# Theory and Practice of High Resolution Seismic HRS-Geo Technology Application

Vladimir L. Trofimov • Fanil' F. Khaziev
Alisa V. Trofimova

# Theory and Practice of High Resolution Seismic HRS-Geo Technology Application

## Optimizing Environment Model Selection

Springer

Vladimir L. Trofimov
Faculty of Geology
Lomonosov Moscow State University
GSP-1, Leninskie Gory, Moscow, Russia

Fanil' F. Khaziev
Faculty of Geology
Lomonosov Moscow State University
GSP-1, Leninskie Gory, Moscow, Russia

Alisa V. Trofimova
Mechanical, Aerospace and Nuclear
Engineering Department
School of Engineering
Rensselaer Polytechnic Institute
Troy, NY, USA

ISBN 978-3-031-41589-0 ISBN 978-3-031-41590-6 (eBook)
https://doi.org/10.1007/978-3-031-41590-6



This Springer imprint is published by the registered company Springer Nature Switzerland AG
The registered company address is: Gewerbestrasse 11, 6330 Cham, Switzerland

Paper in this product is recyclable.

This book outlines the scientific, methodological, and practical foundations for the application of the HRS-Geo technology, the purpose of which is to build detailed 2D and 3D seismoacoustic models in the form of effective acoustic impedance (AI), reflection coefficients (RC), and the most important geological indicators for prospecting and industrial exploration of real geological environment.

Particular attention is paid to improving the process of searching for the AI and RC model using the vector of objective functions, for which various types of residuals between real and model data are iteratively calculated. Such residuals are estimated from the energy (amplitudes) and the shape of the seismic signal. For these discrepancies, the information content and sensitivity of the amplitude (energy) and phase spectra of the seismic record are analyzed in the frequency domains in which the contribution of the medium model parameters perturbations (elastic wave velocity, medium density, and wave absorption) has a significant effect on the result of seismic record inversion.

The well-known criteria for predicting the oil and gas content of productive deposits are analyzed from the standpoint of the principles of system analysis and the set of geological indicators found from the HRS data. Specific examples are used to consider weak seismic signals in the initial wave field against the background of a high level of noise and reflected waves from strong interfaces.

The features of solving the main seismic and geological problems using the HRS-Geo technology are shown. Numerous examples are illustrated using a special (optimal) processing graph and a set of additional record regularization procedures that ensure the resolution of seismic data with a seismic record discretization step in time.

An assessment of various results of automated interpretation of well logging data obtained through the widespread use of the method of normalization of well logs and their functional transformations into information systems are given. On this basis, the construction of lithological-stratigraphic columns and detailed geoacoustic models in the section of wells and predictive lithological-stratigraphic columns is carried out according to HRS data in the inter-well space.

Examples of the use of the HRS-Geo technology on materials obtained in various seismogeological conditions of the Timan-Pechora oil and gas province with high efficiency of the performed studies are considered.

For scientists and specialists involved in deep complex geological interpretation of seismic data in the search and exploration of oil and gas deposits, as well as students of geophysical and geological specialties, graduate students of these specializations, and developers.

# Preface

A method of oil and gas fields exploration is known, including seismic exploration and drilling of a number of deep wells open up productive deposits. In this way, it is possible to outline the productive zones (sections) of the studied territory by changing the kinematic and dynamic characteristics of reflected waves, determined by seismic observations that are made within the zones linking the exploration wells – standards, as well as the locations of the projected deep boreholes. At the same time, the width of such zones is limited by the condition of constancy of the real environment geological model in the area of tracing reflections from the target productive horizon, and the length of the sections is set at least twice as wide [1].

Due to the imposed limitation on the constancy of the real environment models and the area of reflection tracing, the data of this method do not provide a sufficiently reliable classification of wave field anomalies caused by changes in the composition and properties of the studied deposits – lithological composition, reservoir properties, and oil and gas saturation. As a result, newly designed deep drilling wells may turn out to be outside the oil and gas potential, and the funds spent on their drilling and research may be spent unproductively.

There is also known a method of exploration of oil and gas fields, including seismic exploration and deep drilling of a number of boreholes, in which broadband acoustic logging and vertical seismic profiling are carried out in the wells of the first stage, according to which a model of the studied section is built. The choice of holes for vertical seismic profiling is determined by specific seismogeological conditions. The features of the reflected waves generating in the productive and non-productive intervals of the formation under explore are studied. The form of record reflected waves that carry information about the presence (in productive intervals) and absence (in unproductive intervals) of an oil and gas deposit is chosen as standards. These standards are compared with the help of pattern recognition methods with the reflected wave shapes recorded during land-based seismic observations, performed by methodological and technical means in such a way as to ensure the correlating of reflected waves. For each profile within the study area, a picture of the distribution of lithological-reservoir properties of the study formation and information about the presence or absence of an oil- and gas-saturated interval in the formation is obtained.

This allows you to outline the oil and gas deposit and rationally place subsequent exploration boreholes [2].

However, in the method described above for predicting the lithological-reservoir properties of the investigated reservoir and information about the presence or absence of an oil- and gas-saturated interval in it, according to the form of reflected wave recording, a number of issues remain unresolved, which ultimately negatively affect the reliability and unambiguity of geological results. We list the main of these problems, the solution of which is not implemented within the framework of the method described above [2]:

1. The use of only the structure of the wave field significantly limits the possibility of predicting the composition and properties of a thin-layered geological section. In addition to the fact the real thin-layered media are characterized by significant acoustic inhomogeneity in the vertical and horizontal directions, the seismic records of reflected waves recorded from such media also differ in that each of its amplitudes at each of its samples contains geological and geophysical information from the overlying elementary boundaries and thicknesses, i.e., ubiquitous interference of information from a number of layers of the studied section is manifested. At the same time, the length (duration) of such interference along the vertical coordinate significantly increases in comparison with the true (original) thickness of the layer (or stratum) of the section on the seismic record; "stretched" by the length of a seismic pulse that probes the real medium and passes through each specific layer. Such information in the form of a "stretched" local response from acoustic heterogeneity associated with geological indicators (lithology, porosity, and oil and gas saturation) exists both separately and in superposition (interference) with each other. A change in the length and shape of a seismic pulse when it passes through a real medium also leads to a complication of the seismic interference pattern, and this leads to a loss of detail in the studied dynamics of the seismic record and the impossibility of performing a reliable interpretation of a thin-layered section on this basis (including also the impossibility of extracting reliable information about relevant geological indicators).
2. An analysis of the seismic wave field, using the results of seismic modeling to assess the influence of the main geological indicators on the dynamics of the seismic record, shows geological indicators such as porosity and oil and gas saturation respond in the wave field in the form of weak seismic signals – local responses that it is advisable to store in an undistorted view in the dynamics of a seismic record. The method [2] does not use special procedures for saving in the dynamics of a seismic record the required amount of identification information about such geological indicators.
3. Due to the lack of a complex graph in the method of preliminary processing of seismic materials [2], which would ensure, on the one hand, the preservation of the seismic record dynamics, adequate to the actual distribution of acoustic heterogeneities of the studied medium (including the preservation of relatively weak seismic signals associated with porosity and oil saturation in the dynamics

of the record), and on the other hand, it would provide the complete elimination of interference in the structure of identification information associated with lithology, porosity, and oil and gas saturation does not allow this method to be classified as a reliable method that provides unambiguous geological results in the process of predicting geological indicators.

The purpose of the method proposed by the authors, considered in monographs [3, 4] and implemented in the form of HRS-Geo technology, is to increase the efficiency of the method by eliminating the shortcomings of the above methods [1, 2] and reducing material costs associated with reducing the volume of unproductive drilling.

This goal is achieved by the fact that the proposed method for exploration of oil and gas fields implements the following:

(a) In the first stage holes, well logging is carried out using a set of methods, according to which the processing and automated interpretation of logging diagrams are performed. At the same time, the substance composition, porosity, the content of bound water, and the useful capacity of the reservoir are determined, the oil and gas saturation of the reservoirs is estimated, and a number of very important physical characteristics of the geological section are calculated, on the basis of which lithological-stratigraphic columns and detailed geoacoustic models along vertical profiles of the borehole are built.
(b) According to the results of seismic modeling, the influence is estimated – the contribution to the interference seismic record of lithological composition, porosity, and oil and gas saturation in numerical terms and in the form of graphs for the amplitudes of each seismic record sample.
(c) The inverse dynamic (amplitude) seismic problem (IDSP) is solved. Based on this, 2D sections and 3D cubes of effective acoustic impedance (AI) and reflection coefficient (RC) are built with a vertical resolution equal to the seismic record sampling step in time. By solving the inverse dynamic seismic problem, the seismic information associated with the source of elastic oscillations and with the real geoenvironment of the studied geological section is realized, and the interference of geological indicators is completely eliminated on the results obtained in the form of AI and RC values. When solving the IDSP, the traces of the seismic wave field are converted into an impulse response of the medium, in particular, into the traces of effective seismoacoustic logging, each of which is some equivalent of a deep hole section in which such logging was performed with a seismic record sampling step along the time.
(d) Using the developed system of petrophysical equations and the found correlations between the effective acoustic impedance and the main geological indicators, a number of geological parameters related to lithological composition, reservoir properties, and oil and gas saturation (such as clayiness, sandiness, porosity, water saturation, effective and oil-saturated thicknesses, density of hydrocarbon distribution, etc.) are determined by high-resolution seismic data.

(e) According to the set of such predictive geological indicators, areas and zones of possible productivity of the studied sediment complexes are identified based on the prediction of structural, tectonically sealed, and non-structural hydrocarbon traps with a quantitative assessment of the composition and properties of porous facies, and the points of subsequent exploration wells are oriented.

To implement the method, the following operations are sequentially performed.

1. Within the oil and gas basin, seismic exploration is carried out – 2D/3D survey using the common depth point method (CDP) with the length of a number of 2D profiles or 3D area 2–3 times the size of the structure, according to which exploratory drilling is performed, which makes it possible to discover at least one field.
2. In each of the drilled holes, geophysical surveys of wells are performed. Processing and automated interpretation of data from standard well logging methods is carried out by the method of functional transformations of geophysical parameters by integrating them into information systems. The main stages of well logging data processing are: determination of the lithological composition; finding of porosity; estimation of oil and gas saturation; and assessment of the filtration properties of the studied reservoirs. On this basis, a thin-layer geological model is formed in the vertical section of the borehole, lithological-stratigraphic columns and detailed geoacoustic well models are built.
3. Using the data of acoustic and density logging of a well set, their lithological-stratigraphic columns, and the seismic wavelet extracted from real seismic records in the vicinity of the wellbore, based on the results of seismic modeling, the information content of the elements of a thin-layer section is analyzed by quantifying the contribution of elementary boundaries and strata into the interference wave field; using the "interference matrix of contributions," the contributions of local responses from lithology, porosity, and oil and gas saturation to the seismic wave field and to the results of its various transformations are estimated.
4. Taking into consideration the presence of weak seismic signals in the seismic record – local responses associated with such geological indicators as porosity and oil and gas saturation, complex processing of seismic data is performed according to some optimal graph, in which only those procedures are used in the processing that allow to save the dynamics of the seismic record, adequate to the actual distribution of acoustic inhomogeneities in the studied medium. One of the distinguishing features of the technological scheme of seismic data processing is the inclusion in its composition of the procedure for inverting the original seismic traces (with some use of the main preprocessing procedures) into the impulse response of the medium, i.e., into the traces of effective reflection coefficients (RC) and acoustic impedances (AI). This procedure makes it possible to maximize the spectrum band of a seismic record, significantly increase its resolution, eliminate a significant amount of regular and irregular noise, and regularize the dynamic parameters of the record while maintaining weak seismic signals associated with the above target geological indicators.

5. The construction of 2D sections and 3D cubes of AI and RC is carried out on the basis of solving the inverse dynamic seismic problem by a numerical method, which is found by the optimization method, and consists in defining AI and RC models for a given wave field structure using known formulas for solving a direct (forward) problem for calculating a seismic wave field. In this case, a convolutional model algorithm is used, in which it is possible to account the noise level, the residual background of multiples, and the regularization factor. To optimize the solution, a system of objective functions has been developed, in which various types of metrics between real and model data are iteratively calculated. A feature of such a system is that it separately calculates and analyzes residuals in terms of energy and waveforms and makes a decision as a whole on optimizing the model. At the same time, the problem of "selecting weak seismic signals" is directly linked to the problem of "solving the inverse dynamic seismic problem," since on this basis the nature of reflected waves weak in intensity becomes clear, which, depending on the values of the reflection coefficients changing along the section, "respond" appropriately to low-contrast seismoacoustic inhomogeneities. It is with such heterogeneities the required geological and geophysical parameters are most often associated – porosity and oil and gas saturation and the corresponding reduced information content of the initial seismic data.
6. Based on the joint interpretation of sections and cubes of AI and RC and well logging data, detailed seismogeological models are built in order to identify potential oil perspective objects in these models. In combination with AI and RC fragments and with sections of the analyzed wells, these models highlight the most informative relationships of structural elements and forms of the studied deposits, relatively thin structural-lithological and combined oil and gas traps. They also record zones of replacement and wedging out of productive horizons, organogenic structures, stratigraphic unconformities, and, under favorable seismogeological conditions, the location of oil-water (OWC) and gas-water (GWC) contacts between fluids. Based on the use of sections and cubes of AI and RC, acoustic inhomogeneities of the real environment with high resolution make it possible to investigate the ancient paleochannels of river systems, elements of the structures of their filling, channel incisions of river paleovalleys, erosion depths, etc.
7. The developed system of petrophysical equations and previously found relations between the effective acoustic impedance and the main geological indicators for the intervals of the studied reservoirs in the well allow us to solve the problem of predicting geological data from high-resolution seismic data. Based on this system and dependencies, for each 2D profile or 3D cube of AI, the most probable values of geological and geophysical parameters of the reservoirs are calculated, such as clay and sand contents, porosity, water saturation, effective and oil-saturated thicknesses, as well as hydrocarbon distribution densities within productive reservoirs. Thus, a picture of the distribution of lithological-reservoir properties for the studied reservoirs is obtained, and, under favorable seismogeo-

logical conditions, information about the presence or absence of an oil- and gas-saturated interval in the reservoir under study is obtained. The conclusion about the deterioration of the reservoir properties and the absence of an oil- and gas-saturated interval along a number of 2D profiles or corresponding sections (zones) of the 3D cube within the study area allows to delineate the oil and gas deposit and optimally place subsequent exploration holes.
8. Correlation dependencies and petrophysical parameters system of equations obtained in the explored field can be used to identify oil and gas deposits in neighboring areas within the oil and gas basin. This makes it possible to classify the structures prepared for deep drilling and prospective objects into perspective and unpromising ones, to determine the sequence of their introduction into prospective drilling and rationally place exploratory wells in a new area. This method makes it possible to detect hydrocarbon deposits in non-structural (non-anticlinal) traps as well. In general, this leads to a reduction in the cost of deep drilling at various stages (prospecting, exploration) of geological exploration for oil and gas.

The provisions (principles) listed above formed the basis for the development of the HRS-Geo technology and are described in the book [3, 4]. This book is devoted to the further development of detailed complex interpretation of seismic data, i.e., direction of research using software tools and methods of high-resolution seismic (HRS).

Moscow, Russia — Vladimir L. Trofimov
Moscow, Russia — Fanil' F. Khaziev
Troy, NY, USA — Alisa V. Trofimova

# Introduction

Reseaches using the HRS-Geo technology, as noted above, are aimed at studying the internal structure of the real geological environment (productive and potentially productive thin-layer section) using high-resolution seismic data, creating detailed seismic-geological models of the studied real geomedium, and quantifying the composition and properties of porous facies and hydrocarbon saturation, and on the forecast of the most important geological indicators for prospecting and industrial exploration of oil and gas deposits.

The authors, being the developers of the software and methodological tools of the HRS-Geo technology, in their work [3, 4] reveal the main possibilities of its informal application in various geological conditions. In contrast to this development, other production service geophysical organizations (both domestic and foreign), using software and technological tools of development companies (mainly foreign) in the field of approximately the same kind of research, most often use them quite formally, often without taking into account all capabilities and features of the software used. The effectiveness of the use of such technologies is often not high enough.

The main obstacle to solving the set geological problems by traditional seismic exploration is the limited possibility of using the structure of the seismic wave field, due to the phenomenon of the ubiquitous interference of this field in the conditions of a thin-layer section (in which the thickness of the studied layers is actually 1–5 m). Numerous modern methods of seismic inversion of wave fields used either completely fail to decipher wave interference in such thin-layer sections or work satisfactorily only in the vicinity of wells (within a radius of a few hundred meters from their vertical sections). At the same time, in the area where there are no wells, the errors can reach unacceptable values for the correct (adequate) geological interpretation of seismic data.

Let's note the main advantages of HRS-Geo technology [3, 4]:

Its **first advantage** is the achievable research detail, which in the limiting case is the sampling step of the seismic survey (this is 1 or 2 ms in time, i.e., ~ 2–6 m along depth).

The **second advantage** is the technology not strictly tied to well data – approximately the same vertical resolution is achieved even in the absence of well data.

The **third advantage** is the interpretation (construction) of a thin-layer model of the medium, i.e., tracing of horizons and tectonic faults, is based on the resulting thin-layer sections of reflection coefficients (RC) and acoustic impedance (AI), but not on that of wave fields (WF). Unlike modern inversion methods, in which the structure of wave fields is first interpreted, after which the inversion procedure itself is performed within the selected (already interpreted) layers, which significantly limits the extraction of detailed information about the medium, in the HRS Geo method, the inversion procedure is performed first. At the same time, modern approaches to inversion do not work without well data used as a priori information.

The **fourth advantage** of the technology is the sensitivity of the technology to weak seismic signals and the possibility of using them to extract detailed geological information from them. Such weak signals are usually associated with such important indicators as lithology, porosity, and oil and gas saturation of thin-layer real media. The problem of detecting and using weak seismic signals is complex, and, as a rule, existing algorithms are not able to extract useful and reliable information from them about a thin-layer and acoustically low-contrast environment.

The **fifth advantage** is the achievement of a maximum lateral resolution due to a significant reduction in the size of the Fresnel zones and a significant weakening of their influence.

These features of the HRS-Geo technology determine its optimal, efficient, and successful solution at this stage of development of seismic exploration and all other traditional seismic tasks, such as:

- Tracing of boundaries and faults in sections of reflectivities and acoustic impedances
- Building a seismogeological model using the resulting sections of reflection coefficient and acoustic impedance
- Various structural constructions of perspective areas
- Detection and evaluation of perspective objects
- Predicting of the main geological indicators of oil and gas objects and a number of others

In addition, it is possible to solve such tasks, which are not characteristic of modern traditional methods of seismic exploration, as:

- Construction of an acoustic thin-layer (as well as low-contrast) model of a real environment with a detailed sampling step in depth.
- Achievement in the constructed model of maximum horizontal detail, due to a significant weakening of the Fresnel zones influence.
- Detection and mapping of oil-water (OWC), gas-oil (GOC), gas-water (GWC) contacts (in favorable seismo-geological conditions).
- Quantitative assessment of the oil and gas saturation of the object and its detailed contouring, etc.

For this kind of research, a high-resolution seismic technique is used, which is based on a numerical method for solving an inverse dynamic seismic problem developed by the authors [3, 4]. The software and methodological implementation of this technique is made in the form of the HRS-Geo technology, which provides for the separation of seismic information associated with the source of elastic oscillations and the environment of the studied geological section, which provides a significant increase in the oil and gas reservoir structure detail. The study of real environments is focused on the use of dynamic parameters of the seismic record, the implementation of the maximum possible resolution of these parameters. In particular, the construction of 2D sections and 3D cubes of effective acoustic impedance and effective reflection coefficients is implemented, having a vertical resolution equal to the seismic record sampling step in time.

In general, the use of high-resolution seismic data – HRS-Geo technology – is focused on the implementation of a number of provisions-principles, such as:

1. Dynamics saving
2. Separation of source and medium information
3. Resolution improvement within the limits of seismic survey modifications
4. Adequacy of measurement scales, optimal matching of data from seismic, and well logging methods
5. Physical conditionality of the reconstructed and used geological and geophysical parameters

In this monograph, for the most effective use of the HRS-Geo technology software tools and for its further improvement, the authors examined in more detail the theoretical and methodological aspects and the mechanism of searching for the parameters of the real environment target model through the seismoacoustic characteristics of the studied section, namely through the effective acoustic impedance (AI) and reflection coefficients (RC). For this purpose, numerical algorithms were developed for determining the residuals between real and model data and using them in the vector of objective functions. In addition, the improvement of the methodology for applying the technology in various seismogeological settings predetermined the need to give a more reasonable definition of the geological basis and the specifics of the complex geological interpretation of the reconstructed HRS from seismic data, as well as to consider the detailed features of the development and application (in combination with HRS data) of the technique and some algorithms for processing and interpreting well logging data using methods of normalization and functional transformations of well logs.

These issues are considered in more detail in the following sections (chapters).

So, in Part I it is shown in the process of searching for the AI and RC model using the vector of objective functions, various types of metrics between real and model data are iteratively calculated, which are estimated from the energy (amplitudes) and the shape of the seismic signal. These discrepancies are used to analyze the information content and sensitivity of the amplitude (energy) and phase spectra of the seismic record. Such an analysis is carried out in the frequency ranges, in which the contribution of the medium model parameter modifications (velocity of elastic waves, density of the medium, and wave attenuation) affects the result of inversion of a seismic record, which is proportional to the frequency of the waves. The results of modeling and numerical experiments show that the main contribution of medium model parameter perturbations (velocity, density, and absorption) is displayed in the low-frequency band of the amplitude spectrum. In this case, the phase spectrum is most responsive to perturbations in its high-frequency region.

Part II shows that adequate preparation of field observation materials, considering the specifics and goals of solving the problem, includes, in addition to preliminary special processing of seismic data, assessing the quality and quantity of information in the original seismic data, also the optimal parameterization of the model and the parameterization of the main characteristics procedures for solving the inverse dynamic (amplitude) seismic problem (IDSP) by a numerical method. Parameterization of the medium model using the characteristics of the amplitude and phase spectra of the original digital data makes it possible to determine the initial and final detail of the seismic data inversion results, i.e., in general, the solution of the IDSP.

In Part III, based on the use of well-known theoretical provisions, principles and criteria for the formation of oil and gas accumulations, i.e., principles of system analysis, substantiate the geological basis and the specifics of the complex geological interpretation of high-resolution acoustic characteristics (HRS). Specific examples show the selection of weak seismic signals in the original wave field against the background of a high level of noise and reflected waves from strong boundaries. Features and examples of complex geological interpretation of data based on the use of HRS-Geo technology, production geophysics, and drilling data are demonstrated. Processing and interpretation of HRS-Geo and logging data is implemented using a special (optimal) processing graph and a set of seismic record regularization procedures.

Part IV shows that to study deep well sections, the principles of continuous functional data transformations are used, in which the physical parameters of the section, extracted from well logging data, are optimally matched with the corresponding high-resolution seismic parameters obtained using the HRS-Geo technology. The main features of the development and improvement of the methodology for processing and interpreting data from standard well logging methods according to the method of integrating geophysical data by their functional transformations into information systems are considered. These features of log data transformation are aimed at quantitative and qualitative study of well sections with complex lithological composition. Positive results of automated interpretation of well logging data are obtained due to the wide application of the well log normalization method and their functional transformations.

Part V presents the results of the HRS-Geo technology application on materials obtained in complex seismogeological conditions in the exploration areas of the Timan-Pechora Province (TPP).

At the first of these areas, located in the southern part of the Mechayu-Pashnin swell and the Tebuk step of the Izhma-Pechora saddle of the TPP, the geological structure of local uplifts was refined based on the results of 3D AI and RC cubes.

The results of reinterpretation of structural constructions are analyzed taking into account drilling data. Based on the classification of seismic facies using the shapes and values of the reference curves – clusters of 3D cubes of AI and RC, oil perspective objects of structural, tectonically, and lithologically shielded types were identified in the deposits of the upper part of the Famennian stage.

In the junction zone of the northeastern part of the Kolvin megaswell and the northwestern part of the Khoreyver depression, a detailed assessment of the most important geological indicators of the real environment is given, primarily the material composition and reservoir properties in the sediments of the Devonian and Upper Silurian strata. Oil promising objects of a structural type in the reservoirs of the target horizons were identified.

In the central part of the Kolvin megaswell of the Pechero-Kolvin aulacogen and in the junction zone of the large Vozeya and Kharyagin swells, using 3D cubes and 2D AI and RC sections, the nature of the layering of sediments along productive and promising horizons was studied. Detailed structural constructions were made for these horizons and various sedimentation environments were considered by seismic facies classification in order to identify zones of facies replacement, zones of heterogeneity of the geological section in the process of studying the detailed internal structure and lithofacies variability of target deposits.

## References

1. Galagan, Ye. A., Demidovich, L. A., Kuznetsov O. L., Litvinov A. Ya., & Tal'virskiy, D. B. (1978). *Sposob razvedki neftyanykh i gazovykh mestorozhdeniy* (The method of exploration of oil and gas fields). BelNIGRI, VNIIYAGG, VNIGNI (Avtorskoye svidetel'stvo SSSR (USSR Author's certificate) № 677448, kl. G 01 V 1/00).
2. Galagan, Ye. A., Demidovich, L. A., Kuznetsov, O. L., Litvinov, A. Ya., & Tal'virskiy, D. B. (1979). *Sposob razvedki neftyanykh i gazovykh mestorozhdeniy* (The method of exploration of oil and gas fields). BelNIGRI, VNIIYAGG, VNIGNI (Avtorskoye svidetel'stvo SSSR (USSR Author's certificate) № 819771, kl. G 01 V 1/00).
3. Trofimov, V. L., Khaziyev, F. F., & Trofimova, A. V. (2019). *Tekhnologiya vysokorazreshayushchey seysmiki pri poiskakh i razvedke neftegazovykh zalezhey* (High-resolution seismic technology in prospecting and exploration of oil and gas deposits) (p. 343). M., EAGE Geomodel LLC.
4. Trofimov, V. L., Khaziev, F. F, & Trofimova, A. V. (2022). *Oil and gas reservoir prospecting and exploration. High-Resolution Seismic (HRS) techniques and technology*. Springer (Hardcover Book, eBook). https://doi.org/10.1007/978-3-030-84389-2

# Contents

# List of Notations

| Parameter | Description | Measure unit |
|---|---|---|
| $v$ | Longitudinal wave velocity | m/s |
| $d$ | Medium density | $g/cm^3$ |
| $h$ | Thickness of the elementary layer | m |
| $\alpha$ | Absorption coefficient | $m^{-1}$ |
| $m$ | The absorption parameter in the formula $\alpha = m \cdot \omega$ | $m^{-1}rad^{-1}s$ |
| $f$ | Frequency | Hz |
| $\Delta f$ | Frequency interval | Hz |
| $\omega$ | Cyclic (circular) frequency $\omega = 2 \cdot \pi \cdot f$ | rad/s |
| $R_{n+1}(\omega, \alpha)$ | Reflection coefficients in the frequency domain | – |
| $K_{n+1}$ | Reflection coefficients at normal incidence of waves to the layer interface | – |
| $z_{in}^{n}$ | Input acoustic impedance of the *n* layer system | $(m \cdot g)/(s \cdot cm^3)$ |
| $s(t)$ | Elementary impulse | c.u. |
| $r(t)$ | Reflection coefficients | – |
| $*$ | Convolution operator | – |
| $V_a$ | Seismic wave average velocity | m/s |
| $\sigma_v$ | Velocity standard deviation | m/s |
| $\sigma_d$ | Density standard deviation | $g/cm^3$ |
| $\sigma_m$ | Standard deviation of the absorption parameter | $m^{-1}rad^{-1}s$ |
| $\sigma_a$ | Absorption coefficient standard deviation | $m^{-1}$ |
| $\sigma_{ai}$ | Standard deviation of acoustic impedance | $(m \cdot g)/(s \cdot cm^3)$ |
| $\sigma_n$ | Random noise standard deviation | c.u. |
| $\mu_v$ | Average value of velocity (mathematical expectation) | m/s |
| $\mu_d$ | Density average | $g/cm^3$ |
| $\mu_m$ | Average absorption parameter | $m^{-1}rad^{-1}s$ |
| $\mu_{ai}$ | Average acoustic impedance | $(m \cdot g)/(s \cdot cm^3)$ |
| $\mu_n$ | Average of random noise | c.u. |
| $\Sigma A$ | Total change in seismic trace amplitudes, $\Sigma A = \Sigma \Delta A_i$ | c.u. |
| $\Delta A_a$ | The average of the change in the amplitudes of the seismic trace | c.u. |

| Parameter | Description | Measure unit |
|---|---|---|
| $\Delta A$ | The relative value of the change in the amplitudes of the seismic trace | % |
| $\Sigma E$ | The total change in the amplitude spectrum, $\Sigma E = \Sigma \Delta E_i$ | c.u. |
| $\Delta E_a$ | The average of the change in the amplitude spectrum of the seismic trace | c.u. |
| $\Delta E$ | The relative value of the change in the amplitude spectrum of the seismic trace | % |
| $\Sigma P$ | The total change in the phase spectrum, $\Sigma P = \Sigma \Delta P_i$ | rad |
| $\Delta P_a$ | Average value of phase spectrum change of the seismic trace | rad |
| $\Delta P$ | The relative value of the change in the phase spectrum of the seismic trace | % |
| $S/N$ | Signal-to-noise ratio | – |
| $\Delta fq$ | Range (interval) of frequencies | Hz |
| $\Delta F_{Amp}$ | Bandwidth variation across the amplitude spectrum | Hz |
| $\Delta F_{Phs}$ | Bandwidth variation across the phase spectrum | Hz |
| $\delta F$ | The ratio of the frequency bands of the phase and amplitude spectra | – |
| $S_{amp}$ | Relative band change over the amplitude spectrum | % |
| $S_{phs}$ | Relative band change across the phase spectrum | % |
| $C_{amp}$ | Changing the frequency band and energy in the amplitude spectrum | % |
| $C_{phs}$ | Change in bandwidth and phase in the phase spectrum | % |
| $dh$ | Absorbing layer thickness, depth interval | m |
| $dt$ | Sampling step along the time axis, time interval of the absorbing layer | ms |
| $m_a$ | The average of the absorption parameter in the layer | $m^{-1}rad^{-1}s$ |
| $H$ | Depths of the top and bottom of the absorbing layer | m |
| $\lambda$ | Wavelength | m |
| $\lambda_{60}$ | Wavelength at 60 Hz | m |
| $\Delta t_a$ | Average resolution along the time axis in the reservoir | s |
| $\Delta h_a$ | Average depth resolution in the layer, reservoir | m |
| $k_{bn}$ | Bound water factor | – |
| $k_p$ | Porosity coefficient | – |
| $K_p$ | Porosity | % |
| $K_{p.bn}$ | Porosity for bound water | % |
| $K_o$ | Oil saturation factor | % |
| $k_a$ | Absolute permeability | mD |
| $k_w$ | Phase permeability to water | mD |
| $k_o$ | Oil phase permeability | mD |
| $P_p$ | Porosity parameter | – |
| $P_s$ | Saturation parameter | – |
| $P_{sf}$ | Fictive saturation parameter | – |
| $\rho_r$ | Resistance of oil-bearing rocks | Ohm·m |
| $\rho_{wr}$ | Resistance at full water saturation | Ohm·m |

# Abbreviations

| | |
|---|---|
| AI | Acoustic impedance, effective acoustic impedance |
| BRS | Buried river system |
| CAL | Caliper (borehole diameter) |
| CDP, CMP | Common depth (middle) point method |
| CMRW | Correlation method of refraction waves |
| DDSP | Direct (forward) dynamic (amplitude) seismic problem |
| FCP | Filtration and capacitive properties |
| GBS | Geophysical borehole (well) studies (well logging) |
| GeoGraphix | Seismic data interpretation system |
| GGL | Density gamma-gamma logging |
| GOC | Gas-oil contact |
| GR | Gamma ray logging |
| GSP | Geological section prediction |
| GWC | Gas-water contact |
| HRS | High-resolution seismic |
| HRS-Geo | HRS technology in prospecting and exploration of oil and gas deposits |
| ICM | Interference matrix of contributions |
| IDSP | Inverse dynamic (amplitude) seismic problem |
| IL | Induction logging |
| LL | Lateral logging |
| LLS | Lateral logging sounding |
| MLL | Microlateral logging |
| MRW | Reflected wave method |
| NL | Neutron gamma ray logging |
| NLET | Neutron-neutron logging on epithermal neutrons |
| NML | Nuclear magnetic logging |
| NNT | Neural network technology |
| NLT | Neutron-neutron logging on thermal neutrons |
| OMEGA | Seismic data processing system |
| OGP | Oil and gas province |

| | |
|---|---|
| OGR | Oil and gas region |
| PAL | Pseudoacoustic logging |
| PM VSP | Polarization VSP method |
| RC | Reflection coefficients, effective reflection coefficients |
| RZ | Electrical apparent resistivity |
| SKOR | Software package for VSP (PM VSP) data processing |
| SN | Sonic logging |
| SP | Spontaneous potential |
| VSP | Vertical seismic profiling |
| WF | Wave field (seismic) |
| WPS | Well measurement (logging) data processing package |
| 2D, 3D | Two-dimensional, three-dimensional measurement, surveys, models |

# Chapter 1
# Optimization of the Medium Model in Solving the Inverse Dynamic Problem of Seismics in the HRS-Geo Technology Based on the Spectral Characteristics of Seismic Data

**Abstract** To improve the algorithm for solving the inverse dynamic seismic problem (IDSP), the process of searching for acoustic impedance (AI) and reflectivity (RC) models using objective functions (objective function vector) is studied in detail, in which various types of residuals between real and model data are iteratively computed. Such residuals from the energy (amplitudes) and the shape (power distribution in space) of the seismic signal are estimated. For these discrepancies, the information content and sensitivity of the amplitude and phase spectra of the seismic record are analyzed. The analysis is performed in the frequency ranges, in which the contribution of parameter perturbations of the medium models (velocity of elastic waves, density of the medium, and wave attenuation in the geological environment) affects the result of a seismic record inversion, which is proportional to the frequency of the waves. Numerical experiments have shown the main contribution of the medium model parameter variations is displayed in the low-frequency band of the amplitude spectrum. The phase spectrum responds to such changes to the greatest extent in its high-frequency domain. An increase in the acoustic contrast of the velocity and density of the medium raises the information content of the spectra, and the highest sensitivity is found in the phase spectrum of the seismic record at higher frequencies. An increase in the absorption factor leads to a change in the spectra, and the largest contribution in this case falls on the phase component of the spectrum. In the seismogram unwrapped phase spectrum, there are signs by which it is possible to limit the operating frequency range (to determine the limits of the parameter applicability), as well as to detect low-contrast sections and make a conclusion about the absence of a certain thickness layers.

The solution of the inverse dynamic (amplitude) seismic problem (IDSP) in the HRS-Geo technology is found by the optimization method, which consists in the selection of models of AI and RC to the given structure of the wave field according to the known formulas of the direct (forward) problem solution for the seismic wave field calculation



V. L. Trofimov et al., *Theory and Practice of High Resolution Seismic HRS-Geo Technology Application*, https://doi.org/10.1007/978-3-031-41590-6_1

[1]. At the same time the convolution model algorithm for solving the direct dynamic seismic problem is used, in which it is possible to take into account the noise level, the residual background of multiple waves and the regularization factor.

To optimize the solution, a system of objective functions (vector of objective functions) has been developed, in which various types of residuals between real and model data are iteratively calculated [2, 3]. A feature of this system is that it separately calculates and analyzes the residuals in energy and in the form of signals and makes a decision on the whole model optimization. This separate consideration of the wave field characteristics for two main reasons:

- Energy (amplitude) residuals allow you to control the stability of the iterative convergence process of the original model to the desired solution and analyze the sign for optimal dynamic regularization of the finding a solution process;
- Phase residuals allow you to more successfully manage the solution of the problem, selecting the optimal from a variety of solutions. The process of searching for the AI and RC model is completed when the vector of the objective functions system is stabilized and/or when they reach specified limit level.

There is another reason for the separate estimation of residuals (metrics vector). The phase characteristics of the wave field are more sensitive to changes in the seismic properties of the medium and contain much more information about the subsurface medium than the energy ones. In turn, the use of energy factors give stability to the inversion system. The combined use of the described characteristics enables to extract the distribution of the medium properties with maximum detail in the process of the stable wave field data inversion.

## 1.1 Informative Content of Amplitude and Phase Spectra of Seismic Signals

Using wave field modeling, we will consider the properties of seismic signals in different geological situations and evaluate the characteristics are necessary to justify the choice of certain methods for solving the inverse dynamic problem of seismics. It is convenient to investigate the characteristics of signals that vary in time and space in the spectral domain by decomposing the studied complex signal (function) according to some basis of elementary functions with well-known and controlled properties. Considering functions in the spectral domain simplifies the evaluation of some important characteristics, makes it possible to describe time-dependent processes in the simplest way, and also provides us with a way to effectively compare different types of parameters. For these purposes, the Fourier transform technique is usually used, where trigonometric functions are chosen as basis functions. To translate digital data into the frequency domain, there are efficient and fast, discrete analogs of the Fourier transform.

Consideration of seismic signals in the spectral domain has another advantage. In the frequency domain, the complex spectrum can be represented, in general, by means of two independent characteristics: amplitude and phase spectra, where the first describes the distribution of energy over a frequency range (energy), the second determines the shape of the signal (energy distribution of the components of the spectrum in time). These two frequency functions have different properties and contain different amounts of information. They have different sensitivities to the inhomogeneities of the medium, and when solving inverse problems, they show different degrees of stability. Thus, it is possible to optimally combine the useful properties of each of the functions in the formulation and solution of the inverse dynamic seismic problem in the form of a complex multi-criteria optimization problem. We investigate these properties from the point of view of justification, selection and application of methods in solving the inverse dynamic problem of seismics.

Let's consider the theoretical and experimental aspects that characterize the chosen methods in the development of high-resolution seismic technology.

For discussion, consider a simple convolution of two functions, in which, if one function is interpreted as a probing impulse and the other as a characteristic of the medium, then this model will describe in a simplified version the process of a wave pulse propagation in the geological medium

$$f(t) = \int_0^t s(\tau)\cdot r(t-\tau)\,d\tau = s(t) * r(t), \tag{1.1}$$

where $s(t)$ is the sounding pulse, $r(t)$ is the reflection coefficients of the medium, $f(t)$ is the seismic trace.

Using the known properties of the Fourier transform and the convolution function we have

$$F(j\omega) = \int_{\infty}^{-\infty} f(t)e^{-j\omega t}dt,\quad S(j\omega) = \int_{\infty}^{-\infty} s(t)e^{-j\omega t}dt,\quad R(j\omega) = \int_{\infty}^{-\infty} r(t)e^{-j\omega t}dt, \tag{1.2}$$

where $j$ is an imaginary unit, $F(j\omega)$, $S(j\omega)$, $R(j\omega)$ are the complex amplitude-phase spectra of the seismic trace, the probing impulse and the pulse characteristic of the medium, respectively.

Taking into account the properties of the convolution operation, we present this model in the spectral domain

$$F(j\omega) = S(j\omega)\cdot R(j\omega), \tag{1.3}$$

where $F(j\omega) = F_{re}(\omega) + jF_{im}(\omega)$, $S(j\omega) = S_{re}(\omega) + jS_{im}(\omega)$, $R(j\omega) = R_{re}(\omega) + jR_{im}(\omega)$.

Function with the index *re* is real part, with the *im* is the imaginary part of the corresponding complex-valued functions of a real variable ω.

In exponential form, this write can be continued as

$$F(j\omega) = S(j\omega)\cdot R(j\omega) = |F(j\omega)|\cdot e^{jP(\omega)}, \tag{1.4}$$

where $A(\omega) = |F(j\omega)| = \sqrt{F_{re}^2(\omega) + F_{im}^2(\omega)}$ is the amplitude spectrum of the function $F(j\omega)$; $P(\omega) = \tan^{-1}(F_{im}(\omega)/F_{re}(\omega))$ is the phase spectrum of the function $F(j\omega)$.

### 1.1.1 Seismic Data Phase Spectrum Unwrapping

One of the important points in digital signal processing is finding an unambiguous continuous phase spectrum. In particular, such a frequency function is used in the calculation of a complex kepstra to determine the shape of a seismic signal, in the calculation of Hilbert transformations for the analysis of instantaneous characteristics of seismic data and in many other applications.

The calculation of the continuous phase spectrum is also an important step in this work. Let's give a brief overview of the methods of unwrapping the phase spectrum, based on the work [4].

The complex spectrum of the signal *f*(*t*) in polar coordinates can be written as

$$F(j\omega)=\left|F(j\omega)\right|\cdot e^{jP(\omega)}, \tag{1.5}$$

where $|F(j\omega)|$ is the amplitude spectrum, $P(\omega)$ is the phase spectrum.

The main value of the phase spectrum $P(\omega)$ for frequencies in the range $[-\pi, \pi]$ is determined by the function

$$P(\omega)=arctg\frac{F_{im}(\omega)}{F_{re}(\omega)}, \tag{1.6}$$

If the signal is not minimum delayed in phase (not minimum phase), when the limits $[-\pi, \pi]$ are crossed, the function (1.6) breaks, acquiring ambiguity due to its multiplicity of $2\pi$. The operation of removing the limits of the main value interval, or otherwise, mapping the limit $[-\pi, \pi]$ to the interval $[0, \omega_N]$, is the finding of a continuous frequency function, called the unwrapping of the phase spectrum.

There are several methods for unwrapping the phase spectrum. The peculiarity of the continuous phase calculation procedure is it has the property of incorrectness, namely instability, i.e. small changes in the frequency step and small changes in the main phase value in the interval $[-\pi, \pi]$ can lead to unpredictable changes in the phase value in the interval $[0, \omega_N]$. We will briefly discuss the main methods of phase unwrapping according to [4].

***Schafer Algorithm*** [5] The phase unwrapping procedure consists in comparing the phases of neighboring frequencies, detecting phase changes beyond $|2\pi|$ and eliminating this gap over the entire frequency range $[0, w_N]$. The disadvantage of this algorithm is instability with respect to the sampling step in frequency.

***Tribollet Algorithm*** [6, 7] The method is successfully used in the processing of seismic data. This method uses adaptive integration of the derivative of the phase spectrum. Briefly, the algorithm is as follows:

1. The complex spectrum of the signal *f*(*t*) is calculated using the fast Fourier transform:

$$F(j\omega)=F_{re}(\omega)+jF_{im}(\omega), \tag{1.7}$$

2. The derivative of the phase spectrum through the real and imaginary parts of the complex spectrum (1.3) is calculated in the form:

$$P'(\omega)=\frac{dP(\omega)}{d\omega}=\frac{F_{re}(\omega)F'_{im}(\omega)-F_{im}(\omega)F'_{re}(\omega)}{F_{re}^2(\omega)+F_{im}^2(\omega)}, \tag{1.8}$$

where $P'(\omega)$, $F'_{re}(\omega)$ и $F'_{im}(\omega)$ are frequency derivatives of ω.

3. By setting the initial phase condition at ω = 0 in the form of $P(0) = 0$, a continuous phase spectrum is determined using an integral (if necessary, the frequency step is condensed):

$$P(\omega)=\int_{\omega}^{0}P'(\omega)d\omega, \tag{1.9}$$

By the time the algorithm was published in 1977, its disadvantage was large iterative calculations. Currently, with modern computing resources, this disadvantage is minimized.

***Median Filtering Algorithm*** [8] The phase unwrapping consists of the following main points.

1. The phase difference calculation between adjacent phases by the formula (1.6):

$$dP(k)=P(k)-P(k-1), \tag{1.10}$$

2. Determination of the median correction of three consecutive values of differences:

$$\Delta P(k)=median\left[dP(k-1),dP(k),dP(k+1)\right], \tag{1.11}$$

3. Iterative calculation of the unwrapped phase taking into account the found correction:

$$P0(k)=P(k),\ k=1,\ P0(k)=P(k-1)+\Delta P(k),\ k=2,3\ldots, \tag{1.12}$$

***Krajnik Algorithm*** [8] It differs from the median filtering algorithm [8] by the method of determining the correction, where the phase difference between neighboring frequencies is calculated, but their ratio. The advantage of the method is there is no need to search and detect discontinuities (*Shafer algorithm*), and there is no need to interpolate phases between discrete frequencies (*Tribollet algorithm*).

***Factorization Algorithm*** [9] The method uses the z-transform of the signal. The method analyzes the location of z-transform zeros (the polynomial roots) in a narrow band near the unit circle. If the root (zero) of multiplicity 1 is close to the circle, then the phase jump in formula (1.6) will be ± π. The sign is "–" if zero falls inside the circle and "+" if it is outside it. If there is zero on the circle line, then the correction will be 2π. The problem at the time of development in 1980 was the need for a large number of calculations when searching for zeros of the z-transformation. To date, there are effective algorithms for finding the roots of high-order polynomials.

In this study, to assess the influence and sensitivity of the phase spectrum, the Schafer algorithm [5] was used as the fastest in terms of execution speed and relatively simple in practical implementation. The accuracy and stability of this method is sufficient to describe numerical experiments and obtain the necessary results in this work.

### 1.1.2 Seismic Trace Spectral Features Relative to Their Information Content

It is known the maximal energy of a seismic pulse is located in a narrow, low-frequency band. This part of the frequency band in the application to the geological environment is a thick-layer component. This component is usually well described by the energy spectrum $A(\omega)$. It is usually determined by the operating limited frequency band $[\omega_1,\omega_2]$.

The thin-layer component of media is usually located in the high-frequency bands of the spectrum, where the signal energy is low and represented by weak signals – sometimes comparable to the level of noise and even lower in level.

What signal characteristic can reliably represent these weak signals in a seismic trace?

Since the phase of the signal in general is not directly related to the energy of the signal, but is a time characteristic (determined by the time of appearance, distribution and disappearance of the signal), it is logical to assume the phase spectrum contains the most information about the thin-layer component of the medium, i.e. the nature of weak signals. The phase of the signal does not decrease with increasing frequency in the spectrum – on the contrary, it becomes more sensitive to changes, because it describes faster signal changes. At the same time, due to the fact the phase spectrum is practically not determined by the energy of the signal, its working band is much wider than the one in the energy spectrum, under the most favorable conditions, this band can be maintained as a working band up to the Nyquist frequency in the interval $[0,\omega_N]$ or close to it (for the most part, this is the main problem of the methodology of field observations and preliminary processing of seismic data). Thus, it can be assumed the sensitivity to changes in the parameters of the medium is most reflected in the phase characteristics, i.e. the greatest detail can be achieved using the phase spectra of seismic traces (preferably taking care of preserving weak signals in the input data). The latter, as is known, are closely related to the expected geological indicators of the real medium, such as porosity and oil and gas saturation [2, 3].

For clarity, when rationale the above theoretical assumptions, the amplitude and phase spectra are schematically depicted in Fig. 1.1.

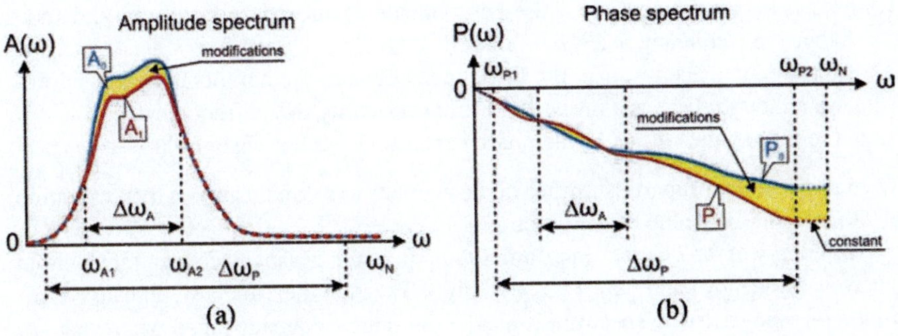


**Fig. 1.1** Components of the signal spectrum: (**a**) amplitude; (**b**) phase (continuous). $A_0$, $P_0$ are amplitude and phase spectra of the original signal, respectively; $A_1$, $P_1$ are the amplitude and phase spectra of the modified signal, respectively

These figures show the changes in the amplitude (Fig. 1.1a) and phase (Fig. 1.1b) spectra of the recorded wave field when the acoustic characteristics of the medium are modified.

The sensitivity of the spectra will determine the potentially possible detail of the seismic data, i.e., the informativeness of the source materials. Let's define expressions for the theoretical assessment of the sensitivity of each of the characteristics.

The sensitivity (informativeness) of the spectra is estimated by the relative change of them. The changes in the spectra for a separate component of the spectrum are given in the form:

$$\delta A(\omega) = \frac{A_1(\omega) - A_0(\omega)}{A_0(\omega)} \text{ – for the amplitude spectrum;} \tag{1.13}$$

$$\delta P(\omega) = \frac{P_1(\omega) - P_0(\omega)}{P_0(\omega)} \text{ – for the phase spectrum.} \tag{1.14}$$

Figure 1.1 shows there are frequency intervals for the amplitude (energy) and phase (determining the shape of the signal) spectra, where the greatest amount of information is contained, i.e. the areas where the parameters of seismic signals are most sensitive to acoustic changes in the characteristics of the medium.

It is possible to give a physical meaning to signal spectra: the amplitude spectrum determines the distribution of energy over frequency components, the phase spectrum sets the allocation of energy components over time.

For the amplitude spectrum, this is the frequency range $\Delta\omega_A = [\omega_{A1}, \omega_{A2}]$, for the phase spectrum is $\Delta\omega_P = [\omega_{P1}, \omega_{P2}]$. At the same time, the amplitude spectrum in this range manifests itself in such a way there are significant changes in the spectral components (the greatest difference) in the interval $\Delta\omega_A$ (Fig. 1.1a), and outside the frequency range, the changes are insignificant (Fig. 1.1a).

The behavior of the phase spectrum in the frequency range $\Delta\omega_P$ is also determined by the difference in the spectral components, however, according to the graph of the phase spectrum, it can be seen there are three features (Fig. 1.1b):

1. there is a proportionality of the components of the phase spectrum and their changes in frequency $\omega$ ($P(\omega) \sim \omega$, $\delta P(\omega) \sim \omega$);
2. at the frequencies $\omega < \omega_{P1}$, the difference between the components is insignificant, although there is a property of proportionality to the frequency $\omega$;
3. at the frequencies $\omega > \omega_{P2}$, the phase spectrum does not change $P(\omega) \equiv const.$

We will use these properties further in the analysis and conclusion on the evaluation of the informative value of the spectra.

The energy of the seismic signal amplitude spectrum is usually concentrated within a narrow frequency band $\Delta\omega_A = [\omega_1,\omega_2]$ (Fig. 1.1a). And therefore a significant contribution of changes to the spectrum is also in the same frequency range $\Delta\omega_A$. Changes in the phase spectrum usually occupy a much wider frequency band, almost up to the boundary frequency of digital data - the Nyquist frequency $\omega_N$, i.e., $\Delta\omega_P = [0,\omega_N]$ (based on the need for an unambiguous description of the waveform) (Fig. 1.1b). At the same time, weak signals, being located in the high frequency band, should also respond significantly to phase changes due to independence from the energy component (the phase distribution is uniquely determined only by the shape of the signal).

The following property of the phase spectrum should be noted: the higher the frequency, the stronger the constituents of the signal phase component respond to changes in the characteristics of the medium, and the more it reacts to weak signals. The informativeness for the entire signal is determined by the integration of the functions $\delta A(\omega)$ and $\delta P(\omega)$ over the frequency range, where there are significant changes in the spectra.

The integration intervals for the spectra are defined as $\Delta\omega_A = [\omega_1,\omega_2]$, $\Delta\omega_P = [0,\omega_N]$ for the amplitude and phase, respectively. Then the informativeness at thus defined frequency intervals for the amplitude and phase spectra is calculated as:

$$\Delta A = \int_{\omega_{A1}}^{\omega_{A2}} \delta A(\omega) d\omega, \quad \Delta P = \int_{\omega_{P1}}^{\omega_{P2}} \delta P(\omega) d\omega, \text{ respectively.} \tag{1.15}$$

Note the relative changes in the phase spectrum $\delta P(\omega)$ are proportional to the frequency $\omega$: $\delta P(\omega) \sim \omega$. The relative variations in the amplitude spectrum $\delta A$ are proportional to the frequency band $[\omega_1,\omega_2]$, where the main part of the signal energy is concentrated: $\delta A \sim \Delta\omega_A$. In this case, $\delta A(\omega) \rightarrow 0$, with $\omega_1 \rightarrow 0$ and $\omega_2 \rightarrow \omega_N$ (Fig. 1.1a). Then $\Delta P > \Delta A$. This means the sensitivity in the phase spectrum is greater than in the amplitude that with given changes in the characteristics of the medium (the contribution of changes in the medium to the phase part of the spectrum is greater, especially in the high-frequency part). That is, the phase spectrum of a seismic signal may contain information about a thin-layer section with appropriate preprocessing of field data.

We show, this conclusion is also true for the parameters of the medium (reflection coefficients and acoustic impedance) in the wave field model described by the convolutional function.

In this case, the seismic trace model is described in the time domain by the convolution function of the seismic pulse $s(t)$ and reflection coefficients $r(t)$. In the frequency domain, the convolution is represented by the product of the corresponding spectra. Writing the spectra in exponential form highlights the amplitude and phase spectra in the expression (1.4). This form of record the spectrum allows you to work independently and systematically with these functions. In accordance with the convolution properties in the frequency domain, the amplitude spectra of the pulse and reflection coefficients are multiplied, and their phase spectra are added.

Let $A(\omega)$, $P(\omega)$ are the amplitude spectra of the pulse and reflection coefficients, respectively. $A_S(\omega)$ and $A_R(\omega)$ are the amplitude spectra of the pulse and reflection coefficients, respectively. $P_S(\omega)$ and $P_R(\omega)$ are the phase spectra of the pulse and reflection coefficients, respectively. Then, in accordance with the convolution properties in the frequency domain, we have:

$$F(j\omega) = A(\omega)e^{jP(\omega)} = A_S(\omega)A_R(\omega)e^{j[P_S(\omega)+P_R(\omega)]}, \tag{1.16}$$

where $A(\omega) = A_S(\omega)\bullet A_R(\omega)$ and $P(\omega) = P_S(\omega) + P_R(\omega)$.

Let's define the original model of the environment and its modification, and the corresponding characteristics of the fields, denoting them with indexes 0 and 1:

$$F_0(j\omega) = A_0(\omega)e^{jP_0(\omega)} = A_{S0}(\omega)A_{R0}(\omega)e^{j[P_{S0}(\omega)+P_{R0}(\omega)]},$$

$$F_1(j\omega) = A_1(w)e^{jP_1(\omega)} = A_{S1}(\omega)A_{R1}(\omega)e^{j[P_{S1}(\omega)+P_{R1}(\omega)]}. \tag{1.17}$$

Then the relative changes for each frequency of the amplitude spectrum of such a model are described as:

$$\begin{aligned}\delta A(\omega) &= \frac{A_1(\omega)-A_0(\omega)}{A_0(\omega)} = \frac{A_{S1}(\omega)A_{R1}(\omega)-A_{S0}(\omega)A_{R0}(\omega)}{A_{S0}(\omega)A_{R0}(\omega)} \\ &= \frac{A_{R1}(\omega)-A_{R0}(\omega)}{A_{R0}(\omega)} = \frac{\delta A_R(\omega)}{A_{R0}(\omega)}\end{aligned} \tag{1.18}$$

here it is assumed, the characteristics of the pulse do not change, i.e. $A_{S0}(\omega) = A_{S1}(\omega)$.

Integrating the expression for $\delta A(\omega)$ we obtain:

$$\Delta A_R = \int_{\omega_{A1}}^{\omega_{A2}} \delta A(\omega)d\omega = \int_{\omega_{A1}}^{\omega_{A2}} \frac{\delta A_R(\omega)}{A_{R0}(\omega)}d\omega. \tag{1.19}$$

The subintegral expression in (1.19) depends only on the amplitude spectrum of the reflection coefficients, and the energy of the seismic signal is concentrated in a limited frequency band. As shown above, $\delta A \sim \Delta\omega_A$. Then the obtained expression implies the proportionality of the change in the amplitude spectrum of the reflection coefficients $\Delta A_R$ in the frequency band $\Delta\omega_A = [\omega_1, \omega_2]$.

Let's make similar estimates for the phase spectrum, taking into account the features for its transformations:

$$\delta P(\omega) = \frac{P_1(\omega) - P_0(\omega)}{P_0(\omega)} = \frac{P_{S1}(\omega) + P_{R1}(\omega) - P_{S0}(\omega) - P_{R0}(\omega)}{P_{S0}(\omega) + P_{R0}(\omega)}$$
$$= \frac{P_{R1}(\omega) - P_{R0}(\omega)}{P_{S0}(\omega) + P_{R0}(\omega)} = \frac{\delta P_R(\omega)}{P_0(\omega)} \quad (1.20)$$

The equality $P_{S0}(w) = P_{S1}(w)$ is also taken into account here.

Integrating the resulting expression by frequency, we obtain:

$$\Delta P_R = \int_{\omega_{P1}}^{\omega_{P2}} \delta P(\omega) d\omega = \int_{\omega_{P1}}^{\omega_{P2}} \frac{\delta P_R(\omega)}{P_{R0}(\omega)} d\omega. \quad (1.21)$$

At the same time, $\delta P_R(\omega)$~$\omega$ and the subintegral function (1.21) are integrated in a wider range $\Delta\omega_P = [0,\omega_N]$, than in the case of a narrow-band amplitude spectrum concentrated in the range $\Delta\omega_A = [\omega_1,\omega_2]$ (Fig. 1.1). This determines the information content of the phase spectrum relative to sensitivity to modifications in the parameters of the medium model (RC and AI).

Let's consider the informational content of seismic data on concrete examples using numerical modeling of wave fields.

## 1.2 The Choice of the Wave Field Modeling Method

In the seismic method of studying the geological environment, various techniques are used to calculate synthetic seismograms, differing in the dimension of the generated wave fields and the domains of their calculation. Each of them has its own advantages and disadvantages.

There are calculations in the time and spectral domains. The main disadvantage of calculating seismic traces in the time domain (for example, the Baranov-Kunetz method) is the difficulty of taking into consideration inelastic absorption and insufficiently complete simulation of multiples waves.

For the research tasks of this section, the method of calculating wave fields in the frequency domain was chosen. In the software implementation, the theory of plane waves for complex absorbing media to calculate the seismic trace in the implementation of the Brekhovsky method [10] was used. At the same time, the unaccounted-for sphericity of the fronts of real waves in no way limits the applicability to practical data, since usually real data are always given as a result of preprocessing to one-dimensional seismic traces according to the location of common depth points in the CMP method. The software implementation of this algorithm is described in a number of papers [11–13].

The method of calculating wave fields in the frequency domain allows the use of more complex models of the source data when solving a direct (forward) dynamic (amplitude) seismic problem (DDSP). In particular, by setting the absorption parameters, it is possible to study the effects of inelastic absorption of seismic waves by the medium. At the same time, it is possible to set different patterns of wave velocity dispersion. Theoretical models of absorbing media and reference data on absorption coefficients for different rocks are given in [14]. It is also possible to study the dynamic effects of a probing pulse more effectively by considering the change in the shape and energy of an elementary pulse over time. The chosen method makes it possible to more fully calculate the multiple waves that occur when a wave passes through the medium.

In brief, at this study the method of calculating waves in a thin-layer medium is as follows. As an initial model, equal-time distributions of velocity, density, absorption coefficients and geometric characteristics of the medium in the form of equitime thicknesses of layers are given. The dependence of the absorption coefficient on frequency is given, the dispersion relation is a function of the seismic wave velocity on frequency, in this case, is the Futterman model.

***Geophysical Formulation of the Problem*** The medium bounded by two half-spaces (Fig. 1.2), consisting of a system of equitime layers, is determined by distribution models: velocity $v$, density $d$, absorption parameter $m$ and layer thicknesses $h$. The dependence of the absorption coefficient on the frequency is linear: $\alpha = m \bullet \omega$. The velocity dispersion relation is given by the formula

$$v(f_i) = v(f_r) \cdot \left[1 - \frac{2mv(f_r)}{\pi} \ln \frac{f_i}{f_r}\right]^{-1}, \tag{1.22}$$

where $f_i$ is

$$f_i \le f_r \exp\left\{\frac{\pi\left[1 - \frac{v(f_r)}{c}\right]}{2mv(f_r)}\right\}, \tag{1.23}$$

and where $f_i$, $f_r$ are arbitrarily selected frequencies, $c$ is maximum wave speed in the rocks of the earth crust.

A probing elementary seismic pulse $s(t)$ in the time domain is given.

It is necessary to solve a direct problem – to calculate the seismic trace for a given model.

In accordance with the approach [10], a complex spectrum of reflection coefficients is calculated by sequentially determining the input impedance of the layer system (Fig. 1.2):

$$R_{n+1}(\omega) = \frac{z_{in}^n(\omega) - z_{n+1}(\omega)}{z_{in}^n(\omega) + z_{n+1}(\omega)}, \tag{1.24}$$

where $z_{n+1}(\omega)$ is the acoustic impedance of the n+1st layer, $z_{in}^n(\omega)$ is input acoustic impedance of the $n$-layer system and presented by

$$z_{in}^n(\omega) = \frac{z_{in}^{n+1}(\omega) - iz_n \tan\varphi_n(\omega)}{z_n(\omega) - iz_n \tan\varphi_n(\omega)}, \tag{1.25}$$

where $\varphi_i = \frac{h_i}{v_i}\omega$ is the phase change within the formation.

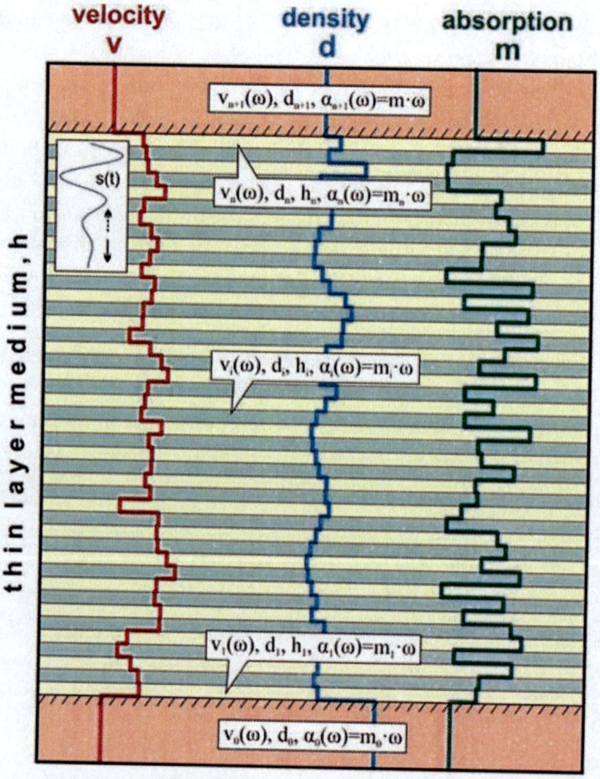


**Fig. 1.2** An equitime model of the medium bounded by half-spaces and represented by the distributions of velocity, density, absorption coefficients and layer thicknesses

Taking into account the reflection coefficients at the normal incidence on the boundary of the layers

$$K_{n+1} = \frac{z_n - z_{n+1}}{z_n + z_{n+1}}, \tag{1.26}$$

we obtain an iterative formula for calculating the spectrum of reflection coefficients of an $n$ – layer medium accounting multiple waves:

$$R_{n+1}(\omega) = \frac{K_{n+1} + R_n \exp(2i\varphi_n)}{1 + K_{n+1}\exp(2i\varphi_n)} = \frac{K_{n+1} + R_n \exp(2ip_n h_n)}{1 + K_{n+1}\exp(2ip_n h_n)}, \tag{1.27}$$

where $p_i(\omega) = \frac{\varphi_i}{h_i} = \frac{\omega}{v_i}$ is wave number.

To account for the absorption of waves in the medium, the wave number $p_i(\omega)$ in (1.27) is replaced by a complex value:

$$P(\omega,\alpha) = p(\omega) + i\alpha(\omega), \tag{1.28}$$

where $\alpha(\omega)$ is wave absorption function.

Substituting (1.28) into (1.27), we obtain a formula for calculating the reflection coefficients in the frequency domain, which describes the model of the medium accounting absorption:

$$R_{n+1}(\omega,\alpha)=\frac{K_{n+1}^{\alpha}+R_{n}^{\alpha}\exp(2i\varphi_{n})\exp(-h_{n}\alpha_{n})}{1+K_{n+1}^{\alpha}R_{n}^{\alpha}\exp(2i\varphi_{n})\exp(-h_{n}\alpha_{n})}, \tag{1.29}$$

and

$$K_{n+1}^{\alpha}(v,d,\alpha)=\frac{\omega(z_{n}-z_{n+1})+iv_{n}v_{n+1}(\alpha_{n+1}d_{n}-\alpha_{n}d_{n+1})}{\omega(z_{n}+z_{n+1})+iv_{n}v_{n+1}(\alpha_{n+1}d_{n}+\alpha_{n}d_{n+1})}. \tag{1.30}$$

There are various functions of the absorption coefficient versus frequency depending on the mechanism and parameters of relaxation (viscosity) of the medium: Kelvin-Voigt, Maxwell, Zener models (standard linear body), generalized linear body and optimal linear body [14]. These models differ in that the equation of state in an ideally elastic non-absorbing medium $\sigma = C\cdot\varepsilon$ is modified by adding stress and/or strain relaxation parameters or combinations thereof. The corresponding wave equations will describe the process of seismic wave propagation in a frequency-dependent absorbing medium. In the Kelvin-Voigt absorption model, the frequency dependence is quadratic $\alpha \sim \omega^2$, in the Maxwell model it is power–law in the form of $\alpha \sim \omega^{1/2}$, the standard linear body combines the first two variants – at low frequencies the contribution of the quadratic dependence is greater, at high frequencies it is power-law $\sqrt{\omega}$. Due to the presence of medium viscosity, simultaneously with the appearance of wave energy absorption, their velocities become frequency-dependent (the velocity dispersion emerges) and decrease in comparison with velocities in ideally elastic media. Without going into more complex wave absorption schemes (for example, when parameterizing discrete media), in practice, an intermediate transient absorption model between Kelvin-Voigt and Maxwell bodies for the seismic range – the Zener model is used, where the absorption dependence on frequency can be approximated by a linear law due to the weak dependence of velocity versus frequency [14].

In the case of the linear absorption law $\alpha(\omega) = m \cdot \omega$ formula (1.30) is written as:

$$K_{n+1}^{\alpha}(v,d,m)=\frac{(z_{n}-z_{n+1})+iv_{n}v_{n+1}(m_{n+1}d_{n}-m_{n}d_{n+1})}{(z_{n}+z_{n+1})+iv_{n}v_{n+1}(m_{n+1}d_{n}+m_{n}d_{n+1})}. \tag{1.31}$$

The inverse Fourier transform of expression (1.29) will give the distribution of the complex impulse response of the medium $r(t)$, which takes into account the absorption of the medium and the background of multiples:

$$r(t)=\frac{1}{\pi}Re\int_{0}^{\infty}R_{n+1}^{\alpha}(\omega)e^{iwt}d\omega. \tag{1.32}$$

By convolution of the obtained function $r(t)$ with a seismic pulse of a given shape, we obtain the desired wave field – a one-dimensional seismic trace (IDSP solution):

$$f(t) = s(t) * r(t) = \int_{t-l}^{\infty} s(\tau) \cdot r(t - \tau) \cdot d\tau, \tag{1.33}$$

where $l$ is the time length of the seismic pulse.

The technology for determining the sensitivity of seismic data consisted in the initial parametrization of the model, the change (modification) of its studied characteristics, quantitative and qualitative estimates of parameters in the time and frequency domains, and their comparison with each other.

As initial data, acoustic and density measurements obtained from one of the wells in Western Siberia were taken, which were approximated by a equitime model, and absorption coefficients were defined by software generation of pseudorandom numbers (Fig. 1.3).

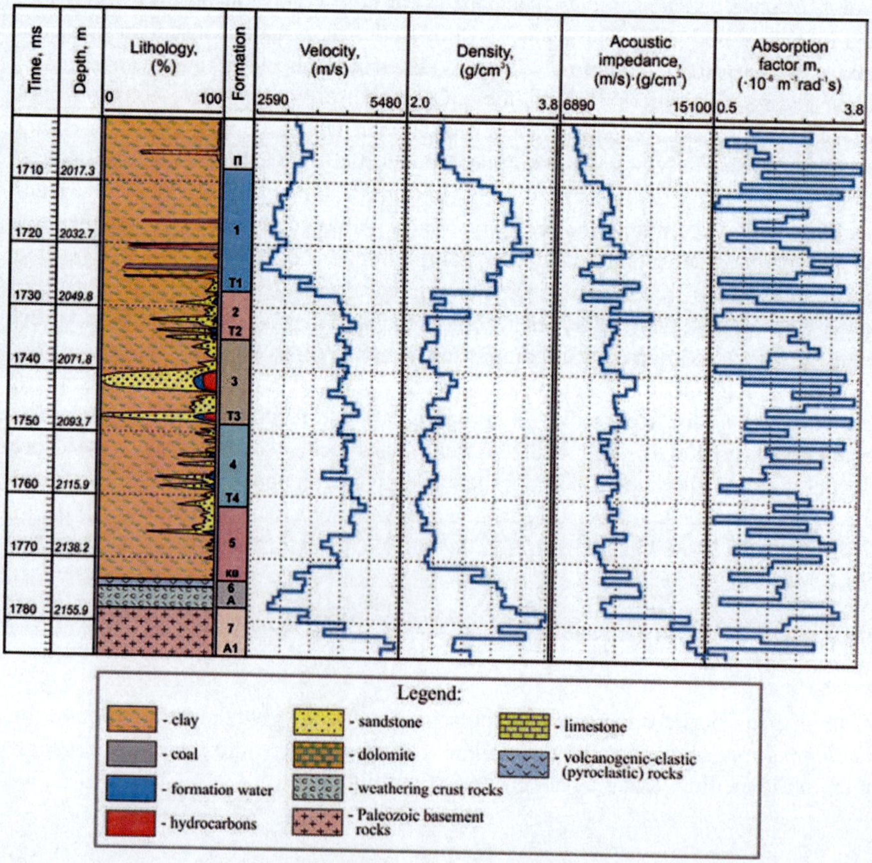


**Fig. 1.3** Initial velocities, densities, and absorptions of a thin-layer medium model

The study was carried out by sequential and combined modification of velocity, density and absorption distributions, modeling of fields by the method described above, calculation of spectral components, evaluation and comparison of their characteristics. The characteristics of the specified models and the results of quantitative estimates of the obtained parameters are summarized in Table 1.1. At the same time, the Berlage pulse with controlled parameters was used as a model of an elementary seismic pulse: with a maximum of the component amplitude spectrum at a frequency of $F_{max} = 60$ Hz, and the attenuation parameter $R_{ext} = 0.5$ (Fig. 1.4).

## 1.3 Spectral Characteristics Sensitivity Evaluation Under Geological Medium Wave Velocity Modification

To modify the initial velocity model, a series of random velocity corrections with a uniform distribution was generated.

Figure 1.5 shows a histogram of one of the velocity change realizations with a uniform distribution with an average value of $V_a = 0.0$ m/s and a standard deviation of $\sigma_V = 250$ m/s.

This correction was added to the initial velocity distribution (Fig. 1.3) and the seismic paths for the initial and modified models were calculated. The results are shown in Fig. 1.6 and Table 1.1. It can be seen from the figure the velocity modifications led to a corresponding change in the distribution of acoustic impedance, which in turn changed the shape of the seismic trace.

Similarly, corrections for the modification of the initial density and absorption distribution were formed.

To study the effect of interference, a random noise with a normal distribution was also programmatically generated using the Box-Muller method (Figs. 1.9 and 1.10).

The construction of a random variable with a normal distribution by the Box-Muller method consists in the transformation of uniformly distributed random variables by formulas:

$$y_1 = \cos(2\pi x_1)\sqrt{-2\ln x_2}, \; y_2 = \sin(2\pi x_1)\sqrt{-2\ln x_2},$$

where $x_1$ and $x_2$ are independent, random variables with a uniform distribution over the interval (0, 1); $y_1$ и $y_2$ are independent, random variables with a standard normal distribution: $(\mu, \sigma^2) = (0, 1)$. The transition for a quantity with arbitrary parameters $(\mu, \sigma^2)$ is performed according to the well-known formula: $\xi = \mu + \sigma \cdot y$. The method is chosen because there is no need to summarize a large number of realizations of a random variable and it is more accurate than the method traditionally proposed, according to the central limit theorem.

The amplitude and unwrapped phase spectra are calculated from these seismic traces (Fig. 1.7). According to the data of the amplitude spectrum in the frequency range, 4 intervals are allocated according to the signal to noise (S/N) ratios, as well

**Table 1.1** Model parameters, seismic traces and their spectra

| No | Model name | Model parameters | $\sum$A, c.u. | $\Delta A_a$, c.u. | ΔA, % | $\sum$E, c.u. | $\Delta E_a$, c.u. | ΔE, % | $\sum$P, rad | $\Delta P_a$, rad | ΔP, % |
|---|---|---|---|---|---|---|---|---|---|---|---|
| 1 | $V_{100}$ | $\sigma_v = 100$ m/s | 1.172 | 0.008 | 25.5 | 3.989 | 0.016 | 16.6 | 853.6 | 3.40 | 41.9 |
| 2 | $V_{250}$ | $\sigma_v = 250$ m/s | 2.993 | 0.021 | 65.1 | 8.094 | 0.032 | 33.6 | 1330.5 | 5.30 | 76.8 |
| 3 | $V_{400}$ | $\sigma_v = 400$ m/s | 4.718 | 0.033 | 102.7 | 10.192 | 0.041 | 42.3 | 1894.4 | 7.55 | 82.9 |
| 4 | $D_{0.1}$ | $\sigma_d = 0.1$ g/cm$^3$ | 1.550 | 0.011 | 33.7 | 4.969 | 0.020 | 20.6 | 1049.2 | 4.18 | 74.7 |
| 5 | $D_{0.15}$ | $\sigma_d = 0.15$ g/cm$^3$ | 2.355 | 0.016 | 51.2 | 6.974 | 0.028 | 28.9 | 1142.7 | 4.55 | 91.4 |
| 6 | $D_{0.3}$ | $\sigma_d = 0.3$ g/cm$^3$ | 4.691 | 0.033 | 102.1 | 9.483 | 0.038 | 39.4 | 1450.2 | 5.78 | 107.6 |
| 7 | $M_1$ | $m_a = 1.55 \cdot 10^{-6}$ m$^{-1}$rad$^{-1}$s, $\sigma_m = 10^{-6}$ m$^{-1}$rad$^{-1}$s | 0.510 | 0.004 | 11.1 | 1.812 | 0.007 | 7.5 | 2186.7 | 8.71 | 94.0 |
| 8 | $M_2$ | $m_a = 2.0 \cdot 10^{-6}$ m$^{-1}$rad$^{-1}$s, $\sigma_m = 10^{-6}$ m$^{-1}$rad$^{-1}$s | 0.618 | 0.004 | 13.4 | 2.174 | 0.009 | 9.0 | 2824.2 | 11.25 | 94.1 |
| 9 | $M_3$ | $m_a = 3.0 \cdot 10^{-6}$ m$^{-1}$rad$^{-1}$s, $\sigma_m = 10^{-6}$ m$^{-1}$rad$^{-1}$s | 0.841 | 0.006 | 18.3 | 2.919 | 0.012 | 12.1 | 3262.0 | 13.00 | 127.3 |
| 10 | $M_4$ | $m_a = 4.0 \cdot 10^{-6}$ m$^{-1}$rad$^{-1}$s, $\sigma_m = 10^{-6}$ m$^{-1}$rad$^{-1}$s | 1.043 | 0.007 | 22.7 | 3.577 | 0.014 | 14.8 | 5297.5 | 21.11 | 144.2 |
| 11 | $M^5$ | $m_a = 5.0 \cdot 10^{-6}$ m$^{-1}$rad$^{-1}$s, $\sigma_m = 10^{-6}$ m$^{-1}$rad-1 s | 1.229 | 0.009 | 26.7 | 4.144 | 0.017 | 17.2 | 5672.7 | 22.60 | 367.6 |
| 12 | $M_6$ | $m_a = 1.0 \cdot 10^{-5}$м-1рад-1с, $\sigma_m = 10^{-6}$м$^{-1}$рад$^{-1}$с | 1.961 | 0.014 | 42.7 | 6.110 | 0.024 | 25.4 | 9222.6 | 36.74 | 454.1 |
| 13 | $M_7$ | $m_a = 2.0 \cdot 10^{-5}$ m$^{-1}$rad$^{-1}$s, $\sigma_m = 10^{-6}$ m$^{-1}$rad$^{-1}$s | 2.835 | 0.020 | 61.7 | 8.030 | 0.032 | 33.3 | 11331.4 | 45.14 | 722.0 |
| 14 | $M_8$ | $m_a = 4.0 \cdot 10^{-5}$ m$^{-1}$rad$^{-1}$s, $\sigma_m = 10^{-6}$ m$^{-1}$rad$^{-1}$s | 3.611 | 0.025 | 78.6 | 12.600 | 0.050 | 52.3 | 13566.6 | 54.05 | 23466.2 |
| 15 | $V_{250} + D_{0.15}$ | $\sigma_v = 250$ m/s, $\sigma_d = 0.15$ g/cm$^3$ | 5.248 | 0.036 | 114.2 | 9.349 | 0.037 | 38.8 | 1502.0 | 5.98 | 114.9 |
| 16 | $V_{250} + D_{0.15} + M_2$ | $\sigma_v = 250$ m/s, $\sigma_d = 0.15$ g/cm$^3$, $m_a = 2.0 \cdot 10^{-6}$ m$^{-1}$rad$^{-1}$s, $\sigma_m = 10^{-6}$ m$^{-1}$rad$^{-1}$s | 4.669 | 0.032 | 101.6 | 9.392 | 0.037 | 39.0 | 4424.8 | 17.63 | 146.7 |

Parameters description:
No is the model number
$\sigma_v$ is the velocity variance, m/s
$\sigma_d$ is the density variance, g/cm$^3$
$\sigma_m$ is the variance of the absorption factor, m$^{-1}$rad$^{-1}$s
$m_a$ is the average absorption factor, m$^{-1}$rad$^{-1}$s
$\sum$A is the total change at the seismic trace amplitudes, c.u
$\Delta A_a$ is the average value of the change at the seismic trace amplitudes, c.u
$\sum$E is the total change in the amplitude spectrum, c.u
$\Delta E_a$ is the average value of the amplitude spectrum change, c.u
ΔE is the relative average value of the change in the amplitude spectrum, %
$\sum$P is the total change in the phase spectrum, rad
$\Delta P_a$ is the average value of the phase spectrum change, rad
ΔP Is the relative average value of the change in the phase spectrum, %

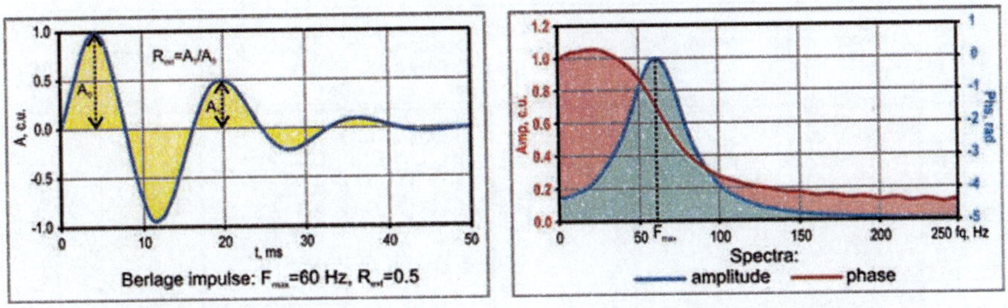


**Fig. 1.4** Berlage impulse and spectra

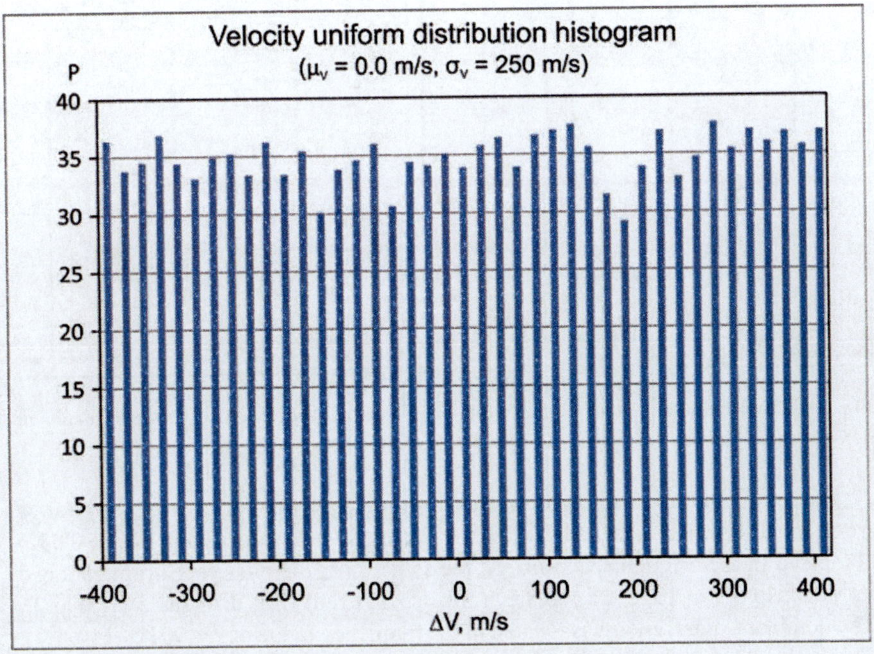


**Fig. 1.5** The histogram of velocity variation for uniform distribution: $\mu_v = 0.0$ m/s, $\sigma_v = 250$ m/s

as taking into account the energy of the components of the spectrum (Fig. 1.11, Tables 1.2 and 1.3).

For models with different parameter perturbations, quantitative estimates of the spectra characterizing their sensitivity to changes in the medium model have been performed. Parameter estimates are given in Table 1.4.

Estimates of the signal spectra characteristics for the initial, velocity-modified and random noise over frequency ranges are summarized in Tables 1.2 and 1.3.

A change in velocity of the model with a standard deviation $\sigma_V = 250$ m/s led to a change in the amplitudes of the seismic trace by a total $\sum A = 2.993$ c.u. in absolute magnitude (on average by $\Delta A_a = 0.021$ c.u.) and amounted to 65.1% relative to the initial one (Figs. 1.6 and 1.9, Table 1.1, No 2).

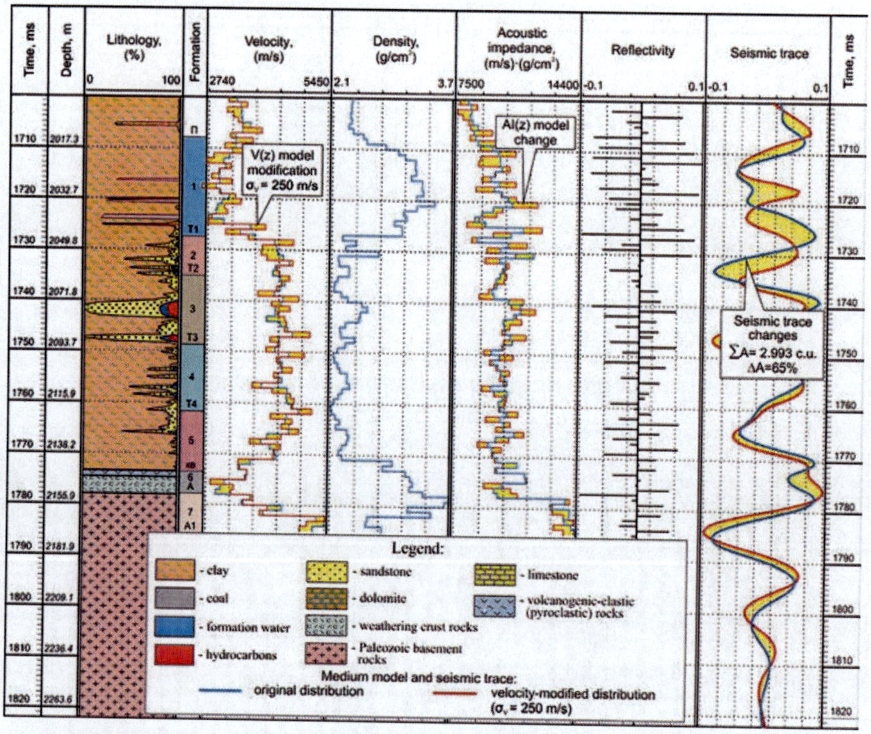


**Fig. 1.6** The detail influence analysis of a thin-layer medium velocity model on the formation of a wave field

The amplitude spectrum of the seismic trace over the entire frequency range from 0 to 500 Hz has changed by $\sum E = 8.093$ c.u. (33.6%) (on average $\Delta E_{av} = 0.032$) compared to the original spectrum. At the same time, the phase spectrum changed its values to $\sum P = 1330$ rad (76.8%) ($\Delta P_a = 5.301$ rad) (Fig. 1.8, Table 1.1, No 2). The dominant signal energy is located in the frequency range of 25–90 Hz (Fig. 1.7). The main changes in the signal energy also occur in the same frequency band (Fig. 1.8).

The amplitude spectra for both models decrease their intensities with increasing frequency and tend to zero from a frequency of about 270 Hz (visually the changes are no longer noticeable). The phase spectra of the models behave differently – the changes become noticeable with increasing frequency and are most significantly manifested in the high frequency band. That is, in the frequency range where the thin-layer medium manifests itself in the geophysical description.

Consider the distribution of the trace spectra characteristics in different parts of the frequency spectrum.

According to the constructed random noise with a signal-to-noise ratio S/N = 5 (parameters $\mu_a = 0.0$ и $\sigma_n = 0.008$ c.u.) (Figs. 1.9 and 1.10), four bands are allocated in the spectral domain: 0–100 Hz, 100–160 Hz, 160–270 Hz, and 270–500 Hz

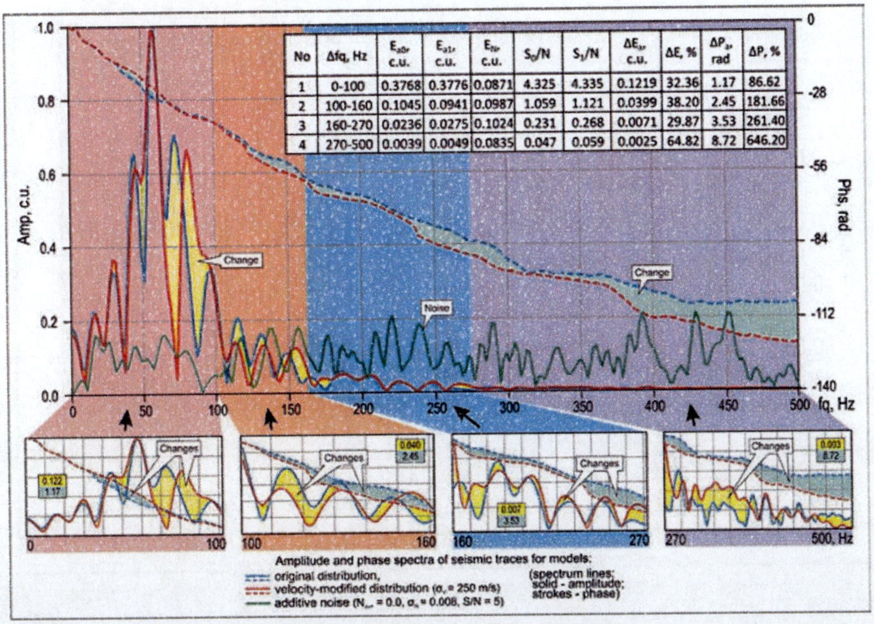


| No | Δfq, Hz | $E_{a0}$, c.u. | $E_{a1}$, c.u. | $E_{N}$, c.u. | $S_0$/N | $S_1$/N | $ΔE_a$, c.u. | ΔE, % | $ΔP_a$, rad | ΔP, % |
|---|---|---|---|---|---|---|---|---|---|---|
| 1 | 0-100 | 0.3768 | 0.3776 | 0.0871 | 4.325 | 4.335 | 0.1219 | 32.36 | 1.17 | 86.62 |
| 2 | 100-160 | 0.1045 | 0.0941 | 0.0987 | 1.059 | 1.121 | 0.0399 | 38.20 | 2.45 | 181.66 |
| 3 | 160-270 | 0.0236 | 0.0275 | 0.1024 | 0.231 | 0.268 | 0.0071 | 29.87 | 3.53 | 261.40 |
| 4 | 270-500 | 0.0039 | 0.0049 | 0.0835 | 0.047 | 0.059 | 0.0025 | 64.82 | 8.72 | 646.20 |

**Fig. 1.7** The medium velocity-modification influence on the structure of the amplitude and phase spectra of the seismic trace at different frequency ranges

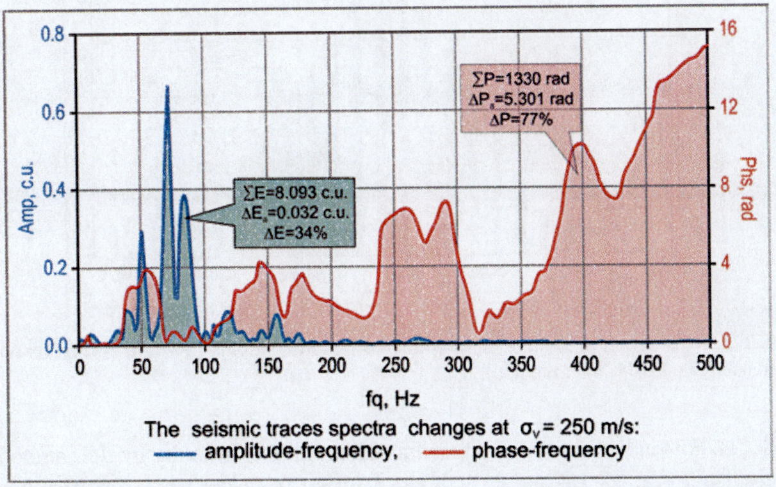


**Fig. 1.8** The difference in the amplitude and phase spectra of the seismic trace relative to the initial ones with variations in the wave velocity in the medium

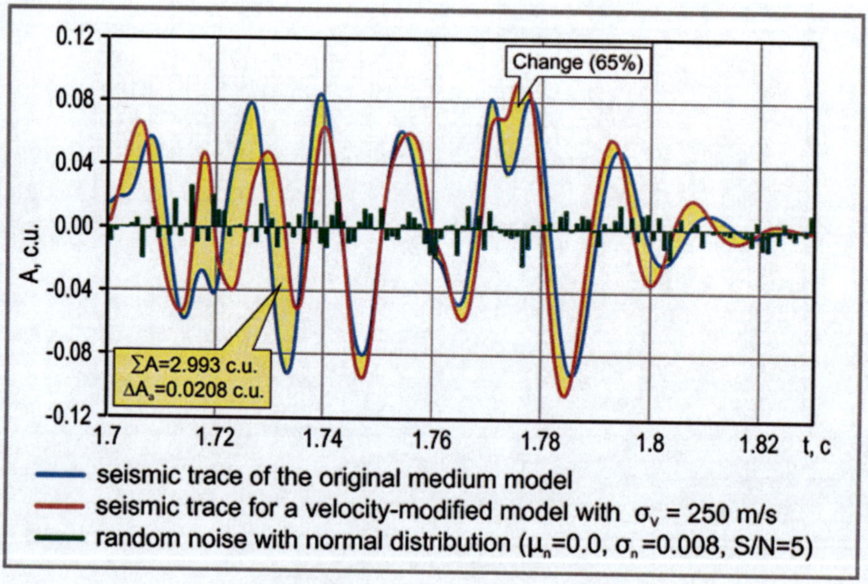


**Fig. 1.9** The effect of changes in the wave velocity of medium on the structure of the seismic trace

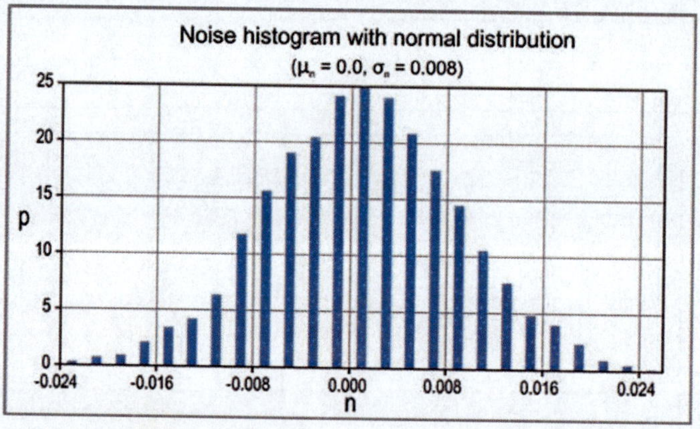


**Fig. 1.10** The histogram of random noise with a normal distribution: $\mu_n = 0.0$, $\sigma_n = 0.008$ (noise realised using the Box-Muller method)

(Fig. 1.11). For each of the bands, the parameters for both velocity models and noise are calculated (Tables 1.2 and 1.3). It can be seen from the tables that for the 1st section, where the main energy components of the frequencies are located, the signal-to-noise ratio is S/N = 4.33, the energy change is $\Delta E_a = 0.12$ c.u. (32%), the phase component of the spectrum is $\Delta P_a = 1.17$ rad (87%) (here the signal significantly exceeds the noise). In the second section: S/N = 1.06, $\Delta E_a = 0.04$ c.u. (38%), $\Delta P_a = 2.45$ rad (182%) (the signal level is approximately the same as the noise

**Table 1.2** Parameter distribution of the trace spectrum along frequency range

| No | Δfq, Hz | $E_{a0}$, c.u. | $E_{a1}$, c.u. | $E_N$, c.u. | $S_0/N$ | $S_1/N$ | $\Delta E_a$, c.u. | ΔE, % | $\Delta P_a$, rad | ΔP, % |
|---|---|---|---|---|---|---|---|---|---|---|
| 1 | 0–100 | 0.3768 | 0.3776 | 0.0871 | 4.325 | 4.335 | 0.1219 | 32.36 | 1.17 | 86.62 |
| 2 | 100–160 | 0.1045 | 0.0941 | 0.0987 | 1.059 | 1.121 | 0.0399 | 38.20 | 2.45 | 181.66 |
| 3 | 160–270 | 0.0236 | 0.0275 | 0.1024 | 0.231 | 0.268 | 0.0071 | 29.87 | 3.53 | 261.40 |
| 4 | 270–500 | 0.0039 | 0.0049 | 0.0835 | 0.047 | 0.059 | 0.0025 | 64.82 | 8.72 | 646.20 |

Parameters description:
No is the frequency range number
Δfq is the frequency range, Hz
$E_{a0}$ is the energy of the original signal, c.u
$E_{a1}$ is the energy of the velocity modified signal ($\sigma_V = 250$ m/s), c.u
$E_N$ is the random noise energy, c.u
$S_0/N$ is the signal-to-noise ratio for the original model
$S_1/N$ is the signal-to-noise ratio for a velocity -modified model
$\Delta E_a$ is the energy change when model is modified, c.u
ΔE is the relative change in energy when model is modified, %
$\Delta P_a$ is the phase change when model is modified, c.u
ΔP is the relative change in phase when model is modified, %

**Table 1.3** Signal and noise characteristics across frequency range

| No | Δfq, Hz | $E_{a0}$, c.u. | $E_N$, c.u. | $S_0/N$ | ΔEa, c.u. | ΔE, % | ΔPa, rad |
|---|---|---|---|---|---|---|---|
| 1 | 0–100 | 0.3768 | 0.0871 | 4.325 | 0.2897 | 76.9 | 4.97 |
| 2 | 100–160 | 0.1045 | 0.0987 | 1.059 | 0.0058 | 5.6 | 4.65 |
| 3 | 160–270 | 0.0236 | 0.1024 | 0.231 | 0.0788 | 333.3 | 15.43 |
| 4 | 270–500 | 0.0039 | 0.0835 | 0.047 | 0.0796 | 2046.7 | 58.01 |

Parameters description:
No is the frequency range number
Δfq, Hz is the frequency range, Hz
$E_{a0}$ is the energy of the original signal, c.u
$E_N$ is the energy of the noise, c.u
$S_0/N$ is the signal-to-noise ratio for the original model
$\Delta E_a$ is the difference in signal and noise energy, c.u
ΔE is the relative difference in signal and noise energy, %
$\Delta P_a$ is the difference in signal and noise phase, rad

level). Ha In the third section: S/N = 0.23 (the level of the useful signal is less noise), $\Delta E_a = 0.007$ c.u. (30%), $\Delta P_a = 3.53$ rad (261%). On the fourth, S/N = 0.05 (the level of the useful signal is significantly less than noise), $\Delta E_a = 0.003$ c.u. (65%), $\Delta P_a = 8.72$ rad (646%) (Table 1.2).

Note the signal-to-noise ratio decreases with increasing frequency band numbers. Frequency band 2 can be attributed to moderate weak signals. Bands 3 and 4 are for very weak signals. At the same time, with an increase in the band number, the changes in the values of the phase spectra increase from 1.17 on the 1st band (0–100 Hz) to 8.72 on the 4th band (270–500 Hz). But it is the 2nd, 3rd and 4th bands that determine the thin layer nature of the medium. The increase in $\Delta P_{S01}$ indicates the phase component of the spectrum contains the largest information about the fine details of the geological section. Thus, as predicted theoretically above in formulas (1.1)–(1.21), it is the waveform that is most sensitive to changes in the parameters of the section.

Let's consider the effect of changes in the velocity model on the structures of the seismic trace, amplitude and phase spectra. To study this question, thin-layer velocity modifications with standard deviations $\sigma_v = 100$ m/s, $\sigma_v = 250$ m/s и $\sigma_v = 400$ m/s were set (Table 1.1).

The seismic traces of the models, spectra and their changes are shown in Figs. 1.12, 1.13 and 1.14, respectively.

Changes in the structure of seismic traces when variation of the model with parameters $\sigma_v = 100$ m/s, $\sigma_v = 250$ m/s, and $\sigma_v = 400$ m/s, respectively, are shown in Fig. 1.12. It can be noted with an increase in the differentiation of the medium, the distinction relative to the original model also increases: at $\sigma_v = 100$ m/s, the change is $\sum A = 1.172$ c.u. in absolute magnitudes (or on average $\Delta A_a = 0.008$ c.u.) (25%), at $\sigma_v = 250$ м/с, the change is $\sum A = 2.993$ c.u. ($\Delta A_a = 0.021$ c.u.) (65%) and at $\sigma_v = 400$ m/s, $\sum A = 4.718$ c.u. ($\Delta A_a = 0.033$ c.u.) (103%) (Table 1.1, No 1–3).

Table 1.4 shows the parameters of the spectra characterizing the sensitivity (and hence the quantity of information about the medium) for the models of velocity variation $\sigma_v = 50$ m/s, $\sigma_v = 100$ m/s and $\sigma_v = 250$ m/s. With increasing differentiation of the medium, the frequency band $\Delta F_{Amp}$ in the amplitude spectrum changed from 22.5 to 54.0 Hz, in the phase spectrum $\Delta F_{Phs}$ increased from 52.5 to 400.0 Hz, respectively (Table 1.4). The ratio of frequency bands $\delta F$ changed from 2.33 to 8.13

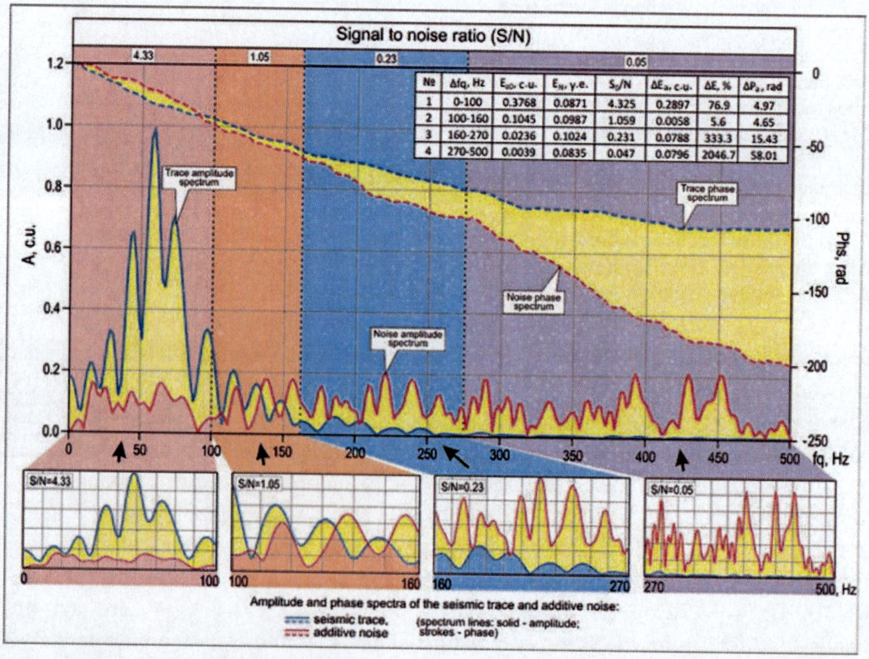


**Fig. 1.11** The amplitude and phase spectra of the signal and noise and distribution of the signal-to-noise ratio over frequency ranges

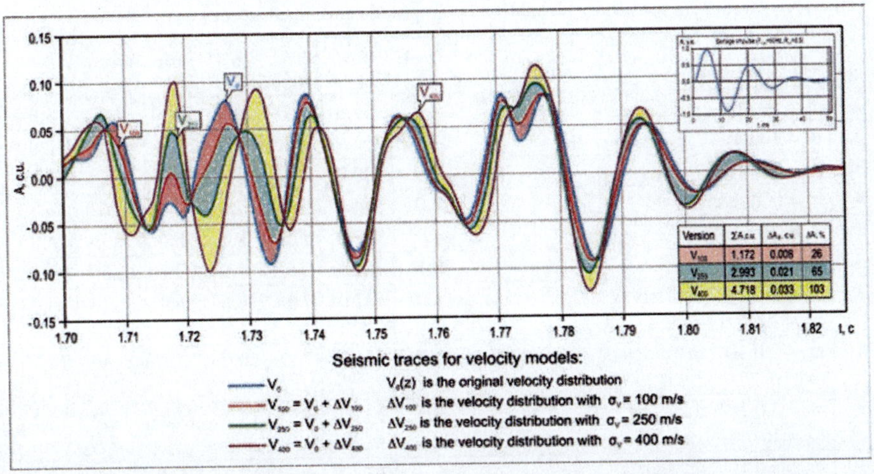


**Fig. 1.12** The synthetic seismograms for various models of wave velocity distribution in the medium

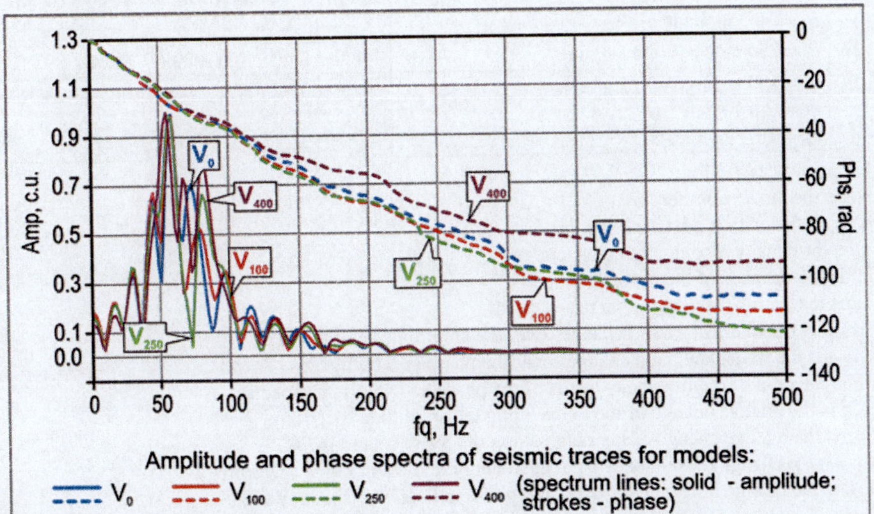


**Fig. 1.13** The amplitude and phase spectra of seismic traces for various models of medium velocity

(Table 1.4). The relative frequency band changes were: from 4.5% to 10.8% for the amplitude spectrum of $S_{amp}$ and from 10.5% to 80.0% for the phase one $S_{phs}$. Relative changes in frequency band and signal energy are expressed in the parameter $C_{amp}$ with a change from 2.6% to 8.37%. Relative changes in the frequency band and seismic signal phase are represented by the parameter $C_{phs}$ with a change from 3.94% to 79.49% (Table 1.4). All the given estimates of the spectra parameters show

**Table 1.4** Parameters sensitivity estimation by amplitude and phase spectra

| No | Model parameters | $\Delta F_{Amp}$, Hz | $\Delta F_{Phs}$, Hz | $\delta F$ | $S_{amp}$, % | $S_{phs}$, % | $C_{amp}$, % | $C_{phs}$, % |
|---|---|---|---|---|---|---|---|---|
| 1 | $\sigma_v = 50$ m/s | 22.5 | 52.5 | 2.33 | 4.5 | 10.5 | 2.60 | 3.94 |
| 2 | $\sigma_v = 100$ m/s | 31.0 | 252.0 | 8.13 | 6.3 | 50.4 | 4.21 | 42.80 |
| 3 | $\sigma_v = 250$ m/s | 54.0 | 400.0 | 7.41 | 10.8 | 80.0 | 8.37 | 79.49 |
| 4 | $\sigma_d = 0.10$ g/cm$^3$ | 38.0 | 244.0 | 6.42 | 7.6 | 48.8 | 5.63 | 46.75 |
| 5 | $\sigma_d = 0.15$ g/cm$^3$ | 44.0 | 378.0 | 8.59 | 8.8 | 75.6 | 6.76 | 74.35 |
| 6 | $\sigma_d = 0.30$ g/cm$^3$ | 70.1 | 466.0 | 6.65 | 14.0 | 93.2 | 11.50 | 93.03 |
| 7 | $m_a = 1.0 \cdot 10^{-6}$ m$^{-1}$ rad$^{-1}$s, $\sigma_m = 10^{-6}$ m$^{-1}$ rad$^{-1}$s | 18.0 | 132.0 | 7.33 | 3.6 | 26.4 | 1.14 | 25.23 |
| 8 | $m_a = 2.0 \cdot 10^{-6}$ m$^{-1}$rad$^{-1}$s, $\sigma_m = 10^{-6m-1}$ rad$^{-1}$s | 50.0 | 394.0 | 7.88 | 10.0 | 78.8 | 5.40 | 78.70 |
| 9 | $m_a = 3.0 \cdot 10^{-6m-1}$ rad$^{-1}$s, $\sigma_m = 10^{-6}$ m$^{-1}$ rad$^{-1}$s | 66.0 | 394.0 | 5.97 | 13.2 | 78.8 | 8.52 | 78.70 |
| 10 | $m_a = 5.0 \cdot 10^{-6}$ m$^{-1}$rad$^{-1}$s, $\sigma_m = 10^{-6}$ m$^{-1}$rad$^{-1}$s | 86.0 | 394.0 | 4.58 | 17.2 | 78.8 | 13.23 | 78.69 |
| 11 | $m_a = 1.0 \cdot 10^{-5}$ m$^{-1}$rad$^{-1}$s, $\sigma_m = 10^{-6}$ m$^{-1}$rad$^{-1}$s | 98.0 | 466.0 | 4.76 | 19.6 | 93.2 | 16.55 | 93.18 |
| 12 | $m_a = 4.0 \cdot 10^{-5}$ m$^{-1}$rad$^{-1}$s, $\sigma_m = 10^{-6}$ m$^{-1}$rad$^{-1}$s | 480.0 | 494.0 | 1.03 | 96.0 | 98.8 | 95.82 | 98.79 |
| 13 | $\sigma_v = 250$ m/s; $\sigma_d = 0.15$ g/cm$^3$ | 70.0 | 448.0 | 6.40 | 14.0 | 89.6 | 11.58 | 89.25 |
| 14 | $\sigma_v = 250$ m/s, $\sigma_d = 0.15$ g/cm$^3$, $m_a = 2.0 \cdot 10^{-6}$ m$^{-1}$rad$^{-1}$s, $\sigma_m = 10^{-6}$ m$^{-1}$rad$^{-1}$s | 86.0 | 492.0 | 5.72 | 17.2 | 98.4 | 14.94 | 98.38 |

Parameter description:
No is the model number
$\sigma_v$ is the velocity model variance, m/s
$\sigma_d$ is the density model variance, g/cm$^3$
$\sigma_m$ is the variance of the absorption factor, m $^{-1}$rad$^{-1}$s
$m_a$ is the average absorption factor, m $^{-1}$rad$^{-1}$s
$\Delta F_{Amp}$ is the frequency band changes over the amplitude spectrum, Hz
$\Delta F_{Phs}$ is the frequency band changes over the phase spectrum, Hz
$\delta F$ is the ratio of the frequency bands of the phase and amplitude spectra
$S_{amp}$ is the relative change of the range along the amplitude spectrum, %
$S_{phs}$ is the relative change of the range along the phase spectrum, %
$C_{amp}$ is the change of frequency band and energy in the amplitude spectrum, %
$C_{phs}$ is the change of frequency band and phase in the phase spectrum, %

a greater sensitivity of the phase characteristics relative to the energy ones, which means it is possible to extract detailed information about the velocity distribution of a thin-layer medium.

The structures of the phase and amplitude spectra of the traces for these models, representing different degrees of velocity contrast, are shown in Fig. 1.13. The general dependencies for amplitude and phase spectra are preserved: with increasing frequency, the sensitivity in energy decreases, and in the form of the signal increases. This is most clearly seen on the graphs of the differences in the amplitude and phase spectra relative to the original spectra ($V_0$) (Fig. 1.14, Table 1.1). Quantitative

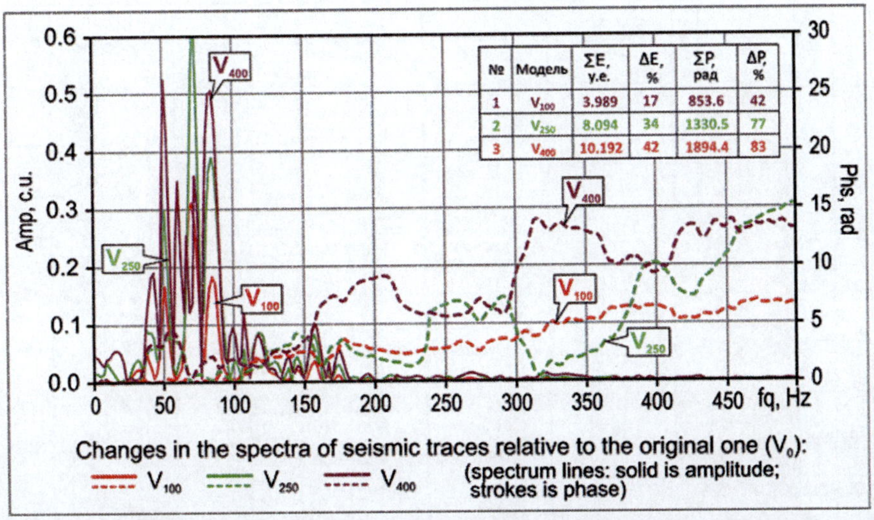


**Fig. 1.14** The difference in the amplitude and phase spectra of seismic traces relative to the original ones for different models of wave velocity distribution in the medium

estimates of changes are also given on the graphs. For the model $V_{100}$ ($\sigma_v = 100$ m/s) the total change of the amplitude spectrum is $\sum E = 3.989$ c.u. (17%), of the phase is $\sum P = 854$ rad (42%); for the model $V_{250}$ – $\sum E = 8.094$ c.u. (34%), $\sum P = 1331$ rad (77%); and for the model $V_{400}$ – $\sum E = 10.192$ c.u. (42%), $\sum P = 1884$ rad (83%) (Fig. 1.14, Table 1.1). Note if you increase the value of the standard deviation of the velocity model in both spectra increase changes, however, the phase spectrum reacts more sharply, and his contribution at high frequencies is much higher than that of the amplitude spectrum.

## 1.4 Spectral Characteristics Sensitivity Evaluation Under Density Modification in the Medium

The change in the original density model was calculated similarly to the formation of a perturbed velocity model: a model of random density corrections with a uniform distribution was constructed and added to the initial density model (Fig. 1.2). The calculated seismic traces with the initial and modified density models are shown in Fig. 1.15. The spectra and their changes are shown in Figs. 1.16 and 1.17.

Variations of the density model with parameters $\sigma_d = 0.1$ g/cm$^3$, $\sigma_d = 0.15$ g/cm$^3$, and $\sigma_d = 0.3$ g/cm$^3$, respectively, are shown in Fig. 1.18.

The amplitude and phase spectra for the described density modification models are shown in Fig. 1.19. The corresponding changes in the amplitude and phase spectra are shown in Fig. 1.20. With their own features, the general depencies found in the previous numerical experiment with respect to velocity are confirmed in the case

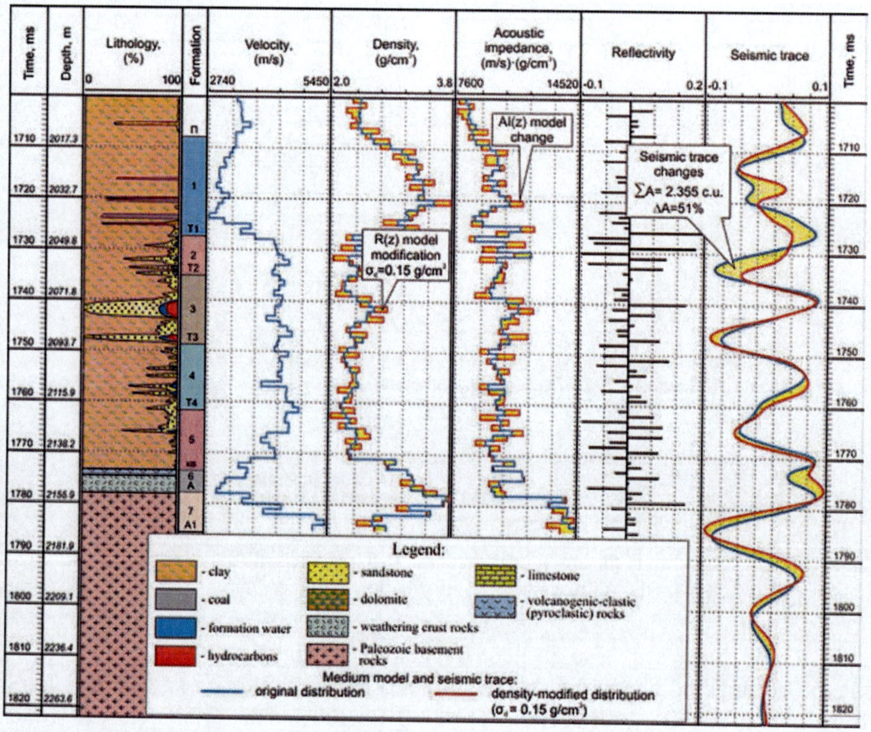


**Fig. 1.15** The detail influence analysis of a thin-layer medium density model on the formation of a wave field

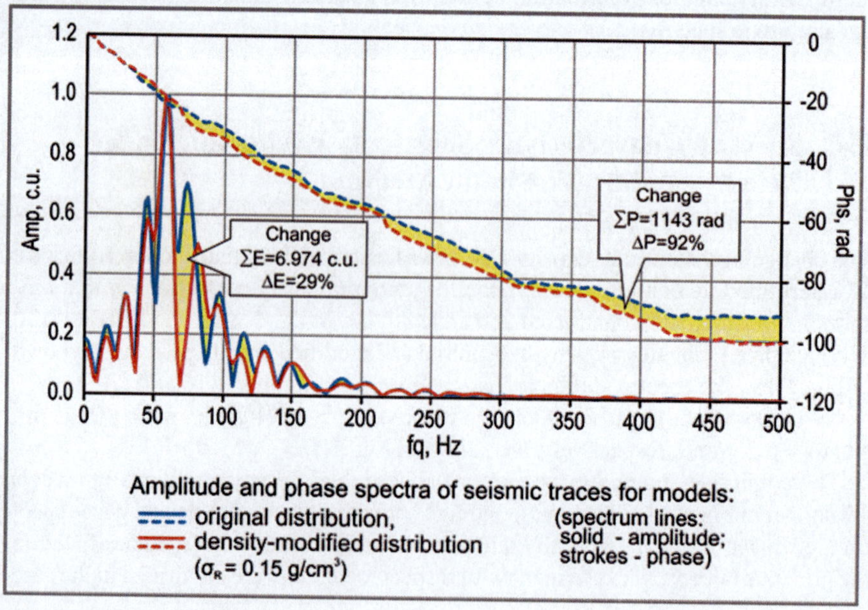


**Fig. 1.16** The medium density modification effect on the structure of the amplitude and phase spectra of the seismic trace

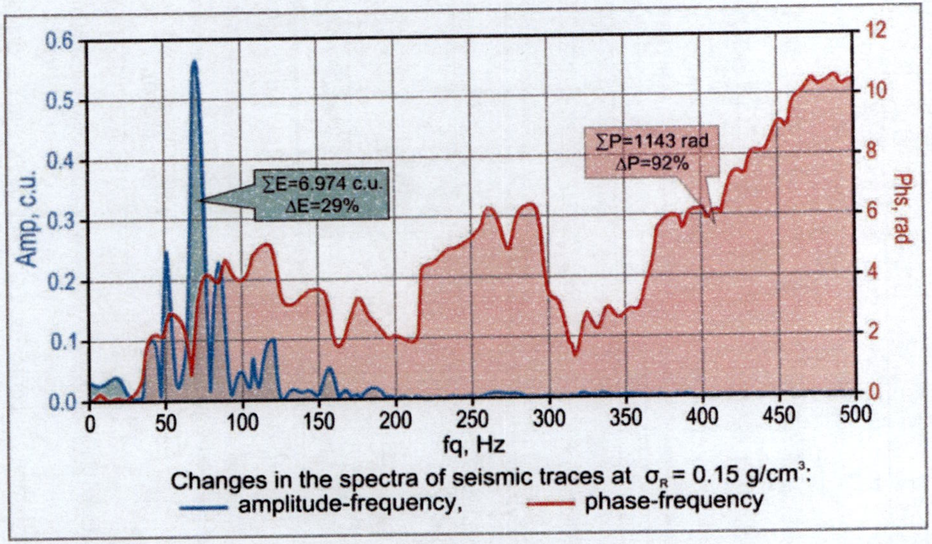


**Fig. 1.17** The difference in the amplitude and phase spectra of the seismic trace relative to the original ones with a variation in the density of the medium

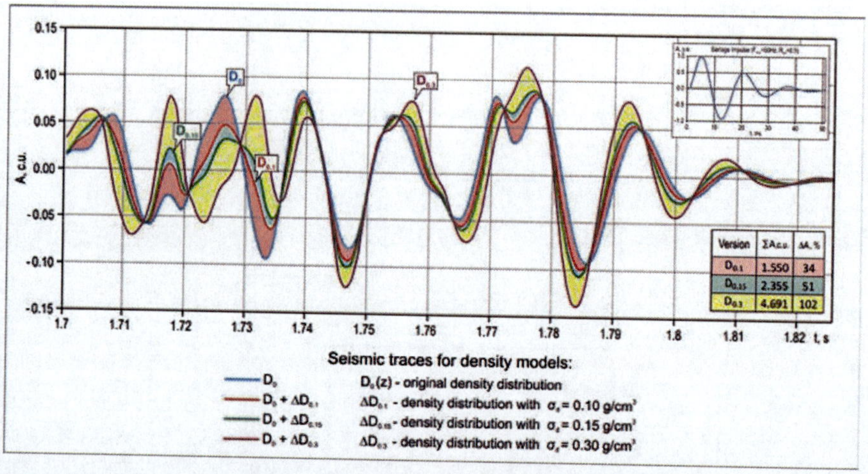


**Fig. 1.18** The synthetic seismograms for various models of medium density

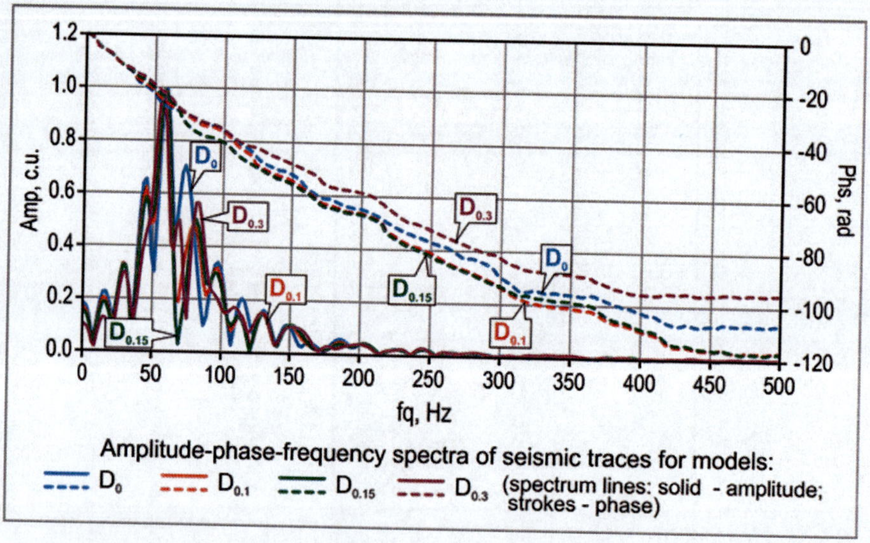


**Fig. 1.19** The amplitude and phase spectra of seismic traces for various models of medium density

of the density distribution. Specifically, for models $D_{0.1}$ ($\sigma_d$ = 0.1 g/cm³), $D_{0.15}$ ($\sigma_d$ = 0.15 g/cm³), $D_{0.3}$ ($\sigma_d$ = 0.3 g/cm³), respectively, $\sum E$ = 4.969 c.u. (21%); $\sum E$ = 6.974 c.u. (29%); $\sum E$ = 9.483 c.u. (39%) (Fig. 1.20, Table 1.1, No 4–6).

Increasing the differentiation of the medium in density increases the total change in amplitudes from $\sum A$ = 1.55 c.u. (34%) in absolute (and relative) magnitude to $\sum A$ = 4.691 c.u. (102%), which is numerically comparable to the change in amplitudes with variations in wave velocity in the medium from 100 to 400 m/s.

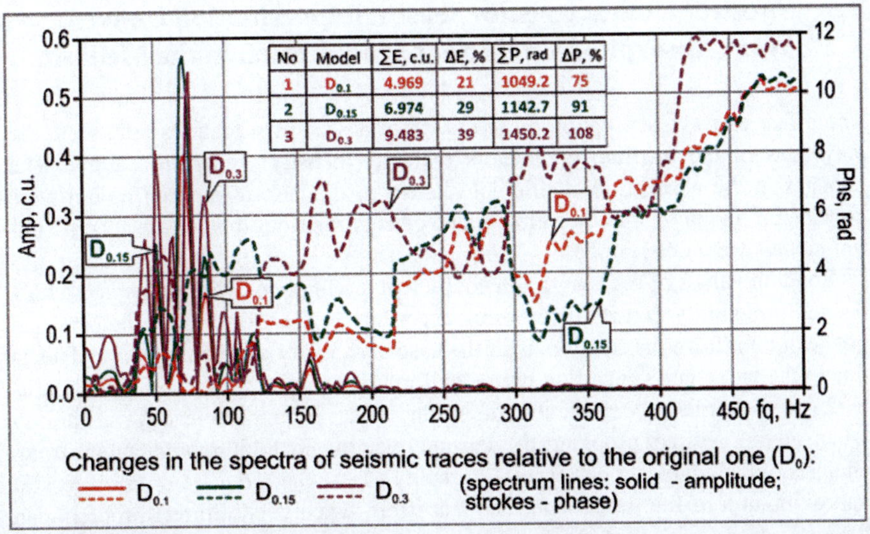


**Fig. 1.20** The difference in the amplitude and phase spectra of seismic traces relative to the original ones with different medium density models

Table 1.4 shows estimates of the spectra parameters characterizing the sensitivity for models of changes in the density distribution of the medium $\sigma_d = 0.10$ g/cm$^3$, $\sigma_d = 0.15$ g/cm$^3$ and $\sigma_d = 0.30$ g/cm$^3$. With an increase in density contrast of the medium, the frequency band $\Delta F_{Amp}$ in the amplitude spectrum changed from 38.0 to 44.0 Hz, in the phase spectrum $\Delta F_{Phs}$ increased from 244.0 to 466.0 Hz, respectively (Table 1.4). The ratio of frequency bands $\delta F$ changed from 6.42 to 8.59 (Table 1.4). The relative frequency band changes were: from 7.6% to 14.0% for the amplitude spectrum of $S_{amp}$ and from 48.8% to 93.2% for the phase that $S_{phs}$. Relative changes in frequency band and energy are expressed in the parameter $C_{amp}$ with a change from 5.63% to 11.5%. Relative changes in the frequency band and signal phase are represented by the parameter $C_{phs}$ with a change from 46.75% to 93.03% (Table 1.4).

Again, all the given estimates of the parameters of the spectra show a greater sensitivity of the phase characteristics relative to the energy ones, which means it is possible to extract detailed information about the density distribution of a thin-layer medium.

Observations along the frequency axis of the spectra also show the greatest changes for the phase spectrum are manifested in the band of higher frequencies that determine the thinness of the medium. Moreover, the changes are proportional to the frequency $\omega$. Conversely, in the amplitude spectrum, the most significant changes are in the limited low frequency band $[\omega_1, \omega_2]$, where the largest energy components are manifested. This also corresponds to the theoretical reasoning described above: sensitivity to changes in the density of the medium is most reflected in the phase spectrum.

## 1.5 Spectral Characteristics Sensitivity Evaluation Under Wave Absorption Coefficient Modification in the Medium

The effect of wave absorption by the medium has its own features relative to the variations of the medium parameters (velocity, density) considered above. The changes in the absorption coefficient relative to the zero-absorption model were considered, in contrast to the velocity and density distributions, where specific initial models were determined.

Random values of the absorption coefficient modification were generated with a normal probability density distribution according to the Box-Muller method, and not with a uniform distribution as in the case of velocity and density ones. This is due to the fact when generating with a uniform distribution, sharp individual outliers of the absorption coefficient value occur, which sometimes leads to instability when solving a direct problem (the seismic trace may exhibit unpredictable strong changes, which does not correspond to reality). Here the rare case of unsuccessful approximation of the model can manifest itself, when even a direct problem can have an unstable solution (the instability of direct and inverse problems was mentioned in the previous monographs [2, 3]).

However, giving and controlling this subtlety, it is possible to apply a more complex, uniform distribution of probability density, while expecting the same conclusions.

An example of the absorption coefficient construction and its histogram are shown in Figs. 1.21 and 1.22.

As mentioned above, when modeling the wave field, the dependence of the absorption coefficient $\alpha(\omega)$ vs the frequency $\omega$ was approximated by the linear function $\alpha(\omega) = m_a \cdot \omega$, where $\alpha$ is the absorption coefficient in $m^{-1}$, $m_a$ is the absorption

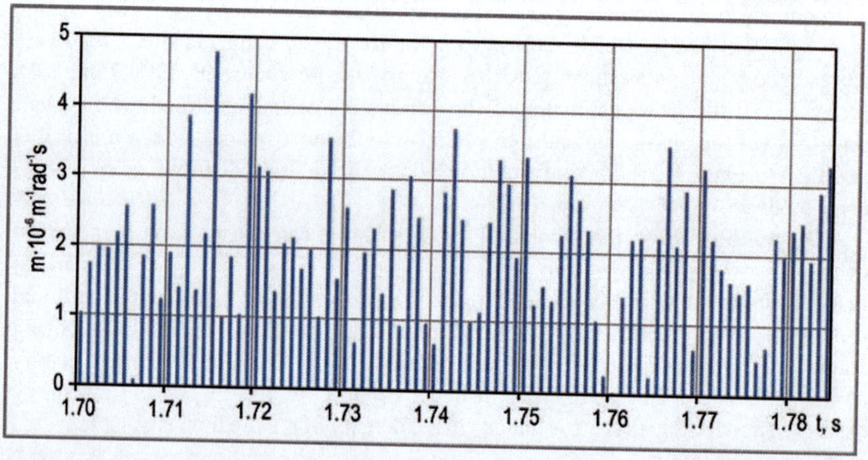


**Fig. 1.21** Absorption coefficient with normal distribution (constructed by the Box-Muller method: $\mu_m = 2 \cdot 10^{-6}$ $m^{-1}rad^{-1}s$, $\sigma_m = 10^{-6}$ $m^{-1}rad^{-1}s$)

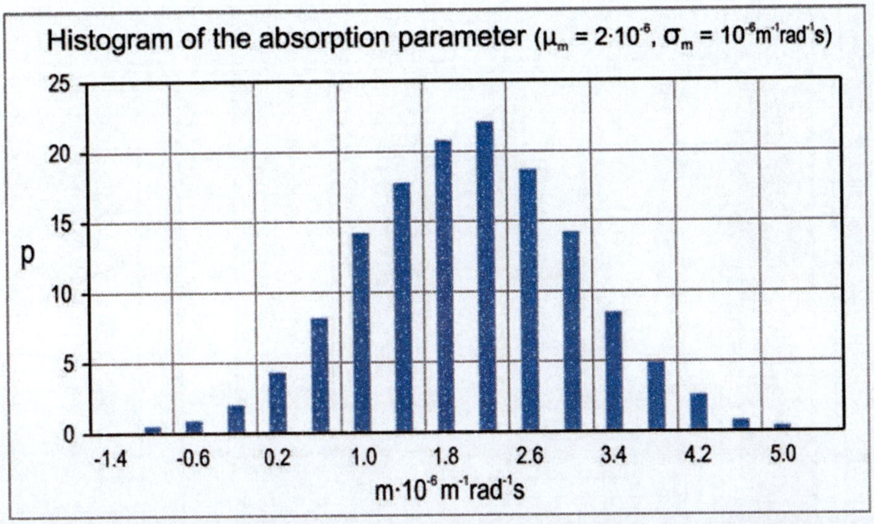


**Fig. 1.22** The absorption parameter *m* histogram with a normal distribution: $\mu_m = 2{\cdot}10^{-6}\ m^{-1}rad^{-1}s$, $\sigma_m = 10^{-6}\ m^{-1}rad^{-1}s$

parameter, $\omega = 2\pi f$ is the cyclic (angular) frequency rad/s, f is frequency in Hz. In this case, the dependence $\alpha(\omega)$ is determined by setting the absorption parameter $m_a$, the unit of which is $m^{-1}\ rad^{-1}s$. That is, the perturbation of the absorption coefficient $\alpha(\omega)$ is reduced to the modification of the absorption parameter $m_a$.

Figure 1.23 shows a detailed analysis of modeling the formation of seismic traces for three distributions of the absorption coefficient. At the same time, the parameters were set in such a way that the standard deviations are the same for all versions ($\alpha(\omega) = m_a{\cdot}\omega = 10^{-6}{\cdot}\omega\ m^{-1}$), and the average values (mathematical expectations) $\mu_a$ changed: $2{\cdot}10^{-6}$, $6{\cdot}10^{-6}$, $1{\cdot}10^{-5}\ m^{-1}\ rad^{-1}$ s, respectively. The amplitudes of the seismic traces for these variants vary in accordance with the specified values of the absorption functions: with an increase in the average absorption value, the attenuation increased.

Figure 1.24 shows graphs of the amplitude and phase spectra when the absorption model is perturbed by $\alpha(\omega) = m_a \cdot \omega = 2{\cdot}10^{-6}{\cdot}\omega\ m^{-1}$.

It is possible to notice minor changes in the amplitude spectrum in the low-frequency operating band and significant changes in the phase spectrum in the high-frequency range with a continuous monotonic increase with the frequency of the contribution of changes in the absorption model. Up to a frequency of 100 Hz, the phase components do not show noticeable changes. This function gives the attenuation of the spectrum components with frequency increasing. The distribution of sensitivity of the considered functions is also confirmed (Fig. 1.24).

Figure 1.25 shows a quantitative assessment of the changes in the spectra during the model modification: amplitude $\sum E = 2.174$ c.u. (9%), phase – $\sum P = 2824$ rad (94%) (Table 1.1, No 8).

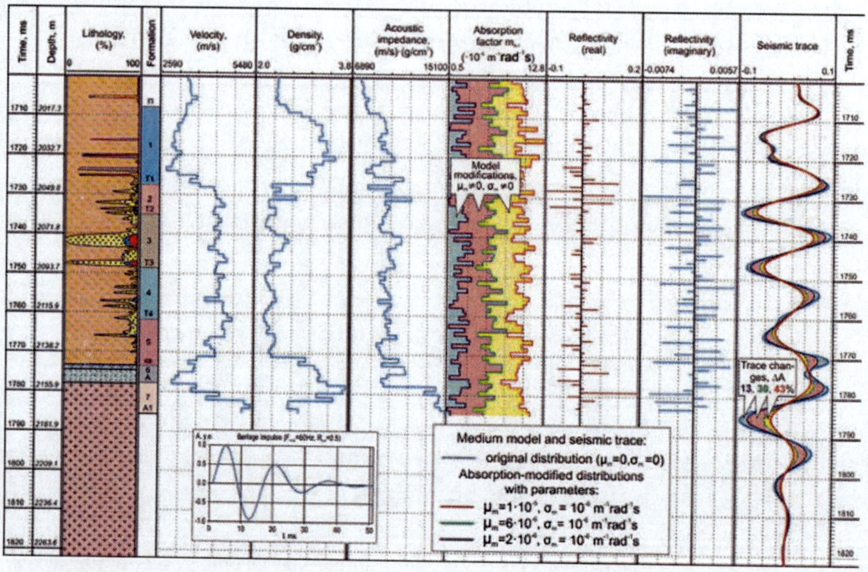


Fig. 1.23 The detail influence analysis of a thin-layer medium velocity, density, and absorption models on the formation of a wave field

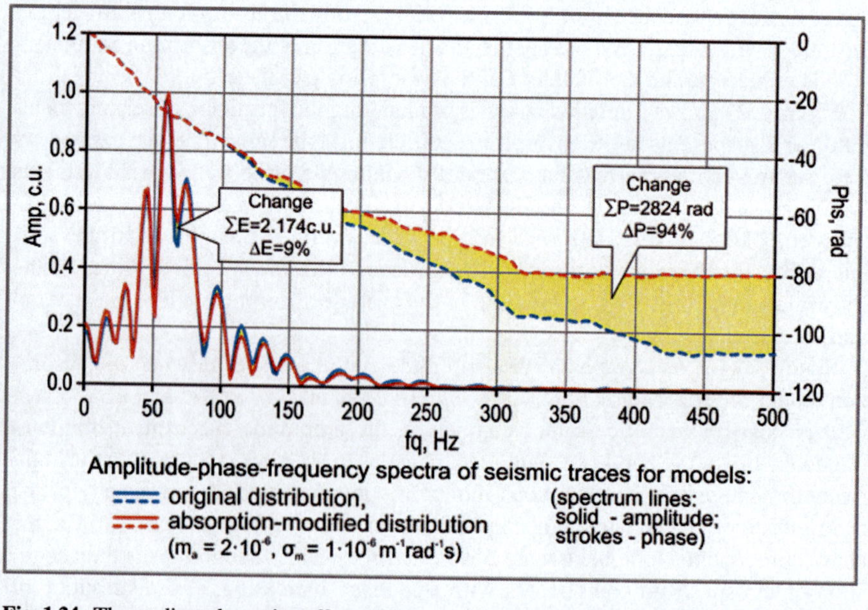


Fig. 1.24 The medium absorption effect on the structure of the amplitude and phase spectra of the seismic trace

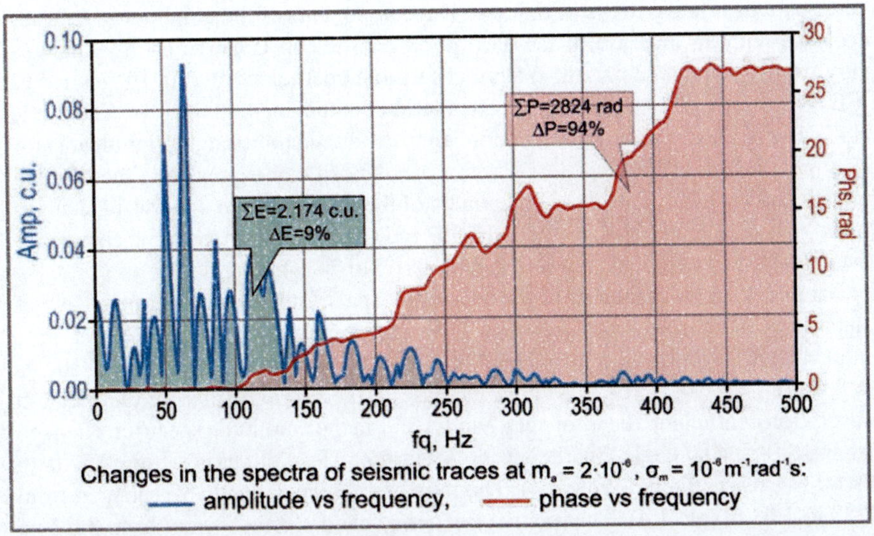


**Fig. 1.25** The difference in the amplitude and phase spectra of the seismic trace relative to the original ones when taking into account the medium absorption

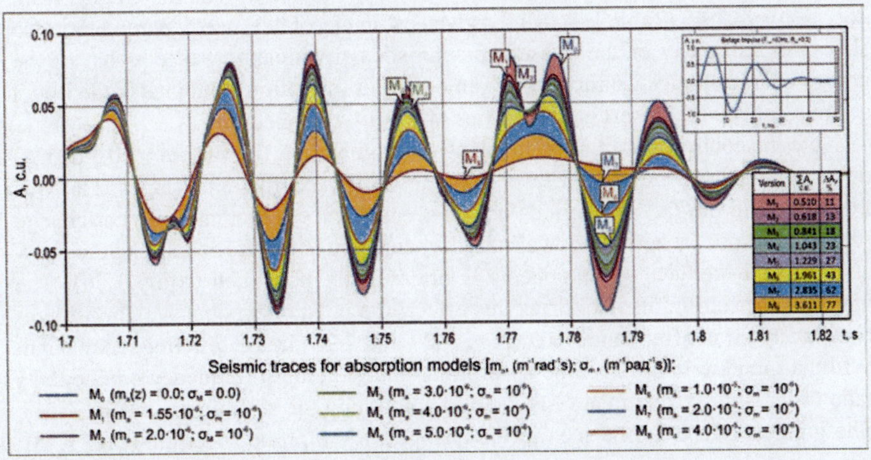


**Fig. 1.26** The synthetic seismograms for various models of environmental absorption

Figure 1.26 shows the seismograms for the largest number of absorption models (8 variants), when the average absorption parameters vary from 0 to $4{\cdot}10^{-5}$ m$^{-1}$ rad$^{-1}$ s with the same varience.

The similarity of the variance means the absorption contrast is the same for all models, although the implementation of random variables was different. Figure 1.28 represents changes in the spectra of eight variants of the absorption coefficient perturbations relative to the initial one. The components of the amplitude spectrum decrease with an increase in the absorption coefficient. Relative energy changes vary from $\sum E = 1.812$ c.u. (8%) at an absorption coefficient of $1.55{\cdot}10^{-6}{\cdot}\omega\ m^{-1}$ to $\sum E = 12.6$ c.u. (52%) at the highest attenuation coefficient $4{\cdot}10^{-5}\ m^{-1}{\cdot}\omega$ (Table 1.1, No 7–14). With large absorption coefficients, the attenuation can be significant (for example, the wave field registered over a gas deposit can express like the M6-M8 models in Fig. 1.26). The phase component of the spectrum also underwent changes from $\sum P = 2187$ rad (94%) to $\sum P = 13{,}567$ rad (23466%) with the same changes in the absorption coefficient (Fig. 1.28, Table 1.1, No 7–14).

Table 1.4 shows estimates of the sensitivity spectrum parameters for models of changes in the absorption distribution of the medium $\alpha(\omega) = m_a{\cdot}\omega = 1{\cdot}10^{-6}{\cdot}\omega\ m^{-1}$, $\alpha(\omega) = 2{\cdot}10^{-6}{\cdot}\omega\ m^{-1}$, and $\alpha(\omega) = 3{\cdot}10^{-6}{\cdot}\omega\ m^{-1}$, $\alpha(\omega) = 5{\cdot}10^{-6}{\cdot}\omega\ m^{-1}$, $\alpha(\omega) = 1{\cdot}10^{-5}{\cdot}\omega\ m^{-1}$ и $\alpha(\omega) = 4{\cdot}10^{-4}{\cdot}\omega\ m^{-1}$. With an increase in the average value of the medium absorption coefficient, the frequency band $\Delta F_{Amp}$ in the amplitude spectrum changed from 18.0 to 480.0 Hz, in the phase spectrum $\Delta F_{Phs}$ increased from 132.0 to 494.0 Hz, respectively (Table 1.4). The ratio of frequency bands $\delta F$ changed from 7.33 to 1.03 (Table 1.4). The relative frequency band changes were: from 3.6% to 96.0% for the amplitude spectrum of $S_{amp}$ and from 26.4% to 98.8% for the phase one $S_{phs}$. Relative changes in frequency band and signal energy are expressed in the parameter $C_{amp}$ with a change from 1.14% to 95.82%. Relative changes in the frequency band and phase are represented by the parameter $C_{phs}$ with a change from 25.23% to 98.79% (Table 1.4). All the given estimates of the spectra parameters also show the sensitivity of the phase characteristics significantly exceeds that of the energy characteristics, confirming the amount of information contained in the initial seismic data on the absorbing properties of a thin-layer medium.

Note the contrast distribution in the absorption models ($\sigma_\alpha = \sigma_m{\cdot}\omega = 10^{-6}{\cdot}\omega\ m^{-1}$) did not change, but the average level of the absorption coefficient changed. This was manifested in the fact the periods on the seismic traces practically do not change (the traces cross the zero level at almost the same points for all models and the influence of high-frequency components is not visually noticeable) (Fig. 1.26). The amplitude spectrum also confirms this effect: the components behave almost synchronously on the frequency axis (Figs. 1.27 and 1.28) in the low frequency band. With an increase in the average absorption value, the high-frequency components fade faster – this effect manifests itself in smoothing the seismic trace (Fig. 1.26). The greatest effect can be seen in the changes in the phase spectrum (Fig. 1.28): significant changes are concentrated in the high frequency region, at the same time, the regularity of the change in the dependence of the phase spectrum to the frequency remains ($\delta P(\omega) \sim \omega$) (see the above assumption).

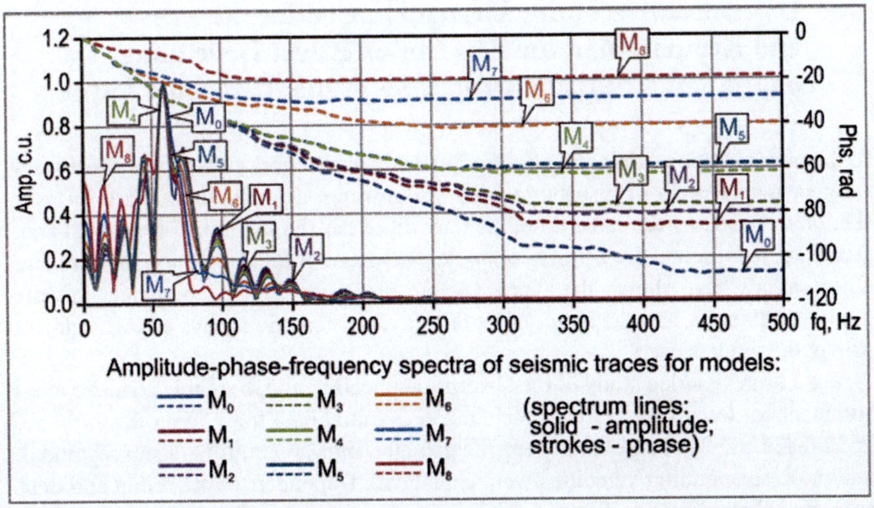


**Fig. 1.27** The amplitude and phase spectra of seismic traces for various models of medium absorption

| No | Model | ΣE, c.u. | Δ E, % | ΣP, rad | ΔP, % |
|---|---|---|---|---|---|
| 1 | $M_1$ | 1.812 | 7.5 | 2186.7 | 94.0 |
| 2 | $M_2$ | 2.174 | 9.0 | 2824.2 | 94.1 |
| 3 | $M_3$ | 2.919 | 12.1 | 3262.0 | 127.3 |
| 4 | $M_4$ | 3.577 | 14.8 | 5297.5 | 144.2 |
| 5 | $M_5$ | 4.144 | 17.2 | 5672.7 | 367.6 |
| 6 | $M_6$ | 6.110 | 25.4 | 9222.6 | 454.1 |
| 7 | $M_7$ | 8.030 | 33.3 | 11331.4 | 722.0 |
| 8 | $M_8$ | 12.600 | 52.3 | 13566.6 | 23466.2 |

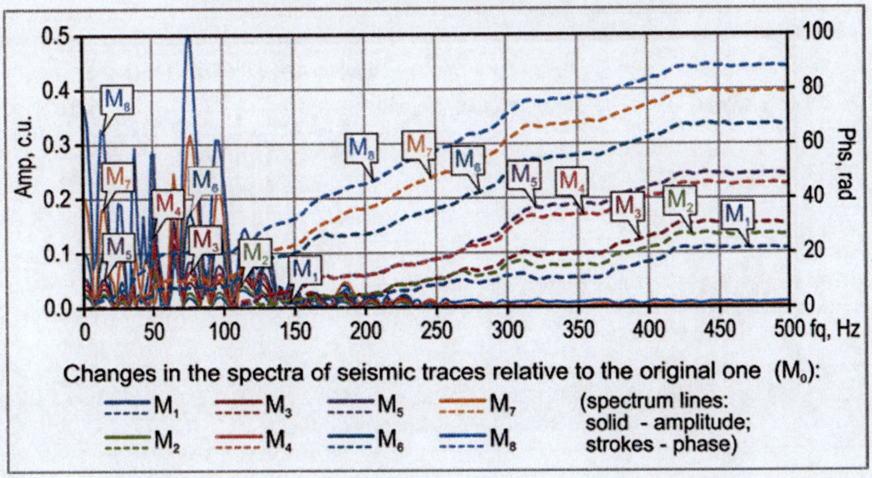


**Fig. 1.28** The difference in the amplitude and phase spectra of seismic traces relative to the original ones with different medium absorption models

## 1.6 Determination of the Absorption Coefficient and Evaluation of the Absorption Effect Depending on the Thickness and Properties of the Absorbing Layer

In the previous subsection, the effects of seismic wave energy absorption were considered, where random distributions of the absorption coefficient in a thin-layer section were set. At the same time, the interval of the studied model was small and, therefore, insufficient for a more complete disclosure of the absorption mechanism. Observations have shown the average value of the absorption coefficient has the greatest influence, and the effect of its thin-layer distribution plays an insignificant role in the models used.

For a more detailed study of the absorption phenomenon, borehole acoustic measurements of longitudinal waves were taken as initial data for a larger depth range: 2632–3835 m. The sonic data were recalculated into an equitime acoustic model, and the corresponding velocities were calculated. Dependences are found and densities and absorption coefficients are computed.

To simulate the absorption effect, the fact that absorption depends on the density of rocks was used. Namely, when a wave passes through less dense rocks, the wave fades more, the speed of waves passage decreases. To overcome a less dense medium, the wave spends more energy, leaves more of its energy in it than when passing through a medium with denser rocks, according to the law of energy conservation during wave processes. Using this fact, the absorption coefficient distribution was generated from the density distribution, applying the linear law of the change in the absorption coefficient from the density of rocks (1.34). Figure 1.29 shows the calculated distribution of the absorption coefficient using a linear relationship:

$$\alpha(\omega) = m \cdot \omega = (-2.5 \cdot d + 7.0) \cdot 10^{-6} \cdot \omega, \quad \mathrm{m}^{-1}, \tag{1.34}$$

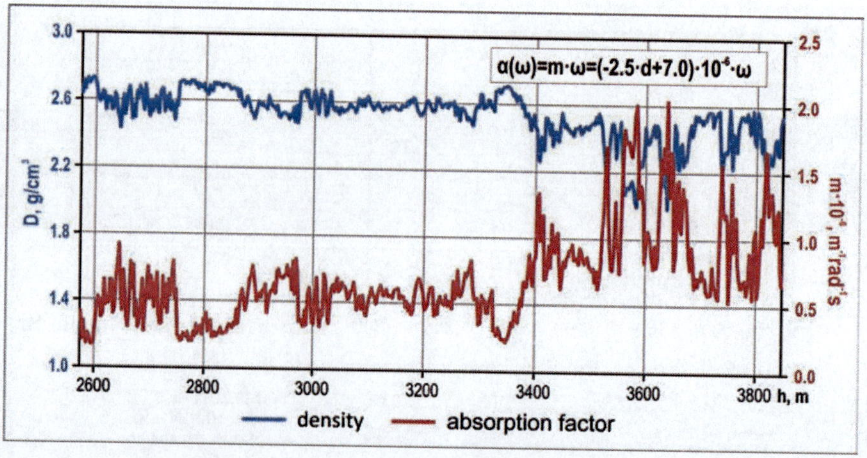


**Fig. 1.29** Calculation of the absorption coefficient from the density

where *d* is the density, g/cm$^3$; α(ω) = m·ω, *α* is the absorption coefficient, m$^{-1}$; *m* is the absorption factor, m$^{-1}$rad$^{-1}$s; ω = *2πf* is the angular frequency, rad/s; *f* is frequency, Hz.

One of the seismic wave amplitudes attenuation features is the absorption effect depends on how far the wave travels. The longer the distance moved by the wave, the greater the effect of the wave amplitude damping, i.e., the property of accumulation of the absorption effect should manifest itself here. In this case, the wave crosses abnormally absorbing layers on its way, leaving in them a certain part of the energy generated in the source at the initial moment of its path. Next, in a numerical experiment, let's see if this is the case.

Another feature is the uneven dependence of attenuation on frequency, i.e. the dispersion of waves. This effect is taken into account when choosing and using the method of modeling the propagation of seismic waves and is described in the corresponding section above. In this paper, the Futterman dispersion relation was used: Eqs. (1.22 and 1.23). In this case, a linear approximation of the dependence of the absorption function on frequency was used: α(ω) = m·ω, where m is the coefficient in the linear dependence of absorption on frequency, ω is the angular frequency.

In Fig. 1.30 the comparison of model data of seismic traces with and without absorption is given. It can be seen from the figure with increasing depth, the attenuation of amplitudes intensifies. Energy changes become more significant (Fig. 1.30). The changes in amplitudes when taking into consideration the absorption of waves in the medium were ∑A = 9.582 c.u. (28%) (Table 1.5, version No 12).

The analysis of spectral characteristics is shown in Figs. 1.31 and 1.32 and in Table 1.5. The amplitude spectrum shows slight variations in the operating frequency band. The energy changes there are ∑E = 3.964 c.u. (18%) (Table 1.5, No 12). Significant variations are reflected in the phase spectrum and amount to ∑P = 31,180 rad (82%) (Table 1.5, No 12).

In Table 1.7 (No. 12), you can find estimates of the spectra parameters by sensitivity. The change in the frequency band $\Delta F_{Amp}$ in the amplitude spectrum was 60.0 Hz, in the phase spectrum was $\Delta F_{Phs}$ = 456.0 Hz, respectively (Table 1.7). The ratio of frequency bands δF is 7.6 (Table 1.7). Relative frequency band changes were 12.0% for the amplitude spectrum $S_{amp}$ and 92.1% for the phase one $S_{phs}$. The relative change in frequency band and energy represents the $C_{amp}$ parameter with 8.56%. The relative change in frequency band and phase is expressed by the $C_{phs}$ parameter with a value equal to 91.19% (Table 1.7).

These estimates show the phase spectrum contains the largest amount of information that could be extracted with the use of adequate digital processing technology. Another effect of the simulation experiment is an increase in the effect of attenuation with depth (cumulative property of the absorption effect).

To observe the dependence of the phenomenon of attenuation of anomalously absorbing media on the thickness of the absorbing layer, a wave field with different thicknesses of absorbing layers was simulated.

The numerical experiment was performed for 11 variants with different thickness of the active absorbing layer. The initial parameters of the anomalous wave absorption models and the results of quantitative estimates are given in Table 1.6.

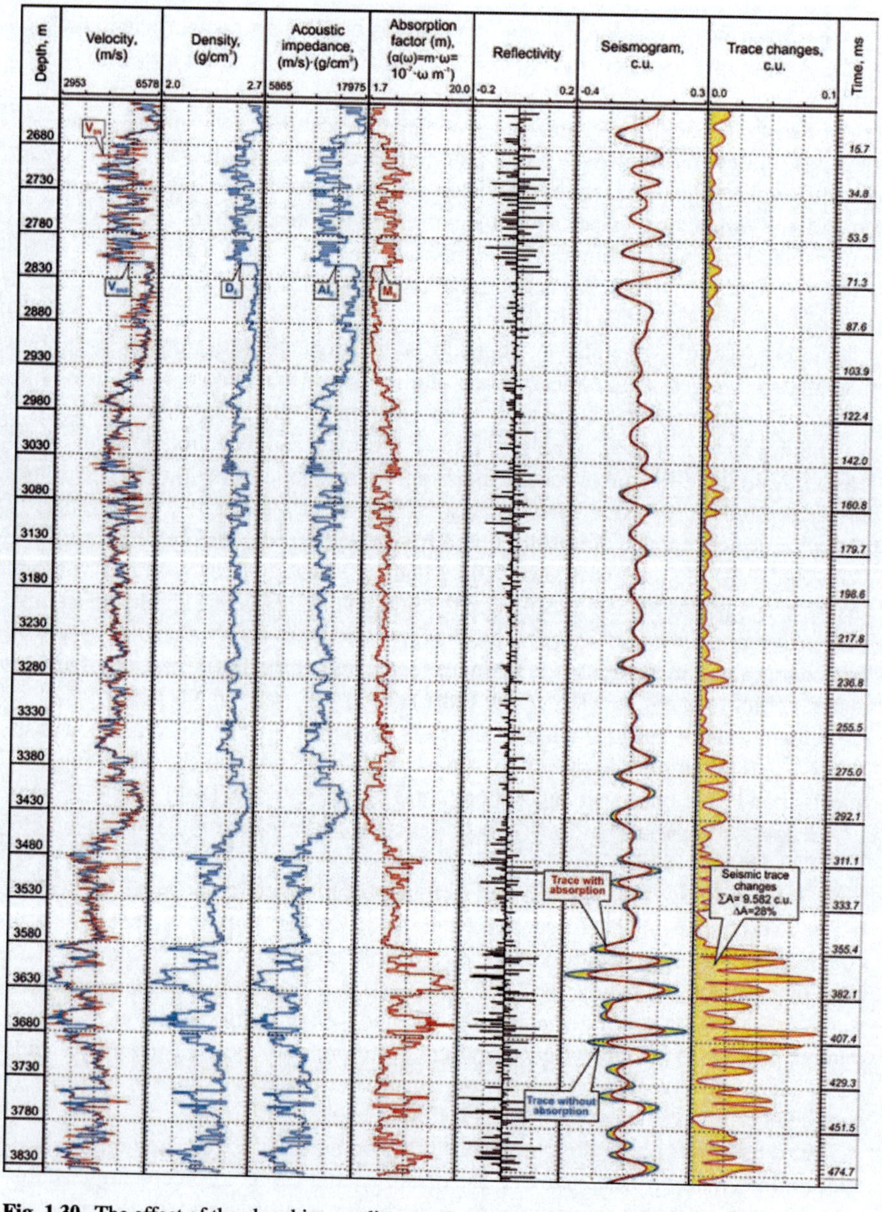


**Fig. 1.30** The effect of the absorbing medium on the structure of the wave field

For visual observation and data analysis, Fig. 1.33 presents 5 different variants from the total number of versions of the work performed.

The thickness of the absorbing layers varied from 13.4 to 261.8 m. At the same time, the average value of the absorption coefficient in the layers remained approximately the same $\mu_m = 5.75 \cdot 10^{-6}$ m$^{-1}$rad$^{-1}$s, with a standard deviation

**Table 1.5** Parameters of models, seismic traces and their spectra

| No | Model name | Model parameters | | | $\sum A$, c.u. | $\Delta A_a$, c.u. | $\Delta A$, % | $\sum E$, c.u. | $\Delta E_a$, c.u. | $\Delta E$, % | $\sum P$, rad | $\Delta P_a$, rad | $\Delta P$, % |
|---|---|---|---|---|---|---|---|---|---|---|---|---|---|
| | | dh, m | dt, ms | $m_a \cdot 10^{-6}$, $m^{-1}$ $rad^{-1}$s | | | | | | | | | |
| 1 | H5 | 13.41 | 5 | 5.142 | 1.904 | 0.0036 | 5.46 | 0.807 | 0.0032 | 3.62 | 6244.7 | 24.88 | 47.92 |
| 2 | H10 | 25.74 | 10 | 5.967 | 3.658 | 0.0070 | 10.50 | 1.583 | 0.0063 | 7.10 | 12567.0 | 50.07 | 56.20 |
| 3 | H20 | 52.13 | 20 | 5.706 | 6.325 | 0.0121 | 18.15 | 2.344 | 0.0093 | 10.52 | 17727.2 | 70.63 | 63.28 |
| 4 | H30 | 77.23 | 30 | 5.948 | 8.889 | 0.0170 | 25.51 | 3.550 | 0.0141 | 15.92 | 31055.1 | 123.73 | 81.50 |
| 5 | H40 | 101.53 | 40 | 6.236 | 11.276 | 0.0216 | 32.36 | 4.283 | 0.0171 | 19.21 | 36255.6 | 144.44 | 88.63 |
| 6 | H50 | 129.88 | 50 | 5.823 | 12.486 | 0.0239 | 35.83 | 4.781 | 0.0190 | 21.45 | 40940.6 | 163.11 | 95.07 |
| 7 | H60 | 157.03 | 60 | 5.687 | 14.005 | 0.0268 | 40.19 | 5.324 | 0.0212 | 23.88 | 43977.9 | 175.21 | 99.24 |
| 8 | H70 | 182.94 | 70 | 5.707 | 15.303 | 0.0293 | 43.92 | 5.843 | 0.0233 | 26.21 | 49133.9 | 195.75 | 106.32 |
| 9 | H80 | 208.90 | 80 | 5.714 | 16.475 | 0.0316 | 47.28 | 6.382 | 0.0254 | 28.63 | 50209.9 | 200.04 | 107.80 |
| 10 | H90 | 235.82 | 90 | 5.644 | 17.405 | 0.0333 | 49.95 | 6.869 | 0.0274 | 30.81 | 52785.3 | 210.30 | 111.32 |
| 11 | H100 | 261.83 | 100 | 5.651 | 18.404 | 0.0353 | 52.81 | 7.283 | 0.0290 | 32.67 | 59729.2 | 237.96 | 120.86 |
| 12 | MP | Absorption calc-n by model | | | 9.582 | 0.0184 | 27.50 | 3.964 | 0.0158 | 17.78 | 31180.5 | 124.23 | 81.71 |

Parameters description:
No is the model number
$\sigma_v$ is the velocity variance, m/s
$\sigma_d$ is the density variance, g/cm$^3$
$\sigma_m$ is the variance of the absorption factor, $m^{-1}rad^{-1}s$
$m_a$ is the average absorption factor, $m^{-1}rad^{-1}s$
$\sum A$ is the total change at the seismic trace amplitudes, c.u
$\Delta A_a$ is the average value of the change at the seismic trace amplitudes, c.u
$\Delta A$ is relative changes in seismic trace amplitudes, %
$\sum E$ is the total change in the amplitude spectrum, c.u
$\Delta E_a$ is the average value of the amplitude spectrum change, c.u
$\Delta E$ is the relative average value of the change in the amplitude spectrum, %
$\sum P$ is the total change in the phase spectrum, rad
$\Delta P_a$ is the average value of the phase spectrum change, rad
$\Delta P$ is the relative average value of the change in the phase spectrum, %

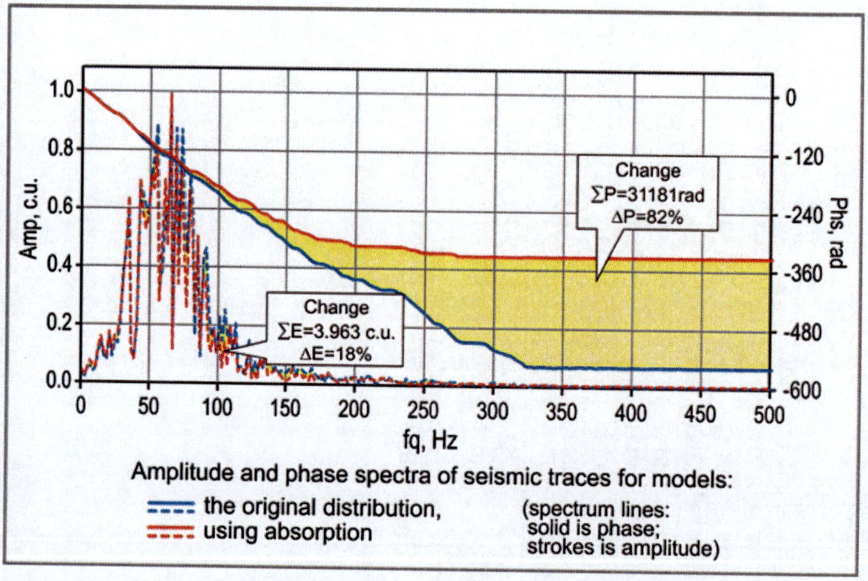


**Fig. 1.31** The effect of medium absorption on the structure of the amplitude and phase spectra of the seismic trace

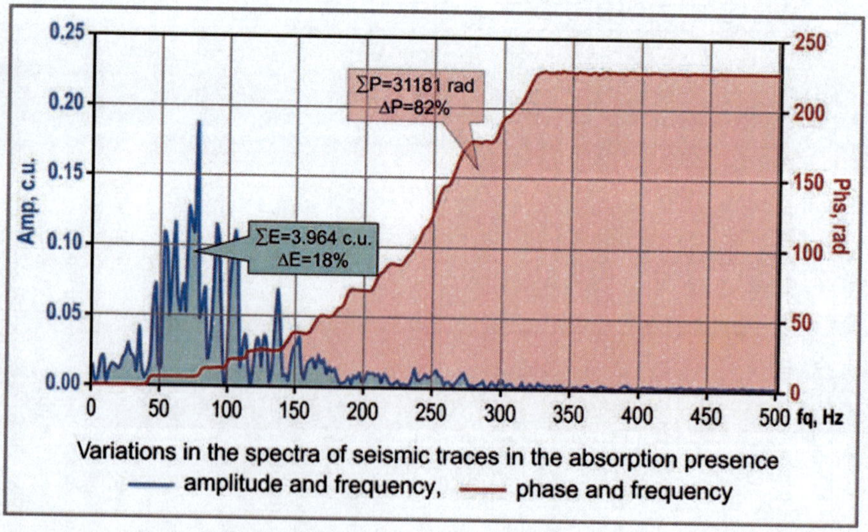


**Fig. 1.32** The difference in the amplitude and phase spectra of the seismic trace relative to the original ones in the presence of wave absorption in the medium

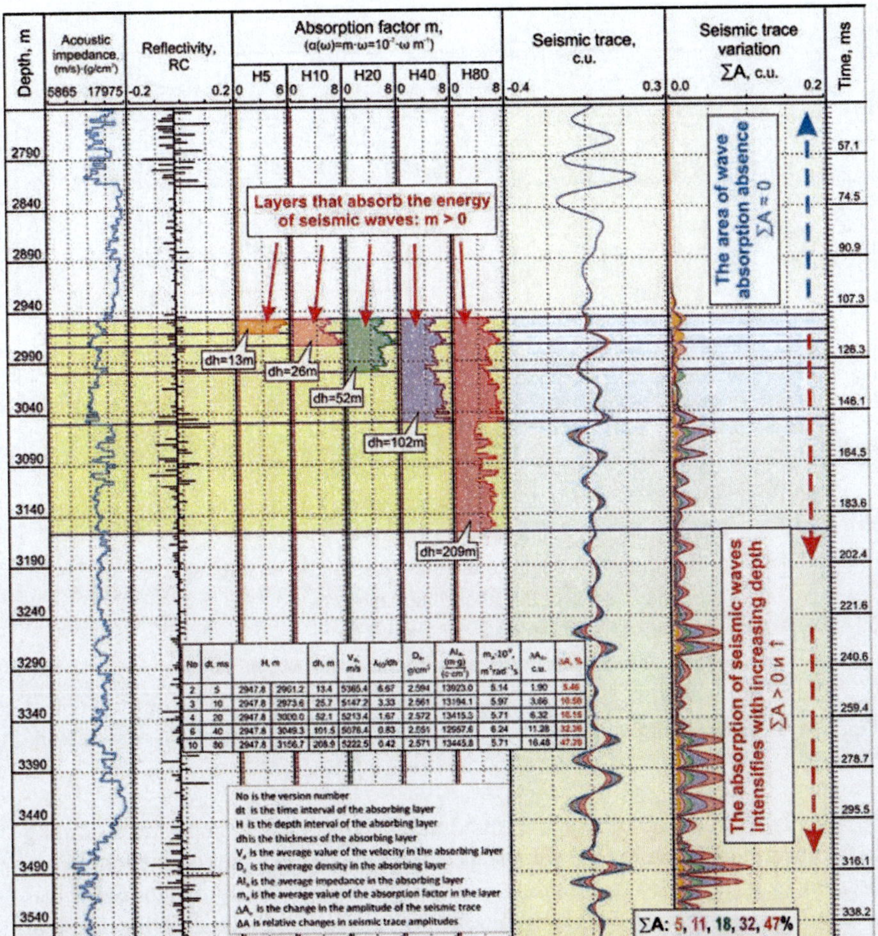


**Fig. 1.33** The effect of the absorbing layer thickness on the structure of the seismic trace

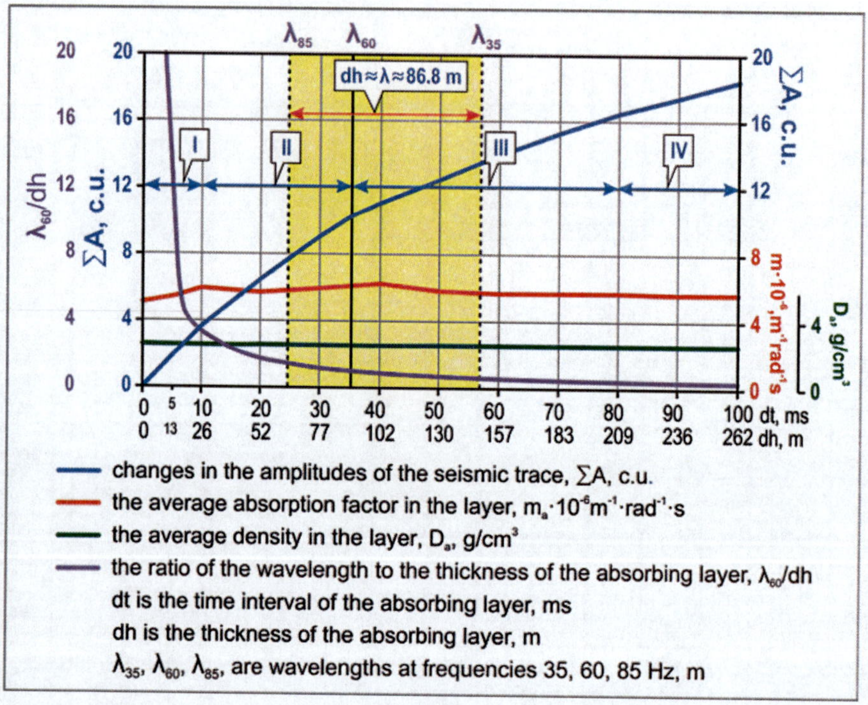


**Fig. 1.34** Dependence of the seismic trace amplitudes (energy) vs the thickness of the absorbing layer

$\sigma_m = 0.27 \cdot 10^{-6}$ m$^{-1}$ rad$^{-1}$s, which ensured the homogeneity of the layers according to the absorption parameter. The constancy of all the layers under consideration can be seen from Fig. 1.34 and Table 1.6. Initially, the uniformity of the layers was ensured by density, the average value of which also remained at the level of $\mu_d = 2.57$ g/cm$^3$, with a standard deviation from the average $\sigma_d = 0.01$ g/cm$^3$. The homogeneity of anomalously absorbing layers is also manifested in their acoustic characteristics: velocity, acoustic impedance, respectively, $\mu_v = 5210$ m/s, $\sigma_v = 73$ m/s and $\mu_{ai} = 13{,}403$ (m·g)/(s·cm$^3$), $\sigma_{ai} = 242$ (m·g)/(s·cm$^3$) (Table 1.6). Thus, in the numerical modelling, one investigated parameter of the model remains, that is the thickness of the absorbing object, which will make it possible to draw an unambiguous conclusion about the influence of the absorbing layer thickness on the characteristics of the wave field in the time and spectral domains.

From Fig. 1.33 and Table 1.6 it is seen when the thickness of the actively absorbing layer changes from dh = 13.4 m to dh = 261.8 m, the change in the amplitudes (energy) of the traces was from $\sum A = 1.9$ c.u. (6%) to $\sum A = 18.4$ c.u. (53%). Figure 1.34 clearly shows the dependence of attenuation vs thickness of the absorbing layer while maintaining the same other factors (velocity, density and absorption coefficient).

**Table 1.6** The influence of the absorbing layer thickness on the structure of the seismic trace

| No | dt, ms | H, m | | dh, m | $V_a$, m/s | $\lambda_{60}$/dh | $D_a$, g/cm$^3$ | Al, $\left(\frac{m \cdot g}{s \cdot cm^3}\right)$ | $m_a \cdot 10^{-6}$, м$^{-1}$ рад$^{-1}$с | $\Delta A_a$, у.е. | ΔA, % |
|---|---|---|---|---|---|---|---|---|---|---|---|
| 1 | 0 | 0.0 | 0.0 | 0.0 | 0.0 | | 0.000 | 0.0 | 0.000 | 0.00 | 0.0 |
| 2 | 5 | 2947.8 | 2961.2 | 13.4 | 5365.4 | 6.67 | 2.594 | 13923.0 | 5.142 | 1.90 | 5.46 |
| 3 | 10 | 2947.8 | 2973.6 | 25.7 | 5147.2 | 3.33 | 2.561 | 13194.1 | 5.967 | 3.66 | 10.50 |
| 4 | 20 | 2947.8 | 3000.0 | 52.1 | 5213.4 | 1.67 | 2.572 | 13415.3 | 5.706 | 6.32 | 18.15 |
| 5 | 30 | 2947.8 | 3025.0 | 77.2 | 5148.7 | 1.11 | 2.562 | 13199.2 | 5.948 | 8.89 | 25.51 |
| 6 | 40 | 2947.8 | 3049.3 | 101.5 | 5076.4 | 0.83 | 2.551 | 12957.6 | 6.236 | 11.28 | 32.36 |
| 7 | 50 | 2947.8 | 3077.7 | 129.9 | 5195.4 | 0.67 | 2.567 | 13355.2 | 5.823 | 12.49 | 35.83 |
| 8 | 60 | 2947.8 | 3104.8 | 157.0 | 5234.2 | 0.56 | 2.573 | 13484.9 | 5.687 | 14.01 | 40.19 |
| 9 | 70 | 2947.8 | 3130.8 | 182.9 | 5226.9 | 0.48 | 2.572 | 13460.6 | 5.707 | 15.30 | 43.92 |
| 10 | 80 | 2947.8 | 3156.7 | 208.9 | 5222.5 | 0.42 | 2.571 | 13445.8 | 5.714 | 16.48 | 47.28 |
| 11 | 90 | 2947.8 | 3183.6 | 235.8 | 5240.4 | 0.37 | 2.574 | 13505.6 | 5.644 | 17.40 | 49.95 |
| 12 | 100 | 2947.8 | 3209.6 | 261.8 | 5236.5 | 0.33 | 2.574 | 13492.5 | 5.651 | 18.40 | 52.81 |

Parameters description:
No is the model number
dt is the time interval of the absorbing layer, ms
H is the depth interval of the absorbing layer, m
dh is the thickness of the absorbing layer, m
$V_a$ is the average value of the velocity in the absorbing layer, m/s
$\lambda_{60}$/dh is the ratio of the wavelength to the thickness of the absorbing layer
$D_a$ is the average density in the absorbing layer, g/cm$^3$
$Al_a$ is the average acoustic impedance in the absorbing layer, (m·g)/(s·cm$^3$)
$m_a$ is the the average absorption factor in the layer, m$^{-1}$rad$^{-1}$s
$\Delta A_a$ is the changes in the amplitudes of the seismic trace, c.u
ΔA is relative changes in seismic trace amplitudes, %

Figure 1.34 also presents a graph of the change in the ratio of the wavelength to the thickness of the absorbing layer. The graph highlights the wavelength range for the boundary frequencies of 40 and 80 Hz, where the main energy of the signal is located. The graph demonstrates the wavelength value for the frequency with the maximum component of the spectrum of the elementary probing medium signal – 60 Hz (Berlage pulse). The behavior of the graph indicates at thicknesses significantly smaller than the length of the seismic wave, the absorption contribution to the seismic trace is smaller and with a decrease in the thickness of the formation, the absorption effect decreases. At thicknesses of large wavelengths, the absorbing effect increases. The layer absorbs more energy of the information carrier.

From the graph of changes in the amplitudes (energy) of $\sum A$, 4 sections (zones) can be distinguished using the breaking points (Fig. 1.34). The zones differ in the slope of the graph (the rate of change in the trace energy). The fastest changes correspond to the I zone, the smallest – IV (Fig. 1.34). The boundary of the transition from zone II to zone III coincides to the thickness of the formation corresponding to the wavelength at a frequency of 60 Hz, i.e. the component of the spectrum with maximum energy. The length of the sections increases by a multiple of a quarter of the wavelength. Here, the thickness of the absorbing layer and the length of the seismic wave are equal and amount to λ = 86.8 m. In section I and at the beginning of zone II, where the small thickness of the absorbing layer compared to the wavelength, the absorbing effect is weakly manifested $\sum A$ = 1.9 and 3.66 c.u. (6% и 11%). In the second and following zones, the absorption effect increases.

The dynamics of the spectral characteristics of the seismic trace are presented in the Table 1.5 and Figs. 1.35 and 1.36. The amplitude spectrum changed from $\sum E$ = 0.807 c.u. (4%) to $\sum E$ = 7.283 c.u. (33%), and the phase spectrum, respectively, in the range $\sum P$ = [6244–31180] rad (48–121%) (Fig. 1.35). The differences in spectral characteristics relative to the original one are reflected in Fig. 1.36. And again, the phase characteristic shows excellent potential for the amount of information contained.

The results of the sensitivity assessment (the amount of information contained about the medium) of variants with thicknesses are shown in Table 1.7. An increase in the thickness of the absorbing layer led to a change in the frequency band $\Delta F_{Amp}$ in the amplitude spectrum from 0.0 to 94.0 Hz, in the phase spectrum $\Delta F_{Phs}$ from 456.0 to 486.0 Hz, respectively (Table 1.7). The ratio of frequency bands $F_r$ changed from ~456 to 5.17 (Table 1.7). The relative frequency band changes were: from 0.0% to 18.8% for the amplitude spectrum $S_{amp}$ and from 91.2% to 97.2% for the phase one $S_{phs}$. Relative changes in frequency band and energy are expressed in the parameter $C_{amp}$ with a change from 0.0% to 15.7%. Relative changes in the

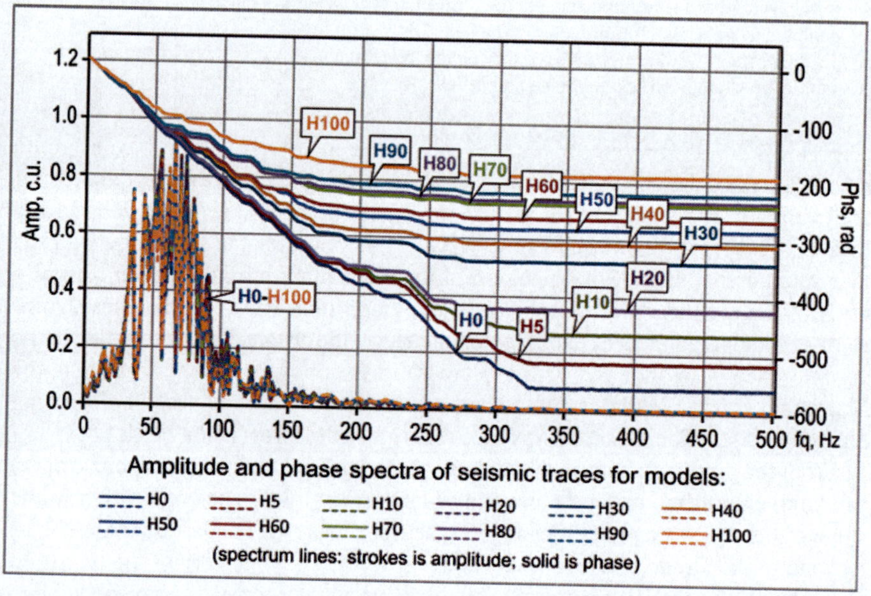


Fig. 1.35 Amplitude and phase spectra of seismic traces for different thicknesses of the absorbing layers

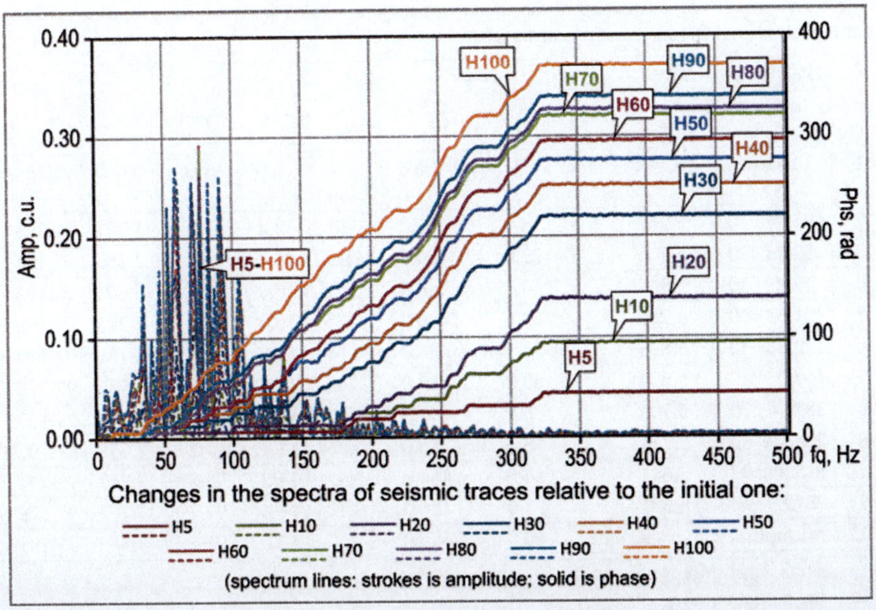


**Fig. 1.36** The difference in the amplitude and phase spectra of seismic traces relative to original ones for different models of the absorbing layer thickness

frequency band and phase are represented by the parameter $C_{phs}$ with a change from 91.2% to 97.2% (Table 1.7).

From the content presented in Fig. 1.33, several more interesting observations can be noted.

First, we detect that above any absorbing layer, the corresponding trace is not subject to change and the influence of the studied layer is practically not noticeable (the absorption contribution is zero) (Fig. 1.33).

Second, the absorption effect leads to a narrowing of the frequency band, especially from the high frequencies and it is most observed on the graphs of the amplitude (power) spectrum. In the phase spectrum, the effect of absorption is highlighted by the appearance and expansion of horizontal sections, but not by a narrowing of the frequency band.

Third, along the depth axis at the level of the formation itself, changes in the dynamics of seismic traces are already being observed. Fourth, at the same time, the influence of the absorbing layer does not stop below the level of the location of the object under study, which also absorbs seismic waves.

Fifth, this influence is also not limited to the deepest parts of the seismic trace, where absorbing objects are completely absent.

Moreover, and sixth, the absorption effect increases with depth, thereby complicating the interference pattern of the wave field, creates obstacles for further unambiguous interpretation of seismic materials (Fig. 1.33). This point must be kept in mind when optimally preparing seismic data for the stage of wave field inversion

**Table 1.7** Sensitivity estimation of the seismic trace parameters by amplitude and phase spectra for models with different thickness of the absorbing layer

| | Параметры моДеди | | | | | | | | | | |
|---|---|---|---|---|---|---|---|---|---|---|---|
| No | dh, m | dt, ms | $m_a \cdot 10^{-4}$, $m^{-1}rad^{-1}s$ | $\Delta F_{Amp}$, Hz | $\Delta F_{Phs}$, Hz | δF | $S_{amp}$, % | $S_{phs}$, % | $C_{amp}$, % | $C_{phs}$, % |
| 1 | 13.41 | 5 | 5.142 | < 1 | 456.0 | > 456 | 0.0 | 91.2 | 0.000 | 91.191 |
| 2 | 25.74 | 10 | 5.967 | 14.0 | 456.0 | 32.57 | 2.8 | 91.2 | 1.241 | 91.193 |
| 3 | 52.13 | 20 | 5.706 | 22.0 | 456.0 | 20.73 | 4.4 | 91.2 | 2.256 | 91.194 |
| 4 | 77.23 | 30 | 5.948 | 50.0 | 486.0 | 9.72 | 10.0 | 97.2 | 6.693 | 97.199 |
| 5 | 101.53 | 40 | 6.236 | 52.0 | 486.0 | 9.35 | 10.4 | 97.2 | 7.295 | 97.199 |
| 6 | 129.88 | 50 | 5.823 | 56.0 | 486.0 | 8.68 | 11.2 | 97.2 | 8.188 | 97.199 |
| 7 | 157.03 | 60 | 5.687 | 60.0 | 486.0 | 8.10 | 12.0 | 97.2 | 9.036 | 97.199 |
| 8 | 182.94 | 70 | 5.707 | 76.0 | 456.0 | 6.00 | 15.2 | 91.2 | 12.074 | 91.193 |
| 9 | 208.90 | 80 | 5.714 | 78.0 | 456.0 | 5.85 | 15.6 | 91.2 | 12.639 | 91.192 |
| 10 | 235.82 | 90 | 5.644 | 84.0 | 464.0 | 5.52 | 16.8 | 92.8 | 13.811 | 92.794 |
| 11 | 261.83 | 100 | 5.651 | 94.0 | 486.0 | 5.17 | 18.8 | 97.2 | 15.704 | 97.199 |
| 12 | Absorption calc-n by model | | | 60.0 | 456.0 | 7.60 | 12.0 | 91.2 | 8.560 | 91.191 |

Parameters description:
No is the model number
dh is the thickness of the absorbing layer, m
dt is the time interval of the absorbing layer, ms
$m_a$ is the the average absorption factor in the layer, $m^{-1}rad^{-1}s$
$\Delta F_{Amp}$ is the frequency band changes over the amplitude spectrum, Hz
$\Delta F_{Phs}$ is the frequency band changes over the phaze spectrum, Hz
δF is the ratio of the frequency bands of the phase and amplitude spectra
$S_{amp}$ is the relative band variation over the amplitude spectrum, %
$S_{phs}$ is the relative band variation over the phase spectrum, %
$C_{amp}$ is the frequency band and energy changes in the amplitude spectrum, %
$C_{phs}$ is the frequency band and energy changes in the phase spectrum, %

(IDSP solutions) and taken into account in the process of interpreting seismic survey materials.

Note also, since as a result of wave absorption, the frequency constituents are weakened when the wave passes through the absorbing object, then information about the characteristics of the object under study can be extracted by the degree of attenuation. However, in the presence of strong absorption of waves by an object, it is possible to lose information about a thin-layer medium (the medium itself disappears along with the complete absorption of high-frequency components). This is the case when it will not be possible to extract information not only about such a high absorbing object, but also below its underlying perspective objects, although they may have characteristics of not greatly absorbing layers. i.e., the component of their contribution to the wave field is negligible (weak signal) or absent. Such deep objects turn out to be seismically invisible from the observation surface. This is also shown by the described experiment in Fig. 1.33. This also defines the limits of the method applicability with respect to thin-layer medium, as well as the adequate use of procedures for digital processing of seismic data.

## 1.7 Spectral Characteristics Sensitivity Evaluation Under the Combined Wave Velocity and the Density Modifications of the Medium

From the point of view of seismic exploration, acoustic heterogeneities in the medium are manifested in changes in acoustic impedance and the presence of wave absorption. Often, absorption in the processing of seismic data is compensated and inhomogeneities of acoustic impedance in the form of a product of wave velocity and medium density ($ai = v{\cdot}\rho$) are for analysis used.

Let us consider the combined effect of velocity and density on the behavior of the seismic trace and spectral characteristics in order to assess their sensitivity (the most common in the practice of seismic exploration).

Figure 1.37 shows a detailed analysis of the effect of a thin-layer medium on the combined modification of velocity and density with parameters $\sigma_v = 250$ m/s and $\sigma_d = 0.15$ g/cm$^3$. This led to a total change in the amplitudes (energy) of the seismic trace by $\sum A = 5.247$ c.u. (114%). The combined action increased the contrast of acoustic impedance (Fig. 1.37) compared to that with a separate effect of velocity (Fig. 1.6) and density (Fig. 1.15). Such an increase in contrast gave to a stronger

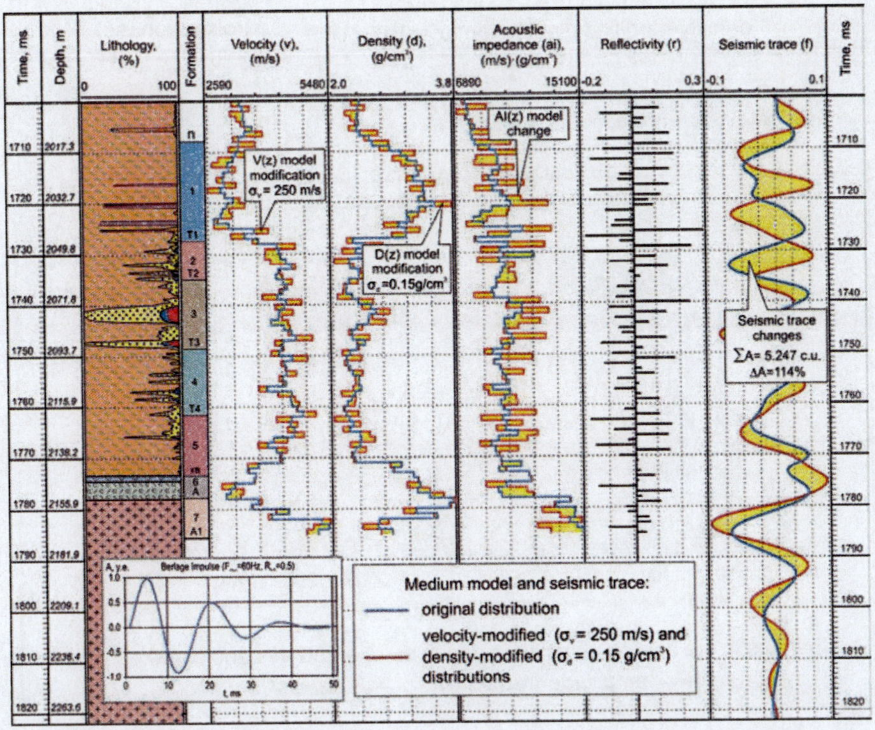


**Fig. 1.37** The detail influence analysis of a thin-layer medium velocity and density models on the formation of a wave field

change in the energy of the seismic trace $\sum A = 5.247$ c.u. (114%) compared to the case of variation in velocity $\sum A = 2.993$ c.u. (65%) and density $\sum A = 2.355$ c.u. (51%) (Table 1.1, No 2, 5, 15).

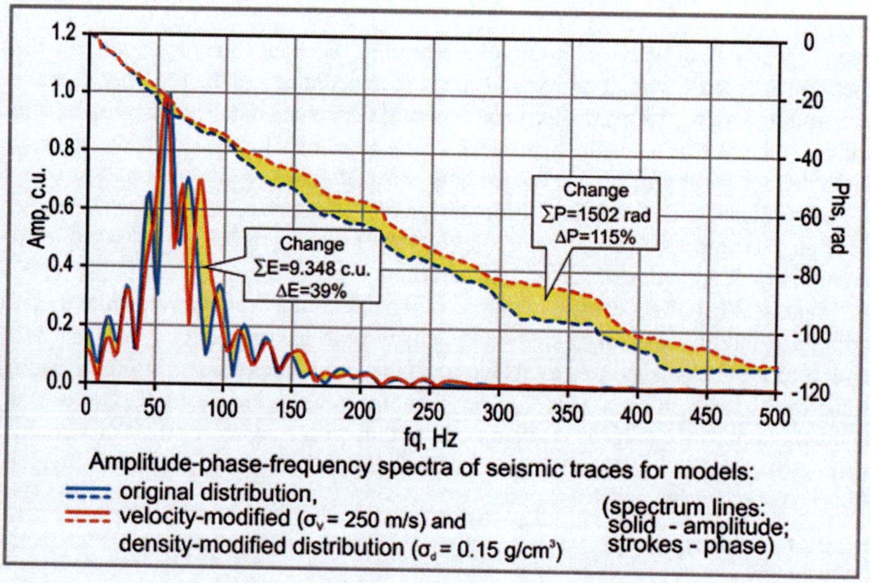


**Fig. 1.38** The effect of changes in the medium velocity and density on the structure of the amplitude and phase spectra of the seismic trace

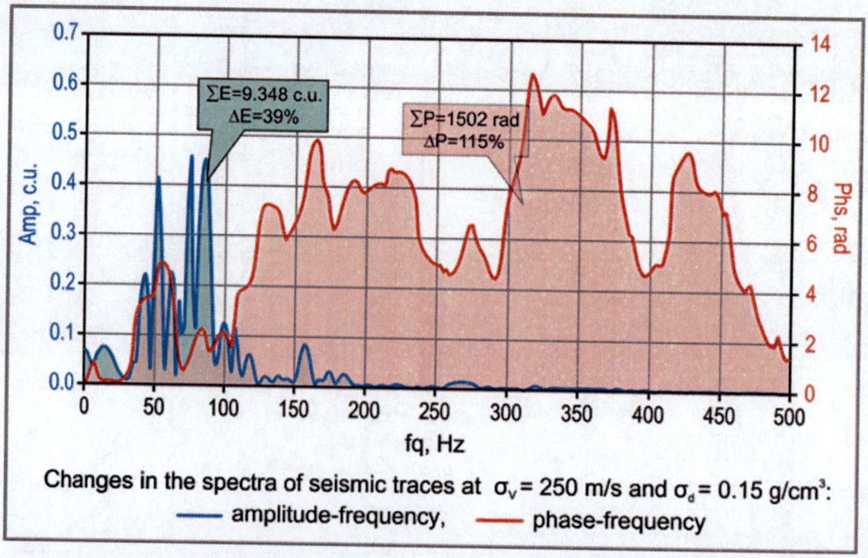


**Fig. 1.39** The difference in the amplitude and phase spectra of the seismic trace relative to the initial ones with variations in the wave velocity and density in the medium

The following changes are observed in the spectra (Figs. 1.38 and 1.39). The amplitude spectrum changed the energy $\sum E = 9.348$ c.u. (39%). The phase spectrum changed by $\sum P = 1502$ rad (115%). Compared with the spectra with individual modifications of velocity ($\sum E = 8.093$ c.u. (34%), $\sum P = 1330$ rad (77%)) and density ($\sum E = 6.974$ c.u. (29%), $\sum P = 1143$ rad (91%)) this change in the spectra was $\sum E = 9.348$ c.u. (39%) for the amplitude and $\sum P = 1502$ rad (115%) for the phase, respectively (Table 1.1). The combined effect of velocity and density on the structures of the spectra shows an increase in the contribution to the components of the spectrum compared to individual disturbances in density and velocity.

Table 1.4 shows the parameters of the spectra characterizing the sensitivity (quantity of information about the medium) for models with a joint change in the distribution of velocity and density $\sigma_v = 250$ m/s, and $\sigma_d = 0.15$ g/cm$^3$. Changing the parameters of the medium model shows the frequency band of $\Delta F_{Amp}$ in the amplitude spectrum has changed by 70.0 Hz, in the phase spectrum of $\Delta F_{Ph}$ by 448.0 Hz (Table 1.4). The ratio of frequency bands $F_r$ has changed 6.4 times (Table 1.4). The relative frequency band changes were 14.0% for the amplitude spectrum of $S_{amp}$ and 89.6% for the phase $S_{phs}$. The relative changes in frequency band and signal energy are expressed in the parameter $C_{Amp} = 11.58\%$. Relative changes in the frequency band and signal phase are presented to the parameters with $C_{phs} = 89.25\%$ (Table 1.4).

The combined effect of variations in the velocity and density distributions increase the sensitivity of the seismic trace spectra. At the same time, the general trend about the quantitative ratio of the information content of the energy and phase trace components is confirmed.

The trend of changing spectra remains the same as in the above described variants of the numerical simulation of the wave field: the contribution to the amplitude spectrum is more evident in the low-frequency band, but to the phase one – in the high-frequency domain.

## 1.8 The Spectral Characteristics Sensitivity Evaluation When a Joint Modifications of the Wave Velocity, Medium Density and the Presence of Wave Absorption

In reality, the propagation of seismic waves depends not only on the acoustic contrast of the medium, but also on its absorbing ability, and this process is always accompanied by wave attenuation in accordance with the physical law of energy conservation.

Let's examine the general case – the joint action of all the main factors: velocity, density and attenuation coefficient.

A detailed analysis of the formation of a seismic trace in such an environment is shown in Fig. 1.40. The selected parameters for the modified model in the direct problem are as follows: $\sigma_v = 250$ m/s for velocity ($V_{250}$), $\sigma_d = 0.15$ g/cm$^3$ for density ($D_{0.15}$), $m_a \cdot \omega = 2 \cdot 10^{-6} \cdot \omega$ m$^{-1}$ and $\sigma_m \cdot \omega = 10^{-6} \cdot \omega$ m$^{-1}$ – for the absorption coefficient ($M_2$) (Fig. 1.40, Table 1.1, verion No 16).

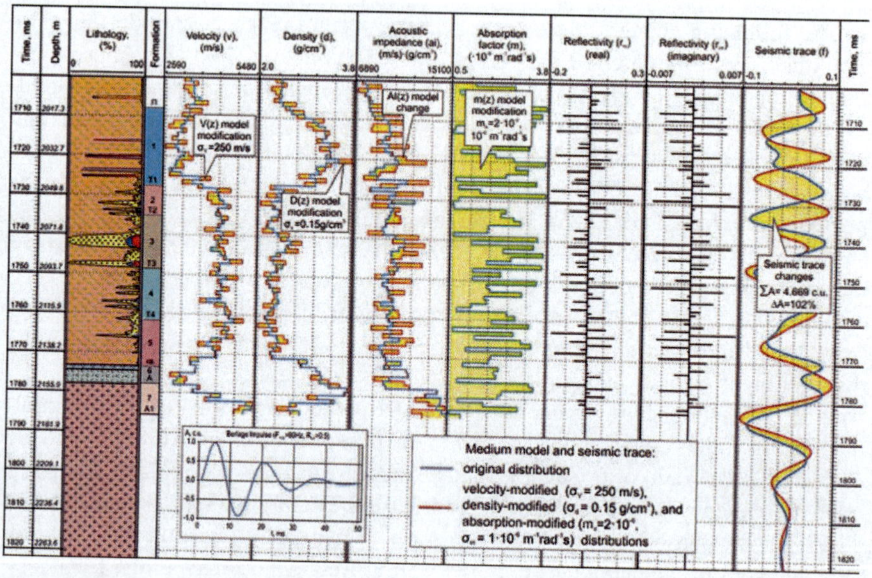

**Fig. 1.40** The detail influence analysis of a thin-layer medium velocity, density, and absorption models on the formation of a wave field

The parameters of the model were chosen in such a way the effect on the seismic trace was approximately the same in absolute magnitude.

The combined effect of all factors on the amplitudes of seismic traces ($\sum A = 4.699$ c.u. (102%)) turned out to be large compared to the individual effects of velocity ($\sum A = 2.993$ c.u. (65%)), density ($\sum A = 2.355$ c.u. (51%)), and absorption coefficient ($\sum A = 0.618$ c.u. (13%)). However, slightly less than with the combined effect of velocity and density ($\sum A = 5.247$ c.u. (114%)) (Table 1.1). This is explained by the fact with this absorption distribution, some compensation of energy occurs due to the effect of interference redistribution of amplitudes. With other values of the absorption coefficient, the patterns may be different.

The spectral characteristics of the traces are shown in Figs. 1.41 and 1.42. The changes in the amplitude spectrum $\sum E = 9.392$ c.u. (39%) (Fig. 1.42) are practically the same as in the previous case study when the velocity and density modifications $\sum E = 9.348$ c.u. (38.8%) (Fig. 1.39, Table 1.1). That is, the contribution to the amplitude spectrum of the absorption coefficient relative to the previous one the case is insignificant with this absorption parameter.

However, the phase spectrum has a completely different effect: its change is $\sum P = 4425$ radians (147%) versus $\sum P = 1502$ rad (115%) than in the previous case, and this means the sensitivity is 2.9 times higher (290%) than that in the option described above.

Table 1.4 shows the parameters of the spectra characterizing the sensitivity (quantity of information about the medium) for models with a complex change in the distribution of velocity, density, and absorption coefficient $\sigma_v = 250$ m/s,

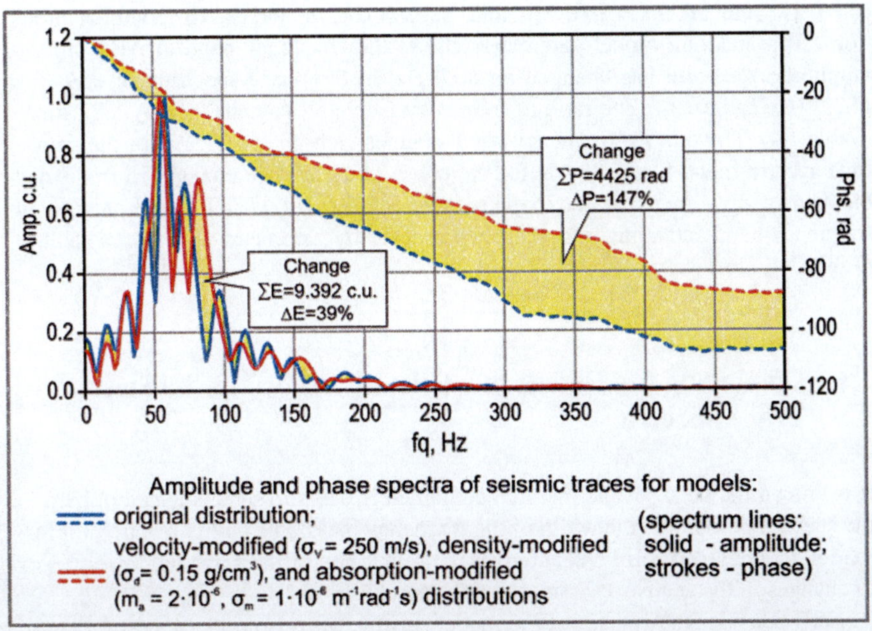


**Fig. 1.41** The effect of changes in the medium velocity, density and absorption on the structure of the amplitude and phase spectra of the seismic trace

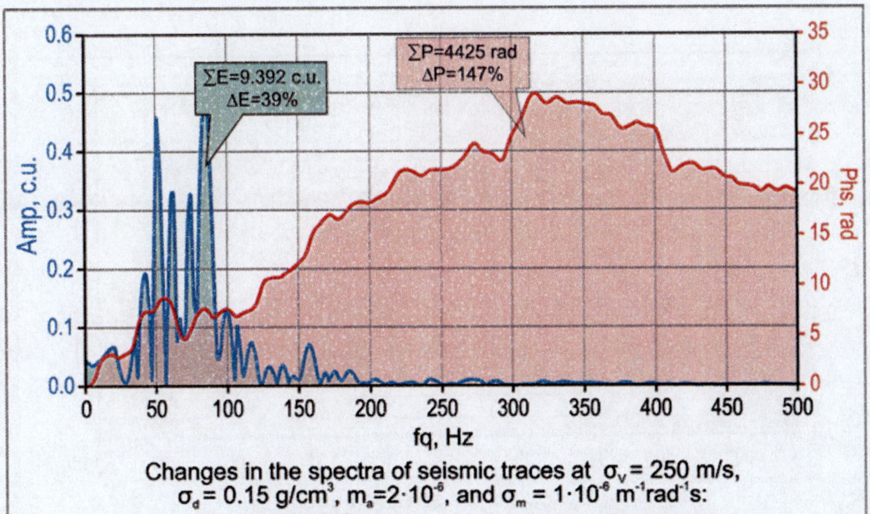


**Fig. 1.42** The difference in the amplitude and phase spectra of the seismic trace relative to the initial ones with variations in the velocity, density and absorption of the medium

$\sigma_d = 0.15$ g/cm³ and $m_a = 2{\cdot}10^{-6}$ m$^{-1}$rad$^{-1}$s, $\sigma_m = 10^{-6}$ m$^{-1}$rad$^{-1}$s. The complex variation of the medium model parameters shows the frequency band of $\Delta F_{Amp}$ in the amplitude spectrum has changed by 86.0 Hz, in the phase spectrum of $\Delta F_{Phs}$ by 492.0 Hz (Table 1.4). The ratio of frequency bands δF has changed by 5.72 times (Table 1.4). The relative frequency band changes were $S_{amp} = 17.2\%$ for the amplitude spectrum and $S_{phs} = 98.4\%$ for the phase one. Relative changes in frequency band and energy are expressed in the parameter $C_{amp} = 14.94\%$. Relative changes in frequency band and signal phase are represented by parameter with $C_{phs} = 98.38\%$ (Table 1.4, No 14).

## 1.9 Displaying the Details of a Seismic Record in the Seismic Trace Spectra

It is known the detail of the medium contained in seismic signals is determined in the spectral domain by the width of the frequency band. The ability to extract information in a detailed form characterizes the sensitivity of the wave field parameters to changes in the acoustic parameters of the medium.

Let's consider the spectra of the model seismic trace from the point of view of displaying the resolution of the seismic record and the potential possibility of restoring a detailed description of the environment. Figure 1.43 shows the amplitude and phase spectra of the model seismic trace.

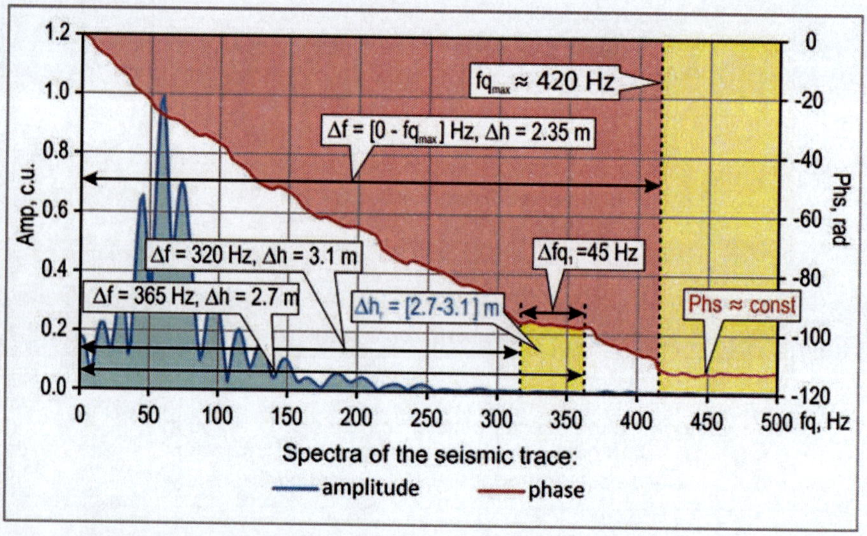


**Fig. 1.43** Resolution mapping in the amplitude-phase spectra of the seismic trace

**Table 1.8** The maximum possible resolution of the IDSP solution depending on the frequency band of the original seismic data

| No | $f_{min}$, Hz | $f_{max}$, Hz | $V_a$, m/s | $\Delta t_a$, s | $\Delta h_a$, m |
|---|---|---|---|---|---|
| 1 | 0.0 | 500.0 | 3940.5 | 0.00100 | 1.97 |
| 2 | 0.0 | 270.0 | 3940.5 | 0.00185 | 3.65 |
| 3 | 0.0 | 160.0 | 3940.5 | 0.00313 | 6.16 |
| 4 | 0.0 | 100.0 | 3940.5 | 0.00500 | 9.85 |
| 5 | 0.0 | 70.0 | 3940.5 | 0.00714 | 14.07 |
| 6 | 5.0 | 500.0 | 3940.5 | 0.00101 | 1.99 |
| 7 | 5.0 | 270.0 | 3940.5 | 0.00189 | 3.72 |
| 8 | 5.0 | 160.0 | 3940.5 | 0.00323 | 6.36 |
| 9 | 5.0 | 100.0 | 3940.5 | 0.00526 | 10.37 |
| 10 | 5.0 | 70.0 | 3940.5 | 0.00769 | 15.16 |
| 11 | 15.0 | 500.0 | 3940.5 | 0.00103 | 2.03 |
| 12 | 15.0 | 270.0 | 3940.5 | 0.00196 | 3.86 |
| 13 | 15.0 | 160.0 | 3940.5 | 0.00345 | 6.79 |
| 14 | 15.0 | 100.0 | 3940.5 | 0.00588 | 11.59 |
| 15 | 15.0 | 70.0 | 3940.5 | 0.00909 | 17.91 |
| 16 | 20.0 | 500.0 | 3940.5 | 0.00104 | 2.05 |
| 17 | 20.0 | 270.0 | 3940.5 | 0.00200 | 3.94 |
| 18 | 20.0 | 160.0 | 3940.5 | 0.00357 | 7.04 |
| 19 | 20.0 | 100.0 | 3940.5 | 0.00625 | 12.31 |
| 20 | 20.0 | 70.0 | 3940.5 | 0.01000 | 19.70 |
| 21 | 0.0 | 320.0 | 3940.5 | 0.00156 | 3.08 |
| 22 | 0.0 | 365.0 | 3940.5 | 0.00137 | 2.70 |
| 23 | 0.0 | 420.0 | 3940.5 | 0.00119 | 2.35 |

No is the version number
$f_{min}$ is the frequency minimum of the range, Hz
$f_{max}$ is the frequency maximum of the range, Hz
$V_a$ is the average velocity in the formation, m/s
$\Delta t_a$ is the average time resolution in the formation, s
$\Delta h_a$ is the average depth resolution in the formation, m

Table 1.8 shows the results of calculating the resolution of seismic record depending on the bandwidth for the reflected wave velocity $V_{av} = 3940.5$ m/s. Here $\Delta t_{av}$ describes the resolution along the time axis, $\Delta h_{av}$ is the resolution in depth.

Figure 1.43 shows the components of the amplitude spectrum fade with increasing frequency. Visually, above the frequency of 300 Hz, changes in the components of the spectrum are not noticeable.

The graph of the unwrapped phase spectrum manifests itself over the entire frequency range $\Delta f = 0$–500 Hz and phase changes range from 0 to 115 radians. There are areas of the spectrum where the phase does not change. For example, these are sections 320–365 Hz and 420–500 Hz. Horizontal intervals in the phase spectrum indicate the absence of perceptible phase changes in the components.

Horizontal lines of the phase spectrum may indicate the absence of information about any structures in the medium or, in general, the absence of information in the data, thereby determining the range of operating frequencies. The horizontal section Δf = 420–500 Hz warns us about the possible lack of information about a thin-layer medium with thicknesses less than 2.35 m. Thus, based on the horizontality in the frequency range Δf = 420–500 Hz, it is possible to determine the operating band Δfw = 0–420 Hz with a frequency limit $fq_{max}$ = 420 Hz.

In Fig. 1.43 and from Table 1.8 (No 21–22), the frequency bands Δf = 0–320 Hz and Δf = 0–365 Hz correspond to a set of layers with thicknesses from Δh = 2.7 to Δh = 3.1 m. The horizontal spectral line indicates that the layers in the range of thicknesses 2.7–3.1 are not represented in a thin-layer section, although the operating frequency extends to 420 Hz, which corresponds to a layer thickness of 2.35 m.

## 1.10 Dependence of the Information Content and Sensitivity of Spectral Parameters on the Seismic Trace Frequency Band

To study the influence of a frequency band on the spectral characteristics, parameter distributions with velocity and density variances $\sigma_v$ = 250 m/s and $\sigma_d$ = 0.15 g/cm$^3$ are given as perturbed relative to the initial model.

Figure 1.44 shows the initial and modified seismic traces with the above parameters, and Fig. 1.45 shows their amplitude and unwrapped phase spectra.

Figure 1.45 shows when the channel is open, when bandpass filtering is not applied, when the frequency band is represented by the range Δf = 0–500 Hz, a thin-layer medium with a detail of Δh = 1.97 m can be recovered from the data. A bandpass filter with limitations Δf = 20–160 Hz allows you to extract data about the medium with a maximum resolution Δh = 7.04 m. Note, outside the range of the operating frequency band Δf = 20–160 Hz, the components of the amplitude spectrum are zeroed, and the values of the phase spectrum are represented by constant values, which is a sign of the information lack about the corresponding layers of the thin-layer medium.

Table 1.9 presents the results of quantitative estimates of the spectra sensitivity at different bands obtained by the frequency bandpass filtering procedure.

For the described model (Figs. 1.44 and 1.45) with parameters $\sigma_v$ = 250 m/s and $\sigma_d$ = 0.15 g/cm$^3$, bandpass filtering in the band Δf = 20–160 Hz led to a decrease the frequency interval in the amplitude spectrum of $dF_{Amp}$ from 70.0 to 54.0 Hz, and in the phase spectrum of $dF_{Phs}$ from 448.0 to 124.0 Hz (Table 1.9, No 1, 18). The ratio of $F_r$ frequency bands decreased from 6.4 to 2.3 (Table 1.9). Relative frequency band variations changed from 14.0% to 10.8% for the $S_{amp}$ amplitude spectrum and from 89.6% to 24.8% for the phase $S_{phs}$. Relative changes in frequency band and energy are expressed in the parameter $C_{amp}$ are from 11.58% to 9.78%. Relative changes in frequency band and phase are represented by parameters with $C_{phs}$ = 89.25% to 24.71% (Table 1.9).

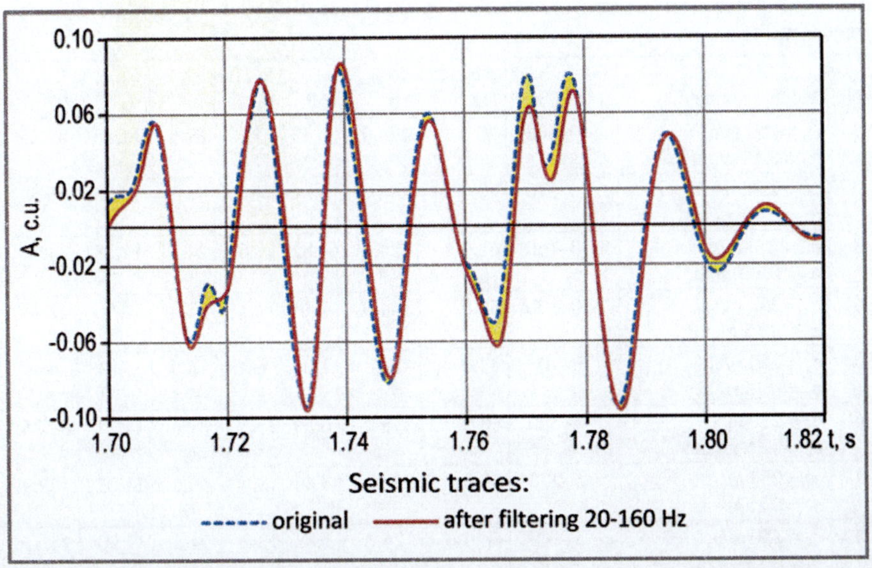


Fig. 1.44 Seismic traces before and after frequency filtering

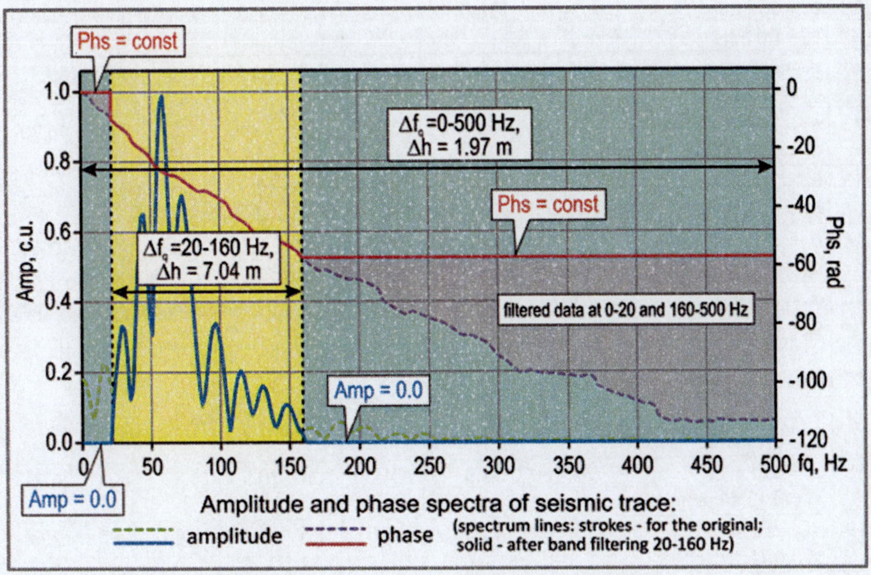


Fig. 1.45 Potential resolution and amplitude and phase spectra of seismic traces before and after filtration

**Table 1.9** Parameters sensitivity estimation by amplitude and phase spectra for different frequency filters

| No | Model parameters | Filter $f_b$–$f_e$, Hz | $\Delta F_{Amp}$, Hz | $\Delta F_{Phs}$, Hz | $\delta F$ | $S_{amp}$, % | $S_{phs}$, % | $C_{amp}$, % | $C_{phs}$, % |
|---|---|---|---|---|---|---|---|---|---|
| 1 | $\sigma_v$ = 250 m/s; $\sigma_d$ = 0.15 g/cm³ | 0–500 | 70.0 | 448.0 | 6.40 | 14.0 | 89.6 | 11.58 | 89.25 |
| 2 | $\sigma_v$ = 250 m/s; $\sigma_d$ = 0.15 g/cm³ | 0–270 | 70.0 | 234.0 | 3.34 | 14.0 | 46.8 | 12.00 | 46.59 |
| 3 | $\sigma_v$ = 250 m/s; $\sigma_d$ = 0.15 g/cm³ | 0–160 | 68.0 | 124.0 | 1.82 | 13.6 | 24.8 | 12.28 | 24.54 |
| 4 | $\sigma_v$ = 250 m/s; $\sigma_d$ = 0.15 g/cm³ | 0–100 | 56.0 | 64.0 | 1.14 | 11.2 | 12.8 | 10.55 | 12.45 |
| 5 | $\sigma_v$ = 250 м/С; $\sigma_d$ = 0.15 г/См³ | 0–70 | 32.0 | 34.0 | 1.06 | 6.4 | 6.8 | 5.81 | 6.47 |
| 6 | $\sigma_v$ = 250 m/s; $\sigma_d$ = 0.15 g/cm³ | 5–500 | 68.0 | 448.0 | 6.59 | 13.6 | 89.6 | 11.22 | 89.25 |
| 7 | $\sigma_v$ = 250 m/s; $\sigma_d$ = 0.15 g/cm³ | 5–270 | 68.0 | 234.0 | 3.44 | 13.6 | 46.6 | 11.63 | 46.60 |
| 8 | $\sigma_v$ = 250 m/s; $\sigma_d$ = 0.15 g/cm³ | 5–160 | 66.0 | 124.0 | 1.88 | 13.2 | 24.8 | 11.90 | 24.55 |
| 9 | $\sigma_v$ = 250 m/s; $\sigma_d$ = 0.15 g/cm³ | 5–100 | 54.0 | 64.0 | 1.19 | 10.8 | 12.8 | 10.16 | 12.46 |
| 10 | $\sigma_v$ = 250 m/s; $\sigma_d$ = 0.15 g/cm³ | 5–70 | 30.0 | 34.0 | 1.13 | 6.0 | 6.8 | 5.42 | 6.48 |
| 11 | $\sigma_v$ = 250 m/s; $\sigma_d$ = 0.15 g/cm³ | 15–500 | 62.0 | 448.0 | 7.23 | 12.4 | 89.6 | 10.27 | 89.37 |
| 12 | $\sigma_v$ = 250 m/s; $\sigma_d$ = 0.15 g/cm³ | 15–270 | 62.0 | 234.0 | 3.77 | 12.4 | 46.8 | 10.66 | 46.70 |
| 13 | $\sigma_v$ = 250 m/s; $\sigma_d$ = 0.15 g/cm³ | 15–160 | 60.0 | 124.0 | 2.07 | 12.0 | 24.8 | 10.90 | 24.68 |
| 14 | $\sigma_v$ = 250 m/s; $\sigma_d$ = 0.15 g/cm³ | 15–100 | 48.0 | 64.0 | 1.33 | 9.6 | 12.8 | 9.12 | 12.64 |
| 15 | $\sigma_v$ = 250 m/s; $\sigma_d$ = 0.15 g/cm³ | 15–70 | 24.0 | 34.0 | 1.42 | 4.8 | 6.8 | 4.41 | 6.65 |
| 16 | $\sigma_v$ = 250 m/s; $\sigma_d$ = 0.15 g/cm³ | 20–500 | 56.0 | 448.0 | 8.00 | 11.2 | 89.6 | 9.23 | 89.39 |
| 17 | $\sigma_v$ = 250 m/s; $\sigma_d$ = 0.15 g/cm³ | 20–270 | 56.0 | 234.0 | 4.18 | 11.2 | 46.8 | 9.59 | 46.73 |
| 18 | $\sigma_v$ = 250 m/s; $\sigma_d$ = 0.15 g/cm³ | 20–160 | 54.0 | 124.0 | 2.30 | 10.8 | 24.8 | 9.78 | 24.71 |
| 19 | $\sigma_v$ = 250 m/s; $\sigma_d$ = 0.15 g/cm³ | 20–100 | 42.0 | 64.0 | 1.52 | 8.4 | 12.8 | 7.96 | 12.68 |
| 20 | $\sigma_v$ = 250 m/s; $\sigma_d$ = 0.15 g/cm³ | 20–70 | 18.0 | 34.0 | 1.89 | 3.6 | 6.8 | 3.29 | 6.68 |

Parameters description:
No is the filter version number
$\sigma_v$ is the velocity variance, m/s
$\sigma_d$ is the density variance, g/cm³

(continued)

**Table 1.9** (continued)

$f_b$ is the lower bandpass filter frequency, Hz
$f_e$ is the higher bandpass filter frequency, Hz
$\Delta F_{Amp}$ is the frequency band changes over the amplitude spectrum, Hz
$\Delta F_{Phs}$ is the frequency band changes over the phase spectrum, Hz
$\delta F$ is the ratio of the frequency bands of the phase and amplitude spectra
$S_{amp}$ is the relative change of the range along the amplitude spectrum, %
$S_{phs}$, is the relative change of the range along the phase spectrum, %
$C_{amp}$ is the change of frequency band and energy in the amplitude spectrum, %
$C_{phs}$, is the change of frequency band and phase in the phase spectrum, %

In general, Table 1.9 shows that the narrowing of the band reduces the sensitivity and information content of the parameters, and therefore the ability to restore a thin-layered medium.

## 1.11 The Effect of the Model Sampling Step on the Amplitude and Phase Characteristics of the Seismic Trace

To study the effect of the detail of the original model on the operating frequency bands in the spectra during parametrization, equitime models with different quantization along time were generated.

Figure 1.46 shows equitime models of the distribution of velocity, density and acoustic impedance with time sampling $\Delta t = 0.001$ s, $\Delta t = 0.002$ s, $\Delta t = 0.004$ s and simulated corresponding seismic traces.

From Fig. 1.46, it can be seen that with an increase in the sampling interval of the model, the number of working layers decreases from $n = 85$ to $n = 22$. At the same time, the seismic traces become more smoothed as the model quantization step increases.

Figure 1.47 presents the spectral characteristics of seismic traces calculated for models of different detail from $\Delta t = 0.001$ s to $\Delta t = 0.008$ s. The figure shows how the operating frequency bands change with variations in the detail of the original model.

For the amplitude spectrum normalized to one at the level of 0.5, frequency bands were allocated, the width of which was 35, 35, 30, 25, 15 Hz for the quantization step of the model 1, 2, 4, 6, 8 ms, respectively (Fig. 1.47). The bands at this level decrease slightly in width with an increase in the quantization interval of the model.

The phase spectrum allocates bands whose upper boundary frequencies are 420, 310, 265, 180, 70 Hz for models with quantization of 1, 2, 4, 6, 8 ms, respectively (Fig. 1.47). Note that with increasing discreteness of model, the horizontal sections of the phase spectrum noticeably and significantly increase. The operating frequency bands, at the same time, are rapidly narrowing, showing the sensitivity of the

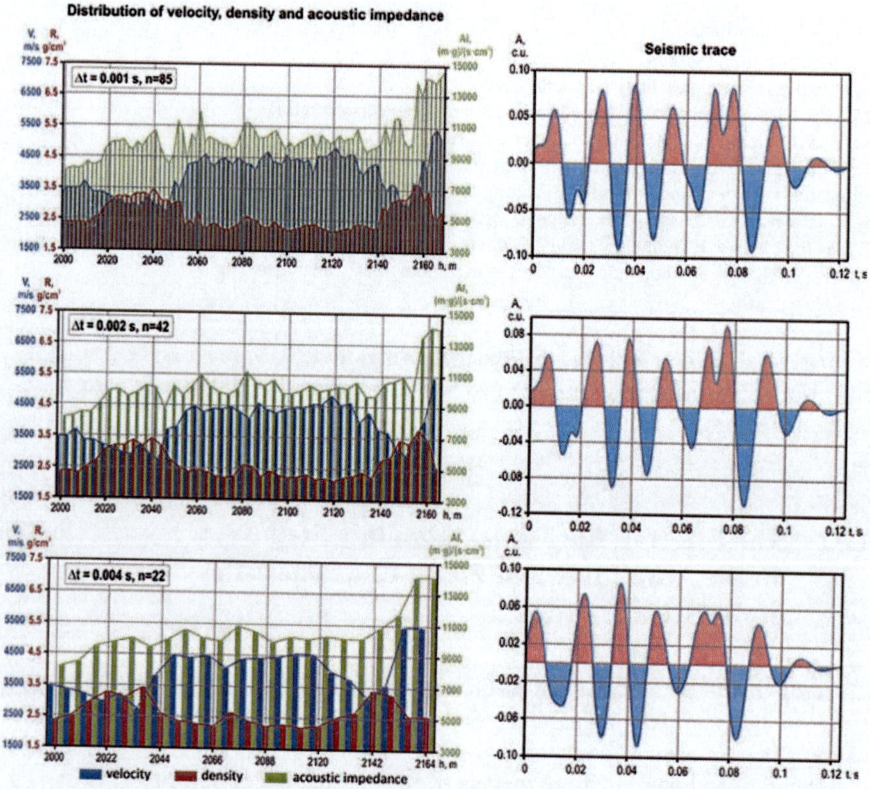


Fig. 1.46 The effect of the sampling step of the medium equitime model on the characteristics of the seismic trace

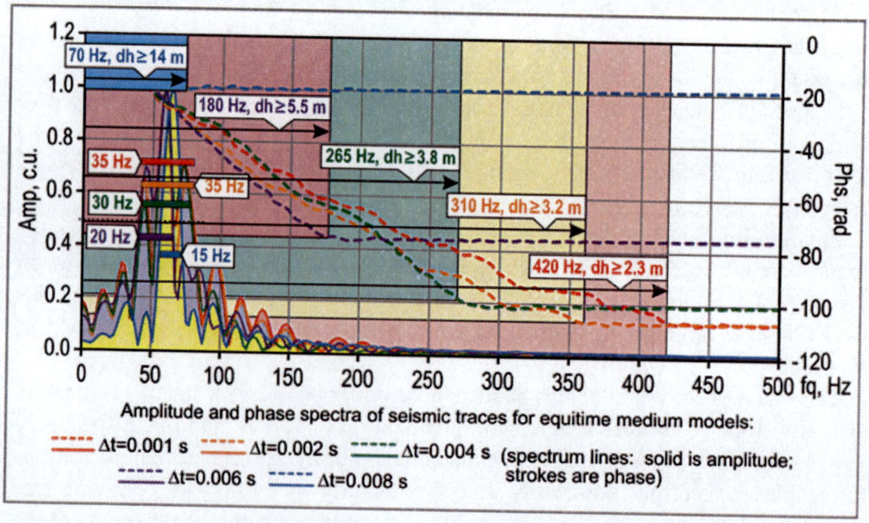


Fig. 1.47 Operating frequency bands with different quantization of an equitime model of the medium

phase spectrum. Thereby confirming the informativeness of the seismic data contained in the phase component.

## 1.12 The Effect of Random Noise to the Amplitude and Phase Characteristics of the Seismic Trace

To observe the noise effect on the amplitude and phase spectra of the seismic trace, random noise with different intensities was program generated by the Box-Muller method.

Figure 1.48 shows one of these implementations of noise with a zero mean of $\mu_n = 0.0$, with a standard deviation $\sigma_n = 0.015$, which corresponds to the signal-to-noise ratio S/N = 3 relative to the considered trace model.

The initial wave trace was calculated according to the velocity and density model without taking into account the absorption effect shown in Fig. 1.2. The distribution of the velocity and density of the medium are taken from the acoustic and density logging of a real well. These field measurements were processed and converted into an equitime form (Fig. 1.2). A theoretical Berlage pulse with controlled parameters $F_{max} = 60$ Hz and $R_{ext} = 0.5$ was used as an elementary pulse probing the medium (Fig. 1.4). The generated random noise was additively added to the calculated synthetic seismic trace.

Figure 1.49 shows the calculated seismic traces with superimposed with different levels of noise. In particular, the traces with the signal-to-noise ratio S/N are depicted = ∞, 50, 10, 5, 3.

Figure 1.50 represents the amplitude and continuous phase spectra of the corresponding model traces. The influence of noise increases with intensity, while with increasing frequency, the contribution of noise in the seismic trace also grows.

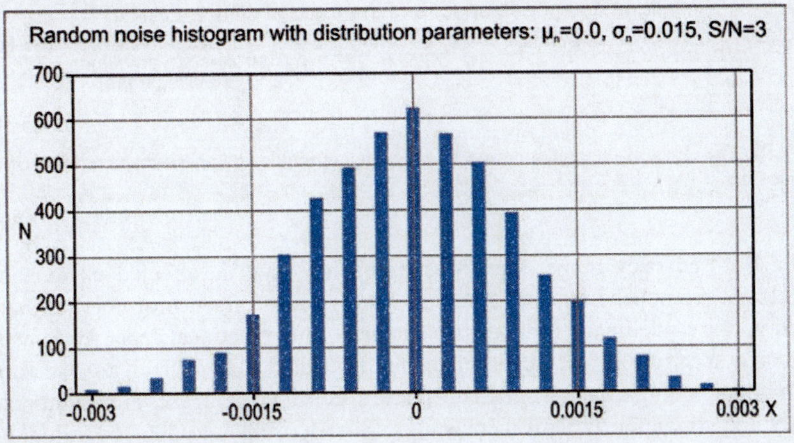


**Fig. 1.48** The histogram of random noise with a normal distribution: $\mu_n = 0.0$, $\sigma_n = 0.015$, S/N = 3 (noise realised using the Box-Muller method)

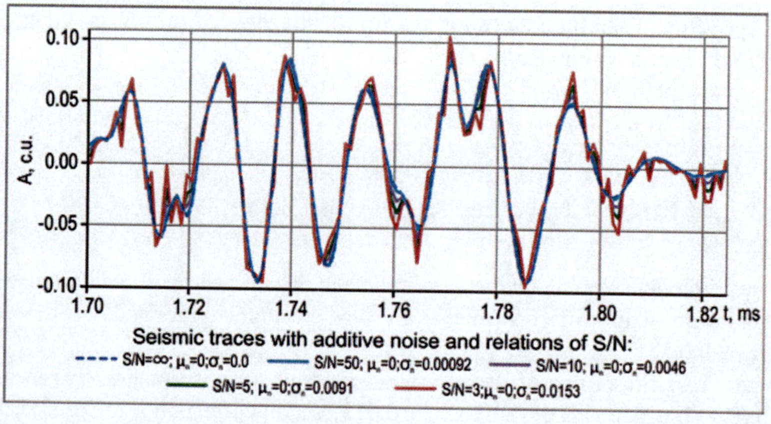


**Fig. 1.49** Seismic traces with different levels of noise

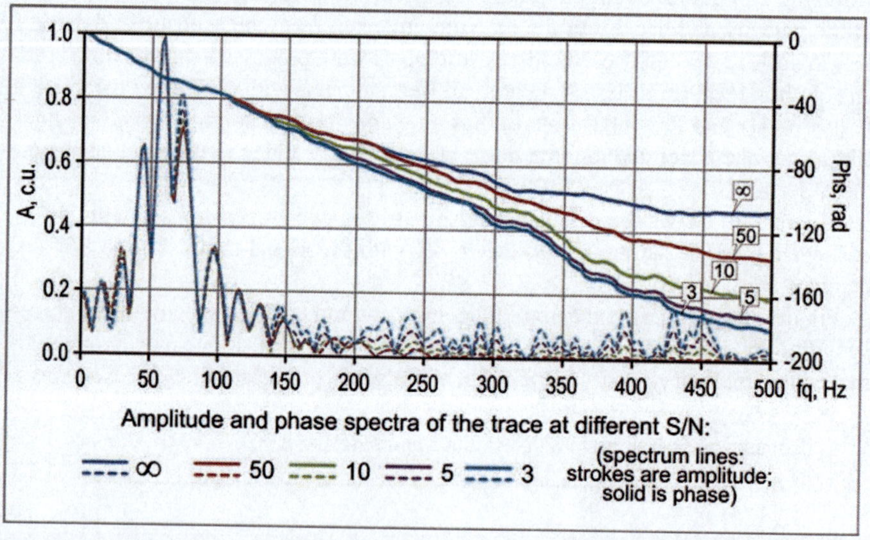


**Fig. 1.50** The effect of the random noise level on the amplitude and phase characteristics of the seismic trace

Along the frequency range, the lowest frequency band is affected up to about 140 Hz. At the upper frequency intervals, the greatest influence of noise is manifested. At the same time, it is not possible to detect any functional dependence in the amplitude spectrum in the upper frequency band and it is difficult to find some approximation of this data, while in the phase spectrum it decreases almost monotonously and continuously with increasing frequency from 0 to 500 Hz and allows you to search and build simpler approximating functions. Although the noise has the greatest effect on the phase spectrum than on the amplitude spectrum, but the structure of such interference in the phase component is more understandable and simple, thereby it becomes possible to selectively suppress them.This creates a basis for removing noise at high frequency bands while maintaining a useful signal there, i.e. a prerequisite for restoring the thin-layer structure of the medium when solving the inverse problem of seismics.

## 1.13 Summary

Further improvement of the numerical algorithm for solving the inverse dynamic seismic problem (IDSP) predetermined the consideration by the authors of the searching process for the AI and RC model using objective functions (vector of objective functions), in which various types of residuals between real and model data are iteratively calculated. Moreover, such discrepancies are estimated from the energy (amplitudes) and the shape of the seismic signal. It is for this kind of discrepancy the information content and sensitivity of the amplitude (energy) and phase spectra of the seismic record were in this chapter analyzed. Such an analysis is performed in the frequency domain in which the contribution of medium models parameters modifications (velocity of elastic waves, density of the medium, and wave attenuation) affects the result of seismic inversion, which is proportional to the frequency of the waves. In their previous work [2, 3], the authors did not state these issues in detail. They are in depth presented here for the first time (in this chapter).

The results of modeling and numerical experiments carried out by the authors in the area of research under consideration allowed the authors to draw the following conclusions:

- The main contribution of medium models parameters changes (wave propagation velocity, medium density and wave absorption coefficient) is displayed in the low-frequency band of the amplitude spectrum. In this case, the phase spectrum responds to such modifications to the greatest extent, primarily in its high-frequency domain.
- The perturbing actions of the medium parameters and the corresponding changes in the amplitude (energy) spectrum $\delta A(\omega)$ of the seismic trace fall on a limited frequency band approximately $[\omega_1, \omega_2] \approx [20, 160]$ Hz and at $\omega_1 \rightarrow 0$. At $\omega_2 \rightarrow \omega_N$ these changes tend to zero ($\delta A(\omega) \rightarrow 0$).
- The phase spectrum exhibits a sensitivity of $\delta P(\omega)$ starting at about 15 Hz. It increases its influence in proportion to the frequency, reaching the greatest contribution of the medium parameters in the high-frequency bands of the spectrum: at $\omega \rightarrow 0$, $\delta P(\omega) \rightarrow 0$ and $\delta P(\omega) \sim \omega$.
- More complex phenomena in the structure of the wave field appear when considering the absorption of seismic waves. An increase in the absorption coefficient in this case leads to a change in the spectra, and the greatest contribution in this case falls on the phase component of the spectrum. However, along with the attenuation of the high-frequency components of the spectrum, accumulative,

interference and other spatial effects are also observed both in depth and in the lateral direction.

- In the unwrapped phase spectrum of the seismogram, there are signs by which it is possible to limit the operating frequency range (determine the limits of applicability of this parameter), as well as to detect low-contrast sections and/or make a conclusion about the disappearance or absence of a certain thickness layers.
- Bandpass filtering of seismic data leads to the appearance of horizontal sections on the graph of the unwrapped phase spectrum; in such areas, the sensitivity of this spectrum is absent (i.e., "does not work").
- Increasing the discreteness (i.e., increasing the time step) of the original seismic record model results in a narrowing of the frequency band and an expansion (increase) of the horizontal sections on the phase spectrum graph. On seismic traces (in the time domain), the effect of smoothing is also clearly manifested.
- The greatest information about the nature and features of the thin-layer structure of a real medium is contained in the phase spectrum.
- The spectra of a seismic record are known to be significantly affected by various interferences and noises. In this case, random noise most shows itself in the high-frequency range of the seismic record. Particular attention to this feature of the record should be paid at the stage of preliminary special processing of seismic observation materials (due to the use of a special graph that preserves the dynamics of the seismic record without significant distortions).
- In addition, it should be noted random noise affects the amplitude and phase spectra of seismic data in different ways. In this case, the composition of the phase spectrum reveals the ability to suppress noise at high frequencies, where noises can exceed the signal level, while preserving the useful signal for its further extraction from the interference structure of seismic data.

The results obtained in this chapter should be considered a very significant prerequisite not only for the most successful solution of the inverse dynamic seismic problem and the effective determination of the acoustic characteristics of the real medium from seismic data, but also for the theoretical and methodological foundations of seismic exploration in general.

## References

1. Khaziev, F. F., & Trofimov, V. L. (2003). *Model'nyye issledovaniya rezul'tatov resheniya obratnoy dinamicheskoy zadachi seysmiki* (Model studies of the solving inverse dynamic seismic problem results). Spetsial'nyy vypusk Geofizika: Tekhnologii seysmorazvedki-II (Geophysics, special edition: Seismic Technologies-II) (pp. 27–37).
2. Trofimov, V. L., Khaziev, F. F., & Trofimova, A. V. (2019). *Technology of high-resolution seismics in the search and exploration of oil and gas deposits* (p. 86). OOO "EAGE Geomodel".
3. Trofimov, V. L., Khaziev, F. F., & Trofimova, A. V. (2022). *Oil and gas reservoir prospecting and exploration. High-Resolution Seismic (HRS) techniques and technology*. Springer. (Hardcover Book, eBook). https://doi.org/10.1007/978-3-030-84389-2

4. Krasnitskiy, Yu. A. (2007). *Nekotoryye algoritmy razvertyvaniya fazovogo spectra* (Some algoritms of phase spectrum deployment). Nauchnyy vestnik MGTU GA, №126, seriya Radiofizika i radiotekhnika (Scientific Bulletin of MSTU GA, No. 126, series Radiophysics and Radio Engineering) (pp. 172–177).
5. Oppenheim, A. V., & Schafer, R. W. (1989). *Discrete time signal processing*. Englewood Cliffs/Prentice Hall.
6. Tribolet, J. M. (1977, April). A new phase unwrapping algorithm. *IEEE Transaction on Acoustic, Speech, and Signal Processing, ASSP-25*, № 2.
7. Tribolet, J. M. (1977). *Seismic application of homomorphic signal processing*. Englewood Cliffs/Prentice Hall.
8. Krajnik, E. (1992). A simple and reliable phase unwrapping algorithm. In J. Vandewalle (Ed.), *Signal processing VI: Theories and application* (pp. 917–919). Elsevier.
9. Steiglitz, K., & Dickinson, B. (1982). Phase unwrapping by factorization. *IEEE Transactions on Acoustics, Speech, and Signal Processing, 30*(6), 984–991.
10. Brekhovskikh, L. M. (1973). *Volny v sloistykh sredakh* (Waves in layered media) (p. 343). Izd-vo M. Nauka (Publishing House M. Science).
11. Losovskiy, Ye. L. (1982). Teoreticheskiye seysmogrammy na idealizirovannykh modelyakh neftegazovoy zalezhi (Theoretical seismograms on idealized models of an oil and gas deposit). *Geofizicheskiy zhurnal, t. 4*(№ 5) (Geophysical Journal, 1982, vol. 4, No. 5.).
12. Gogonenkov, G. N., & Zakharov, Ye. T. (1971). *Teoreticheskiye seysmogrammy v tonkosoistykh pogloshchayushchikh sredakh* (Theoretical seismograms in thin-layer absorbing media.) Izv. AN SSSR. Ser. Fizika Zemli (Physics of the Earth series) (No. 2, pp. 45–54).
13. Trapeznikova, N. A., Averbukh, A. G., Mikhalev, V. M., & Shushakova, N. S. (1975). *Program for calculating theoretical seismograms, spectral characteristics and the level of multiple waves for absorbing thin-layered media*. M. VNIIGeofizika, 35 p. (Collection of programs for processing geoph. data; Issue 23).
14. Kozlov, Ye. A. (2006). *Modeli sredy v razvedochnoy seysmologii* (Medium models in exploration seismology) (p. 480). Tver'. GERS.

# Chapter 2
# Experimental Evaluation of Information Content Regarding the Resolution of the Wave Field and the Choice of Basic Parameters for Solving the Inverse Dynamic Problem of Seismics

**Abstract** Extraction of environmental parameters from HRS data implies adequate preparation of field observation materials, taking into account the features and objectives of solving problem. It includes: special preliminary processing of seismic survey data; assessment of the quality and quantity of information in the original seismic data; optimal parametrization of the environment model; parametrization of the main procedure characteristics for solving the inverse dynamic seismic problem (IDSP) by a numerical method. Parameterization of the medium model based on the use of the input digital data amplitude and phase spectra features makes it possible a prior determine the initial and final detail of the seismic data inversion results and, in general, the results of solving the IDSP. To assess the information content of the solving IDSP process main parameters, it is advisable previously study and highlight the features of residual noise on the seismic record after performing data preprocessing, determine the type and structure of these noises, and construct regularization functions. It is also needed to solve the IDSP with respect to the noise under study, and on this basis, in the final process of searching for the desired medium model, apply the regularization parameters of this already defined noise with certain characteristics. The obtained research results provide the necessary prerequisites for using the properties of the characteristics described above to find a stable and detailed solution to the inverse dynamic seismic problem using multicriteria optimization methods.

Let's consider the above conclusions and note the main features of a detailed and stable finding of the medium model characteristics on a real example.

In order to successful extracting geophysical parameters of the geological medium by solving the inverse dynamic seismic problem on the seismic survey data, it is necessary to carefully prepare the field observation data taking into account the features of the IDSP solution. This is due to the fact that in this case, in order to achieve the task, it is necessary to use the limit of the seismic method

 
V. L. Trofimov et al., *Theory and Practice of High Resolution Seismic HRS-Geo Technology Application*, https://doi.org/10.1007/978-3-031-41590-6_2

possibilities, where any inadequate (incorrect) action leads to instability or forces a return to the search for an unresolved model. Among the main stages of data fitting are: special preliminary processing of seismic data, assessment of the quality and quantity of information in the source data, parameterization of the medium, within which the search for a model of the environment is carried out with the help of IDSP, parameterization of the main characteristics of the IDSP procedure, with the help of which the optimal solution is found.

Special pre-processing of seismic survey data is aimed at optimal preparation of data as input for the application of the procedure for solving IDSP using HRS-Geo technology, where detailed information about the geological medium is mainly preserved, noise waves are suppressed (attenuated) in the observed data without significant amplitude and phase distortion; preservation of true amplitudes; simplification of the seismic record structure relative to the extracting parameters of the medium (AI, RC) for their further using in predicting the main geological characteristics (lithology, reservoir and filtration-capacity properties, the nature and degree of saturation of the reservoir, determination of the fluid type), etc.

At the stage of assessing the quality and quantity of information in the source data for the solution of the IDSP, on the basis of available seismic materials, the potential possible detail of the solutions obtained is predicted, the residual level of background noise is estimated and ways to eliminate and/or consider them.

The obtained results of the above stages are used for optimal parameterization of the medium, within which the maximum possible detailed model (vertical and horizontal resolutions) by solving the IDSP is found.

The next important stage is the parameterization of the main characteristics of the IDSP procedure to ensure a stable process of finding the model characteristics. Based on the study of the wave field structure, testing and noise analysis, optimal dynamic regularization parameters sets are determined within the framework of multi-criteria optimization. The stage is important because balanced achievement of results is carried out with multidirectional criteria: determination of detailed (most resolved) parameters of the environment contained in the input data in the process of their steady search.

It is known that inverse problems of geophysics are incorrect (ill-posed, ill-defined, etc.). There are several definitions of the mathematical physics problems correctness. According to the classical definition of the solution correctness [4–7], a direct or inverse problem is correct if the solution exists, is unique and stable (continuous dependence on the initial data). The absence of any of the three properties leads to incorrectness. The existence of the solution is provided by the initial statement of the problem: the correspondence of the mathematical description to the physical model and the process. Uniqueness is acquired in the process of setting various types of constraints on the desired model. Stability arises when the noise and interference properties and constraints on the objective functions are correctly taken into account when developing a solution search algorithm.

When solving the IDSP, it is important to ensure the stability of the solution because this property is associated with the detail of the desired model. The reduction of an incorrect problem to the correct one is carried out by the regularization process. Regularization is associated with the weakening of the noise influence.

There are various ways to regularize pooly conditioned inverse problems. Regularization can include frequency filtering of data from noise, which leads to a correct problem. Examples of such regularization by the optimal frequency filtering method are given in [1–3]. Smoothing, mixing and other procedures can also regularize the solution of the problem.

Regularization is often used, which consists in replacing the original problem with another one with a biased but stable solution that differs slightly from the true one. At the same time, the variance of the solution relative to the true that decreases compared to the original one due to the weakening of the noise effect. Such a task is called conditionally correct. Numerically, in the classical regularization theory, the operation consists in adding some constant in the matrix of coefficients of a linear equations system describing a direct problem. The added constant is called the regularization parameter. The regularization parameter has the following important properties [4]: when the action of the parameter decreases, the solution in the limit tends to the true solution of $\lim_{\lambda \to 0} r_\lambda(t) = r_0(t)$; the regularization parameter reduces the influence of noise; the variance of the solution with respect to the true one decreases when using regularization: $\sigma^2 r_\lambda(t) < \sigma^2 r_0(t)$.

However, there is also a disadvantage of a regularized solution in the classical problem: in addition to the fact of the result being biased, there is often a loss of detail in solving the inverse problem [4]. In the HRS-Geo technology, multicriteria optimization with multifactor dynamic regularization is used to overcome these shortcomings. In this case, the regularization function instead of regularization parameter is used, which depends on the frequency and dynamically changes during the iterative search for a solution. This function depends on the structure of the initial seismic data prepared for the IDSP solution. The regularization functions are evaluated immediately before applying the seismic data inversion and are set as input processing parameters together with a parameterized model (which defines maximum possible resolution).

In the HRS-Geo technology, to determine the regularization function, the signal-to-noise ratio is evaluated, data on various levels of noise is tested, and a set of amplitude and phase regularization functions is formed for successive iterations when solving the IDSP. At the same time, according to the classical theory of problem regularization, with each iteration, the impact of the regularization function decreases, gradually bringing it closer to a true stable and detailed solution. Thus, the potentially possible vertical and horizontal resolution of the model contained in the source data is achieved.

Let us consider the main characteristics for the successful inversion of the interference seismic wave field into the distributions of acoustic impedances and reflection coefficients (reflectivities) on model and real observation materials. To study the features of the data inversion process, there is a 3D cube of the wave field obtained during special pre-processing of field seismic surveys.

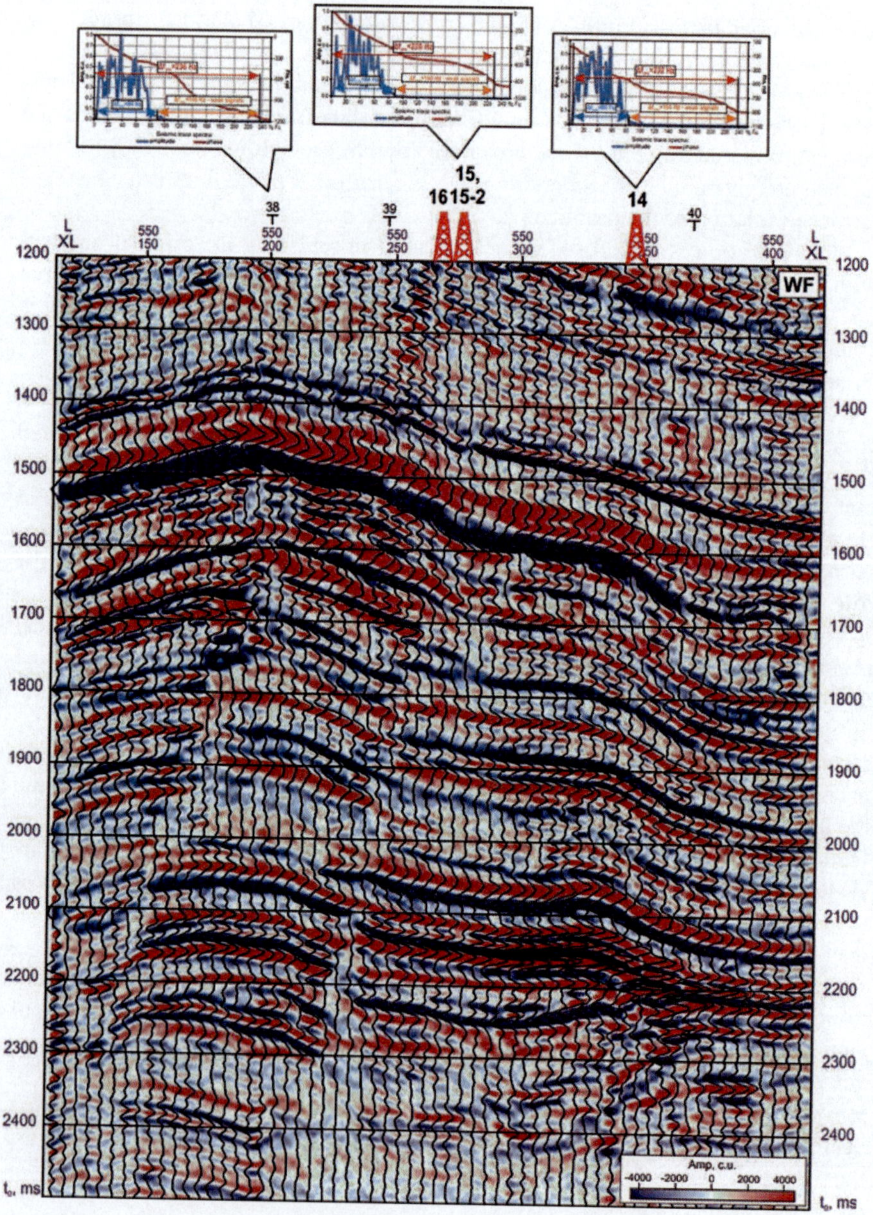


Fig. 2.1 The initial time section along the inline 550 obtained by a special graph of seismic data processing

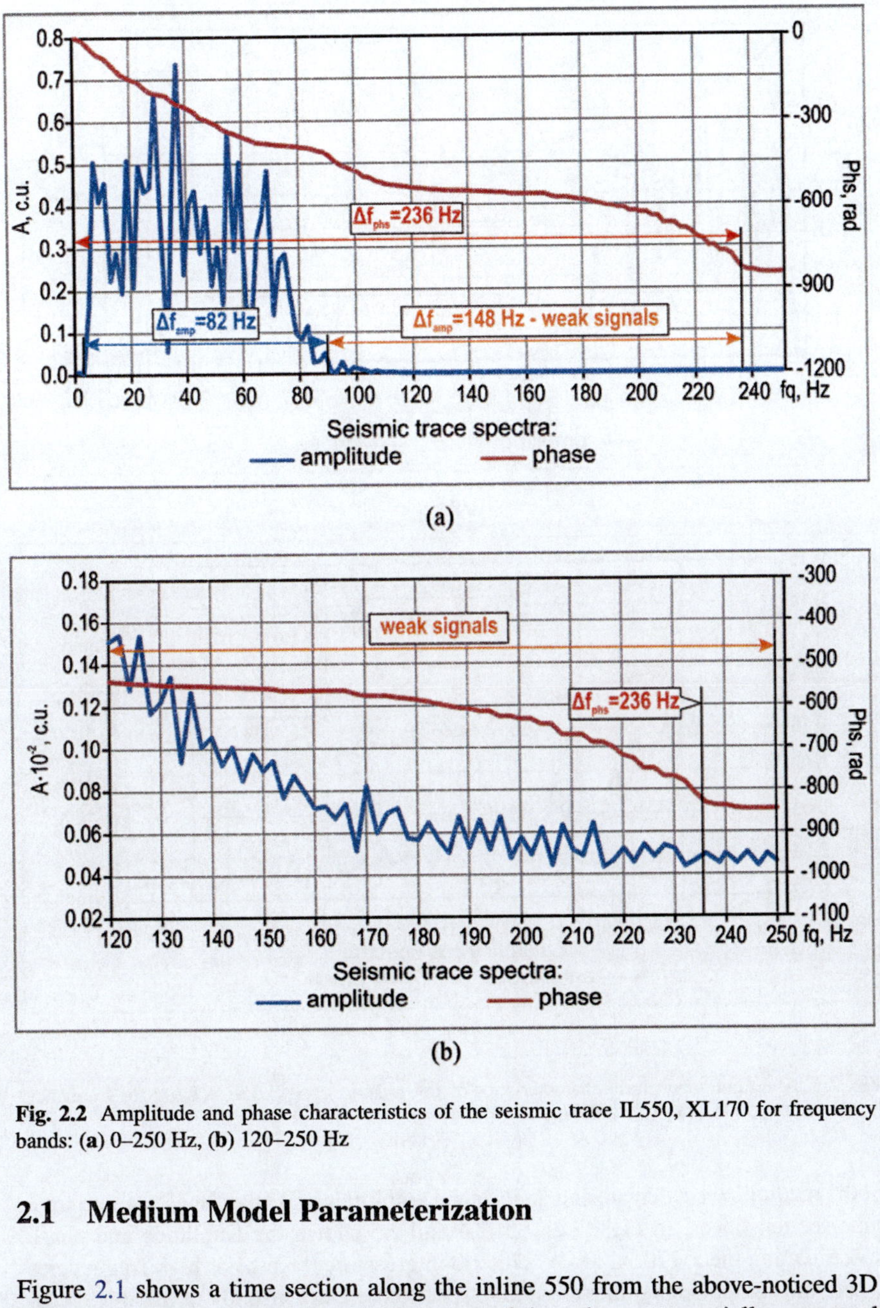


**Fig. 2.2** Amplitude and phase characteristics of the seismic trace IL550, XL170 for frequency bands: (**a**) 0–250 Hz, (**b**) 120–250 Hz

## 2.1 Medium Model Parameterization

Figure 2.1 shows a time section along the inline 550 from the above-noticed 3D cube of the wave field. Previously field data of the cube were specially processed with saving of true amplitudes with the maximum possible control of the frequency band over the entire seismic range of 0–250 Hz. The sampling step of the seismic data was 2 ms. To assess the possibility of extracting the most detailed parameters

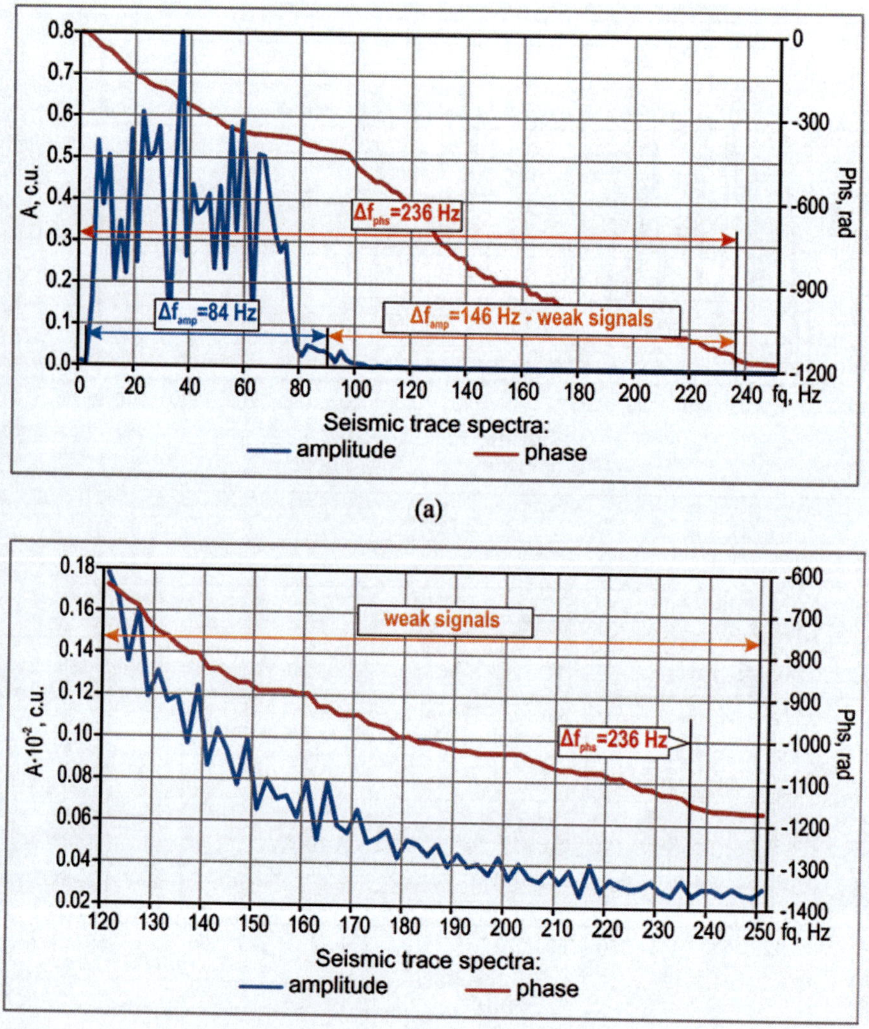


(a)



(b)

**Fig. 2.3** Amplitude and phase characteristics of the seismic trace IL550, XL200 for frequency bands: **(a)** 0–250 Hz, **(b)** 120–250 Hz

of the medium, we calculate and analyze the amplitude and phase spectra of seismic time-section traces. In Figs. 2.2, 2.3, 2.4 and 2.5 shown the amplitude and phase spectra along the xl170, xl200, xl270, xl345 crossline traces. For high frequencies where the components of the amplitude spectrum are not visually noticeable, spectra are constructed separately for the frequency band 120–250 Hz (Figs. 2.2b, 2.3b, 2.4b, and 2.5b). These data show that weak seismic signals are present in the high frequency bands. This indicates satisfactory pre-processing of field seismic materials where information about the thin-layer medium is stored.

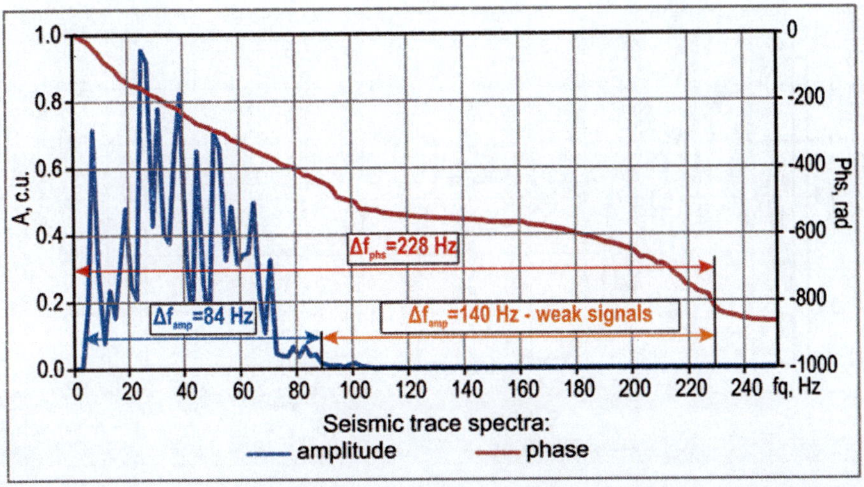


(a)

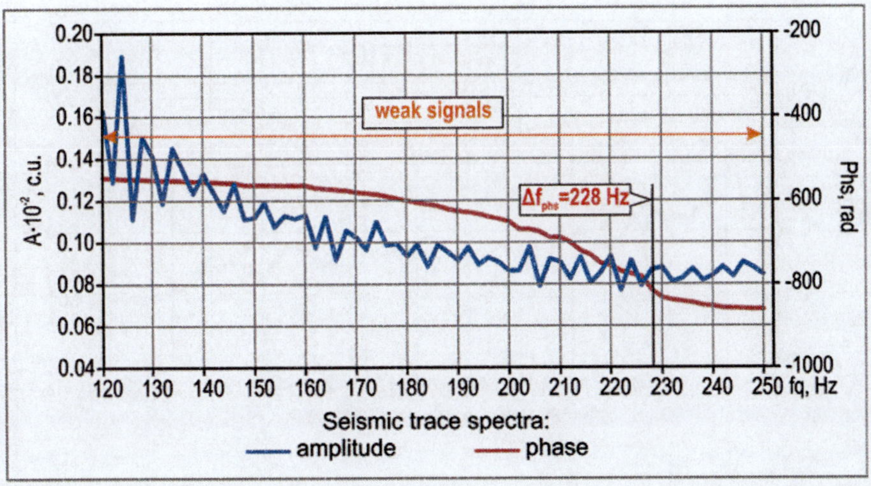


(b)

**Fig. 2.4** Amplitude and phase characteristics of the seismic trace IL550, XL270 for frequency bands: (**a**) 0–250 Hz, (**b**) 120–250 Hz

Let us determine the operating frequency bands from the amplitude and phase spectra. From Fig. 2.2a it can be seen that the main energy of the signal along the xl170 trace in the amplitude spectrum is concentrated in the frequency range 6–88 Hz, the bandwidth is $\Delta f = 82$ Hz (Table 2.1, No 1). This band can provide detail in time $\Delta t_a = 0.0061$ s, in depth $\Delta h_a = 10.15$ m (Table 2.1, No 1). On the phase spectrum graph (Fig. 2.2a), the active frequency band is estimated by the frequency range 0–236 Hz. At the same time, a horizontal section is observed above the frequency of 236 Hz, which indicates the absence of information on the components of

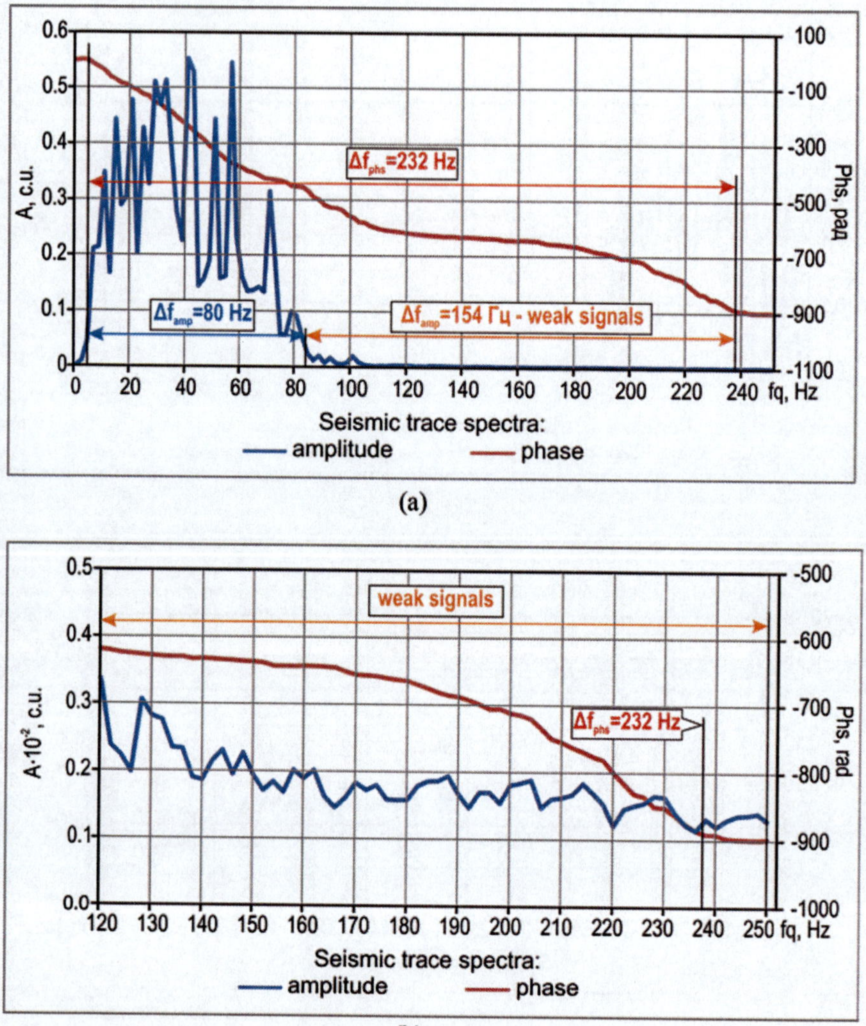


**Fig. 2.5** Amplitude and phase characteristics of the seismic trace IL550, XL345 for frequency bands: (**a**) 0–250 Hz, (**b**) 120–250 Hz

these frequencies. Here, a model can be extracted from the frequency band of the phase spectrum with a potential resolution in time $\Delta t_a = 0.0021$ s, in depth $\Delta h_a = 3.53$ m (Table 2.1). Similar results were obtained for other inline 550 seismic traces (Figs. 2.3, 2.4 and 2.5, Table 2.1, No 2–4). For traces xl200, xl270, xl345, the following resolution estimates were found for the amplitude spectrum $\Delta h_a$: 9.90 m, 9.91 m, 10.27 m, respectively; according to the phase spectrum $\Delta h_a$: 3.53 m, 3.53 m, 3.54 m, respectively (Table 2.1). All the main estimates are summarized in Table 2.1. It can be seen from this table that the frequency band of the amplitude spectrum

**Table 2.1** Prediction of resolution by amplitude and phase spectra during parameterization of the model for the IDSP solution

| No | IL, XL | SP type | $f_{min}$, Hz | $f_{max}$, Hz | $\Delta f$, Hz | $V_a$, m/s | $\Delta t_a$, s | $\Delta h_a$, m |
|---|---|---|---|---|---|---|---|---|
| 1 | 550, 170 | Samp | 6.0 | 88.0 | 82.0 | 3328.0 | 0.00610 | 10.15 |
| 2 | 550, 200 | Samp | 6.0 | 90.0 | 84.0 | 3328.0 | 0.00595 | 9.90 |
| 3 | 550, 270 | Samp | 4.0 | 88.0 | 84.0 | 3331.0 | 0.00595 | 9.91 |
| 4 | 550, 345 | Samp | 4.0 | 84.0 | 80.0 | 3286.0 | 0.00625 | 10.27 |
| Average | | | **5.0** | **87.5** | **82.5** | **3318.3** | **0.00606** | **10.06** |
| Accepted | | | | | | | **0.00600** | **10.00** |
| 1 | 550, 170 | Sphs | 0.0 | 236.0 | 236.0 | 3328.0 | 0.00212 | 3.53 |
| 2 | 550, 200 | Sphs | 0.0 | 236.0 | 236.0 | 3328.0 | 0.00212 | 3.53 |
| 3 | 550, 270 | Sphs | 0.0 | 228.0 | 228.0 | 3331.0 | 0.00219 | 3.65 |
| 4 | 550, 345 | Sphs | 6.0 | 238.0 | 232.0 | 3286.0 | 0.00216 | 3.54 |
| Average | | | **1.5** | **234.5** | **233.0** | **3318.3** | **0.00215** | **3.56** |
| Accepted | | | | | | | **0.00200** | **4.00** |

Parameters description:
No is the version number
IL, XL is the inline and crossline number, respectively
SP type is the spectrum type: Samp is amplitude, Sphs is phase
$f_{min}$ is the band minimum frequency, Hz
$f_{max}$ is the band maximum frequency, Hz
$\Delta f$ is the bandwidth of the operating frequency, Hz
$V_a$ is the average velocity in the formation, m/s
$\Delta t_a$ is the formation average time resolution, s
$\Delta h_a$ is the formation average depth resolution, m

makes it possible to confidently obtain detail on average over time $\Delta t_a = 0.0061$ s, over depth $\Delta h_a = 10.1$ m. The potential of the phase spectrum is much wider: with adequate information extraction the maximum achievable resolution averages $\Delta t_a = 0.00215$ s, $\Delta h_a = 3.56$ m (Table 2.1). The accepted resolution for the amplitude spectrum is $\Delta t_a = 0.006$ s, $\Delta h_a = 10$ m, for the phase spectrum $\Delta t_a = 0.002$ s, $\Delta h_a = 4$ m (depth detail depends on wave velocity). Thus, the model is parametrized with a maximum detail of 2 ms and 4 m. The preliminarily, components of the spectrum indicate that it is possible to extract high resolved models of the study medium from the available data.

## 2.2 Parameterization of the Seismic Data Inversion Process

As noted above, the process of seismic data inversion is an incorrect (ill-posed) task. First, let's see how the result of the wave field inversion appears in an incorrectly posed problem and what the materials look during regularization on a model example. To do this, consider the materials presented in [1–3]. Here we will touch only on the main question of this subsection about the stability of the inverse problem under consideration.

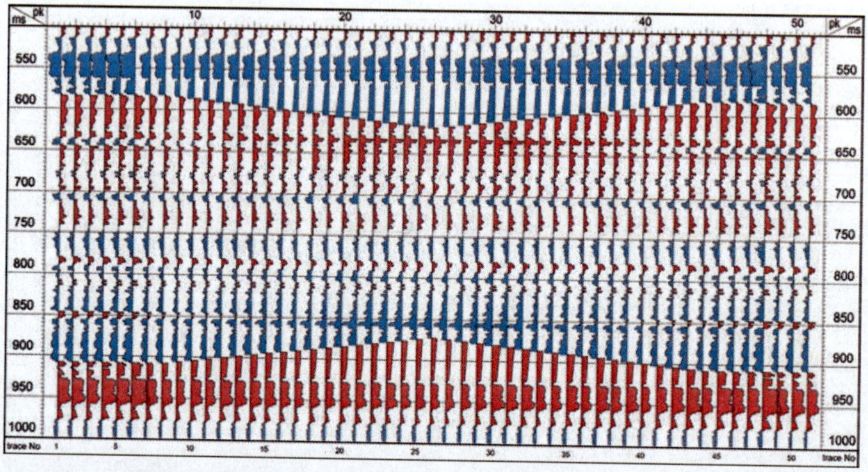

Fig. 2.6 Acoustic impedance model for calculating the wave field

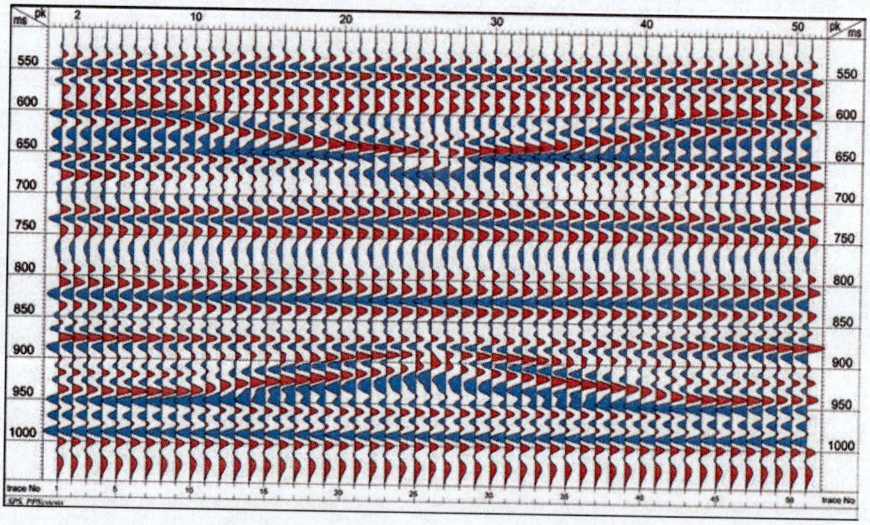

Fig. 2.7 The wave field calculated from the acoustic impedance model presented on the Fig. 2.6

Figure 2.6 shows a thin layer-approximated equtime model formed on the basis of acoustic logging of the well. Figure 2.7 represents the calculated wave field according to the model of Fig. 2.6. To study the effect of noise on the recovery capabilities of the algorithm, noise with a normal probability density distribution with parameters $\mu_N = 0.0$, $\sigma_N = 0.015$ was generated, which corresponds to the signal-to-noise ratio S/N = 3 (Fig. 2.8). The view of the wave field with additive noise looks as shown in Fig. 2.9. The noise energy is strong enough – the wave traces are torn, their smoothness is low. An attempt to solve the inverse problem based on these data

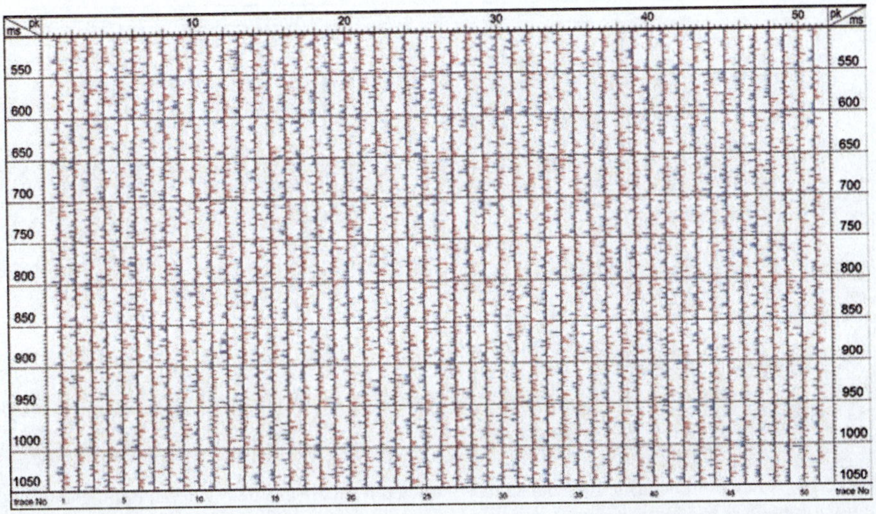

Fig. 2.8 Random noise with normal distribution ($\mu_N = 0.0$, $\sigma_N = 0.015$) for additive to the calculated wave field on the Fig. 2.7

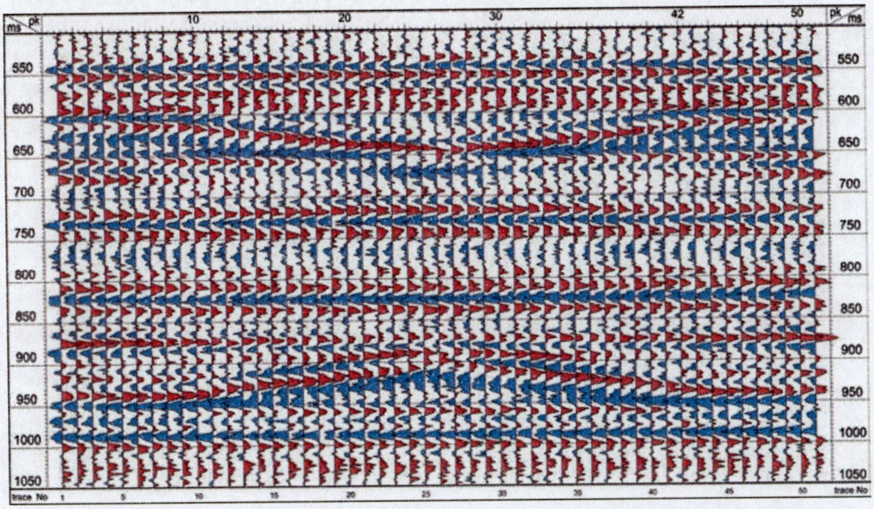

Fig. 2.9 A wave field with added random noise ($\mu_N = 0.0$, $\sigma_N = 0.015$, Fig. 2.8) and signal-to-noise ratio S/N = 3

led to the result shown in Fig. 2.10, where some traces were practically not restored, others were distorted quite strongly and some were partially resolved.

The influence of strong random noise has a rather significant destructive effect on the desired result – the acoustic impedance model. However, the most general kinematic and dynamic patterns of the model being restored have still been

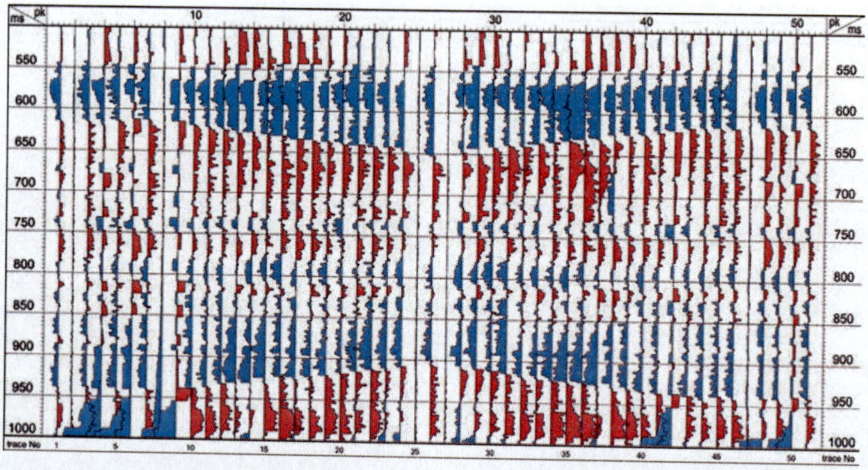

**Fig. 2.10** A model of acoustic impedance reconstructed from a wave field without optimal application of multicriteria regularization

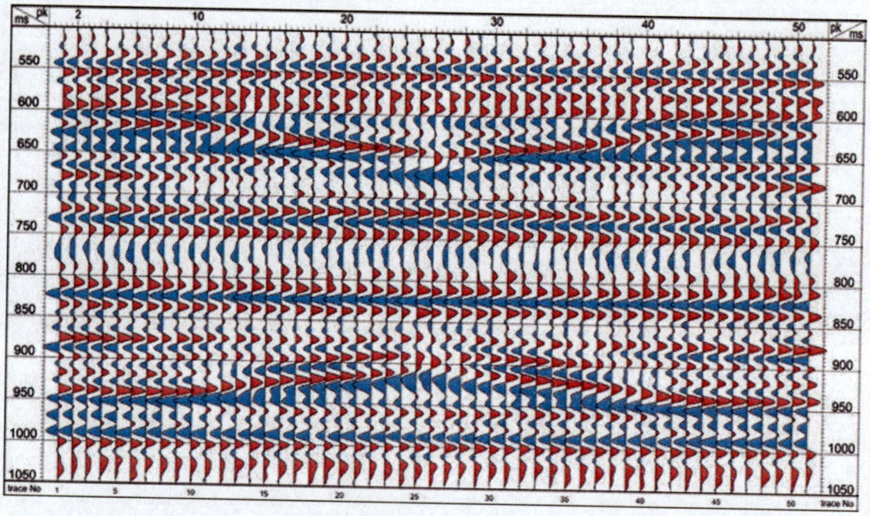

**Fig. 2.11** Optimal filtering of the wave field from random noise

preserved (this indicates the noise immunity of the developed recovery algorithm). At the same time, it can be seen that in some cases the signal-to-noise ratio turned out to be quite unfavorable for recovery, as a result of which, on some traces, distortion led to a sharp decrease in the intensity of the desired acoustic impedance. On separate traces (with numbers No 2, 8, 25, 27, 47) restoration was almost impossible – the stability of the solution was lost here, and on other vertical sections – the resulting traces are only partially distorted (No 4, 6, 9, 42, 50) (Fig. 2.10). In order to reduce the adverse effect of strong random noise on the desired result of accessing the seismic wave field traces (Fig. 2.10), a standard optimal filtration procedure

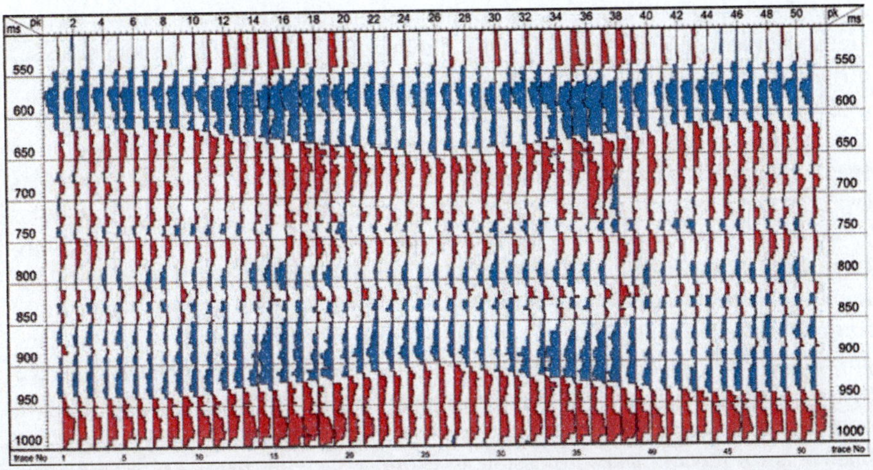


**Fig. 2.12** The model of the acoustic impedance obtained during the regularization of the IDSP solution

was applied. As a result, the initial model of the seismic wave field with superimposed strong random noise, but to a certain extent regularized by optimal filtration, was obtained (Fig. 2.11).

As a result of solving the IDSP using regularization parameters, a reconstructed model of acoustic impedance was obtained (Fig. 2.12). At the same time, the task has acquired stability – all traces have been satisfactorily extracted from the synthetic wave field. The results were quite stable. Although the resolution (vertical and horizontal) of the acoustic model has somewhat deteriorated, the found acoustic model of the medium as a whole retains a thin-layer structure and almost all geological elements are confidently identified and suitable for interpretation.

For a preliminary assessment of the regularizing characteristics of the inversion procedure, we will evaluate the structure of noises and useful signals, quantify the signal-to-noise ratio based on the real data of the 3D cube of the wave field described above.

To determine the type, structure, and characteristics of interference, we will complicate the convolutional model of the seismic trace (2.1), i.e. instead of the traditional:

$$f(t)=\int_0^t s(\tau)\cdot r(t-\tau)\cdot d\tau+n(t)=s(t)*r(t)+n(t), \tag{2.1}$$

we will use a convolutional model of the seismic trace in the form:

$$f(t)=\int_0^t s(\tau)\cdot r(t-\tau)d\tau+\int_0^t s'(\tau')\cdot n_r(t-\tau')d\tau'=s(t)*r(t)+s'(t)*n_r(t) \tag{2.2}$$

where, $f(t)$ is the seismic trace, $s(t)$ is the probing pulse, $r(t)$ is the reflection coefficients of the medium, $s'(t)$ is the pulse filtering the noise, $n_r(t)$ is the sought noise, $n(t)$ is interference noise: $n(t) = s'(t) * n_r(t)$.

Such a description of the seismogram model is justified because often (if not always) noise is also wave-like, participating in the process of wave propagation in space and time (except maybe when the noise does not have an elastic-deformation (wave) nature, for example, noise from electrical networks).

Since noises are involved in the process of wave propagation in space and time (they are of a wave nature), they are subjected to the influence of an elementary impulse. The resulting primary noise is affected by a seismic impulse, which transforms it through filtering: the original noise changes its shape and frequency composition, and participates in the interference process, complicating the overall wave pattern. That is, noise is a constituent element of a more complex interference phenomenon described by convolution. Representation of this process in the form of a mathematical model makes it possible to estimate the contribution of random primary noise $n_r(t)$ to interference noise $n(t)$ (changed by an impulse during wave propagation in space), which, in turn, is included as an element of a seismic record in a general convolutional model, where the components are already the reflection coefficients and the probing impulse (2.2).

Thus, to determine the random noise $n_r(t)$, it is necessary to solve the inverse dynamic seismic problem with respect to $n_r(t)$, having previously estimated the interference noise $n(t)$. In this case, parameters are also set that ensure the solution stability of the problem.

Next, the structure of the noise is evaluated: shape, level in the form of the S/N ratio, type of noise according to statistical characteristics (average, variance, etc.), noise imagine (histogram, type of probability density function, etc.) (Figs. 2.13, 2.14, 2.15, 2.16 and 2.17).

In Fig. 2.13 the results of the main stages of signal and noise assessments are briefly shown. For this purpose, the procedures of bandpass and linear filtering and solving the inverse problem with respect to random noise were applied. Figure 2.13a shows the result of signal and noise evaluation. In this case, the noise itself is further complicated by the interference pattern (Fig. 2.13b). And to determine the "pure" noise, not complicated by interference, we solve the inverse dynamic problem with respect to noise $n_r(t)$. Here, the definition of noise (but not its suppression) is the goal of the inverse problem. In this case, interference involving the signal $s'(t)$ is a "hindrance" for noise $n_r(t)$.

After the random noise was inverted, its internal structure was clearly revealed (Fig. 2.13c). Note that since at this stage, when finding an noise, the inverse problem is solved, which, like all inverse problems, is incorrect, regularizing algorithms are also used here. Figure 2.14 shows the calculated characteristics of the received noise. The constructed histogram and its approximation by a continuous probability density function makes it possible to conclude about the type and characteristics of the noise found. It turned out that this interference, extracted from a real seismic trace, has a random character, the probability density distribution of which is close to normal with the parameters: mathematical expectation $\mu_N = 0$ c.u., standard

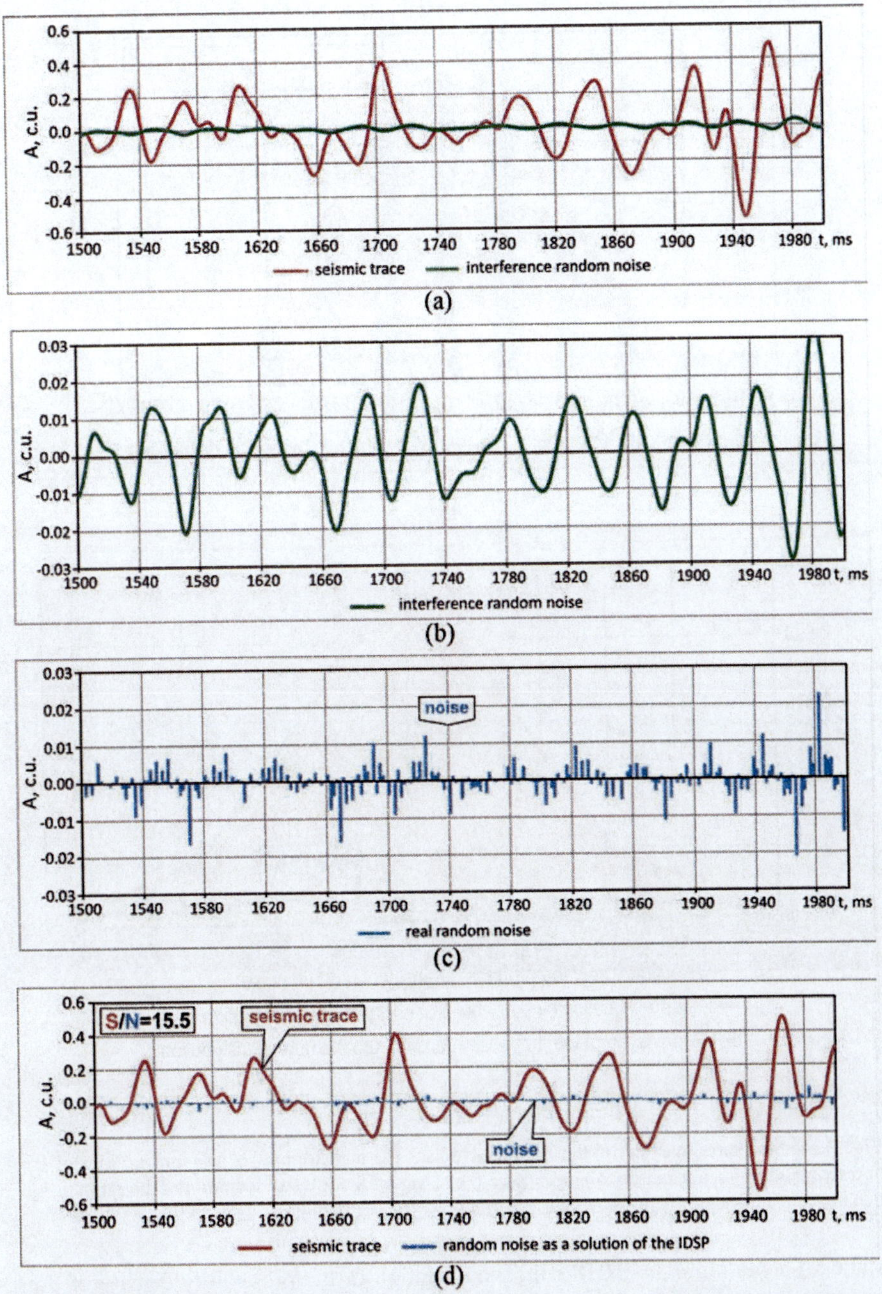


**Fig. 2.13** Extraction of a signal, noise from a seismic trace and determination of the signal-to-noise ratio S/N: (**a**) interference signal and noise; (**b**) interference noise; (**c**) the result of the interference noise inversion by solving the IDSP; (**d**) evaluation of the real signal-to-noise ratio: S/N = 15.5

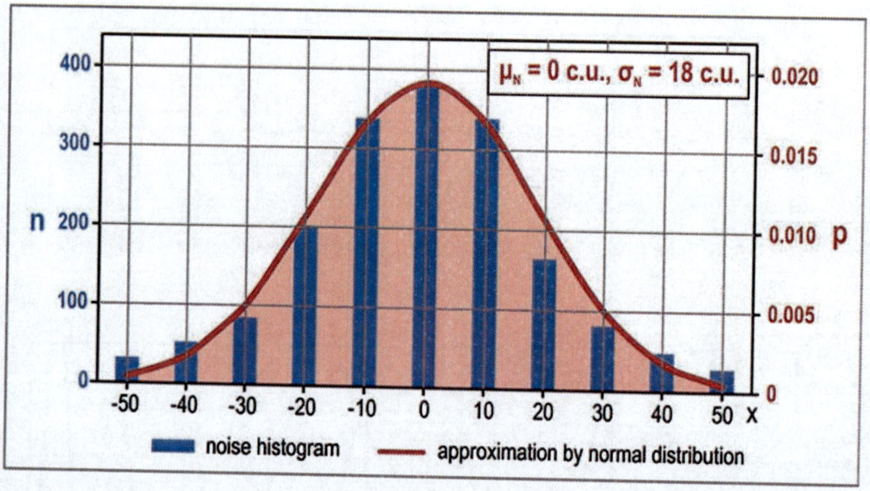


(a)

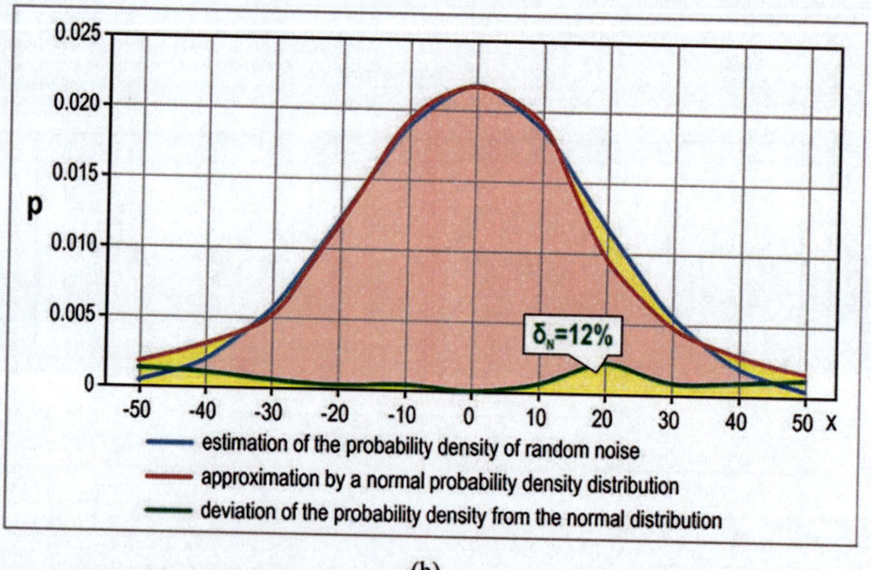


(b)

**Fig. 2.14** Structure analysis of the extracted noise: (**a**) a histogram of random noise and its approximation by a normal distribution ($\mu_N = 0$ c.u., $\sigma_N = 18$ c.u.); (**b**) deviation of the probability density distribution of the noise from the normal law ($\delta_N = 12\%$)

standard deviation $\sigma_N = 0.018$ c.u. The deviation of the probability density of the noise extracted from the seismic trace from the normal law is $\delta_N = 12\%$ (Fig. 2.14b). In the time domain, the estimated signal-to-noise ratio was S/N = 15.5 (Fig. 2.13d). Such a value of S/N indicates successful preliminary special data processing with preservation of true amplitudes and weak signals, which makes it possible to extract and study in detail the fine structure of the medium.

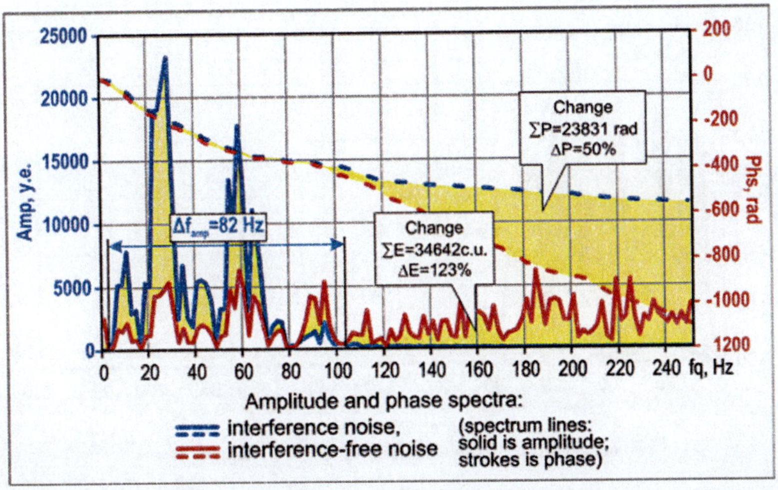


**Fig. 2.15** Change in the amplitude and phase characteristics of the noise when solving the IDSP with respect to random noise

However, noise of such a low level is often a significant obstacle to finding a detailed model: it is required to apply adequate regularization methods when solving IDSP. Figure 2.15 represents the amplitude and phase spectra of interference noise and interference-free noise (noise without "interference") obtained from the outcome of the IDSP solution.

Figures 2.16 and 2.17 show the amplitude and phase spectrum of the initial seismic trace and the separated signal and noise over the entire frequency range, including singly bands of weak signals.

Based on the separated interference signals and the structure and characteristics of noise, sets of regularizing functions were calculated for the amplitude $\xi_{Ai}(\omega)$ (Fig. 2.18a) and phase $\xi_{Pi}(\omega)$ (Fig. 2.18b) spectra. For their effective practical application, these functions were approximated by continuous frequency functions (Fig. 2.18). The peculiarity of the set of these functions is that each of them is dynamically applied at its own stage of iterative computation. At the same time, with an increase in the iteration number, the influence of the regularizing function weakens relative to the previous iteration (although increasing with respect to the initial approximation), thereby reducing the bias of the solution of the IDSP and bringing it closer to an unbiased evaluation of the result (according to the classical property of the regularizing parameter noted above). This makes it possible not only to ensure the stability of the process of finding a solution, but also to maintain maximum detail, unlike traditional approaches to solving incorrect problems, where it is necessary to find the optimum between the variance of the solution and the resolution (see the trade-off curve, Fig. 2.29) [4].

We also note that the above separation of the seismic trace into a signal and a noise, the purpose of which is to determine the regularizing functions, contains

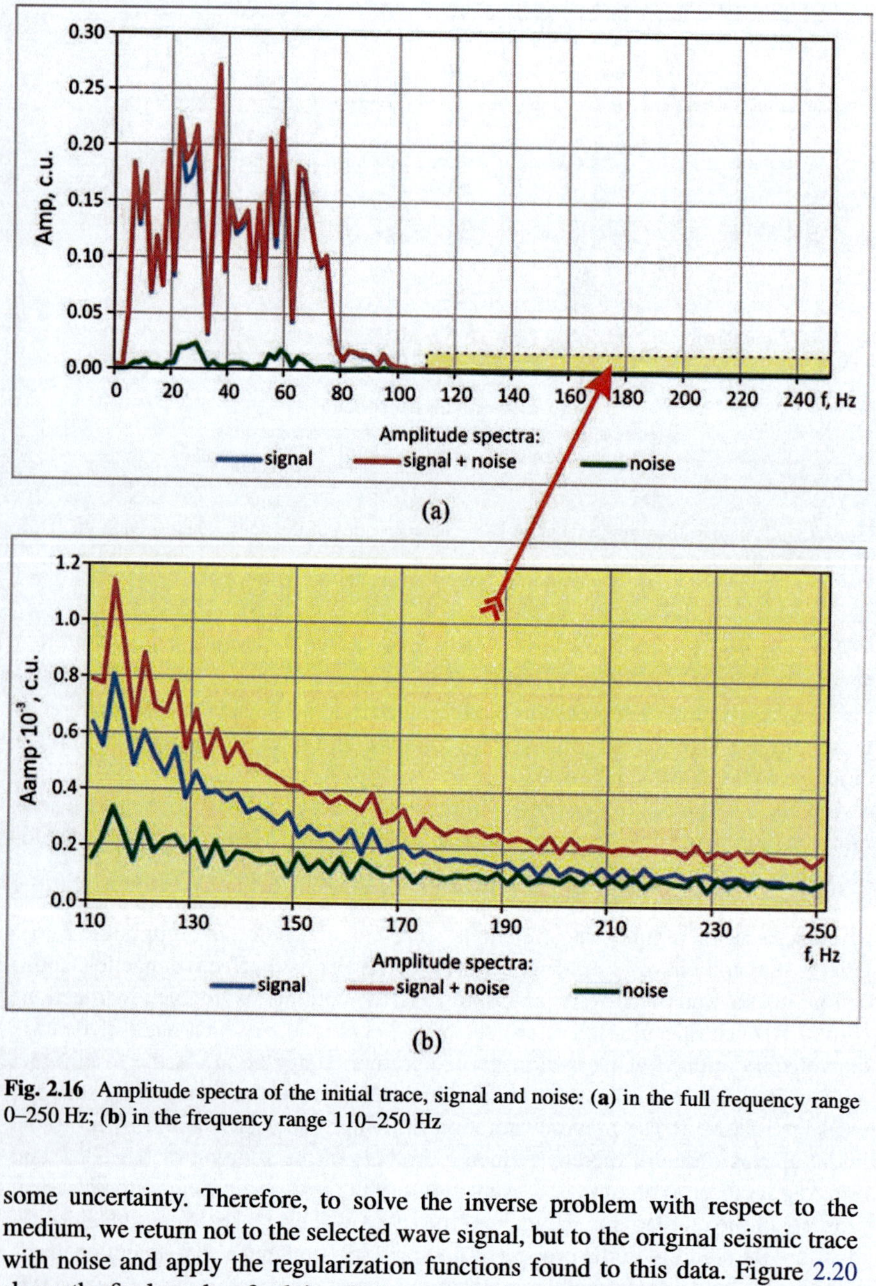


**Fig. 2.16** Amplitude spectra of the initial trace, signal and noise: (**a**) in the full frequency range 0–250 Hz; (**b**) in the frequency range 110–250 Hz

some uncertainty. Therefore, to solve the inverse problem with respect to the medium, we return not to the selected wave signal, but to the original seismic trace with noise and apply the regularization functions found to this data. Figure 2.20 shows the final results of solving the inverse dynamic problem. Figure 2.20a shows the acoustic impedances and the noises contained therein (noises in AI as much as 0.1%). Figure 2.19a shows a histogram of the reflection coefficients found, their approximation by the normal law and parameters: mathematical expectation $\mu_r = 0$,

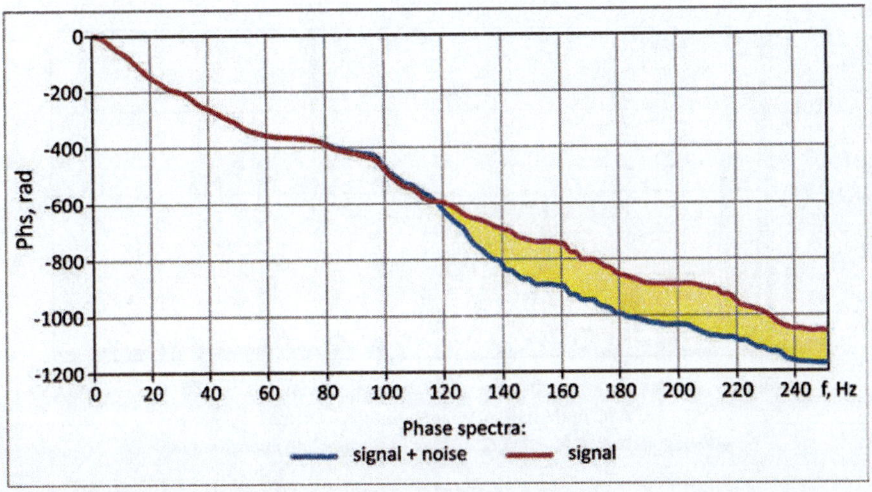


**Fig. 2.17** Phase spectra of the initial trace and the extracted signal

standard deviation $\sigma_r = 0.15$. The deviation of the probability density of the noise defined from the seismic trace is $\delta_r = 5\%$ (Fig. 2.19b). Figure 2.20b shows the reflection coefficients and the noise contained in them. Figure 2.20c shows acoustic impedances and reflection coefficients with remote (or suppressed) noises. The error of the inversion method in this numerical experiment is 0.1% (Fig. 2.20).

Errors in the solution of IDSP are formed from different sources, and difficulties often arise in assessing the contributions of each of the factors. Among the main sources of errors can be noted: the accuracy of determining the seismic pulse, estimating the type and structure of noise, the method of inversion, etc. In general, these estimates fit into the error range of 2–8%.

## 2.3 An Example of Solving the Inverse Dynamic Problem of Seismics

Figure 2.21 presents a general scheme for finding a solution to the IDSP by minimizing the system of objective functions. The figure shows the main characteristics of classical inverse problems: finding the model space within which the model is searched (this stage is described above when parameterizing the model), determining the areas of existence, uniqueness and stability (also shown above how these conditions are provided within the multi-criteria approach), forming a set of objective functions (in HRS-Geo technology energy and phase characteristics in the complex are used to construct objective functions), choosing the starting point for optimizing the solution (with the help of testing various areas of determining objective functions), etc.

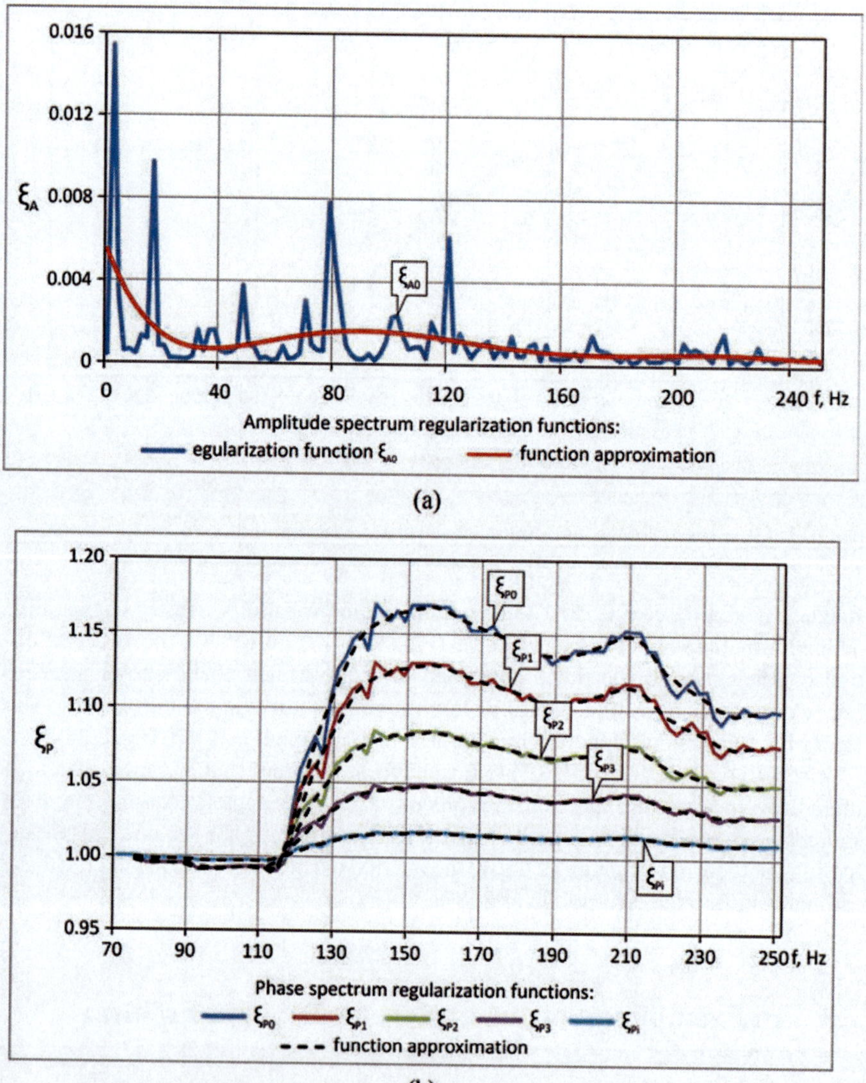


**Fig. 2.18** Regularization functions of the spectra: (**a**) amplitude; (**b**) phase

The peculiarity of the search for a solution to the inverse dynamic problem of seismics for a thin-layer medium also consists in the fact that the processed seismograms with a limited frequency band represent some smooth functions (Fig. 2.22b). From them it is necessary to obtain a solution in the form of discontinuous functions, despite the fact that thin-layer AI and RC have a wide frequency spectrum (Fig. 2.22c, d), in contrast to the PAC traces, where a smooth function is mapped to a smooth one (Fig. 2.22a). In other words, the initial data and the desired

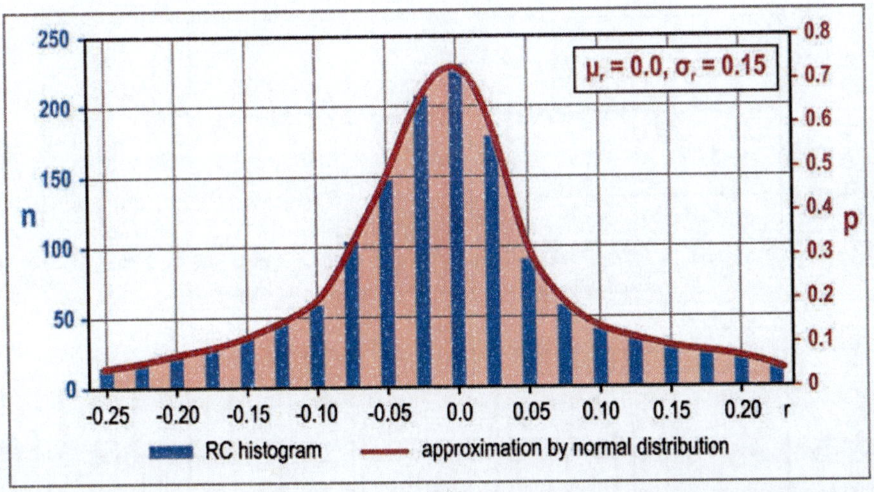


(a)

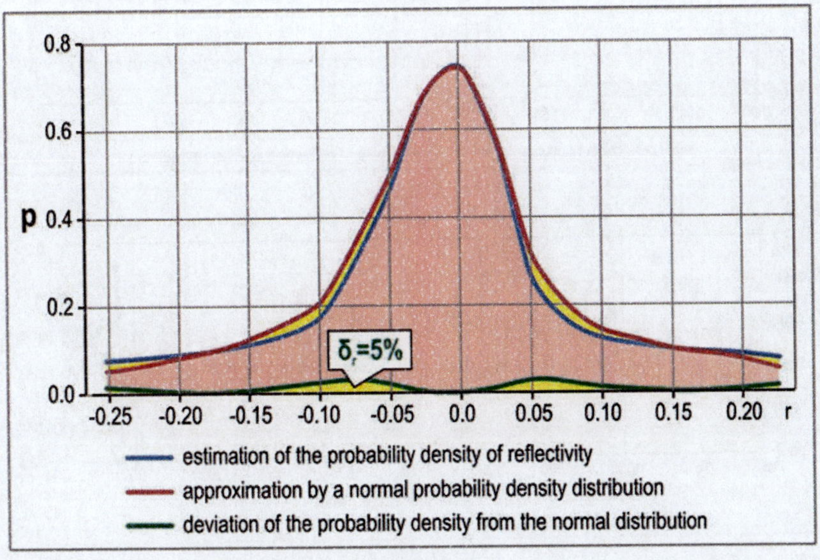


(b)

**Fig. 2.19** Analysis of the structure of extracted reflectivity: (**a**) a histogram of reflectivity and its approximation by a normal distribution ($\mu_N = 0.0$, $\sigma_N = 0.15$); (**b**) deviation of the probability density distribution of the RC from the normal law ($\delta_N = 5\%$)

transformation results are in different classes of functions – the class of continuous (smooth) functions and the class of discontinuous, respectively. In this sense, obtaining a solution is a mapping of finite–dimensional functions from one class to another using the inverse problem solution operator.

The phase characteristics of the wave field are more sensitive to changes in the seismic properties of the medium and contain much more information about the parameters of the desired medium than the energy ones. In turn, adequate use of

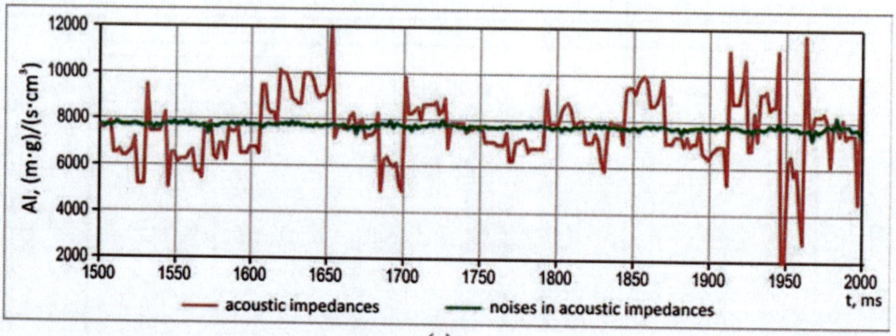


(a)

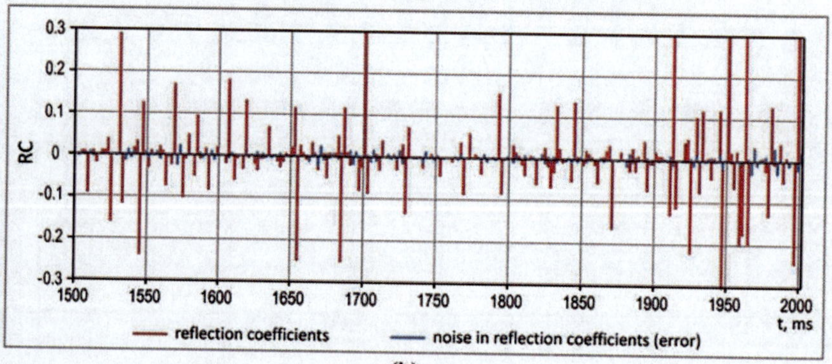


(b)

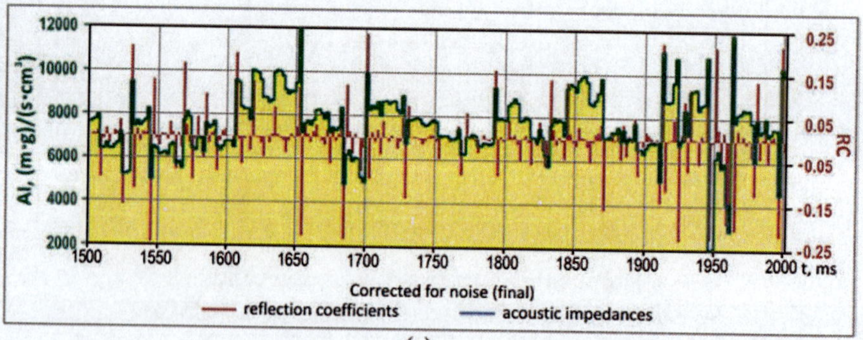


(c)

**Fig. 2.20** Restored acoustic impedance, reflection coefficients and noises: **(a)** acoustic impedance and noise; **(b)** reflection coefficients and noise; **(c)** final reflection coefficients and acoustic impedance

energy factors gives stability to the inversion system. Joint operation with the described characteristics makes it possible to extract the distribution of the properties of the medium with maximum detail in the process of steady inversion of the wave field data.

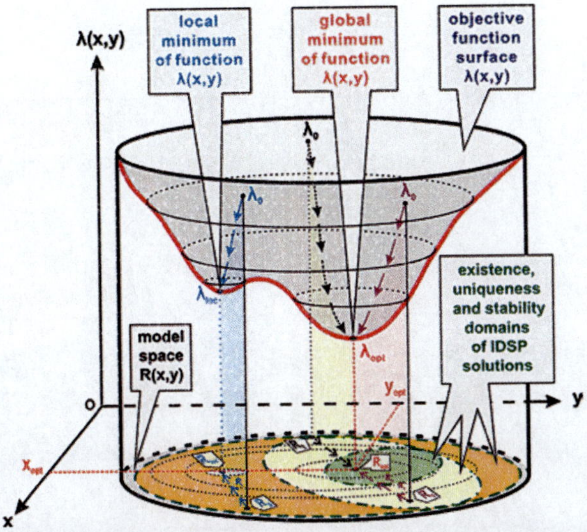


Fig. 2.21 The scheme of the IDSP solution by the optimization method

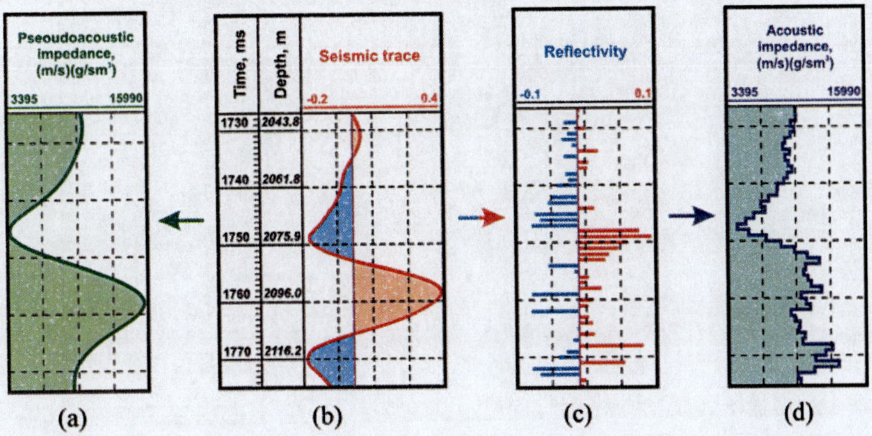


**Fig. 2.22** To the solution of the IDSP in the conditions of a thin-layer model of the medium on fragments: the pseudoacoustic trace (**a**) and the seismic trace (**b**) as smooth functions; the distribution of reflection coefficients (**c**) and acoustic impedance (**d**) as discontinuous functions

System of response functions (vector of functions) with restrictions [2, 3] in a general modified form can be represented as:

$$\begin{cases} \lambda_1(R) = \left\| E_f(R,\omega) - E_m(R,\omega) \right\| + F_1\left(\xi_{Ai}, E_f, R, \omega\right) \Rightarrow \min, \\ \lambda_2(R) = \left\| P_f(R,\omega) - P_m(R,\omega) \right\| + F_2\left(\xi_{Pi}, P_f, R, \omega\right) \Rightarrow \min, \\ S(R) \geq c, \quad R \Rightarrow fix, \\ S(R) < c, \quad R \Rightarrow \text{var}. \end{cases} \tag{2.3}$$

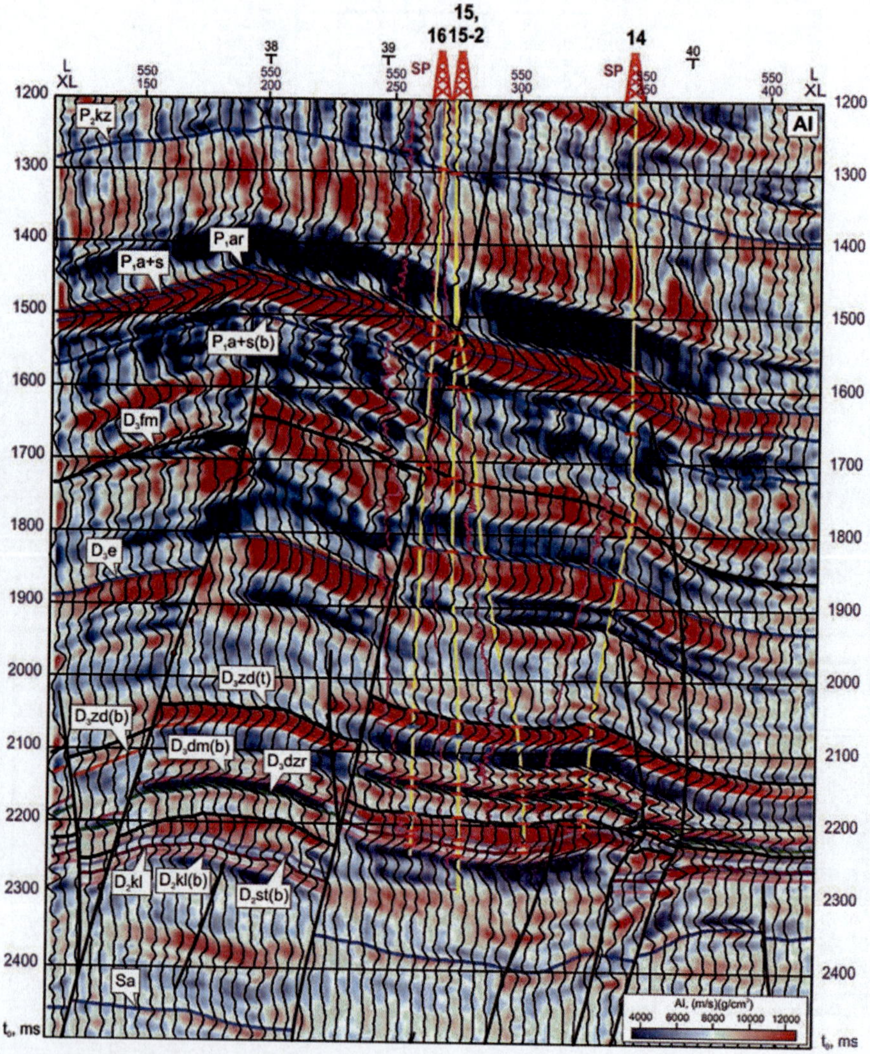


**Fig. 2.23** Section of pseudoacoustic impedance along inline 550 as a result of the solution of the IDSP in the seismic frequency range of 5–88 Hz with interpretation of boundaries and faults and binding to the horizons from wells data

where $\lambda_1$, $\lambda_2$ are objective functions (functionals);

$E_f(R)$, $E_m(R)$ are the energy of real and calculated traces, respectively;

$P_f(R)$, $P_m(R)$ are the phase characteristics of real and calculated traces responsible for the shape of the reflected signal, respectively;

ω is cyclic frequency;

$F_1$, $F_2$ are stabilizing functions;

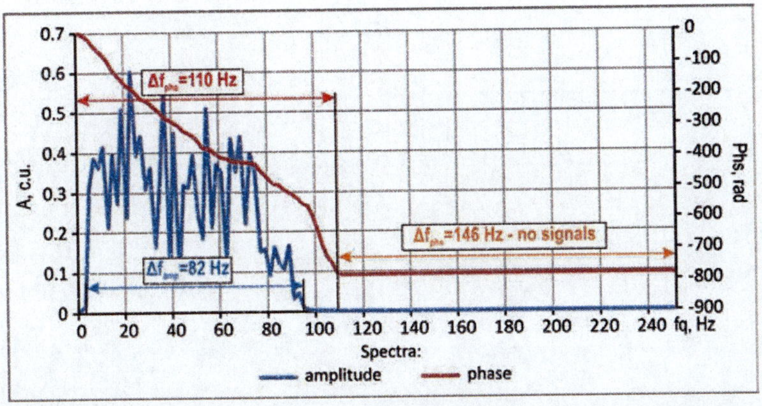


**Fig. 2.24** Amplitude and phase characteristics along the IL550, XI200 seismic trace after band-pass filtration of 5–88 Hz

$\xi_{Ai}$, $\xi_{Pi}$ are regularization functions of the corresponding functionals, depending on the iteration number i;

$S(R) \geq c$, $S(R) < c$ are constraints in the form of inequalities defining fixed and variable parameters;

$c$ is a constant that determines the reliability of the resulting solution;

$R$ is the desired solution of the inverse dynamic problem.

Figure 2.23 shows the traditional method of wave field inversion in the seismic frequency range of 5–88 Hz. The spectra corresponding to this frequency band are shown in Fig. 2.24. It can be seen that weak signals in the higher frequency bands (110–256 Hz) are absent due to filtering of the latter. Such filtering is traditionally carried out in order to give stability to the solution, however, detail is brought to the detriment – interference is preserved at the same time. The frequency band of 82 Hz allows you to achieve a maximum of detail no better than 10 m (Table 2.1), but in practice, as a rule, much worse. It is impossible to see and evaluate the thin layered structure of the medium here.

In the HRS-Geo technology, the working band of the amplitude spectrum is taken as the definition of the initial approximation, in this case, the resolution of the initial approximation is about 10 m (Table 2.1). And since weak signals have been preserved in the phase spectrum and it is most sensitive to them, the active operating band in this case is around 0–236 Hz (Table 2.1), which corresponds to a resolution of 3.5 m. This makes it possible, consistently maintaining stability and applying regularizing functions, to move from the initial model with an amplitude spectrum resolution of 10 m to a potentially possible phase spectrum resolution of 3.5 m (Table 2.1).

Figure 2.25 shows a section of the acoustic impedance of the initial approximation with a detail of 10 m. Figure 2.26 shows a section of acoustic impedance after

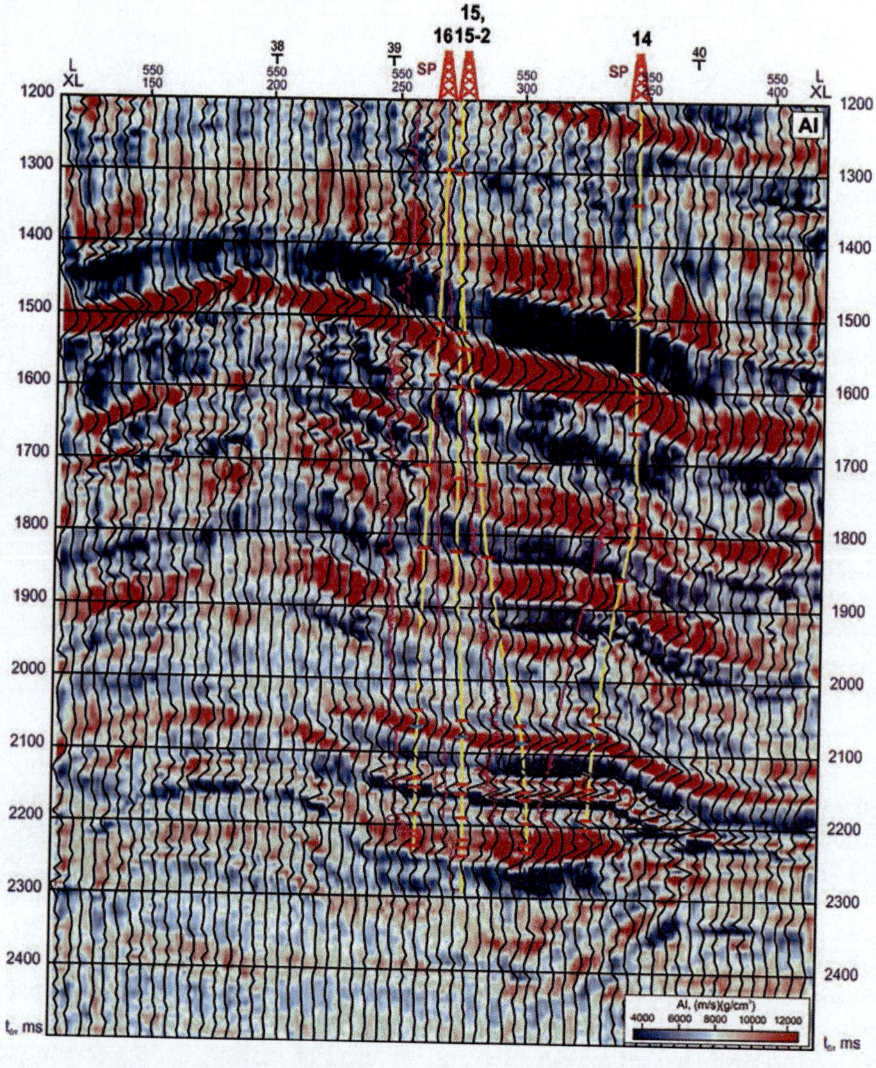


Fig. 2.25 Initial approximation of acoustic impedance along the inline 550

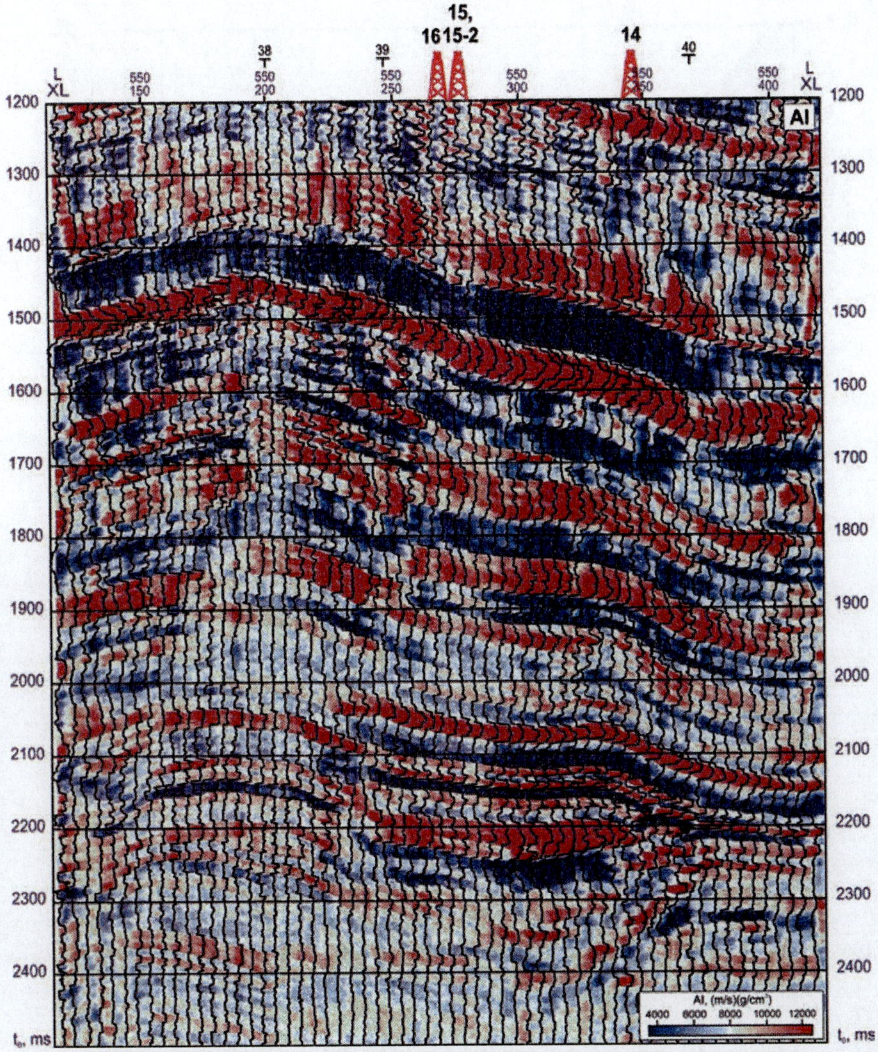


**Fig. 2.26** The optimal section of acoustic impedance according to inline 550, obtained as a result of the IDSP solution

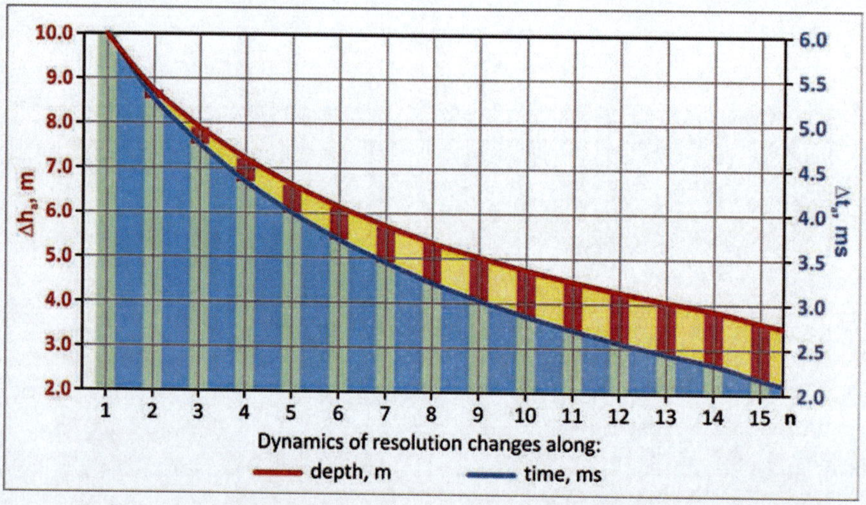


**Fig. 2.28** Dynamics of changes in the resolution of the IDSP result during the iterative solution

**Table 2.2** The resolution change dynamics when solving IDSP

| n | $f_{min}$, Hz | $f_{max}$, Hz | Δf, Hz | $V_a$, m/s | $Δt_a$, s | $Δh_a$, m |
|---|---|---|---|---|---|---|
| 1 | 5.0 | 88.0 | 83.0 | 3318.3 | 0.00602 | **9.99** |
| 2 | 5.0 | 100.0 | 95.0 | 3318.3 | 0.00526 | **8.73** |
| 3 | 5.0 | 110.0 | 105.0 | 3318.3 | 0.00476 | **7.90** |
| 4 | 5.0 | 120.0 | 115.0 | 3318.3 | 0.00435 | **7.21** |
| 5 | 5.0 | 130.0 | 125.0 | 3318.3 | 0.00400 | **6.64** |
| 6 | 5.0 | 140.0 | 135.0 | 3318.3 | 0.00370 | **6.14** |
| 7 | 5.0 | 150.0 | 145.0 | 3318.3 | 0.00345 | **5.72** |
| 8 | 5.0 | 160.0 | 155.0 | 3318.3 | 0.00323 | **5.35** |
| 9 | 5.0 | 170.0 | 165.0 | 3318.3 | 0.00303 | **5.03** |
| 10 | 5.0 | 180.0 | 175.0 | 3318.3 | 0.00286 | **4.74** |
| 11 | 5.0 | 190.0 | 185.0 | 3318.3 | 0.00270 | **4.48** |
| 12 | 5.0 | 200.0 | 195.0 | 3318.3 | 0.00256 | **4.25** |
| 13 | 5.0 | 210.0 | 205.0 | 3318.3 | 0.00244 | **4.05** |
| 14 | 5.0 | 220.0 | 215.0 | 3318.3 | 0.00233 | **3.86** |
| 15 | 5.0 | 236.0 | 231.0 | 3318.3 | 0.00216 | **3.59** |

Parameters description:
n is the iteration number
$f_{min}$ is the band minimum frequency, Hz
$f_{max}$ is the band maximum frequency, Hz
Δf is the bandwidth of the operating frequency, Hz
$V_a$ is the average velocity in the formation, m/s
$Δt_a$ is the formation average time resolution, s
$Δh_a$ is the formation average depth resolution, m

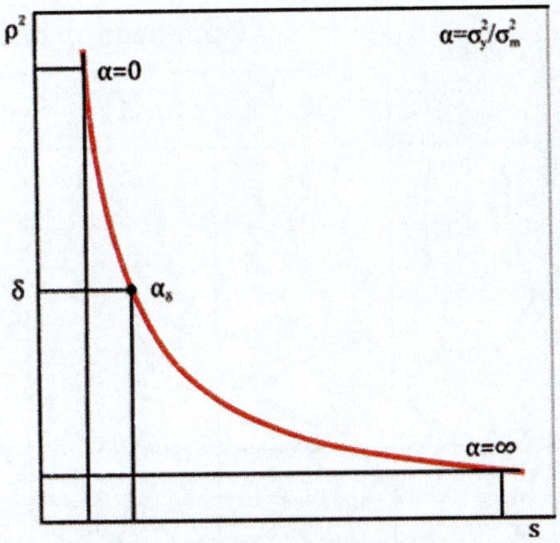


**Fig. 2.29** The trade-off curve between the variance p² and the resolution s. (According to Ref. [4])

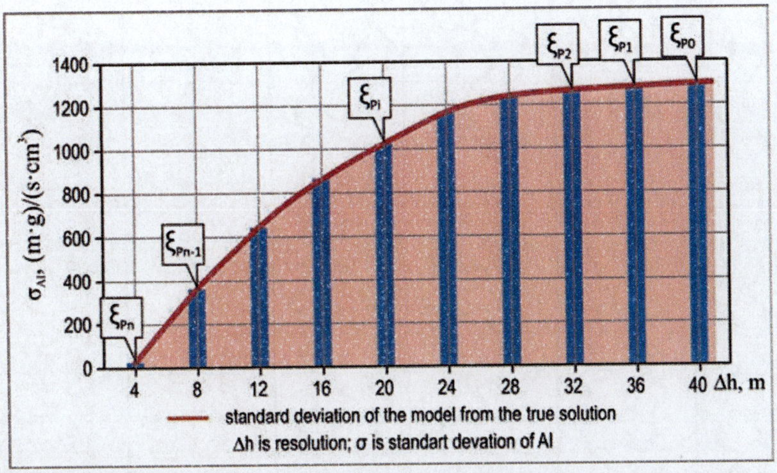


**Fig. 2.30** Dynamics of changes in the standard error of the model in the process of finding the optimal IDSP solution ("trade-off curve" in the HRS-Geo technology)

final optimization with a maximum resolution of 3.5–4.0 m according to estimates using the phase characteristics of the signal (Table 2.1).

Figure 2.27b shows the iteration stages where the resolution changes from the initial approximation to the final result. Also for comparison, the left column (Fig. 2.27a) shows the result of traditional inversion in the seismic frequency range.

Figure 2.28 and Table 2.2 show the dynamics of changes in time and spatial resolutions in the process of solving IDSP. Here, with each iteration, an improvement in the quality of the resulting model is observed, i.e. an increase in the detail of the desired model is manifested.

In practice, according to the classical paradigm of inverse problems of geophysics, "… when constructing a solution to the inverse problem, one has to look for a compromise between dispersion and resolution …" [4]. The relationship between

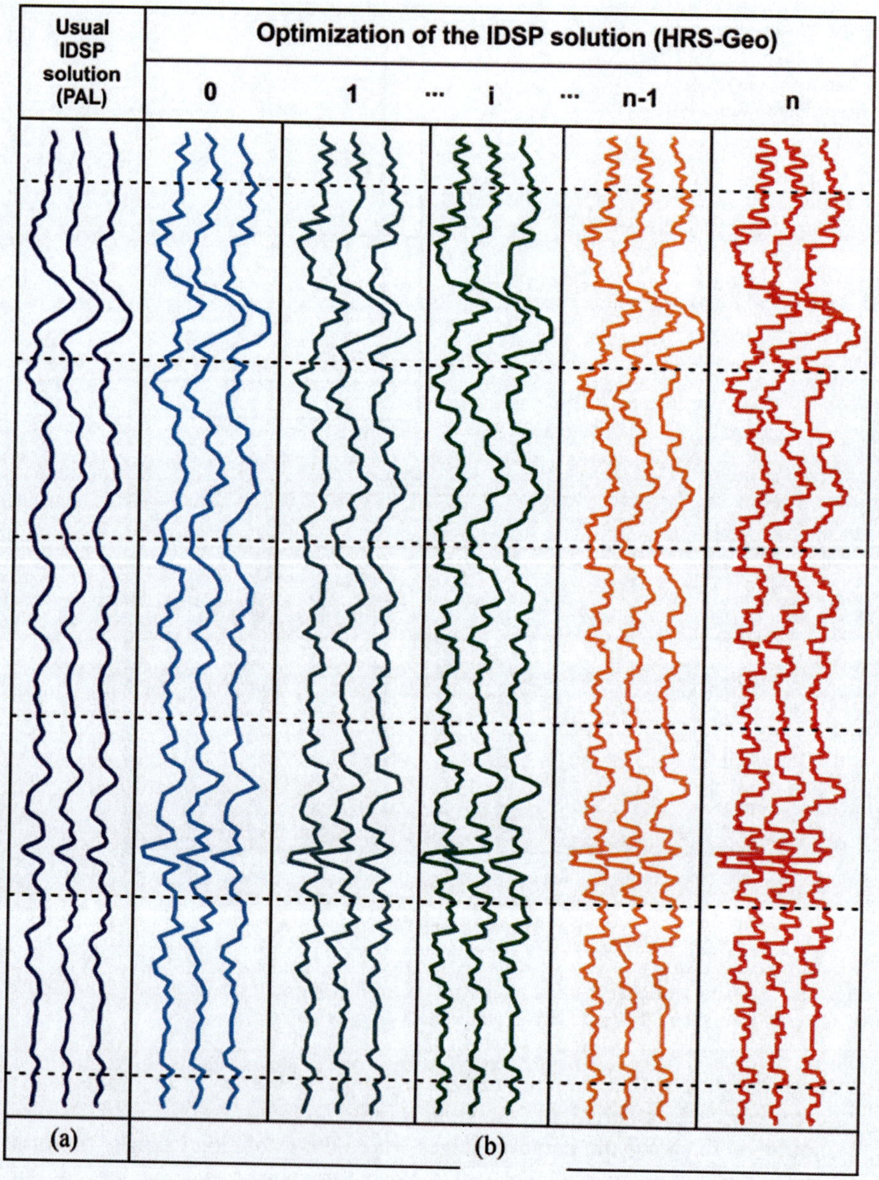


**Fig. 2.27** The change in the detail of acoustic impedance during the iterative IDSP solution: (**a**) the traditional solution method; (**b**) the process of optimizing the IDSP solution in the HRS-Geo technology

**Table 2.3** The model root-mean-square error change dynamics in the process of searching for the optimal IDSP solution

| n | $f_{min}$, Hz | $f_{max}$, Hz | Δf, Hz | $V_a$, m/s | $\Delta t_a$, s | $\Delta h_a$, m | $\sigma_{AI}$, $\frac{m \cdot g}{s \cdot cm^3}$ |
|---|---|---|---|---|---|---|---|
| 10 | 5.0 | 256.0 | 251.0 | 4002.0 | 0.0020 | 3.99 | **25.85** |
| 9 | 5.0 | 130.0 | 125.0 | 4002.0 | 0.0040 | 8.00 | **362.44** |
| 8 | 5.0 | 88.0 | 83.0 | 4002.0 | 0.0060 | 12.05 | **643.85** |
| 7 | 5.0 | 67.5 | 62.5 | 4002.0 | 0.0080 | 16.01 | **867.87** |
| 6 | 5.0 | 55.0 | 50.0 | 4002.0 | 0.0100 | 20.01 | **1023.16** |
| 5 | 5.0 | 46.5 | 41.5 | 4002.0 | 0.0120 | 24.11 | **1168.79** |
| 4 | 5.0 | 40.7 | 35.7 | 4002.0 | 0.0140 | 28.03 | **1244.44** |
| 3 | 5.0 | 36.3 | 31.3 | 4002.0 | 0.0160 | 31.96 | **1266.17** |
| 2 | 5.0 | 32.8 | 27.8 | 4002.0 | 0.0180 | 35.99 | **1278.79** |
| 1 | 5.0 | 30.0 | 25.0 | 4002.0 | 0.0200 | 40.02 | **1294.32** |

Parameters description:
n is the iteration number
$f_{min}$ is the band minimum frequency, Hz
$f_{max}$ is the band maximum frequency, Hz
Δf is the bandwidth of the operating frequency, Hz
$V_a$ is the average velocity in the formation, m/s
$\Delta t_a$ is the formation average time resolution, s
$\Delta h_a$ is the formation average depth resolution, m
$\sigma_{AI}$ is the standard deviation of the IDSP solution from the true sought model, (m·g) / (s·cm³)

the variance of the solution and the resolution is called the trade-off curve. It is argued that these dependencies are in the compromise nature and have the form (Fig. 2.29) [4], i.e. when the variance of the solution decreases, the resolution of the model parameters deteriorates and vice versa (see paragraph 14 of the Annex). However, in the technology of high-resolution seismic HRS-Geo, due to the use of multi-criteria optimization and sequential regularization, a more detailed study of the noise structure and special data processing graph, the "compromise" curve has a different appearance and another pattern (Fig. 2.30). This dependence monotonically increases with decreasing resolution and then the variance stabilizes (Fig. 2.30). At the same time, within certain limits, a decrease in the variance of the solution does not occur due to a deterioration in the detail of the resulting model (Fig. 2.30, Table 2.3) (at least within the limits of applicability of the described method of inversion). Obviously, on model and real examples, a stable transition of the classical curve (Fig. 2.29) to the graph (Fig. 2.30) can be seen with a continuous change in the characteristics of data processing and consistent adequate use of a priori materials within the appropriate limitations listed below.

In order for this pattern to be fulfilled in practice, we outline the limits of applicability of this property. These necessary and sufficient conditions are:

1. low random noise background in the seismic record;
2. it is possible to attenuate noise without significant signal loss;

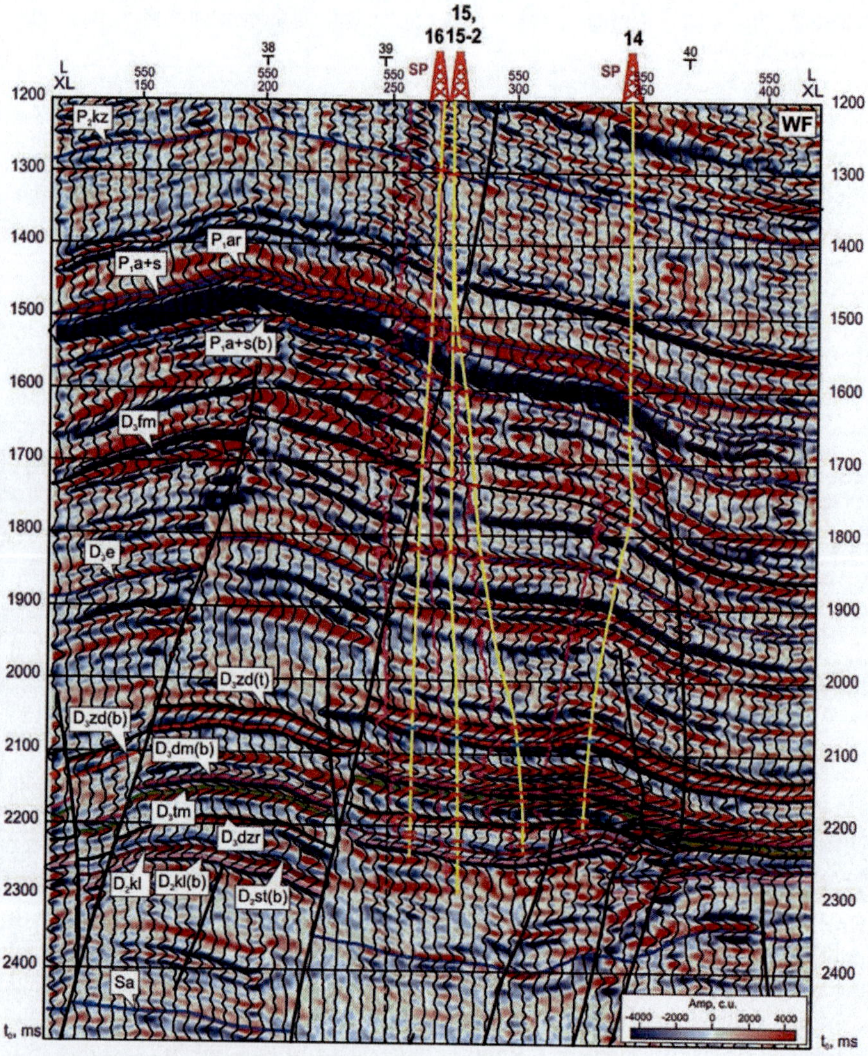


**Fig. 2.31** Wave field along the inline 550 with interpretation of boundaries and faults and binding to horizons from well data

3. the possibility of a detailed description and separation of the main signals and noises;
4. the characteristics and behavior of the noises are known in detail;
5. adequate and optimal parametrization of the model and problem;
6. multi-criteria search for a solution;
7. dynamic application of functions regularizing the problem.

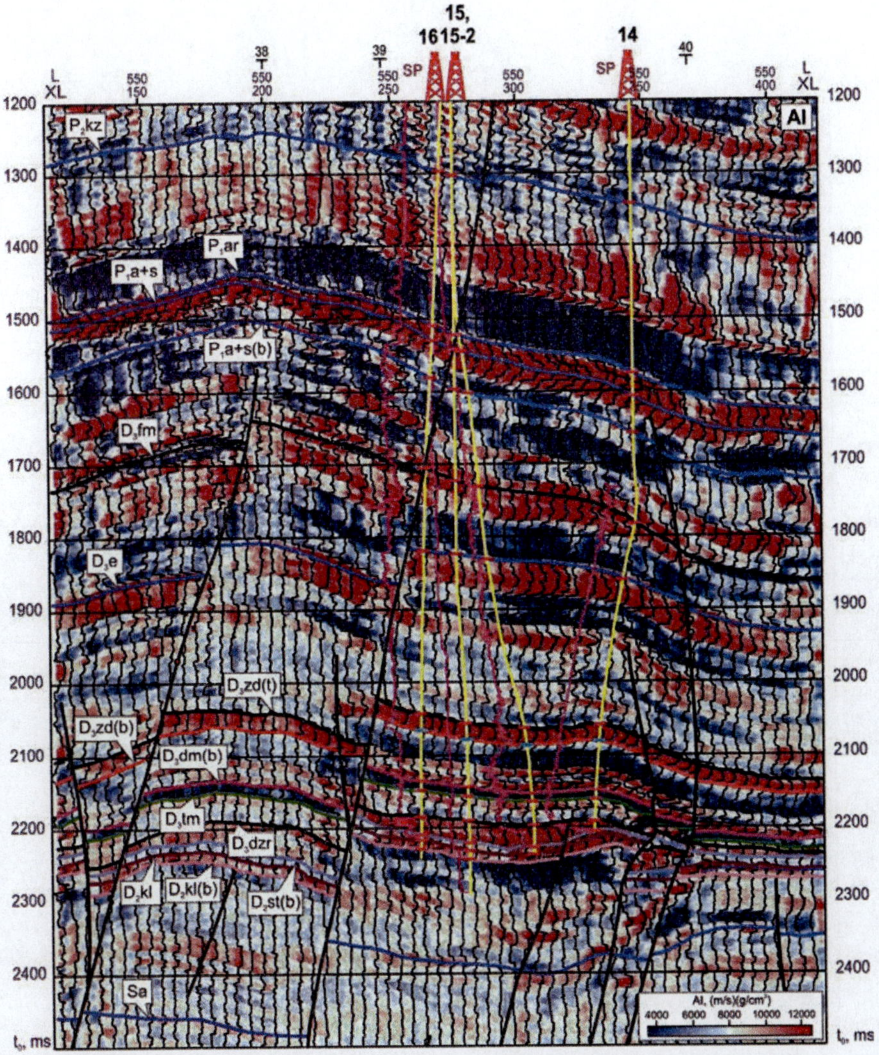


**Fig. 2.32** Sections of acoustic impedance along inline 550 as a result of the IDSP solution with the interpretation of boundaries and faults and binding to the boundaries from well data

Conventionally, the first 2 points refer to the stages of preparing seismic material for inversion (preliminary processing of seismic data). Items 3 and 4 determine the qualitative and quantitative assessment of the seismic record structure. Points 5, 6 and 7 are related to the technology of solving the inverse problem.

Figures 2.32 and 2.33 present the results of solving the inverse problem of seismics in the form of detailed sections of acoustic impedances and reflection coefficients with interpretation of boundaries and faults and linking to the horizons well

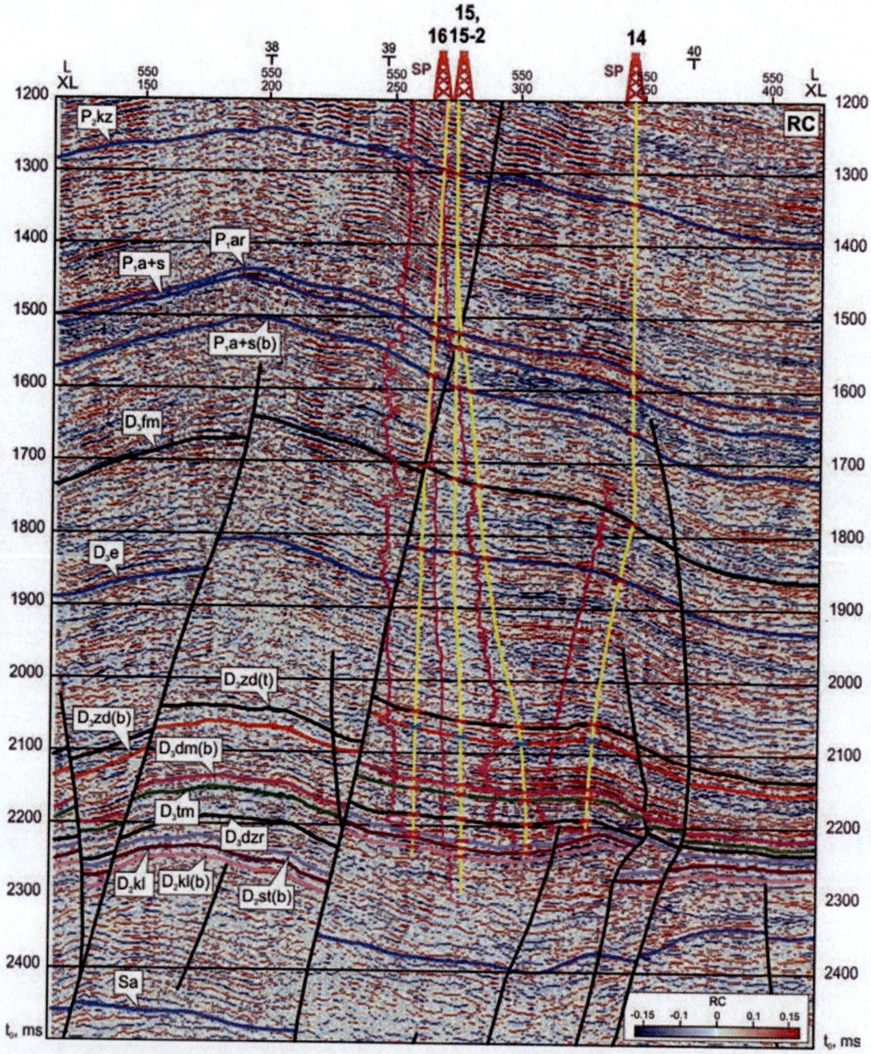


**Fig. 2.33** Section of reflection coefficients along the inline 550 as a result of the IDSP solution with the interpretation of boundaries and faults and binding to the horizons from well data

log data on inline 550 from a 3D cube (Fig. 2.37). The same interpreted data is applied to the initial section of the wave field (Fig. 2.31).

Figures 2.35 and 2.36 show sections of acoustic impedances and reflection coefficients also with interpretation of boundaries and violations and binding to the horizons according to borehole data on a fragment of the crossline 250 from the same 3D cube (Fig. 2.37). The interpreted boundaries and faults are also plotted on the original section of the wave field (Fig. 2.34).

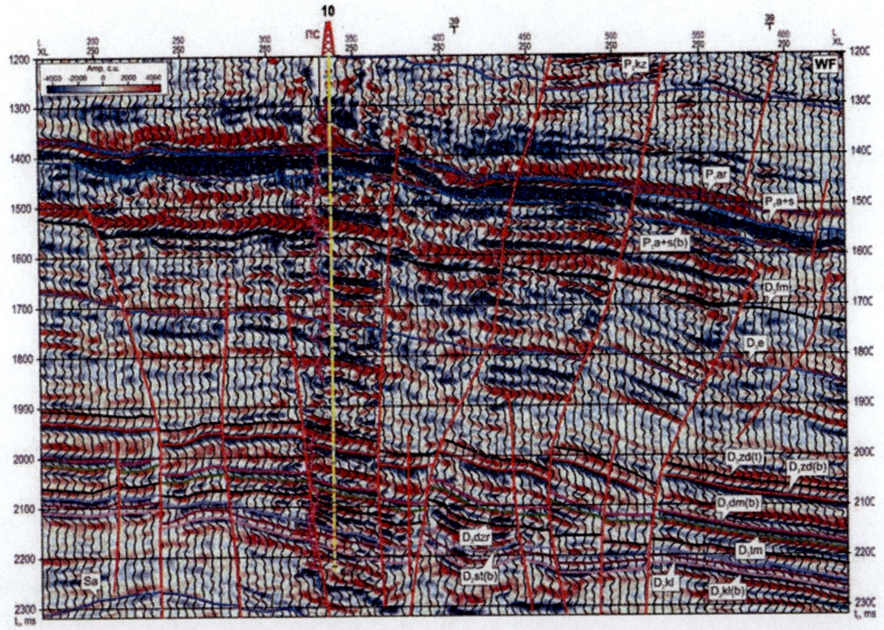


**Fig. 2.34** Wave field along the crossline 250 with interpretation of boundaries and faults and binding to horizons from well data

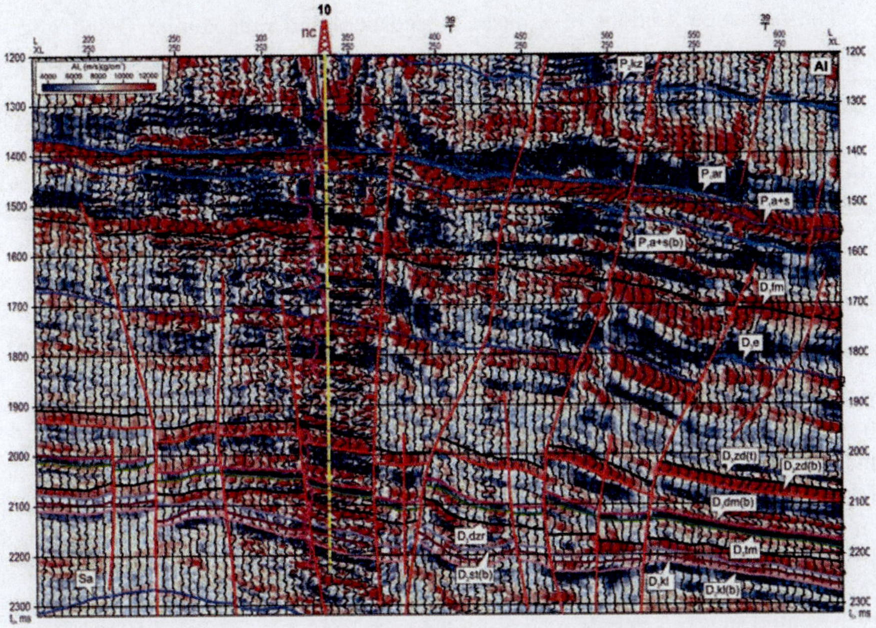


**Fig. 2.35** Sections of acoustic impedance along crossline 250 as a result of the IDSP solution with the interpretation of boundaries and faults and binding to the horizons from well data

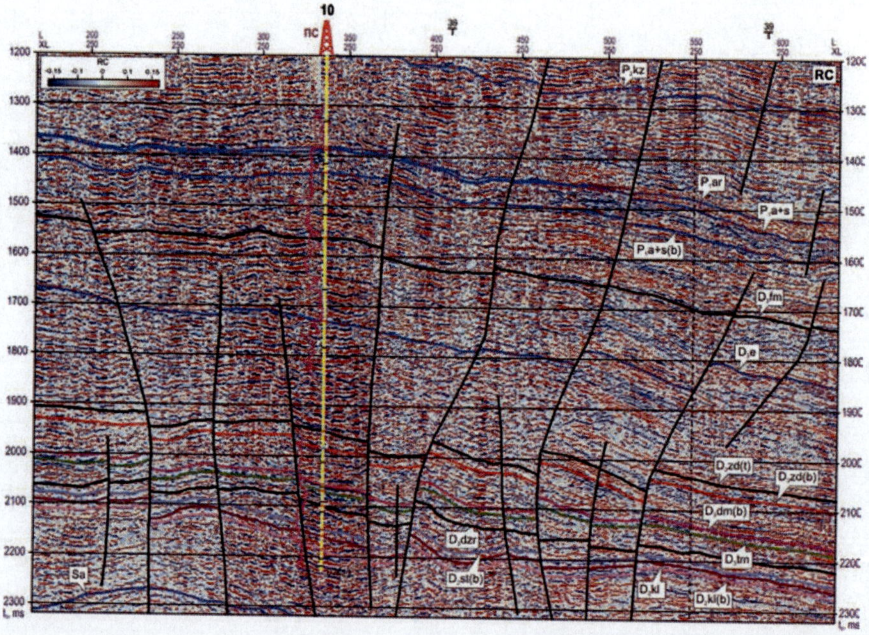


**Fig. 2.36** Section of reflection coefficients along the crossline 250 as a result of the IDSP solution with the interpretation of boundaries and faults and binding to the horizons from well data

The presented sections of acoustic impedances and reflectivities (Figs. 2.32, 2.33, 2.35 and 2.36) received the maximum possible detail of 3.5–4.0 m as a result of the IDSP solution within the framework of the HRS-Geo technology.

Figure 2.37 presents a fragment of the isochron map along the bottom of the domanik horizon $D_3$ dm with an interpretation of a complex system of tectonic disturbances. Here, these faults system control oil and gas deposits and are one of the signs of the search for promising objects. The same figure shows the location of the inline 550 and crossline 250 profiles within a 3D cube, along which the above results of the IDSP solution are shown (Figs. 2.31, 2.32, 2.33, 2.34, 2.35 and 2.36).

Fragments of the wave field and reflection coefficients sections for one of the areas located on the territory of Myanmar as another representative example of the solution of the IDSP are given (Fig. 2.38). Here, the result of the solution of the inverse problem in the form of a reflectivity section represents the most detailed model – a thin-layer model, where the entire internal structure of the medium is revealed: thin layers, their morphology in space, the relationship between thin and small-sized objects and other features. Such materials form the basis for detailed structural constructions and a prerequisite for the prediction of important indicators for geology, such as the lithological composition, objects reservoir properties, the type of fluid and the saturation degree of the reservoir.

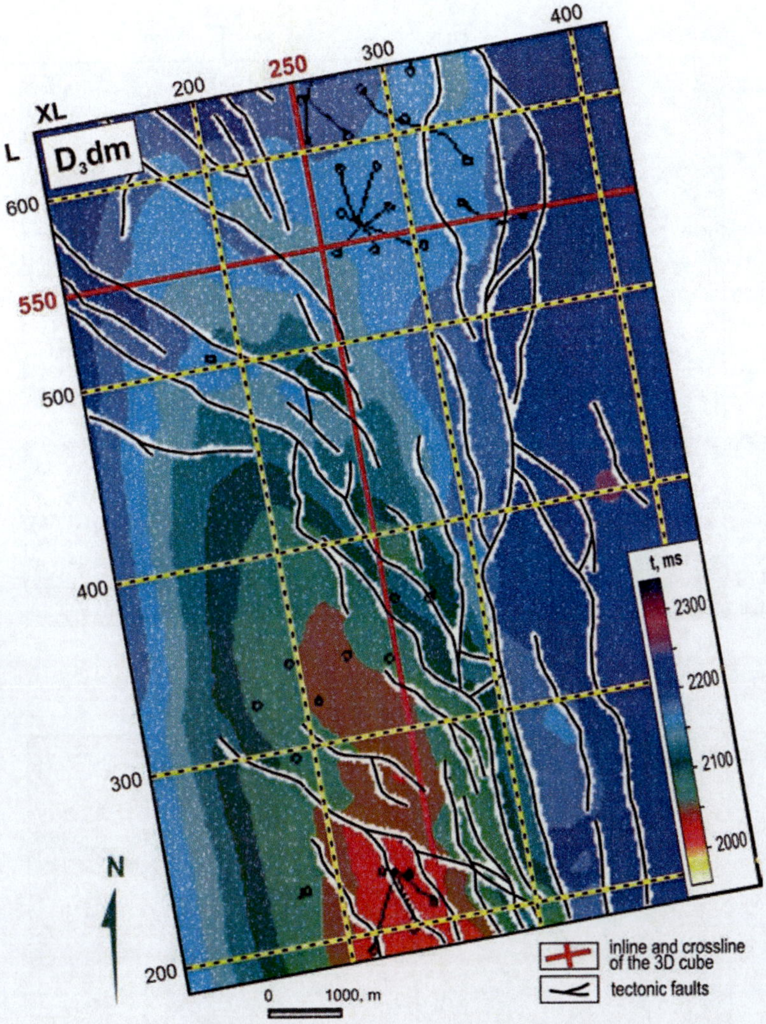


**Fig. 2.37** Fragment of the isochron map along the $D_3$dm horizon with the interpretation of tectonic disturbances and the location of the Xl550 and L250 profiles within the 3D cube of seismic survey

The obtained materials allow us search for features and identify fine details of the internal structure of the section, and interpret the spatial morphology of objects and their genesis in geological history. Based on these data, 3D geological and geophysical models of the studied promising oil and gas deposits are constructed, their most important indicators such as the lithological composition of the composing and enclosing rocks, their reservoir properties and the nature of the filling fluid are predicted.

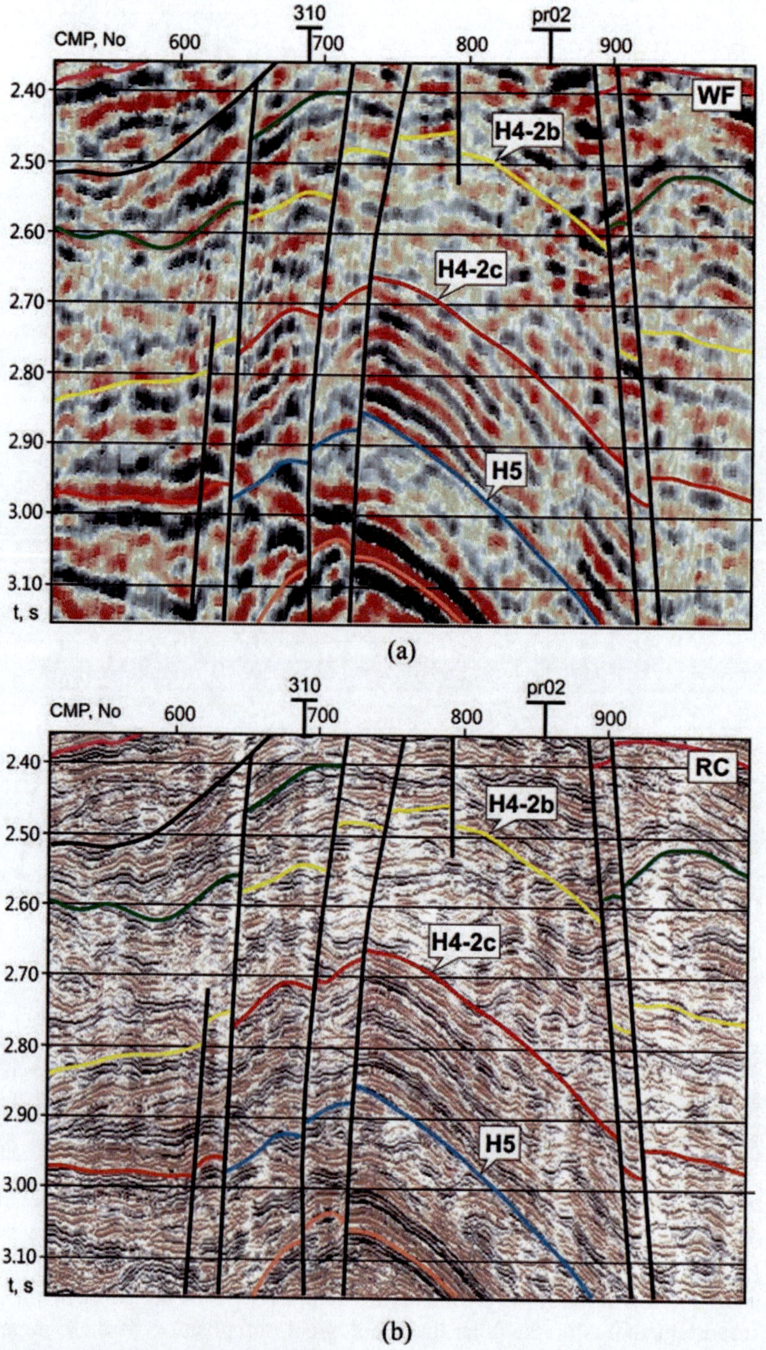


**Fig. 2.38** The example of a IDSP solution with the interpretation of horizons and faults: (**a**) a fragment of the original time section; (**b**) reflectivities as a result of the wave field inversion (Myanmar)

## 2.4 Summary

The construction of an adequate (with the maximum possible detail) model of a real geoenvironment based on the use of high-resolution seismic data of the HRS (i.e., based on the solution of the IDSP) involves the sequential execution of certain stages of a numerical experiment with an information content assessment of the results at each of them. At the same time, successful extraction of environmental parameters from HRS data presupposes adequate preparation of field observation materials, taking into account the features and objectives of solving the problem. It includes: (a) preliminary special processing of seismic survey data; (b) assessment of the quality and quantity of information in the initial data of seismic observations; (c) optimal parametrization of the environment; (d) parametrization of the main characteristics of the procedure for solving the IDSP.

Parameterization of the medium model based on the use of the characteristics of the amplitude and phase spectra of the original digital data makes it possible to predetermine the initial and final detail of the results of the seismic data inversion, i.e., the results of the solution of the IDSP.

To find and evaluate the information content of the main parameters of the process of solving the IDSP, it is advisable to preliminarily study and highlight the features of residual noise on the seismic record after performing preliminary data processing, determine the type and structure of this noise, and build regularization functions on this basis. It is also expedient to solve the IDSP with respect to the noise under study and, on this basis, in the final process of searching for the desired medium model, also apply the regularization parameters of this already known noise with certain characteristics.

As a result, it can be stated that the results of the studies obtained above provide the necessary prerequisites for using the properties of the characteristics described above to find a stable and detailed solution of the inverse dynamic seismic problem (IDSP) using multicriteria optimization methods. For this purpose (i.e., optimizing the solution of the IDSP), the authors, as noted above, used the system of objective functions that implement the iterative calculation of various types of residuals between real and model data. A feature of this system is that it separately calculates and analyzes residuals in terms of energy and waveforms and ultimately makes a decision on optimizing the desired acoustic model of the medium.

## References

1. Khaziev, F. F., & Trofimov, V. L. (2003). *Model'nyye issledovaniya rezul'tatov resheniya obratnoy dinamicheskoy zadachi seysmiki* (Model studies of the solving inverse dynamic seismic problem results) (pp. 27–37). Spetsial'nyy vypusk Geofizika: Tekhnologii seysmorazvedki-II (Geophysics, special edition: Seismic Technologies-II).
2. Trofimov, V. L., Khaziev, F. F., & Trofimova A.V. (2019). *Technology of high-resolution seismics in the search and exploration of oil and gas deposits* (p. 86). OOO "EAGE Geomodel".

3. Trofimov, V. L., Khaziev, F. F., & Trofimova, A. V. (2022). *Oil and gas reservoir prospecting and exploration. High-Resolution Seismic (HRS) techniques and technology*. Springer (Hardcover Book, eBook). https://doi.org/10.1007/978-3-030-84389-2
4. Yanovskaya, T. B., & Porokhova, L. N. (2004). *Obratnyye zadachi geofiziki* (Inverse problems of geophysics) (p. 213). Izdatel'stvo SPbGU (St. Petersburg State University Publishing House).
5. Hadamard J. (1978). *Zadacha Koshi dlya lineynykh uravneniy s chastnymi proizvodnymi giperbolicheskogo tipa* (Hadamard J. The Cauchy problem for linear partial differential equations of hyperbolic type) (p. 351). M.: Nauka.
6. Arsenin, V. Ya., & Tikhonov, A. N. (1982). *Nekorrektnyye zadachi* (Ill-posed problems)/*Matematicheskaya entsiklopediya* (Mathematical Encyclopedia) (pp. 930–935). –M: Sovetskaya entsiklopediya–T.3. (Soviet Encyclopedia, Vol. 3).
7. Morozov, V. A. (1987). *Metody regulyarizatsii neustoychivykh zadach* (Regularization methods of unstable problems). –M.: Izd-vo MGU (Publishing House of Moscow State University).

# Chapter 3
# Features of Integrated Dynamic Interpretation of High Resolution Acoustic Characteristics and Geological Indicators of the Real Environment

**Abstract** Analysis and use in the subsurface study of the known theoretical provisions for predicting the real environment's oil and gas content in the form of a geostructural, lithological, stratigraphic principles system, criteria for the formation and regularities of oil and gas accumulations placement, i.e. principles of system analysis, substantiates the geological basis and the specifics of the complex geological interpretation of high-resolution (HRS) acoustic parameters together with the geological indicators of the real environment extracted from these characteristics. On specific examples, it is shown the selection of weak seismic signals in the original wave field against the background of a high level of noise and reflected waves from strong boundaries is one of the main tasks of the complex seismic data interpretation. Features and examples of complex geological interpretation of data based on use of the HRS-Geo technology, production geophysics and drilling data are demonstrated. The features of processing and interpretation of materials using data from the HRS-Geo and well logging technology are analyzed sequentially formed on the development and application of a special (optimal) processing graph and a set of seismic record regularization procedures in the process of prospecting and exploration of oil and gas deposits. Thus, the maximum vertical and horizontal resolution of seismic data is ensured, which, in turn, makes it possible to identify the features of the detailed internal structure of the productive parts of the formation and their host deposits.

The study of the detailed internal structure of oil-prospective formations, oil and gas deposits, the creation of a thin-layer geological model of the real medium using high-resolution seismic and well logging methods are implemented within the framework of the developed HRS-Geo technology [9, 10]. High-resolution seismic and well logging methods data obtained in various seismic and geological conditions are used to predict the most important geological indicators for prospecting and exploration, such as lithological composition, reservoir properties and oil and gas saturation. The application of the technology makes it possible to increase the

 
V. L. Trofimov et al., *Theory and Practice of High Resolution Seismic HRS-Geo Technology Application*, https://doi.org/10.1007/978-3-031-41590-6_3

reliability of the detection of oil-promising objects and the efficiency of geological exploration in general.

This section highlights the criteria of structural and lithological deposits and the main research directions that the data and the results are guided by using the HRS methodology in combination with logging data. It is shown that the detection of weak seismic signals in the input wave field against the background of a high level of noise and reflected waves from strong acoustic boundaries of the media interface is one of the important tasks of complex interpretation of seismic data. The features of complex interpretation of data based on the use of HRS-Geo technology, field geophysics and drilling are demonstrated. The difficulties encountered in the interpretation of seismic materials using the developed technology and their overcoming in the course of research are analyzed.

## 3.1 The Main Criteria for Predicting Oil and Gas Deposits Based on High-Resolution Seismic, Well Logging and Drilling Data

The establishment of the main criteria for predicting oil deposits in the terrigenous section (for example, in the terrigenous Devonian, characteristic of a significant part of the Volga-Ural oil and gas province) based on a detailed study of oil-prospective sedimentary strata according to seismic data is very relevant. Such a target setting is defined by the authors as finding in a complex of known various search criteria (search prerequisites) for the forecast of oil and gas potential of a place according to the results of the conducted studies, on the basis of which the prediction of the most important geological indicators for petroleum geology is realized using high-resolution seismic data. First of all, this may concern the results obtained with the help of high-resolution seismic technology HRS-Geo [9, 10]. At the same time, it is obviously necessary to find such aggregates of geological and geophysical parameters (geological indicators) that, in specific seismic and geological conditions, can be effectively used for studying and analyzing and in the interrelation of a complex of geological regularities. The latter determines the spatial placement of the productive desired objects, and ultimately determining the location of complex oil-prospective objects (oil and gas traps), oil and gas accumulation zones, local inhomogeneities characterized by a certain set of composition and properties of the studied deposits, etc.

Considering this, in this subsection the authors focused on a brief review and analysis (on some formal basis) of the known criteria for predicting the oil and gas content of productive deposits from the standpoint of the principles of system analysis [1, 3, 4, 7]. They are known to play an essential role in studying a wide range of issues of sedimentary strata accumulation, post-sedimentation transformations, the formation of oil and gas complexes, including permeable and impermeable (shielding) strata, trap structures, interactions of hydrocarbon fluids with host rocks, etc. However, reliable forecast results from the application of these criteria can be obtained with their complex, systematic use and for relatively favorable geological conditions. In addition, they can be obtained under conditions of good drilling of the area, i.e. the presence of a sufficient number of points for calculating representative statistical characteristics, the absence of geological objects due to complex layering, dislocation, poorly differentiated lithological composition of sediments, low magnitude of the anomalous effect of the desired oil-prospective objects. In some cases, including in such geological conditions as the sedimentary situations of terrigenous Devonian deposits in the south-east of the territory of the Republic of Tatarstan, a number of criteria for predicting oil and gas potential for known objective reasons either do not "work" or can only be used in a limited volume [7].

Then, a set of geological indicators extracted from seismic data using the developed high-resolution seismic technology (HRS-Geo) [9, 10] is considered, the nature of the interaction of these indicators with the previously considered geological criteria and the possibility of their application together with a number of "close" (in terms of genetic and morphological features) criteria are analyzed.

### *3.1.1 Criteria for Predicting the Oil and Gas Potential of Productive Sediments Studied on the Basis of System Investigations*

By using system analysis in predicting the oil and gas potential of the subsurface, as is known, the nature of the holistic, multilateral process of oil and gas formation and accumulation as one of the forms of geological matter movement is investigated. On this basis, structural spatial-temporal and genetic relations, patterns of their occurrence and development in time and space are studied [3, 4].

In [1], a set of criteria for predicting the oil and gas content of carbonate deposits for well-studied areas was developed and proposed. According to the intended purpose, the oil and gas content criteria are divided into:

lithological-paleogeographic and petrophysical criteria characterizing the quality of a carbonate reservoir (reservoir and seal);
structural-tectonic and paleotectonic criteria that determine the quality of traps – anticlinal and non-anticlinal, especially traps inherent in carbonates (reef and bioherm uplifts and shell structures);
geochemical and thermodynamic criteria that evaluate the conditions of oil and gas formation, including a quantitative assessment of the generation potential;
hydrogeological and paleohydrogeological (paleofluid-dynamic) criteria, which, in combination with other factors, reveal the mechanism of deposits formation (secondary migration and accumulation of hydrocarbons in traps).

The groups of criteria used for forecasting with a set of indicators inherent in each of them make it possible to solve the problem with sufficient reliability, provided that the indicators meet the following basic requirements.

1. Indicators should be identified and evaluated taking into account the tasks of the prediction, the degree of object knowledge and the nature of the information received.

2. The set of indicators used must meet the conditions of necessity and sufficiency.
3. The set of indicators used for predicting should meet the requirement of comparative simplicity and cheapness and, if possible, be based on objectively measured information.

The group of lithological-paleogeographic and petrophysical criteria includes the following predicting indicators: (1) reservoir characteristics according to the prevailing structural and genetic type of rocks and post-sedimentation processes; (2) reservoir rock thickness, m; (3) open porosity, %; (4) permeability $n \times 10^{-13}$ $m^2$; (5) sealing thickness, m; (6) lithological characteristics of the seal.

The group of structural and tectonic forecasting criteria includes the following indicators: qualitative characteristics of second–order structural elements, specific density of lineaments (km/km$^2$), tectonic prerequisites for the formation of biogerms (reefs) or their enveloping structures, the number of undiscovered uplifts – identified and prepared for exploratory drilling.

Geochemical criteria are an important element of predicting, since the analysis of geochemical conditions highlights the quantitative and qualitative aspects of the processes of oil and gas formation and oil and gas accumulation in the studied deposits.

The analysis of the hydrogeological factor role in the formation and placement of deposits is of paramount importance for the predict of oil and gas potential. For carbonate deposits, for example, the task becomes more complicated due to the expressed lithological heterogeneity and the complex structure of the void space, which worsens the fluid-conducting properties of rocks.

For the purposes of predicting oil and gas potential in the carbonate section, the authors [1] propose to use all groups of criteria in a complex. At the same time, given that individual indicators are not the same in their significance, it is recommended to assign them a certain coefficient of informativeness, the weight of which depends on the representativeness of the factual material, the theoretical validity of the indicator, its impact on the final result of forecasting, etc., then summarize them in a certain way. According to the set of the points, the degree of prospects of the site, territory is determined and they begin to rank them according to the priority degree.

In the work [7], in particular, it is noted that when assessing the prospects of poorly studied locally oil-bearing horizons of the Romashkinsk oilfield, confined first of all to carbonate reservoirs located in dead-end zones and clayed with a low oil recovery coefficient (0.007–0.24), with the absence of a regionally common seal, three main factors were used as criteria:

- structural-tectonic – occurrence conditions of the productive strata;
- lithological – features of the reservoir-layers development, thickness and nature of the caprocks distribution;
- genetic – established oil content and various signs of oil shows in the section.

The resulting assessment of the oil deposits reserves and promising areas in poorly explored reservoirs showed that they are a significant potential in replenishing the oil reserves of the Romashkinsk oilfield.

The work [3] presents geological criteria for predicting and searching for oil and gas accumulation zones in sedimentary formations without separation into carbonate and terrigenous complexes. The main ones are the following:

1. Tectonic criteria. Among the factors controlling the formation and placement of regional oil and gas-bearing territories and oil and gas accumulation zones, the leading role belongs to paleotectonic factors that determine the conditions for the emergence and development of oil and gas formation processes and control the preservation of oil and gas accumulations in the subsurface, as well as structural factors controlling the formation of oil and gas-bearing areas, oil and gas accumulation zones and local hydrocarbon accumulations.
2. Paleogeographic criteria. Paleogeographic conditions clarification of sediment accumulation of the studied lithological-stratigraphic complex is one of the main criteria for predicting the oil and gas potential of the subsurface, since oil and gas formation can occur only in certain paleogeographic conditions. All regional oil and gas complexes, regardless of their lithological properties and facies conditions of their formation, are characterized by one unifying (diagnostic) feature, namely accumulation in an aqueous environment with an anaerobic geochemical medium against the background of relatively stable deflection of the sedimentation basin.
3. Lithological-facies and geochemical criteria. These criteria (conditions) control the material composition of sedimentary formations and the geochemical environment of accumulation and subsequent transformation of the initial organic matter buried in the sediment, and, consequently, the formation of possible oil source sediment complexes in various parts of the sedimentation basin. In addition, the lithofacies conditions of sediment accumulation control the formation of the composition and reservoir rocks involved in the structure of the natural reservoir, their changes in section and space, as well as the distribution of gas-and-oil-permeable strata of seal rocks (caprocks).

   These factors, together with others (paleotectonic and paleohydrogeological), predetermine the conditions for the migration and accumulation of oil and gas and the formation of regional oil and gas levels.
4. Hydrogeological and paleohydrodynamic criteria. It is impossible to establish the regularities of the creation and preservation of oil and gas accumulations without a detailed study of the dynamics and chemistry of formation and fractured waters not only in modern geological conditions, but also in paleogeological terms. The main hydrogeological criteria for assessing the regional oil and gas content of the subsurface include the conditions of distribution in the section and space of zones and areas of difficult and intensive water exchange and hydrogeologically stagnant regime, as well as the degree of flushing by infiltration waters of structural stages sediments in space and time. The hydrogeological closeness of each assessed structural stage is the most important factor in a positive assessment of the prospects for oil and gas subsurface resources.
5. Hydrogeochemical criteria. The hydrogeochemical indirect indicators of the oil and gas subsurface potential include:

   high gas saturation of groundwater with hydrocarbon gases and increased elasticity of water-dissolved gasessaturation pressure;

content of dissolved heavy hydrocarbons of the oil series in groundwater;

specific features of the chemical composition of high mineralized groundwater of reduced sulphate content inherent in oil and gas-bearing territories;

relatively high content of microelements (iodine, bromine, ammonium, etc.) and some organic compounds (naphthenic acids, phenols, etc.) in groundwater.

6. Geothermal criteria. In sedimentation basins with weak heat flow, paleogeological conditions were comparatively less favorable for the development of oil formation processes and the initial migration of petroleum hydrocarbons from producing sediments into reservoirs. In many oil and gas–bearing areas, geothermal conditions are one of the decisive factors in the formation of vertical and areal regional geostructural zonality of the placement of hydrocarbon accumulations, as well as changes in their physical properties in the section.
7. Oil and gas manifestations on the surface and their significance for predicting the subsurface oil and gas potential. Oil and gas shows on the earth's surface always indicate within the study area in the section there are regional oil and gas bearing complexes, which in some places may contain accumulations of oil and gas of industrial importance.
8. Criteria for predicting the safety of formed oil and gas accumulation zones. When predicting the oil and gas potential of the subsoil of the studied territory, along with the factors determining the possibility of the formation of oil and gas accumulation zones, it is necessary to take into account the factors controlling their safety. These include, first of all, the direction and mode of tectonic movements and the nature of their changes in space and time. The most favorable for the preservation of oil and gas accumulations is the predominant development of a stable immersion of the territory, which is accompanied by the accumulation of thick sedimentary formations, including reservoirs and cap rocks, and favorable hydrogeological and geochemical conditions.

Scientifically based and effective forecasting of the prospects for the detection of regionally oil and gas-bearing territories and oil and gas accumulation zones is possible only with a comprehensive study of the totality of the criteria considered in the relationship and the features of their changes in space and time, i.e. in development, in compliance with the basic principles of system analysis [3, 4].

These predicting criteria, based on the analysis of a large geological and geophysical material, formed the basis for assessing the prospects for searching for stratigraphic and lithological types of oil and gas deposits in various oil and gas provinces.

The objects of predicting the oil and gas potential of the subsurface can be distinguished both in the areal distribution and in the section of sediments composing the studied territories and individual areas.

At the regional stage, the objects of forecasting in the areal distribution are large territories, including oil and gas provinces, regions and zones of oil and gas accumulation and the corresponding geostructural, lithological and stratigraphic objects controlling them.

At the search stage, the objects of forecasting in the areal distribution are possible localities and deposits of oil and gas and the corresponding local, structural and lithological-stratigraphic objects controlling them.

The prognostication objects in the context of deposits of individual areas are oil and gas reservoirs that are part of regional oil and gas complexes and are covered with weakly permeable cap rocks.

Thus, for scientifically based predicting of the prospects for the oil and gas content of the studied territories, it is necessary to comprehensively examine the main geological factors in the regional and geological and historical terms, which together and in close interrelation determine the emergence and development of the processes of oil and gas formation and oil and gas accumulation. This also applies to: the tectonics of the research area and its changes in time and space; stratigraphy and lithology of sedimentary formations involved in the structure of the area studied; paleogeographic conditions of sediment accumulation and their changes in time and space; hydrogeological and hydrogeochemical features of the area investigated etc. [3, 4].

### *3.1.2 The Main Criteria for Predicting Oil and Gas deposits Based on a Detailed Study of Target Deposits by HRS Methods*

If the processing and complex geological interpretation of seismic data is carried out within the framework of only "wave" seismic exploration, when identifying various diagnostic signs or criteria for detecting oil and gas traps from seismic data, it is necessary to rely on well-known principles and approaches that are laid down in such areas of interpretation of seismic data as lithological seismic exploration, direct searches and GSP (geological section prediction based on seismic data), seismic stratigraphy and a number of other modern modifications of seismic exploration. In this case, the characteristic anomalies of the seismic wave field are analyzed: a decrease in the average and effective wave velocities reflected from the interfaces below the deposit; an interval velocities decrease and an interval times increase in the deposit area; intense attenuation of energy and a decrease in the dynamic regularity of waves passing through the deposit; a change in the frequency composition of waves passing through the deposit due to frequency-dependent absorption. This may include such as wave amplitudes reflected from oil and gas saturated formations changes ("bright spot" or "dark spot"); the appearance of reflections from the bottom part of the deposit and contact surfaces ("flat spot"); a change in the shape of the recording, a phase change of the wave reflected from the top of the reservoir layer, followed in some cases by diffraction phenomena in the peripheral of the deposit [5]. In this case, the seismic data is also interpreted from the point of view of analysis: the allocation of sedimentation complexes (along the unconformity boundaries); the allocation of seismic facies (according to the geometry of the reflection pattern); the regional sea level relative changes and their correlation with global cycles (according to the schemes of overlap and downlap in the littoral zones of the paleobasin) [8].

The specificity of interpreting the results obtained using high-resolution seismic (HRS) methods is that the interpreter, first of all, does not deal with seismic wave fields and their various derivatives, but with high-resolution sections (cubes) of reflection coefficients (RC) and acoustic impedances (AI). Each trace of such a section (cube) is the equivalent of a deep well section in which seismic and acoustic logging has been carried out. At the same time, only one correlation of structural main features (reflecting boundaries) and their combinations in such sections (cubes) makes it possible to identify and map natural rock associations of various ranks and scales, to establish their confinement to certain stages of the sedimentary basin development. It should be borne in mind that the information content of such results obtained only on the basis of the use of seismic wave fields is approximately an order of magnitude lower. A significant addition to the sections of effective reflection coefficients (RC) and acoustic impedance (AI) (if not the main content) are the results of predicting the lithological composition, reservoir properties and the nature of the fluid saturating the reservoir rock (the most probable distributions of oil, gas and formation water concentrations in the section studied) in the form of appropriate graphic and digital models.

Based on the studies conducted using high-resolution seismic using the HRS-Geo technology, data were obtained that characterize the distribution of the material composition along the studied section, its physical properties, as well as the nature and degree of fluid saturation in the studied real environment [9, 10]. The very process of extracting the necessary information about the geological substance from seismic data acts as a whole as a process of complex geological interpretation of geophysical data and substantiation of a geological structure model of a specifically studied oil-promising object. This process includes both the analysis of various geological and geophysical data and the use of a number of automated procedures performed in a certain sequence. At the same time, the interpretation technology becomes no less complex and branched than the modern technology for processing seismic data (pre-processing in fact).

A feature of the seismostratigraphic approach to the search for non-anticlinal oil and gas deposits is the transfer of data from traditional geological methods to a seismic basis (using the results of the analysis of the seismic record pattern as basic information). In contrast, the peculiarity of using high-resolution seismic methods in the search for oil deposits of various genesis and sizes is, on the contrary, the transfer of data from unconventional seismic methods to a geological basis. This means using as basic information the results of extracting identification information about geological matter from the dynamics of a seismic record using HRS-Geo technology, well logging data and drilling. In those areas where lithological and stratigraphic complexes of rocks have been studied in the main using the HRS-Geo technology, but have not been sufficiently studied by exploratory drilling, the hydrocarbon content in the predicted productive objects (usually such objects are located on an area with proven or assumed oil and gas potential), when calculating the mass of hydrocarbons (oil, condensate or natural gas), the following to standard conditions, it should be called predictive or prospective resources, as is customary [2].

The calculation of forecast resources is done within the framework of the HRS-Geo technology based on the use of the volumetric integral method, taking into account all the necessary calculation principles, quantitative criteria for the allocation of pore reservoirs and standard conditions (taking into account the corresponding coefficients of volumetric expansion of reservoir oil during its degassing). This makes it possible to obtain a quantitative assessment of various components of geological matter, including hydrocarbons.

Thus, on the territory of the western slope of the South Tatar Arch (as for most areas of the Volga-Urals), the geological conditions for deposits of the terrigenous Devonian are generally favorable for the development of the main types of non-anticlinal trap objects adjacent to ancient coastlines. Here, frequent fluctuations in sea level led to the formation of sand layers characterized by sharp lithological variability of rocks, accumulative bodies of various shapes, various objects of lithological and stratigraphic wedging, facies substitution, tectonic shielding, etc. In these conditions, the greatest interest for further increasing industrial oil and gas reserves in areas with proven industrial oil and gas potential of the subsoil are non-arched traps developed outside anticline zones.

Judging by the materials discussed above in [1–4], and by the results that are extracted from seismic data based on the use of HRS-Geo, well logging and drilling technology [9, 10], it can be stated that it is quite acceptable to use the totality of the found geological indicators (parameters) as appropriate to determine the prospects of oil and gas potential factors (criteria). These sets of geological indicators should be used for:

1. study the features of formation of reservoir layers, thicknesses (oil-saturated thicknesses) and the nature of the distribution of fluid-resistant seals – ***lithological criterion***;
2. studying the conditions of occurrence, structure and regularities of alternation of permeable and impermeable sedimentary complexes – ***structural-tectonic criterion***;
3. determination of oil and gas saturation and various signs of oil manifestation in the section and along the area of the study – ***a genetic criterion***;
4. ranking of oil-saturated rock according to filtration and capacitive parameters, screening properties of fluid seals (cap rocks); studying the processes of distribution and migration of fluid systems in relation to the formation and reformation of oil and gas deposits – ***fluid dynamic criterion***;
5. ascertainment of paleogeographic sediment accumulation conditions of the studied lithological-stratigraphic complexes and prediction of the preservation of the formed oil and gas accumulation zones (study of the direction and regime of tectonic movements and the nature of their changes in space and time) – ***paleotectonic criterion***;
6. systematization of materials for calculating the forecast resources of oil and gas, as well as the component of the studied substance at different stages of the deposits study is ***a perspective-resource criterion***.

The listed criteria for predicting oil and gas potential, based on geological indicators that are extracted from high-resolution seismic data based on the use of HRS-Geo technology, are primarily aimed at searching and exploration of hydrocarbon accumulations controlled by relatively small structures, as well as non-anticline traps of various types.

In general, all of the above criteria can be characterized as a single set of geological and geophysical criteria extracted from seismic data based on the use of HRS-Geo technology. The specifics of prospecting and exploration of oil and gas traps using these criteria is they rely mainly on seismic research materials and can be performed with the minimum necessary well logging data for deep wells (i.e. they can be used in geological conditions that are practically little studied by exploratory drilling). In such a geological situation, in addition to the fact that these criteria have independent significance, they can also be used to develop other geological criteria for oil prospecting complex geological and geophysical studies on the territory be investigated.

In geological conditions with a relatively large number of prospecting and exploration wells, the results of the forecast of oil and gas potential obtained using the HRS-Geo technology and the set of geological and geophysical prediction criteria developed on their basis make it possible to generally increase the efficiency of geological exploration for oil and gas. First of all, this is realized by obtaining detailed additional information on: geometrization of the oil and gas trap and establishing its spatial position; mapping of lithological, tectonic and stratigraphic screens in combined and non-anticlinal traps; refinement of the configuration of areas and zones of increased thickness reservoirs with improved filtration and capacitance properties; mapping of water-oil, gas-oil and gas-water contacts (as far as the study detail allows due to a reduced sampling step of seismic data over time and preservation of weak seismic signals responding to low-contrast acoustic contrast ratio), etc. In this case, the whole set of geological and geophysical criteria and the results of the prognosis for oil and gas potential obtained from seismic data can be used in conjunction with all other geological criteria and the results of their application on the objects be specifically studied.

## 3.2 Detection of Weak Seismic Signals in the High Level Noises and Reflected Waves from Strong Interfaces Background

The problem of highliting weak signals on the background of a relatively high level of noises and reflected waves from strong interface boundaries is relevant for almost all oil and gas regions. This issue has been known for a long time – since the moment when it became necessary to solve the problems of petroleum geology with the help of seismic exploration – the task of prognosis various components of geological matter (*this is the period of the 60–80 years of the last century, when the methods of exploration geophysics began to widely develop the physical and geological foundations (prerequisites) for solving the problems of "direct searches"*). The difficulty can be "felt" especially acutely when using relatively "short" field observation systems. Being applied in rather complex deep seismic and geological conditions, these systems do not provide satisfactory quality of seismic results based on the subsequent use of almost any of the graphs for processing the initial seismic data. The need to extract weak "useful" signals from the complex structure of the wave field also arises when studying the detailed internal structure of real geological environments and predicting various components of geological matter.

The main difficulties in solving this problem are predetermined, as is known, by the very nature of the seismic method, more precisely, by the complexity of solving the methods of the inverse dynamic problem of seismics. Based on the use of the obtained adequate results of solving such a problem, it is possible to judge the relatively low acoustic contrast of the section under study and the reflected waves with low intensity relevant to such a section.

As has been repeatedly emphasized above (in Chap. 1), the problem of "***detecting weak signals***" is directly related to the problem of "***solving the inverse dynamic problem of seismics (IDSP)***". On this basis, the nature and mechanism of the formation of the seismic wave field is explained and an adequate solution is found to the problem of separating the different types of information contained in the reflected waves. This approach to the analysis of the seismic wave structure field makes it possible to understand the nature of reflected waves that are weak in intensity, depending on the reflection coefficients and which vary across the section, "respond" accordingly to low-contrast seismic acoustic heterogeneities.

As shown in [9, 10], to solve the inverse dynamic problem of seismics, the authors have developed numerical algorithms for constructing 2D seismic geological models based on the formation of a time sequence of effective reflection coefficients and acoustic impedance. To optimize the solution of the problem, a system of objective functions (objective functions vector) has been designed, in which various types of discrepancies between real and model data are iteratively calculated. The peculiarity of this system is that it calculates and analyzes the energy and waveform differences separately and makes a decision to optimize the model as a whole (*the results of the IDSP solution are given by the authors on test and real materials using the software of the HRS-Geo technology* [9, 10]). At the same time, each of the reconstructed traces of effective seismic acoustic logging (*reflection coefficients and acoustic impedance)* represents some equivalent of a deep well section in which such logging was "carried out" with a sampling step of the seismic record in time. For example, the vertical resolution of the section under exploration at the sampling step of the seismic record in time equal to $\Delta t = 2$ ms will be 3.0–4.5 m for terrigenous sections, and 5.5–6.5 m for carbonate ones (*at elastic wave propagation velocities equal to 3000–4500 m/s in terrigenous and 5500–6500 m/s in carbonate sections*).

This subsection provides examples of extracting information about weak seismic signals from seismic data. Actual materials are in this case used, and some of which are characterized by unfavorable seismogeological conditions for the work (obtained in the area of the southwestern border of the Russian platform and the modern Alpine folding system), and others – relatively favorable seismic geological conditions (on one of the sections of the S-Gubkin area, located in the central part of West Siberia).

The authors in [9, 10] analyzed the informativeness of a thin-layer section elements by quantifying the contribution of elementary boundaries and strata to the interference wave field. The development and application of the concept of the "interference matrix of contributions" at the same time allowed us to evaluate the contributions of local responses from lithology, porosity and oil and gas saturation to the seismic wave field and to the results of its various transformations. Based on the obtained results of seismic modeling, it was found that ***lithology*** is characterized by the highest influence – its contribution to the interference seismic recording is in the range of 60–80%, sometimes reaching 90%. The effect of ***porosity*** on seismic record and its pseudoacoustic transformations is estimated on average in the band of

10–30%. The influence of ***water*** and ***oil saturation*** on the dynamics of the record as a whole, compared with lithology and porosity, is significantly lower and reaches an average of about 10–15%.

Judging by these results already – estimates of the contribution to the seismic record of the main geological indicators, especially such as oil saturation and porosity, confined to real reservoir formations, we can say seismic signals carrying information about these two main indicators (porosity and oil saturation) have the important property that can be characterized as very weak "weak signals" in terms of information content. At the same time, if we focus on the causal relationship that determines the interaction of the geological components of the real environment and the dynamic parameters of the seismic wave field, it is quite obvious in this "bundle" the consequence is "weak" seismic signals, and the "cause-source" is local acoustic inhomogeneities. The latter are presented in the form of corresponding values of reflectivities, their set, which determine the presence of either oil saturation or porosity, affecting the dynamics of the wave field individually with their relatively weak intensities, or acting together.

It should be noted, however, the existing ambiguity in determining such parameters of the environment (lithology, porosity and oil saturation) in the context of each of the wells using borehole measurement methods can be represented as a special "geological interference" of different characteristics (geological indicators) of the real medium [6, 9, 10]. At the same time, according to the set of parameters used, located in a certain system, the material composition, porosity, the content of bound water and useful capacity are determined, the oil and gas saturation of reservoirs is estimated, a number of very important characteristics of the geological section are computed. On this basis, a thin-layer geological model is formed in the vertical section of the well profile, lithological and stratigraphic columns and detailed geoacoustic models of wells are constructed. The main stages of logging data processing are: (1) determination of lithological composition; (2) determination of porosity; (3) determination of oil and gas saturation; (4) evaluation of filtration properties of the studied reservoirs. In such a situation, at each point of the vertical section of the well section, the "geological interference" of geological indicators manifests itself in the form of a certain ratio of the considered set of these indicators – lithological composition, reservoir properties and oil and gas saturation. This ratio, quite obviously, turns out to be proportional to the above-mentioned local contributions (i.e. local responses – weak seismic signals) from the corresponding acoustic heterogeneities associated with them.

It should be pointed out here the "interference" of the components of a geological substance is represented in the form of interacting phases of this substance – various and heterogeneous combinations of the material composition of rocks with their physical chemical features, properties of the mineral matrix, cementing material, pore space structure, etc. These components of a real geological section as a whole reveal themselves in various physical chemical ratios and geological-geophysical properties – they thus "respond" to the continuous nature of the sedimentation process in the studied interval of the section, the conditions of catagenetic transformation of matter, etc. The physical geological properties of such a section, fixed on the diagrams of various well logging methods, act simultaneously and they participate jointly in the consolidated complex interpretation of geophysical measurement materials. These features are well manifested in the results of continuous processing and automated interpretation of well logging diagrams at each point of the studied section [9, 10]. On the results of the final processing of logging diagrams, of course, a superposition of various lithological components is observed (in volume ratios of terrigenous, carbonate, chemogenic, etc. features of the lithological composition), various capacitive and filtration (due to the difference in the structure of the pore space), as well as the different nature of the fluid saturation of reservoir rocks.

Taking into account the above, we can talk about the need for a consistent transition from the parameters of the wave field, namely, from the analysis of the data contained in the interference matrix of contributions (ICM), to the synthesis of the model parameters of the real environment in the form of the results of the IDSP solution (in the form of effective reflection coefficients, acoustic impedance, absorption coefficients, etc.). When implementing such a inversion of the wave field and restoring the model of the medium at the final stage of complex interpretation of seismic data, as a result, the transformation of the initial matrices of the wave field dynamic parameters into matrices of the medium geological indicators such as lithological composition ($C_{cl}$, $C_{sand}$ and $C_{cb}$), reservoir properties ($K_p$), saturation character ($C_o$, $C_g$ and $C_w$) and others [9, 10]. In this work, in particular, it is shown the contributions concerning the nature of reservoir rocks saturation are recalled in the interference wave field primarily in the form of weak seismic signals. In turn, these signals are formed in the form of corresponding responses due to acoustic heterogeneities – their components such as oil, gas and reservoir water. It is quite obvious such important characteristics can be preserved in a seismic wave field only when a special processing and interpretation graph is applied to seismic observation materials that use high-resolution seismic technology procedures. At the same time, as shown in [9, 10], it is advisable to keep local seismic responses undistorted in the dynamics of seismic record. For this purpose, special inadequate procedures are not used in order to preserve the necessary amount of identification information about the above-mentioned basic geological indicators in the dynamics of the seismic record.

With such interacting factors in the processes that form geological information about the real thin-layer medium and the corresponding dynamic structure of the seismic wave field, it is obvious a solution should be sought to the problem of weak signal detection as well as those tasks and issues that are related to this problem. However, it should be borne in mind that it is practically impossible to extract subtle features of the section structure from the seismic information formed in such difficult conditions by simple algorithms for processing and interpreting data (i.e. from the dynamics of seismic recording). It is necessary to radically improve the methods and approaches to the processing and interpretation of seismic data based on the development of fundamentally new algorithms that allow a good understanding and sufficiently fine structure of the real environment using the fine properties of the seismic wave field dynamics.

At present, there are a fairly large number of algorithms for solving inverse dynamic seismic problems, which are based on various approaches to solving the problem of seismic inversion (dynamic interpretation) of seismic data. The authors of [9, 10] solve this problem using the software tools of the developed HRS-Geo technology, the key element of which is the methods for solving the IDSP. Researchs were carried out to study the structure of the seismic wave field (the mechanism of field formation and interpretation of its structure) based on 1D seismic modeling, taking into considiration the contribution of elementary interfaces and strata, as well as to restore the seismic acoustic model of the medium based on the use of test 2D seismic modeling by splitting the structure of the wave field into information about the source of elastic oscillations generation and information about the thin layer model of the medium. It should be emphasized the problem of identifying weak signals is at the same time the problem of noise immunity of processing and interpreting systems. The sensitivity and reliability of the operation of such systems should be many times higher than in the analysis of only powerful reflections, especially when the intensity levels of the useful weak signal and noise are comparable or the level of the first is much lower than the second. In turn, the issue of noise immunity of systems determines the degree of reliability of identified promising objects.

As a result of the carried experiments, conclusions were drawn [9, 10]:

1. the procedure developed by the authors for solving the inverse dynamic seismic problem makes it possible to reliably restore the model of real media in the form of sections and cubes of effective reflection coefficients and acoustic impedances with an accuracy of the seismic record sampling step in time;
2. the results of the restoration of the acoustic impedance model are affected by strong noise (*commensurate in intensity with the useful reflected waves*); their impact on the desired results can be significantly reduced through the use of additional procedures for processing the original seismic records (*for example, optimal filtering procedures*); in general, the recovery procedure is characterized by a fairly high noise immunity;
3. it becames obvious at the stage of seismic data pre-processing, it is necessary to implement a special seismic data processing graph, which, on the one hand, would provide, to a certain extent, an increase in the signal-to-noise ratio (*it is quite possible to do this using the procedures of a traditional standard processing graph*), and with the other, minimally distorted the dynamics of the seismic record, i.e. would retain as much as possible identification information about the main geological indicators (lithology, porosity and oil saturation), which is contained in it.

Below is an example of extracting information about weak seismic signals from reflected waves on actual materials obtained in relatively unfavorable seismic geological conditions. The main unfavorable conditions for seismic surveying here were: the use of a relatively short central spread of seismic receivers – the maximum distance between the explosion source and the recieve device was no more than 675 m; the sources used for excitation of elastic oscillations did not provide sufficient energy of reflections from the target horizons deeper than the time interval $t_0 = 0.7$ s; correlation of reflections from deep boundaries becomes problematic due to their frequent breaks – target boundaries are almost impossible to identify; among a large number of coherent waves, the main part of these is associated with multiple and partially multiple reflections.

For such a situation, in Fig. 3.1, and the result of processing obtained in one of the foreign companies using the procedures of the processing package ProMAX 2D/3D v.98Plus v.98.6 is shown in the form of a CMP time section recorded in the deviations. Here, in the target interval, starting from $t_0 \geq 1.0$ s and deeper, an intense residual background of multiple reflected waves (mainly simple multiples) is recorded, the kinematic formation scheme of which, most likely, is confined to the time interval from $\Delta t_0 = 0.15$–0, 2 s to 0.6–0.75 s. In this case, there is a characteristic, gradual increase in the angles of inclination of the in-phase axes of multiple waves with an increase in the registration time of the seismic record. The result of the wave field inversion in the form of a effective acoustic impedances showed an actually similar situation (*in that, the principle "works": what is the structure of the initial wave field, this is approximately the result of the inversion, i.e. all those distortions of the field dynamics that were at the stage of preliminary processing of seismic data, are practically "transferred" to one degree or another to the result of restoring a detailed seismic acoustic model of the real environment*). It is clear for the existing very "short" field observation system applied in complex deep seismic geological conditions, and the processing graph that was formed based on the use of this processing package, the problem of extracting reflections from the target interval of the section turned out to be practically unsolved.

The use of a standard graph built on the basis of the software and methodological tools of the OMEGA system [9, 10] made it possible to significantly reduce the residual background of multiple waves compared to the previous section – this already turned out to be a significant shift under the unfavorable factors noted above (*the use of a "short" field spacing geophones in conditions of complex geological structure of the section*). However, the resolution of the seismic record was significantly reduced, despite the fact the procedure for deconvolution of the seismic record was widely used in this graph. The result of seismic inversion of the considered wave field is also adequate to the structure of this field – the relatively low resolution of the seismic wave field in the target interval of the section, starting from $t_0 \geq 1.0$ s and deeper, predetermines the result of restoring the detailed internal structure of the studied section with a relatively low resolution (Fig. 3.1b). It is quite obvious the distortions of the dynamics of the seismic record introduced at the preprocessing stage also do not allow one to speak about the possibility of separating weak seismic signals and corresponding inhomogeneities of the section with a relatively low acoustic contrast.

The application of a non-standard pre-processing graph (using the procedures of the HRS-Geo technology at the original seismograms pre-processing stage) can significantly improve the wave pattern (both by suppressing the residual background of multiples and by increasing the resolution of the seismic record).

In particular, the following were obtained: the time section of the CMP, which is shown in Fig. 3.2a, and the result of the inversion – a section of effective acoustic impedance (AI) in the form of variable intensity (Fig. 3.2b). Additionally, the last of them (Fig. 3.2b) also contains time lines – the corresponding stratigraphically identified interfaces (which are used later in the interpretation of seismic data). This processing version seems to the authors to be the most optimal in terms of the choice of parameters and the necessary set of procedures for preliminary processing of observational materials in these very complex seismic geological conditions. From the results of processing and inversion obtained for this variant – the traces of the restored values of effective AI, we can conclude:

1. There is almost no residual background of multiples in the wave field, and the recording resolution is quite high (Fig. 3.2). The structure of the thin-layered section model is characterized by the presence of layers and interlayers with

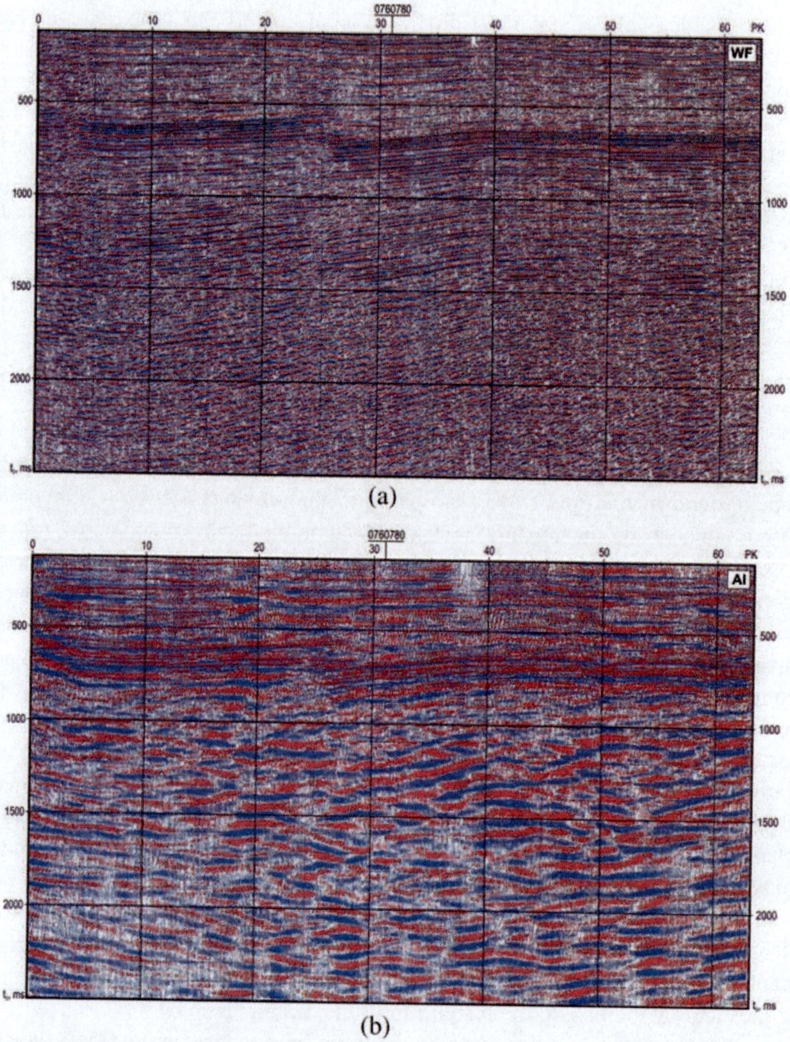


**Fig. 3.1** Results of seismic data processing on profile 0710780 obtained using standard graphs: (**a**) a time section of the MCMP obtained using ProMAX 2D/3D v.98Plus (v.98.6) complex programs; (**b**) a section of effective acoustic impedance reconstructed from a time section obtained using the programs of the OMEGA processing package. Dobrudzhan coal basin. Bulgaria

multidirectional slopes, zones of different dislocation and tectonic fault in the target intervals.

2. Compared with non-optimal field structures and the section structure model (Fig. 3.1), the optimal processing and inversion results (Fig. 3.2) confidently distinguish relatively "weak" seismic signals and corresponding heterogeneities with relatively "low" acoustic contrast. All relatively "small" (*sections with increased and decreased values of acoustic impedance*) and rather "large" struc-

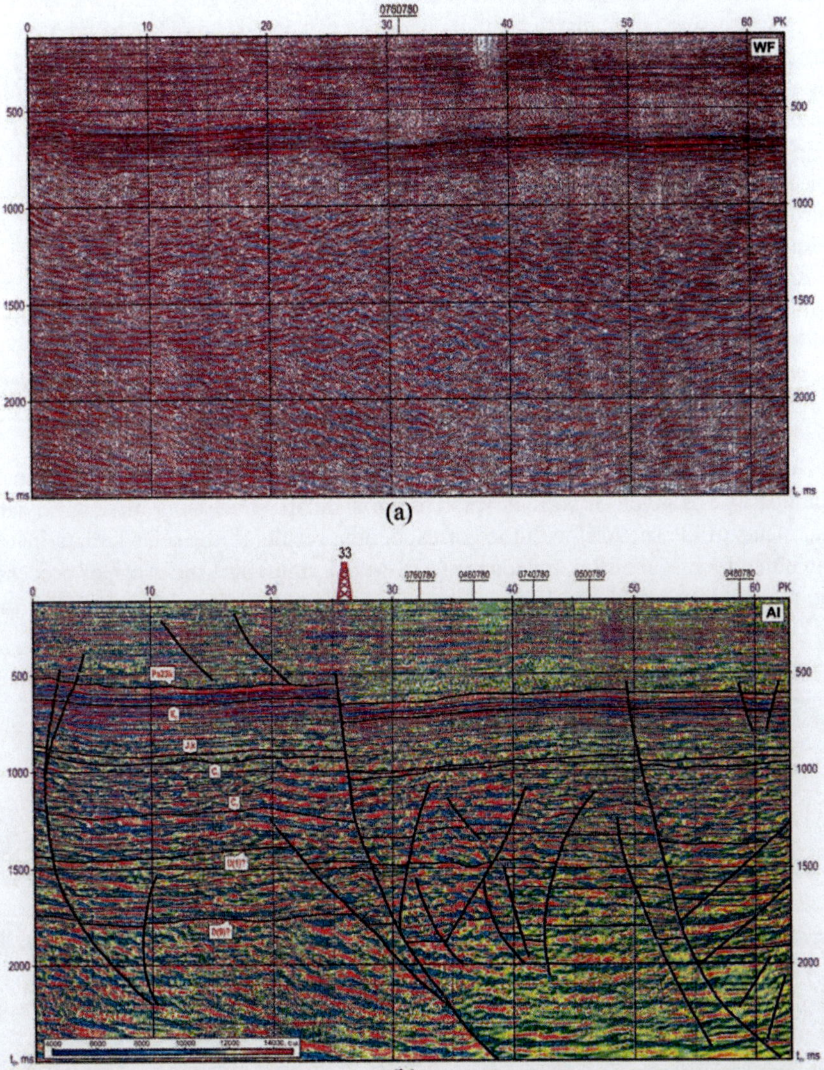


**Fig. 3.2** The results of processing seismic data on profile 0710780 obtained using a special (optimal) graph: (**a**) the time section of the MCMP; (**b**) the section of the effective acoustic impedance, reconstructed from the section (**a**), in the form of variable intensity with correlated reflective boundaries. Dobrudzhan coal basin. Bulgaria

tural and tectonic features of the studied medium are manifested in this case. Despite the significant dislocation (*obviously due to the manifestation of tectonic activity of different intensity and direction in different geological epochs*), manifestations of fracturing of the target section, it is possible to almost unambiguously identify areas of tectonic fault and correlate interfaces within individual blocks (or identify zones of their absence, most likely associated with a break).

Thus, with the help of the work performed on the choice of the graph and parameters of the preprocessing programs, the authors managed to find a solution to the problem of extracting weak seismic signals in a specific complex seismic geological setting. The key element in solving this problem was the development and application of a procedure for solving the IDSP based on the idea of inverting a seismic wave field [9, 10]. The main result in this case are sections and cubes of effective acoustic impedances and reflectivities, on which heterogeneities with a relatively low acoustic contrast are distinguished, among other things. It is on these inhomogeneities that weak seismic signals "respond", which are previously recorded in the original seismic wave field.

As one of the examples of distortions of the fine structure of the sections of the restored acoustic impedance obtained under relatively favorable seismogeological conditions (*the presence of sufficiently representative seismic data obtained, first of all, with the parameters of the observation system close to optimal and relatively "calm" deep geological conditions*), the materials of one of the profiles (5990019) of the S-Gubkin area of the Yamalo-Nenets Autonomous Okrug are below considered (Fig. 3.3).

Here, for a given section of the profile in the selected time window (1350–1700 ms), a comparison is given of various versions for standard processing of seismic data (fragments of the section) with the corresponding results of the seismic data inversion into effective acoustic impedance (AI). At the same time, the graph of processing, the result of which was a fragment of a time section (Fig. 3.3a), includes the

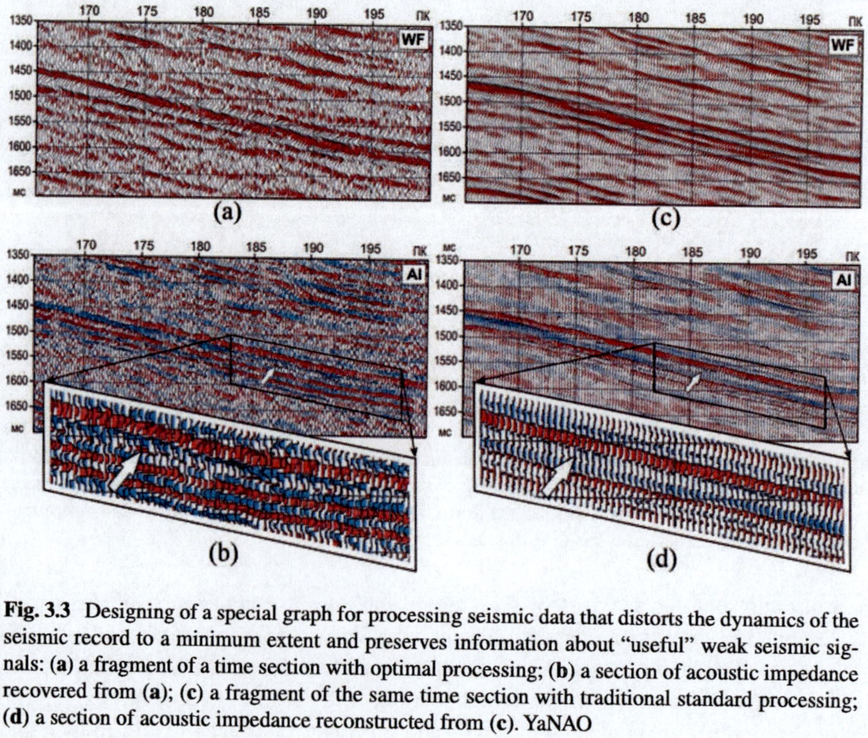


**Fig. 3.3** Designing of a special graph for processing seismic data that distorts the dynamics of the seismic record to a minimum extent and preserves information about "useful" weak seismic signals: (**a**) a fragment of a time section with optimal processing; (**b**) a section of acoustic impedance recovered from (**a**); (**c**) a fragment of the same time section with traditional standard processing; (**d**) a section of acoustic impedance reconstructed from (**c**). YaNAO

most minimal, but at the same time necessary set of processing procedures (in particular, it does not include procedures for correcting residual phase shifts, migration of seismic records, multiple procedures for zero-phase and multichannel FX deconvolutions, as well as bandpass and time–variable filtration).

The results of the inversion of seismic records are presented in Fig. 3.3b. It is quite logical to assume they are restored with minimal distortion. In the section obtained in this way, the most informative part of the section is geologically represented by a set of layers with increased and decreased acoustic impedance, characterized by their various combinations and time thicknesses. This totality, confined to the productive layers of the АП group of the Pokursky and upper part of the Tangalovsky formations of the S-Gubkin deposit (in the middle part of the time interval of the section AI much as (Fig. 3.3b), is the area of tracing in the bottom of the $БП_6$ formation of the $БП_7$ horizon – the so-called Cheusk "chocolate" clays), is well correlated from the smaller pickets at a time $t_0$ = 1600–1650 ms. It is a rather complex stratum of unevenly overlapping sand-siltstone and clay formations. Increased values AI much as (in Fig. 3.3b they are shown in red) correspond to sand-siltstone differences of rocks, and reduced (in Fig. 3.3b they are shown in blue) – clay differences. At the same time, it is necessary to emphasize a number of features in the behavior of structural forms and ratios of intensities of AI restored, rather thin layers (*this is the characteristic interbedding in ≈ 1–3 time samples within each of the selected layers, the duration of their tracing and the nature of the articulation of the layers themselves with increased and decreased AI values*). For visual perception, we will point out only one characteristic feature – in the interval of the section pk 178–200 in the time range $t_0$ = 1515–1610 ms, there is a very thin layer (in 1–2 time discretions) of increased values of the AI intermittent type against the background of a more powerful layer with reduced values of the AI (which, in turn, is directly located on this interval between two layers with well-traceable AI values). It is obvious these characteristic features in the behavior of the above–marked layers and interlayers should be seen in other – the following processing versions and analyze the nature of the differences (distortions) manifested in this case.

The results of the next processing option are shown in Fig. 3.3c for the same section – the time window as in the previous Fig. 3.3a. At the preliminary stage of standard seismic data processing in this case, a complicated graph was applied, including, in addition to the minimum required set of procedures, such procedures as correction of residual phase shifts, migration of seismic records, repeated application of zero-phase and multichannel FX deconvolution procedures, as well as bandpass and time–variable filtration, i.e. the entire powerful "arsenal" of seismic records regularization (*the fragment of the final version of the time section itself is given in* Fig. 3.3c). The results of the seismic records inversion for this variant of the time section are presented in Fig. 3.3d. As can be seen from the figure, the inclusion of additional procedures in the standard graph of seismic data processing, and especially the repeated use of zero-phase and multichannel FX deconvolution procedures, leads to very strong distortions of the acoustic impedance being restored (it should be noted the parameters for AI curve as much as in the other case are preserved without changes).

With a detailed comparison of almost each of the identified layers and thin interlayers presented in Fig. 3.3b, d, there are significant distortions of the second result of the inversion in relation to the first (features of the thin structure of the layers and the intensity ratios of the AI). At the same time, the above–marked very thin layer ($\approx$ 1–2 samples in time) with increased values of AI, lying in the section interval of pk 178–200 in the time range $t_0$ = 1515–1610 ms, practically disappears – only its insignificant (distorted) manifestations can be seen in this area of the section.

The materials obtained in this way show the problem of identifying reliable weak seismic signals exists for both unfavorable and favorable working conditions. It is solved quite effectively on the basis of the selection and use of a special graph of seismic data processing, which ensures the preservation of identification information about the components of geological matter in the dynamics of the record and methods of seismic wave field inversion (HRS-Geo technology), implementing the restoration of the acoustic model of the real environment with the accuracy of the sampling step of the seismic record.

## 3.3 Features of Complex Interpretation of Data Based on the Use of HRS-Geo Technology, Well Field Geophysics and Borehole Drilling

In order to study real thin-layered media, the authors, as noted above, have developed and applied a technique of high-resolution seismics, which is based on a numerical method for solving the inverse dynamic problem of seismics [9, 10]. The software and methodological implementation of this technique is made in the form of HRS-Geo technology. As it was noticed, the study of real environments is focused on the use of dynamic features of seismic record, the implementation of its maximum possible resolution. The construction of 2D sections and 3D cubes of effective acoustic impedance (AI) and reflection coefficients (RC) with vertical resolution equal to the sampling step of the seismic record in time is implemented. The use of the technology is aimed at separating seismic information related to the source of elastic oscillations and the real environment of the studied geological section. This makes it possible to study in detail the structure of reservoir formations, outline the oil and gas deposit, rationally place subsequent exploration wells and reduce the costs associated with reducing the volume of unproductive drilling.

In general, the use of high–resolution seismic data – HRS-Geo technology is focused on the realization of a number of regulations-criteria, such as:

1. maintaining dynamics;
2. separation of source and environment information;
3. increasing the resolution within the limits of seismic exploration modifications;
4. adequacy of measurement scales, optimal coordination of seismic and well logging data;
5. physical conditionality of the restored and used geological and geophysical parameters.

Let us briefly dwell on the features of solving the main seismic geological problems using the procedures of the HRS-Geo technology.

### 3.3.1 Assessment of the Contribution of Boundaries, Strata and Parameters of the Geological Section

The dynamic parameters of the interference wave field are studied using simulation programs for the dynamic characteristics of reflected waves with a quantitative assessment of the contribution of elementary boundaries and strata, i.e. the informativeness of the elements of the thin-layer section under study is evaluated directly by modeling. The influence of lithology, porosity, water and oil saturation on the dynamics of seismic recording and on the results of its pseudoacoustic and other transformations is evaluated. Such geological indicators as porosity and oil and gas saturation are characterized by weak seismic signals – local responses that must remain undistorted in the dynamics of seismic record (which is not often performed when processing seismic data) [9, 10].

The concepts of an elementary formation and an elementary boundary for unambiguous definition of a thin-layer acoustic medium model when the development and use of technology for obtaining seismic data with high spatial resolution are introduced [9, 10].

The analysis of the elements information content of a thin-layer section is realized by quantifying the contribution of elementary boundaries and layers to the interference wave field. In relation to the development and introduction of the concept of "interference contributions matrix" and with its application, the contributions of local responses from lithology, porosity and oil and gas saturation to the seismic wave field and to the results of its various transformations are evaluated.

As an example of seismic modeling in Fig. 3.4a–d (without a detailed description and analysis, which are given with the necessary detail in our works: articles, monographs and scientific and methodological reports) in the interference wave field, an assessment of the contributions of selected formations and elementary boundaries (*a*), lithology (*b*), porosity (*c*) and oil saturation (*d*) is given.

Based on the results of seismic modeling, it was found the ***lithology*** is characterized by the highest influence – its contribution to the interference seismic record is in the range of 60–80%, sometimes reaching 90%.

The effect of ***porosity*** on seismic record and its pseudo-acoustic transformations is estimated on average in the range of 10–30%.

The influence of ***oil saturation*** on the record dynamics in comparison with lithology and porosity is significantly lower and averages about 10–15%.

As shown in [9, 10], it is precisely the developed technology that is most in demand when studying a thin-layer geological section, including reflected waves weak in intensity that respond to low-contrast acoustic heterogeneties of the geological medium (depending on the reflectivities varying along the section).

### 3.3.2 Processing and Interpretation of Well Logging and Vertical Seismic Profiling Data

In vertical sections of deep wells, data from geophysical well surveys (logging) are processed and interpreted. Processing and automated interpretation of these standard well logging methods is carried out according to the methodology described in detail in

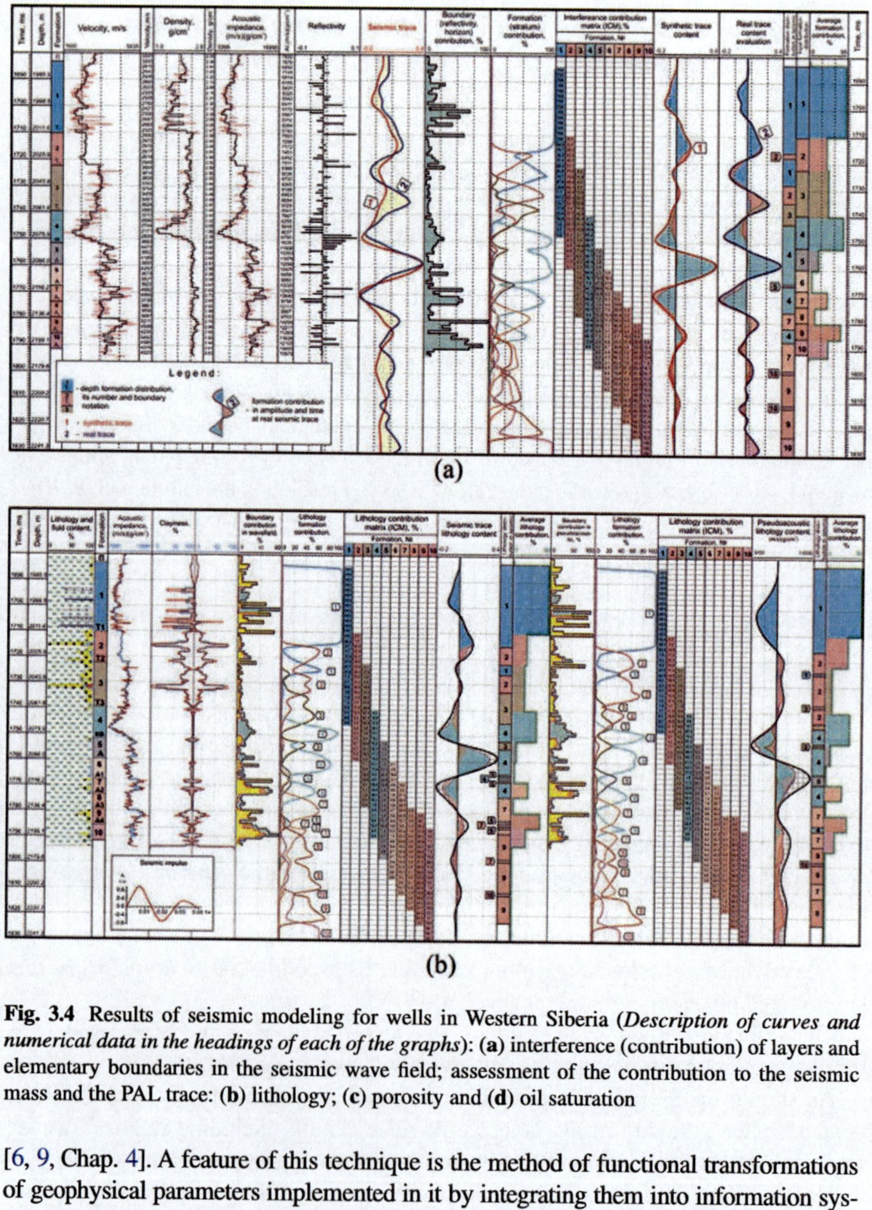


**Fig. 3.4** Results of seismic modeling for wells in Western Siberia (*Description of curves and numerical data in the headings of each of the graphs*): (**a**) interference (contribution) of layers and elementary boundaries in the seismic wave field; assessment of the contribution to the seismic mass and the PAL trace: (**b**) lithology; (**c**) porosity and (**d**) oil saturation

[6, 9, Chap. 4]. A feature of this technique is the method of functional transformations of geophysical parameters implemented in it by integrating them into information systems [6, Chap. 4]. This method, in addition to the continuity of data processing and analysis, provides quantitative interrelationships of all geophysical parameters based on their direct dependencies as well as the integrated use of data and the results of solving individual tasks. The physical prerequisite for solving such problems is the different sensitivity of different geophysical parameters to changes in lithological and reservoir properties. When the indications of individual methods are normalized in rocks of one type, a characteristic discrepancy of indications for rocks of another one is observed according to the lithology and the nature of the fluid. The effect of a particular property

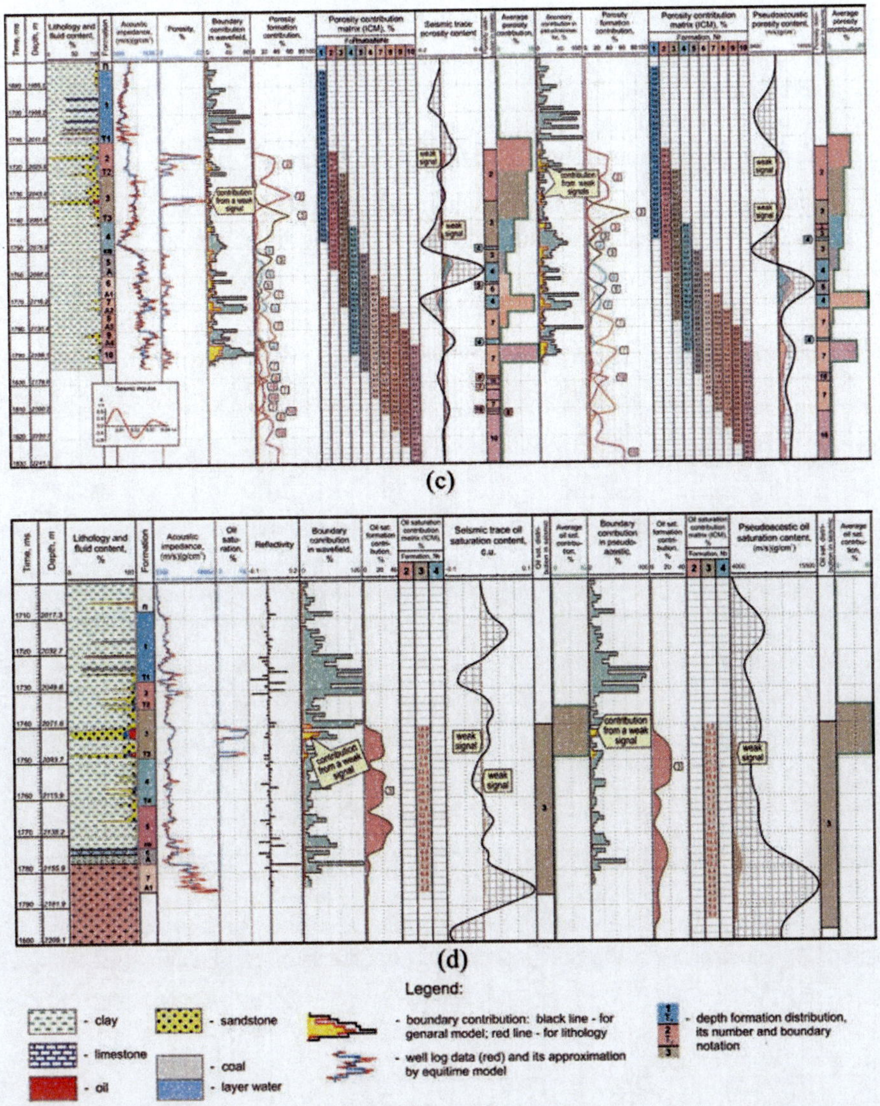


**Fig. 3.4** (continued)

is revealed by sequential compensation of interfering factors by comparative analysis of normalized diagrams of various methods in certain combinations.

When integrating a significant number of these well logging methods, a whole set of rock characteristics is determined. According to the set of parameters used, located in a certain system, the material composition, porosity, content of bound water and useful capacity are determined, the oil and gas saturation of reservoirs is estimated, a number of very important characteristics of the geological section are calculated. On this basis, a thin-layer geological model is formed in the vertical profile of the well, lithological and stratigraphic columns and detailed geoacoustic models of wells are constructed.

The main stages of well logging data processing are:

1. determination of lithological composition;
2. determination of porosity;
3. determination of oil and gas saturation;
4. evaluation of filtration properties of the studied reservoirs.

As an example of such processing and automated interpretation of well logging data Fig. 3.5 shows geological sections drilled by wells in a number of areas in

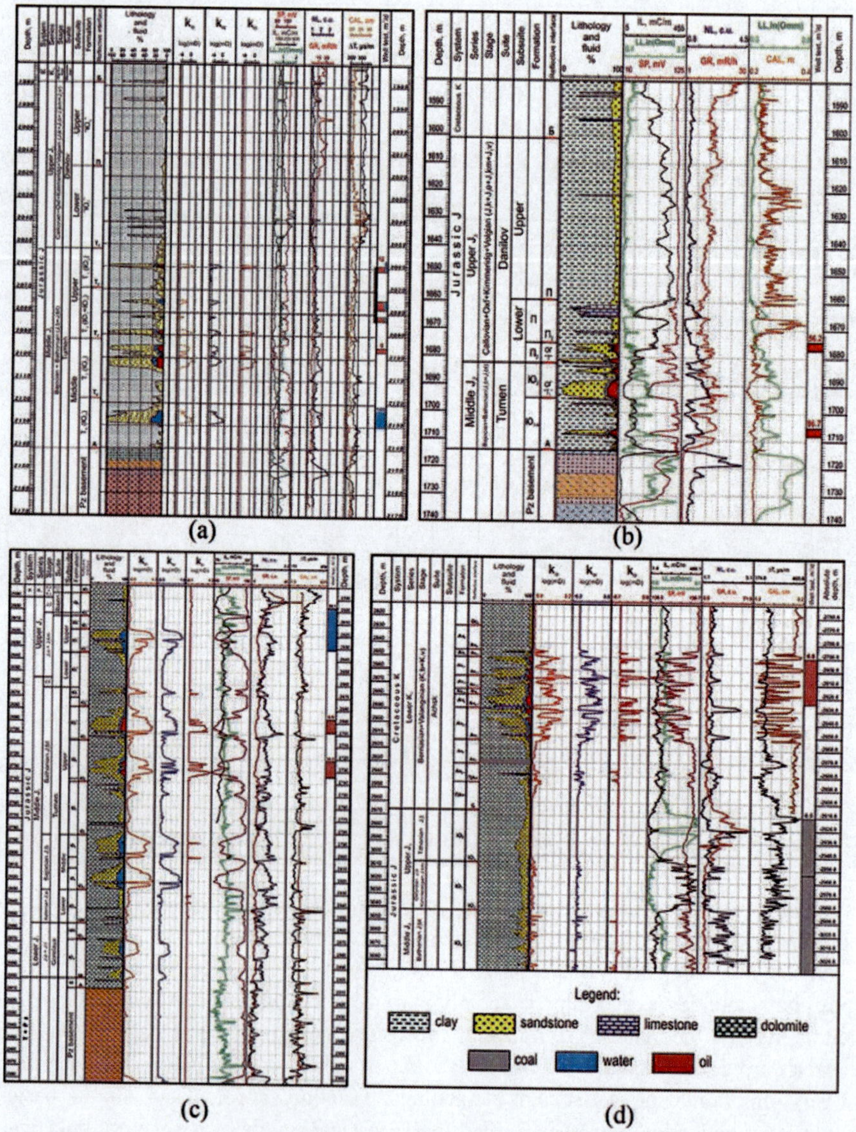


**Fig. 3.5** Lithological and stratigraphic columns for wells: **(a)** 10234 Symoryakh, **(b)** 10351 Uzbek, **(c)** 180 Novoyutym, **(d)** 405 Z-Karen (in the Western Siberia); **(e)** O-6 Okruzhnaya (in the Sakhalin)

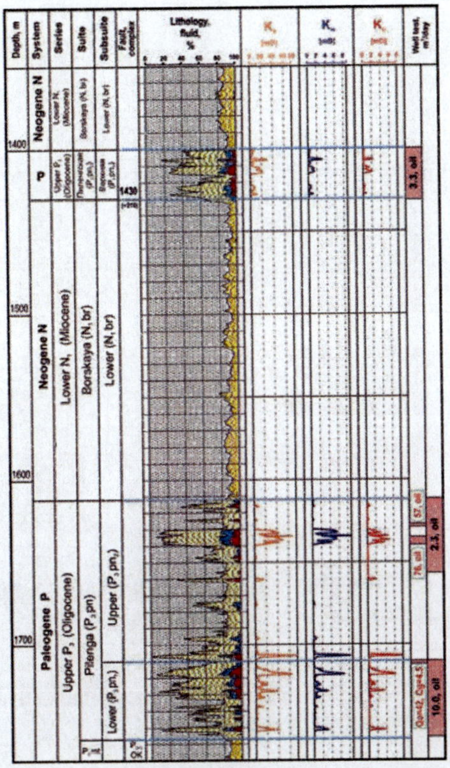


(e)

**Fig. 3.5** (continued)

the Western Siberia and the Sakhalin. Figure 3.6 also presents a correlation scheme for Jurassic deposits along the sections of a number of wells in the S-Yutim prospecting area of the Western Siberia. Together with these geological sections, fragments of effective acoustic impedances (AI) and reflection coefficients (RC) correlated with them are given.

Processing of vertical seismic profiling data by the polarization method (PM VSP) is based on the use of the SKOR software package to determine in the section of the studied well the velocity characteristics of various wave types, effective elastic-strain parameters and the stress state real environment in the natural (in situ) occurrence of a rock mass [9, 10].

### 3.3.3 *Seismic Data Processing by Optimal Graph*

Taking into account the presence of weak seismic signals in the seismic record – local responses associated with such geological indicators as porosity and oil and gas saturation, complex processing of seismic data is performed according to some

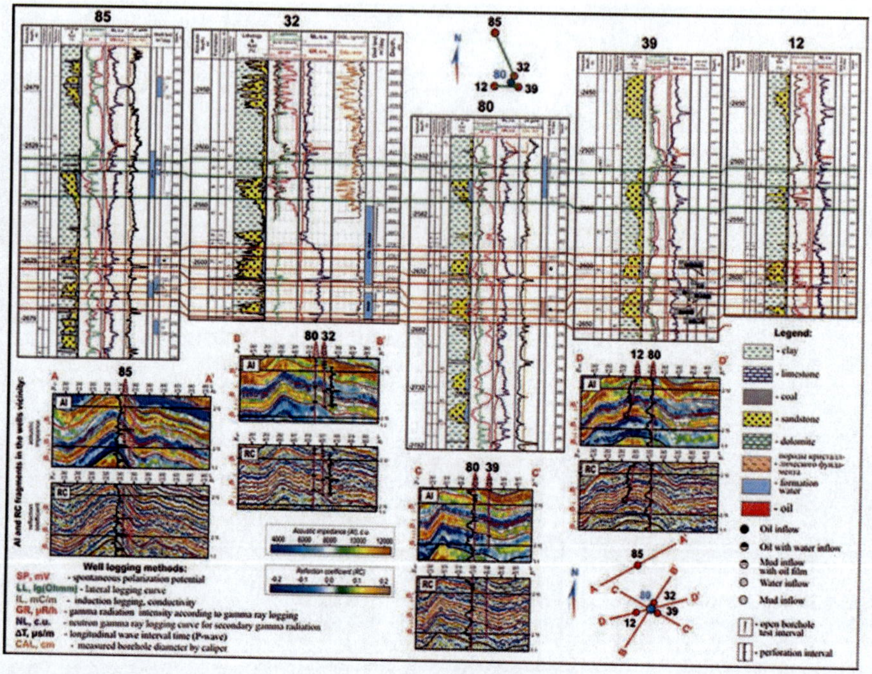


**Fig. 3.6** The correlation scheme of the Jurassic sediments along the wells 185, 302, 180, 309 and 312 together with fragments of sections AI and RC. KhMAO

optimal graph, in which only those procedures are used in the processing that allow maintaining the seismic record dynamics, adequate the actual distribution of acoustic inhomogeneities in the section under study. One of the distinguishing features of the technological scheme of seismic data processing is the inclusion in its composition of the procedure for converting seismic traces into an impulse response of the medium, i.e. into the traces of reflection coefficients (RC) and acoustic impedances (AI). This procedure makes it possible to expand the range of seismic records to the maximum, to significantly increase their resolution, to regularize the dynamic parameters of the record while maintaining weak seismic signals from target geological indicators. Such a procedure is used both in the processing of original seismograms and in that of final migrated 2D sections and 3D cubes of seismic records.

Examples with the results of applying the optimal seismic data processing graph are shown in Figs. 3.7, 3.8 and 3.9. The initial data for the presented results were seismic data obtained in the conditions of the West Siberian, Timan-Pechora provinces and Myanmar.

The use of a complex (optimal) graph in the processing of seismic data ensures the preservation of the seismic record dynamics, which contains identification information about the lithological composition, reservoir properties and oil saturation, the contribution of which to the seismic record, as shown above, is significantly different.

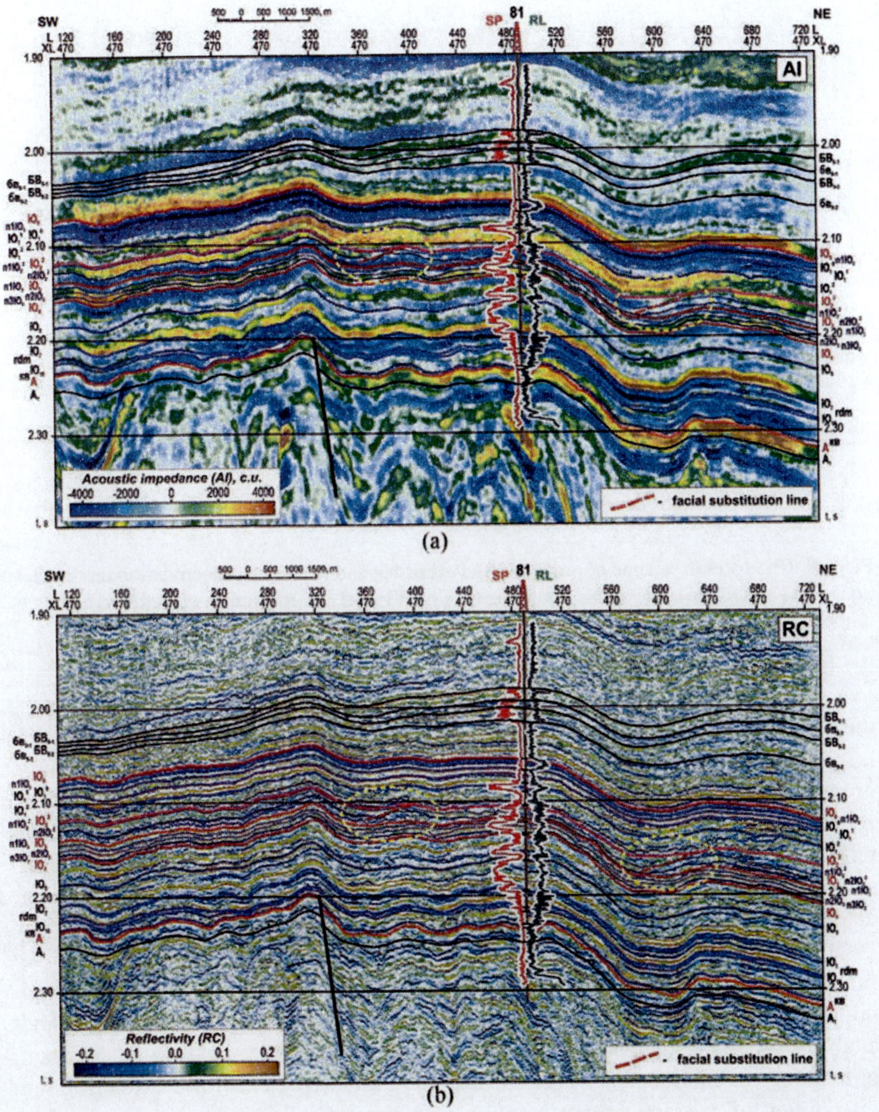


**Fig. 3.7** Sections of effective acoustic impedance AI (**a**) and reflection coefficients (RC) (**b**) along the crossline 470 3D cube of the search area, KhMAO

## *3.3.4 Numerical Solution of the Inverse Dynamic Problem*

The solution of the inverse dynamic seismic problem (IDSP) is found by an optimization method, which consists in selecting the AI and RC models for a given wave field structure using known formulas for solving the direct problem for calculating

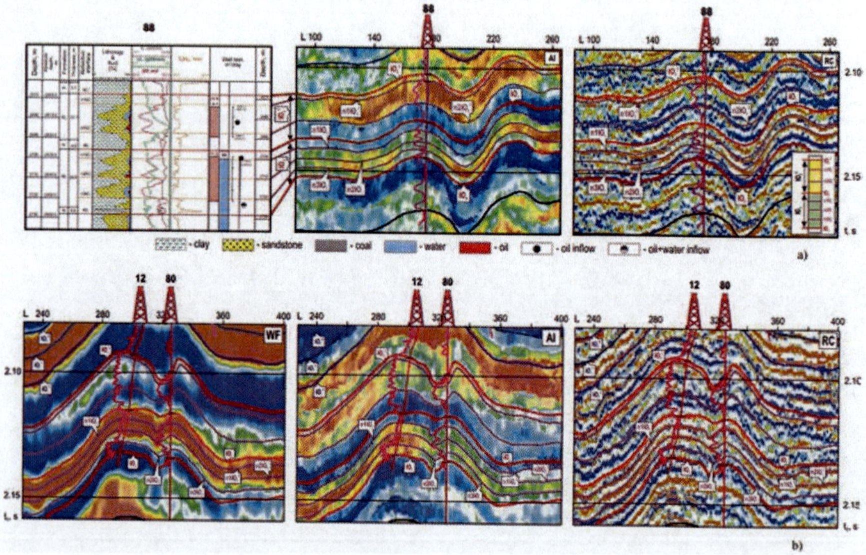

**Fig. 3.8** The patition scheme of productive layers of $IO_2$ and $IO_3$ and their correlation on the sections of AI and RC (**a**); fragments of the wave field (WF), AI and RC and a detailed interpretation of the layers in the vicinity of wells (**b**)

the seismic wave field [9, 10]. In this case, the algorithm of the convolutional model is used, in which it is possible to take into account the noise level, the residual background of multiple waves, and the regularization factor. To search for a solution, the zero-order direct search method (Hooke-Jeeves method) is used, which does not require the calculation of derivatives of the objective function. In the process of solving the inverse dynamic problem, an important role is given to the stage of determining the shape of the initial seismic signal.

To optimize the solution, a system of objective functions (a vector of objective functions) has been developed, in which various types of residuals between real and model data are iteratively calculated. A feature of this system is that it calculates and analyzes the energy and waveform residuals separately and makes a decision on optimizing the model as a whole. Sharing of characteristics allows extracting the distribution of medium properties with maximum detail in the process of stable inversion of wave field data.

Amplitude residuals make it possible to control the stability and sustainability of the iterative convergence process of the initial model to the desired solution and analyze the feature for optimal dynamic regularization of the solution search process.

Phase residuals make it possible to manage the solution of the problem more successfully by selecting the optimal solution from a variety of those. The process

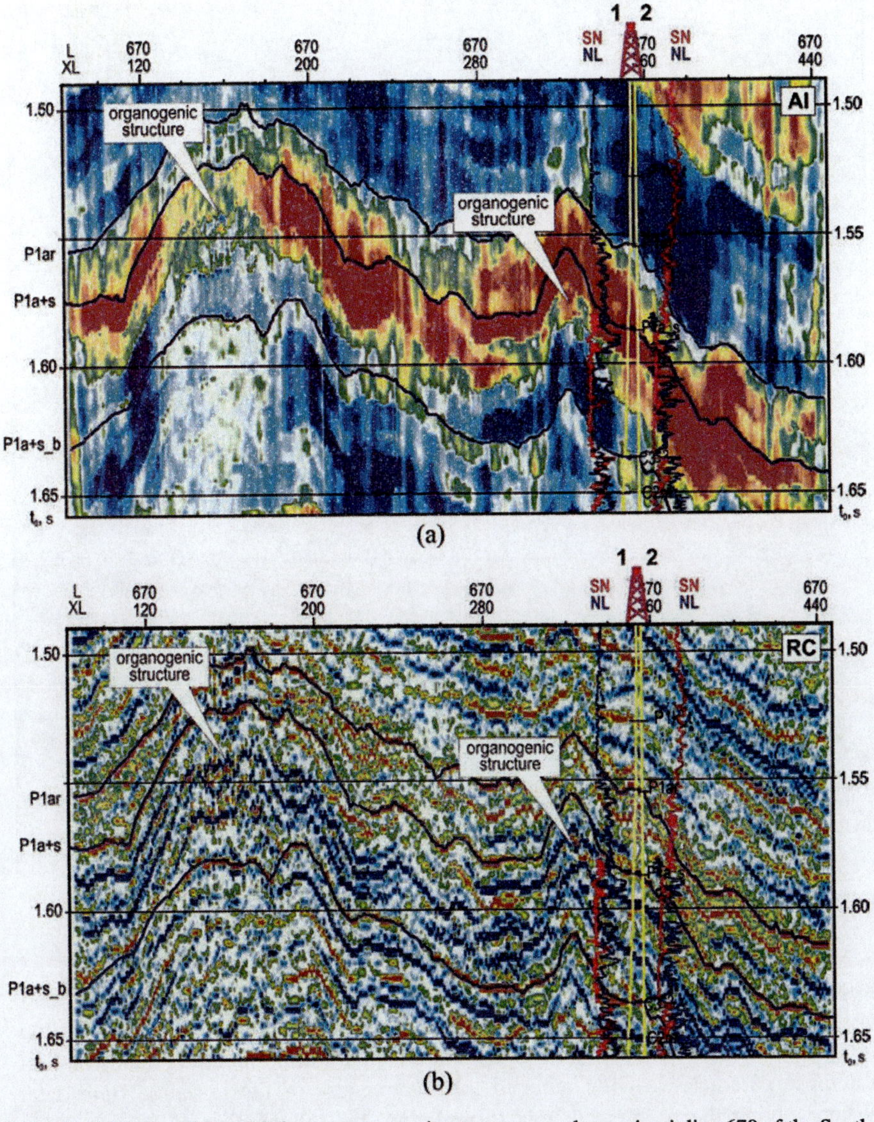


**Fig. 3.9** Mapping of Assel-Sakmar organogenic structures on the section inline 670 of the South Osh section: (**a**) acoustic impedance; (**b**) reflection coefficients; (**c**) a fragment of the section AI along the line 2013–07. Myanmar

of searching for a model AI and RC is completed when the vector of the objective functions system is stabilized and/or when their specified restriction level is reached.

It is worth noting the problem of "*identification of weak seismic signals*" is related to the problem of "*solving the inverse dynamic problem of seismic*", since in such a "bundle" the nature of reflected waves weak in intensity becomes clear, which, depending on the reflection coefficients varying in section, "respond"

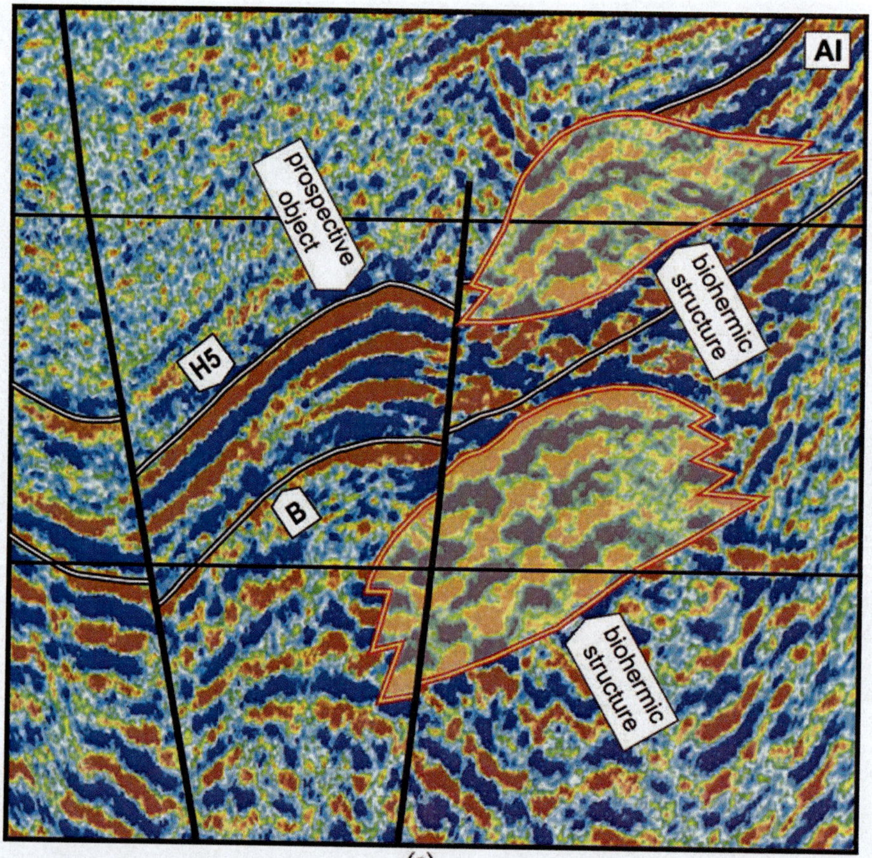


(c)

**Fig. 3.9** (continued)

accordingly to low-contrast seismic acoustic inhomogeneities. It is with such heterogeneities that the desired geological and geophysical parameters are most often associated – porosity and oil and gas saturation and the corresponding manifestation for these parameters of the reduced informativeness of the initial seismic data. When solving the problem in this formulation, the information of wave seismic exploration (in the form of a wave field structure) is modified into that of seismic exploration of material composition (in the form of sections and cubes of high-resolution acoustic seismics).

In general, the numerical algorithm for constructing detailed 2D and 3D seismic acoustic models based on the formation of a sequence of RC and AI is characterized by high vertical and horizontal resolution of the structural elements of a thin-layer section.

Numerous examples with the results of solving the IDSP are given by us in Chap. 5 and in [9, 10].

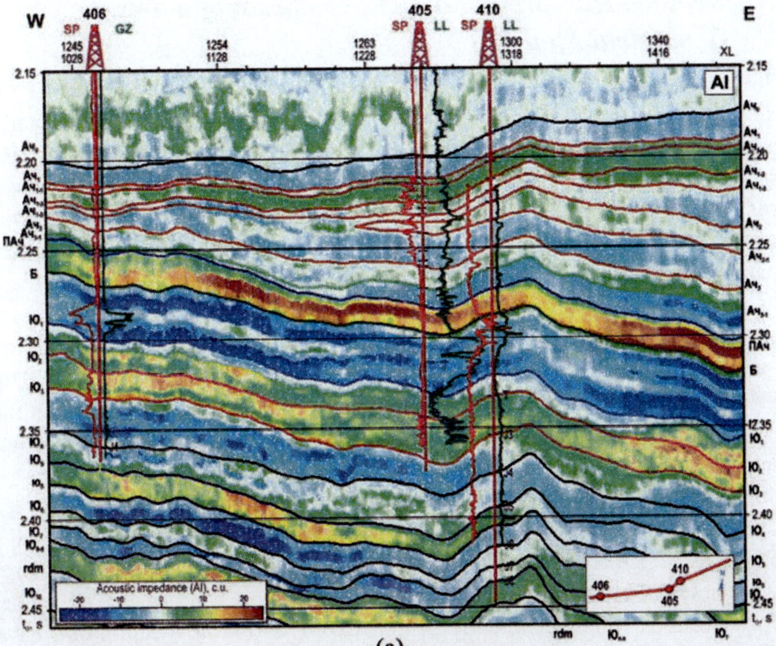


(a)

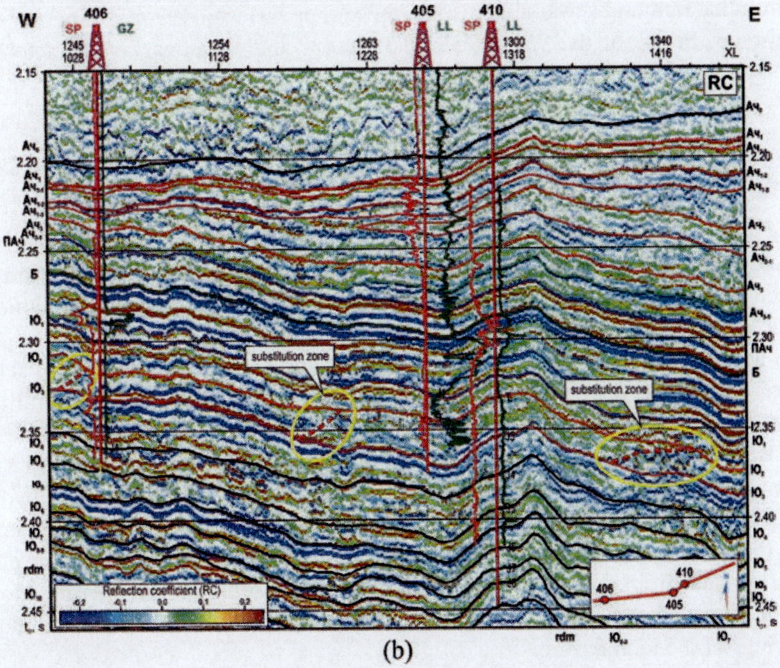


(b)

**Fig. 3.10** Sections of acoustic impedance (**a**) and reflection coefficients (**b**) through wells 406, 405 and 410. KhMAO

### 3.3.5 *Detailed Interpretation and Formation of a Seismic Geological Model*

Detailed interpretation on models of the real environment – 2D sections and 3D cubes of AI and RC is implemented with a vertical resolution equal to the seismic record sampling step in time [9, 10]. Compared to traditional seismic wavefield records, the resolution of 2D sections and 3D RC and AI cubes increases by an order of magnitude on average, that is, the inversion results (RC and AI) turn out to be comparable with well logging data. The practical vertical resolution of the terrigenous section is 3–4 m, the carbonate section is 5.5–6.5 m (with a seismic record sampling step $\Delta t = 0.002$ s and the propagation velocity of elastic oscillations in the terrigenous section $V = 3000–4000$ m/s and in the carbonate one $V = 5500–6500$ m/s).

The construction of detailed seismic geological models based on the joint interpretation of sections and cubes of AI and RC and well logging data is carried out in order to identify potential oil-prospective objects on such models. On these, the most informative ratios of structural elements and forms of the studied sediments, relatively small thickness structural–lithological and combined traps of oil and gas are distinguished, zones of substitution and wedging of productive horizons, organogenic structures, stratigraphic inconformities are fixed, and in the presence of appropriate information (seismic, dynamical expressed) – the location of oil-water (OWC) and gas-water (GWC) contacts between fluids.

From the numerous available results, the authors selected seismic data obtained for the conditions of the West Siberian Province and Myanmar as characteristic examples (Figs. 3.10, 3.11, 3.12 and 3.13).

### 3.3.6 *Structure Constructions of a Thin-Layer Section*

Structural constructions on the main lithologic stratigraphic units of the section are carried out using velocity models of the real medium in the form of structural maps, isochron maps, interval time maps (isopachite) for target horizons and layers.

In a comprehensive analysis of the interpretation results from the entire set of criteria for oil and gas potential, such an important criterion as the "structural factor" is considered of the main ones, since it largely controls the migration routes of hydrocarbons due to the better conductivity of the paths for them in the arched parts of reservoir formations and the best conditions for the creating of traps for hydrocarbon accumulations. At the same time, the studied structural forms are considered from the point of view of the geodynamic model of the structure (geodynamic classification) and from the point of view of the shape (morphological classification), and the faults are classified according to the direction of breaking movement along the wings of the fault fissure.

Reconstructed acoustic heterogeneities of the real environment with high resolution (based on the obtained sections and cubes of AI and RC) also make it possible

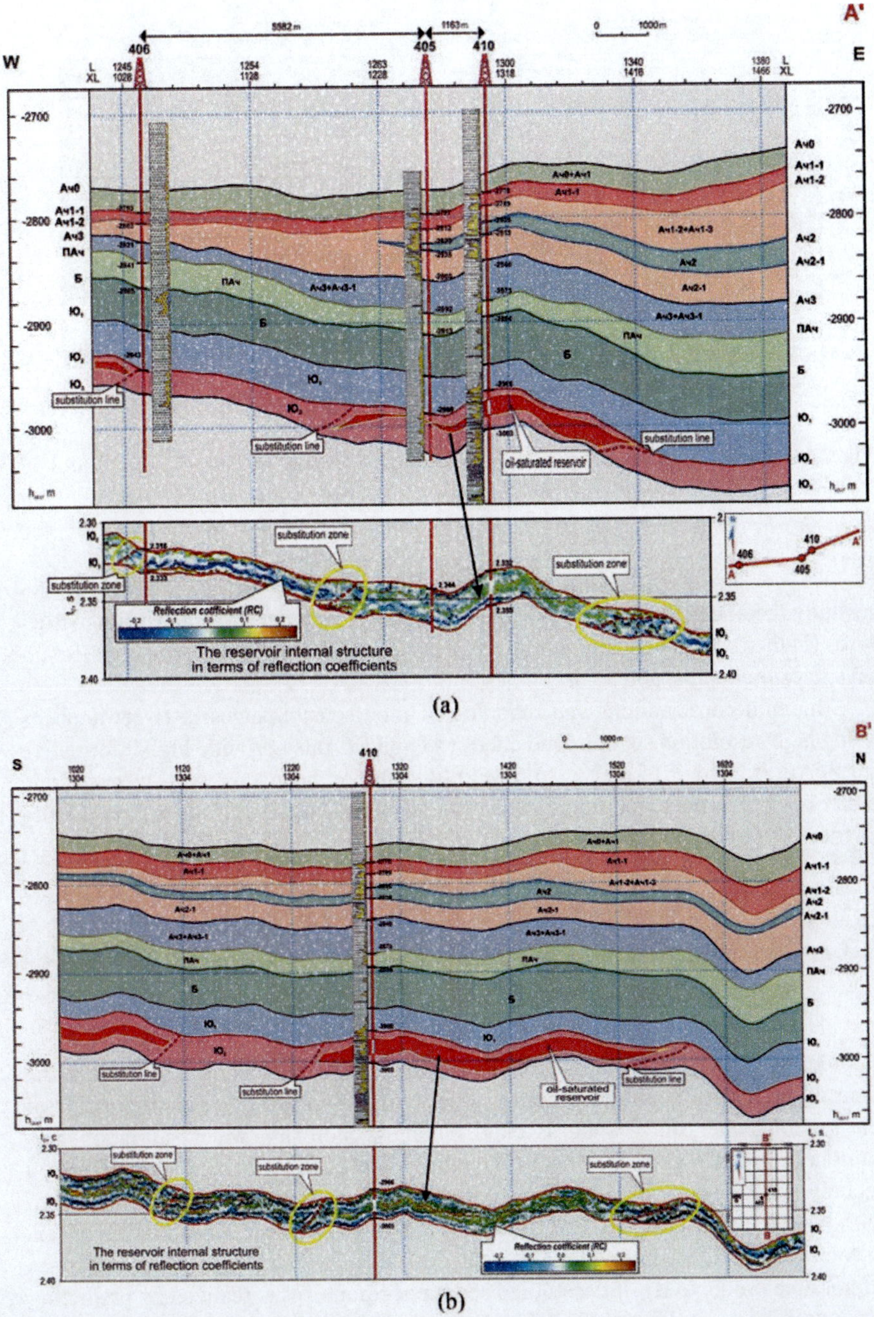


Fig. 3.11 Seismic geological sections along the lines: (a) wells 406, 405 and 410; (b) crossline 1304 through well 410. KhMAO

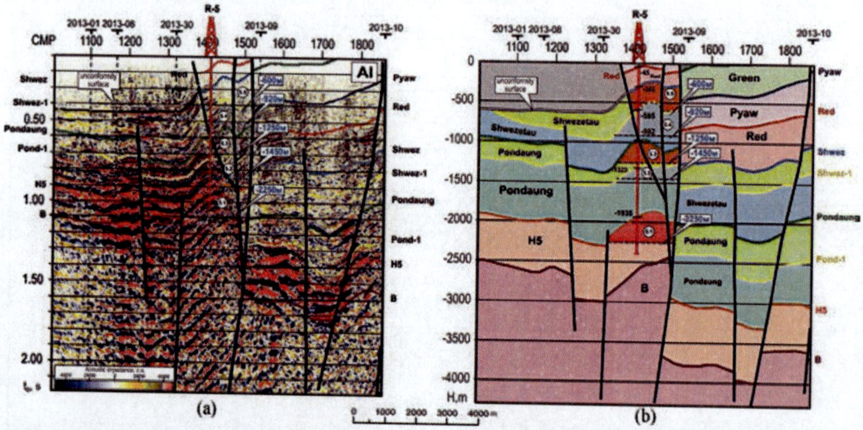

**Fig. 3.12** Acoustic impedance (**a**) and seismic geological section (**b**) along the profile 2013–02 with the recommended well R-5. Myanmar

to study the ancient paleochannels of river systems, elements of their filling structures (in the form of acoustic inhomogeneities), channel incisions of river paleovalleys, erosion depths, and so on.

Structural constructions and mapping of lithofacies changes in target deposits using high-resolution sections and cubes (AI and RC) also provide high informative results (Figs. 3.14, 3.15 and 3.16). The initial data for the given results were seismic data obtained in the exploration areas of the Khanty-Mansiysk Autonomous Okrug of the Western Siberia.

### 3.3.7 *Identification of Oil-Promising Zones and Oil and Gas Deposits*

In order to study in detail the internal structure of the target deposits and determine their oil-prospectives, the distributions of geological and geophysical indicators of the section under study, such as clay, sandiness, porosity, water saturation, oil-saturated thicknesses (capacities) and others, are evaluated using the HRS-Geo technology. Under favorable seismic geological conditions, the prediction hydrocarbons (HC) distribution density, the contours of the possible oil saturation of reservoir rocks in the studied sediments is carried out. The allocation of sites and zones of possible productivity of the studied sediment complexes is realized by predicting structural, tectonically shielded and non-structural hydrocarbon traps with an assessment of the composition and properties of porous facies.

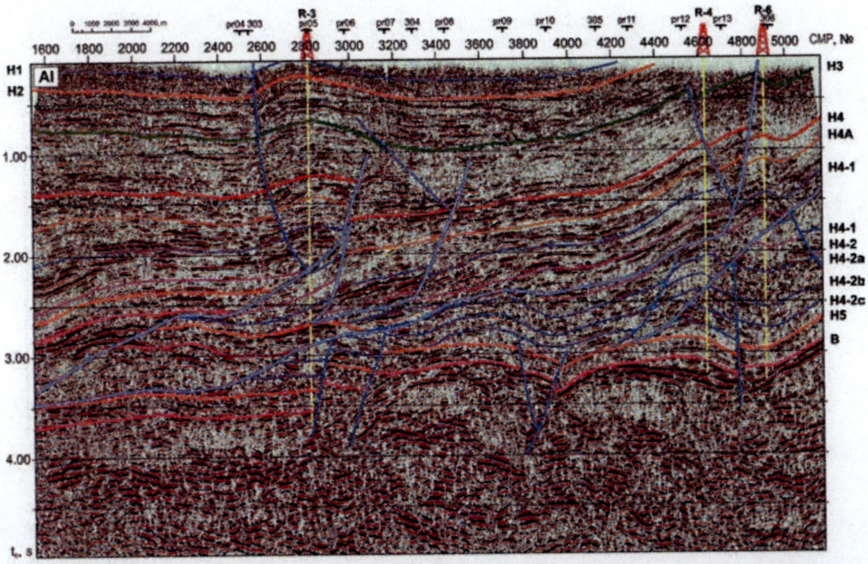

(a)

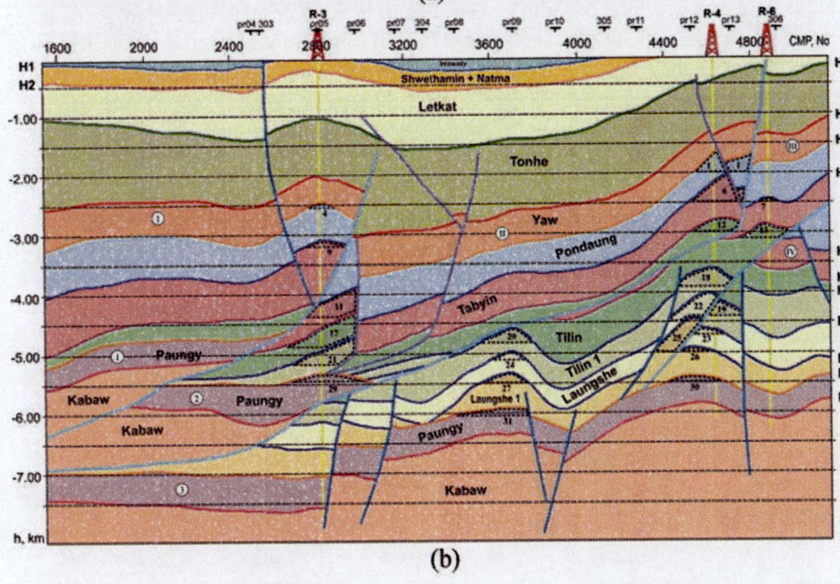


(b)

**Fig. 3.13** Acoustic impedance (**a**) and seimic geological section (**b**) along the profile 310 with recommended wells for optimal opening up of oil-prospective objects. Myanmar

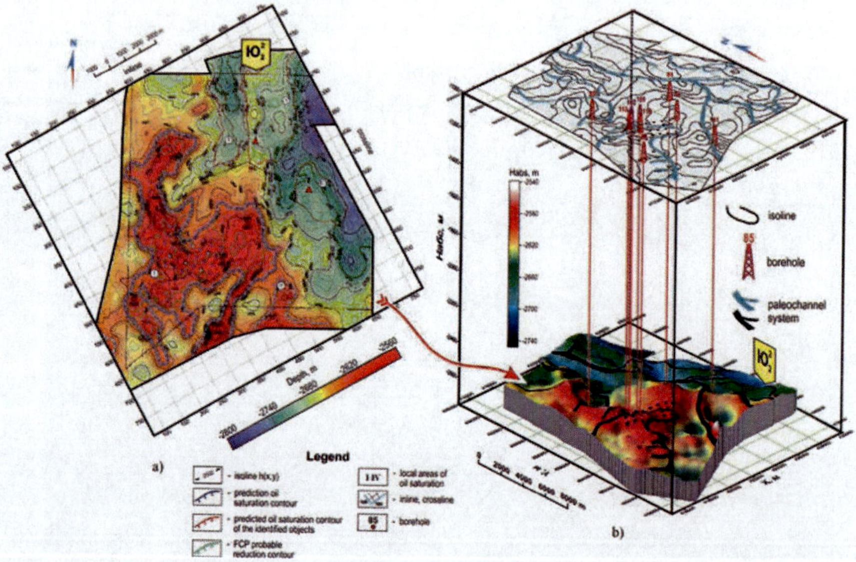


**Fig. 3.14** Results of structural constructions on deposits of the Tyumen formation: (**a**) a structural map along the top of the productive reservoir Ю$_3$; (**b**) a 3D image of the reservoir surface Ю$_2^2$ with a paleo river system. KhMAO

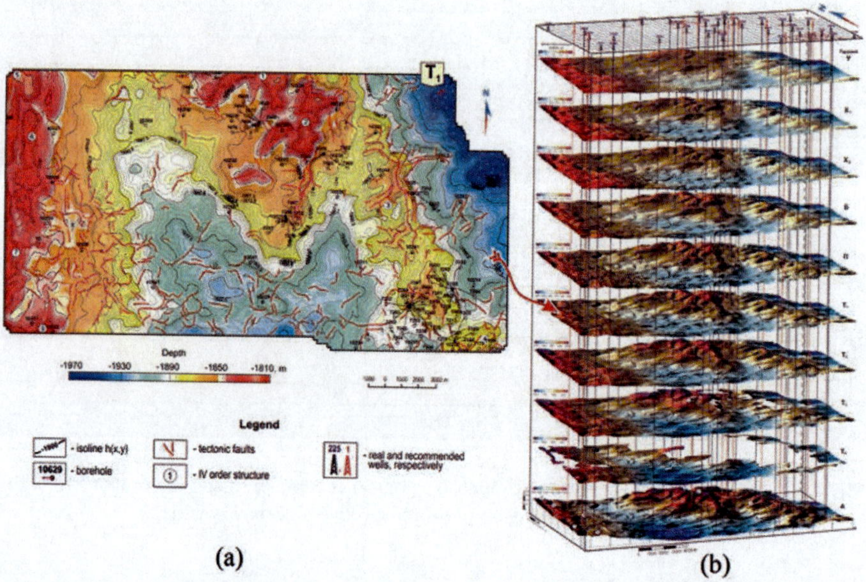


**Fig. 3.15** Results of structural constructions: (**a**) a structural map along the top of the deposits of the Tyumen formation – horizon T$_1$ (Ю$_2$); (**b**) a 3D image of structural maps along the horizons of the sedimentary cover and the surface of the pre-Jurassic base (A). Shaim OGR, KhMAO

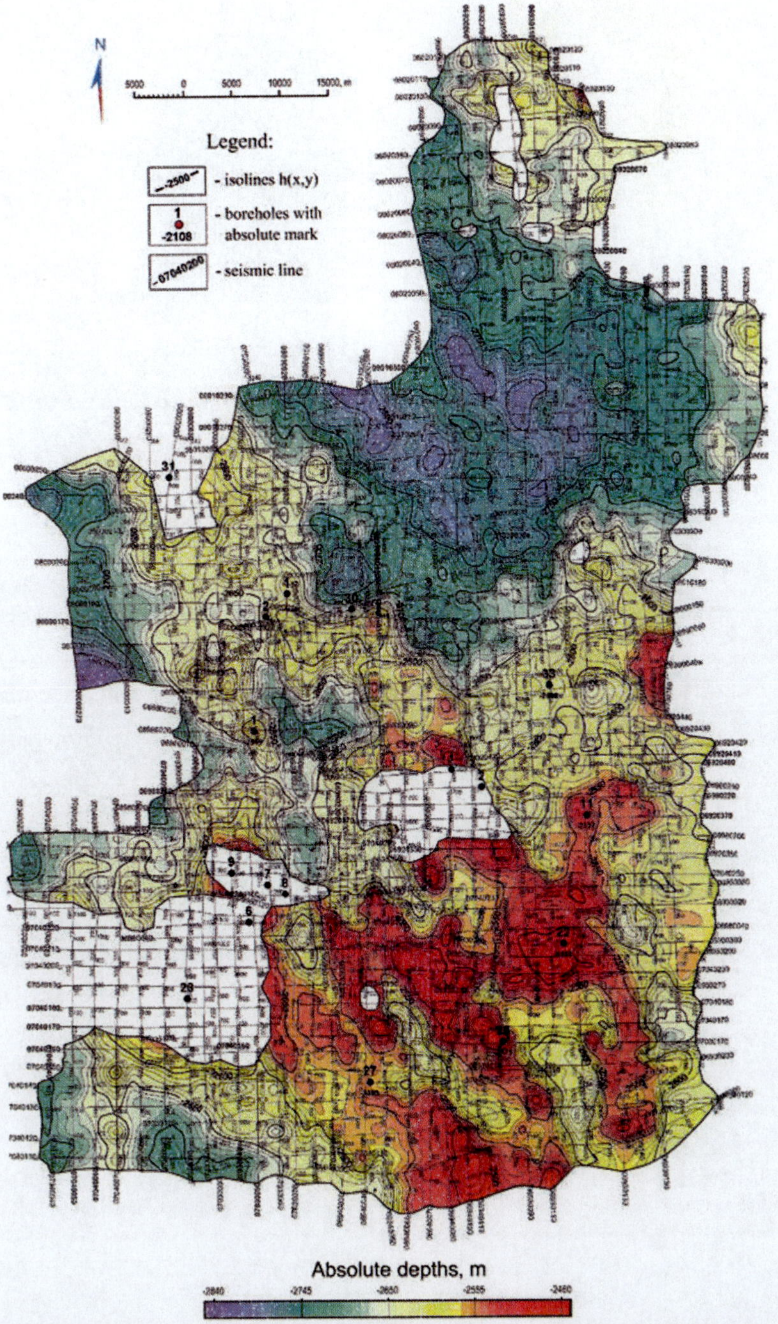


(a)

**Fig. 3.16** Results of structural constructions for deposits: **(a)** the top of the $TIO_{10}$ horizon of the Khudosei formation of the Toar age $J_1t$; **(b)** the surface of the pre–Jurassic basement – horizon A (the bootom of the sedimentary cover). KhMAO

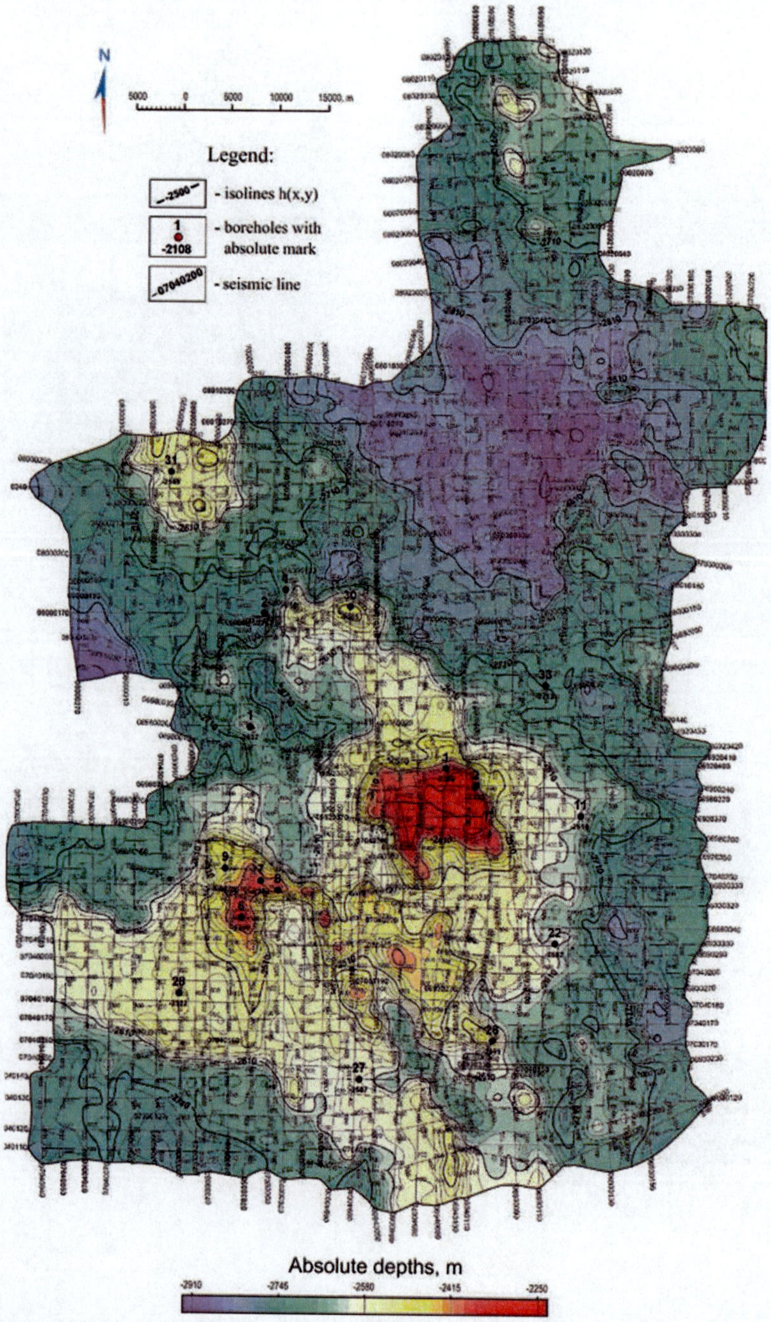


(b)

**Fig. 3.16** (continued)

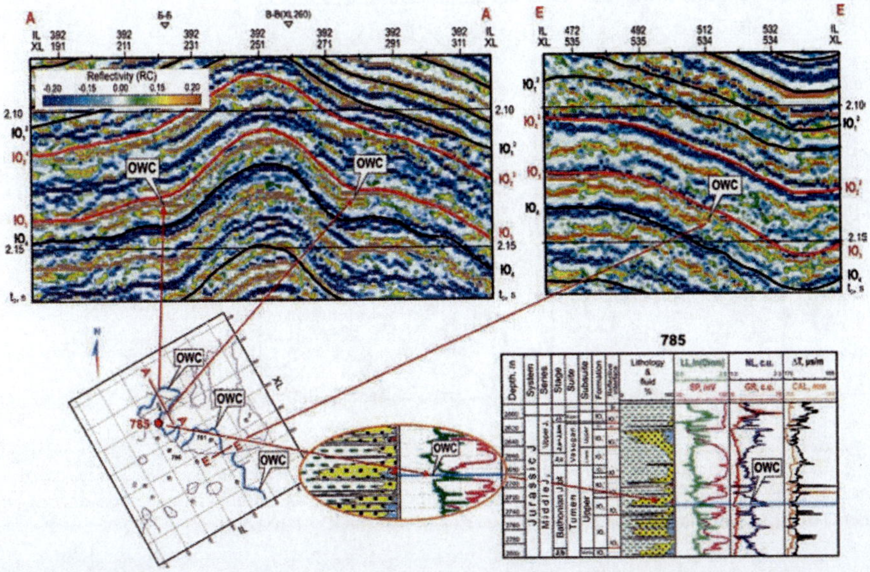


**Fig. 3.17** Fragments of reflection coefficients along the lines A-A and E-E with elements of geological interpretation. KhMAO

In order to study in detail the internal structure of the target deposits and determine their oil-prospectives, the distributions of geological and geophysical indicators of the section under study, such as clayiness, sandiness, porosity, water saturation, oil-saturated thicknesses (capacities) and others, are evaluated using the HRS-Geo technology software. Under favorable seismic geological conditions, the forecast of the hydrocarbons (HC) distribution density, the contours of the possible oil saturation of reservoir rocks in the studied sediments is carried out. The allocation of sites and zones of possible productivity of the studied sediment complexes is realized by predicting structural, tectonically shielded and non-structural hydrocarbon traps with an assessment of the composition and properties of porous facies.

In general, the allocation of sites and zones of possible productivity of the studied sediment complexes is carried out on the basis of the prognosis of structural, tectonically shielded and non-structural hydrocarbon traps with a quantitative assessment of the composition and properties of porous facies.

As examples, Figs. 3.17, 3.18, 3.19 and 3.20 give some results of the identification of oil-prospective zones and deposits, for which the initial data were the materials of seismic observations performed on a number of areas of the West Siberian oil and gas province.

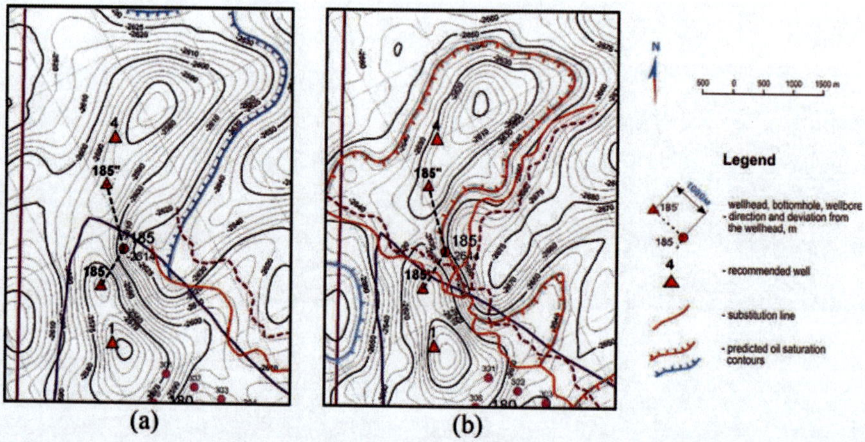


**Fig. 3.18** The deflection scheme of the borehole 185 for opening up productive reservoir layers of the horizons $Ю_2^2$ and $Ю_3$ of the Tyumen formation. The well projections are presented on fragments of structural maps along the horizons: (**a**) $Ю_2^2$ and (**b**) $Ю_3$. KhMAO

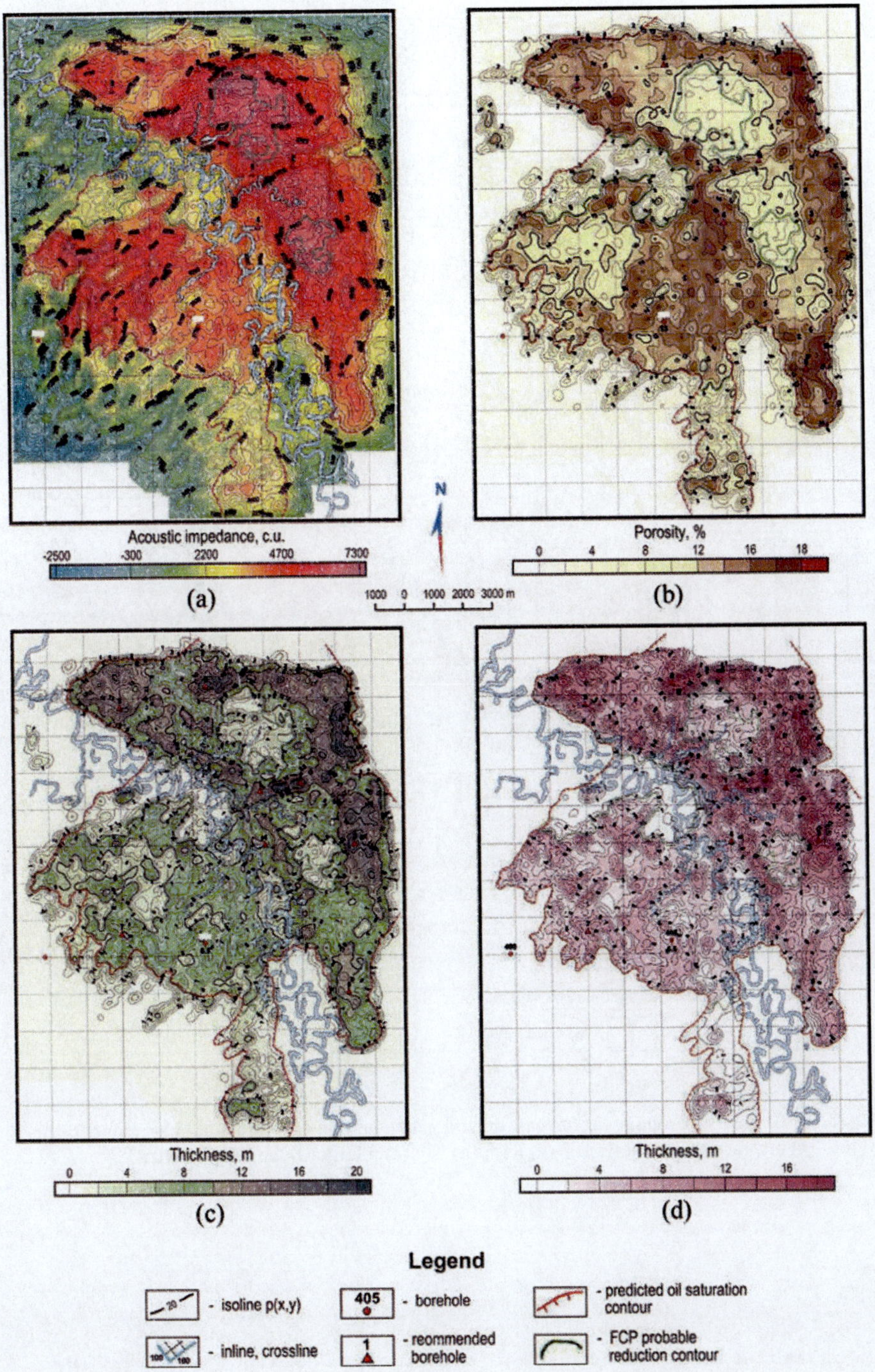


**Fig. 3.19** Maps of the deposits of the $Ach_{1-1}$ horizon (Achimov stratum) with distributions of: **(a)** acoustic impedance; **(b)** porosity; **(c)** effective thicknesses; **(d)** oil-saturated thicknesses. KhMAO

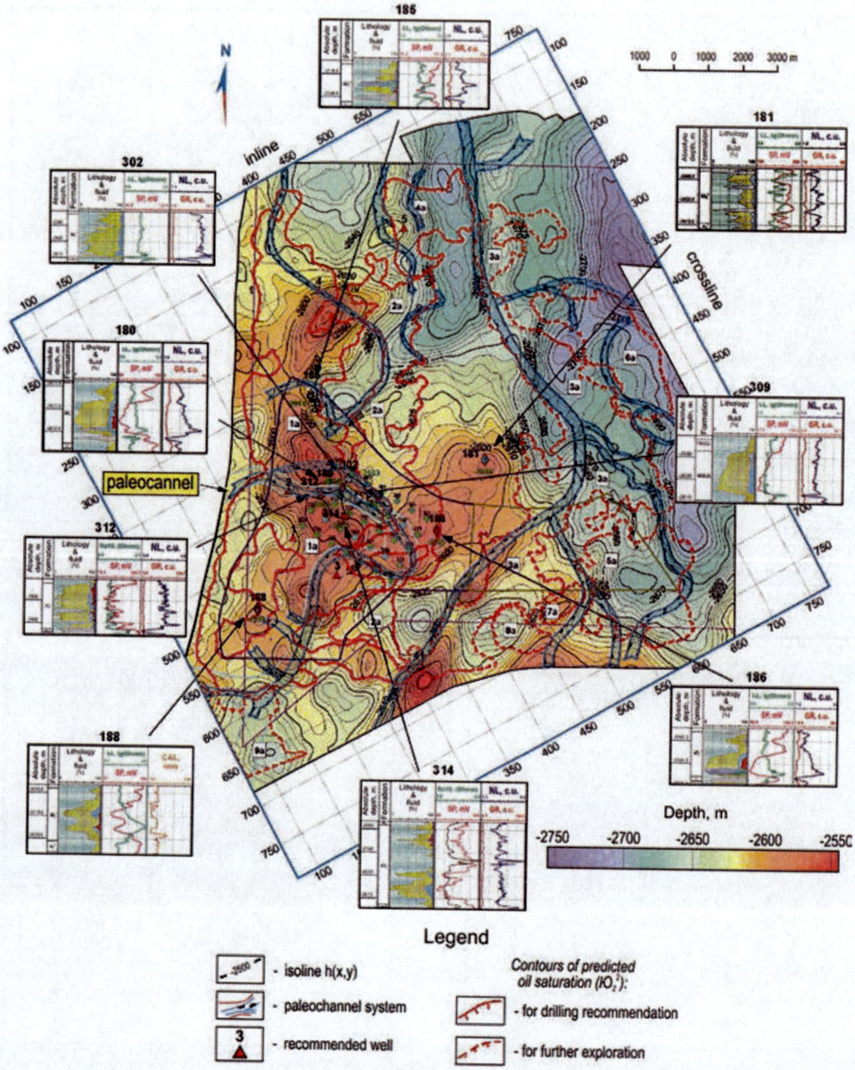


(a)

**Fig. 3.20** Contours of significant forecast oil and gas saturation on a structural basis for deposits of the Tyumen formation – horizons: **(a)** $Ю_2^2$; **(b)** $Ю^3$; **(c)** jointly $Ю_2^2$ and $Ю_3$. KhMAO

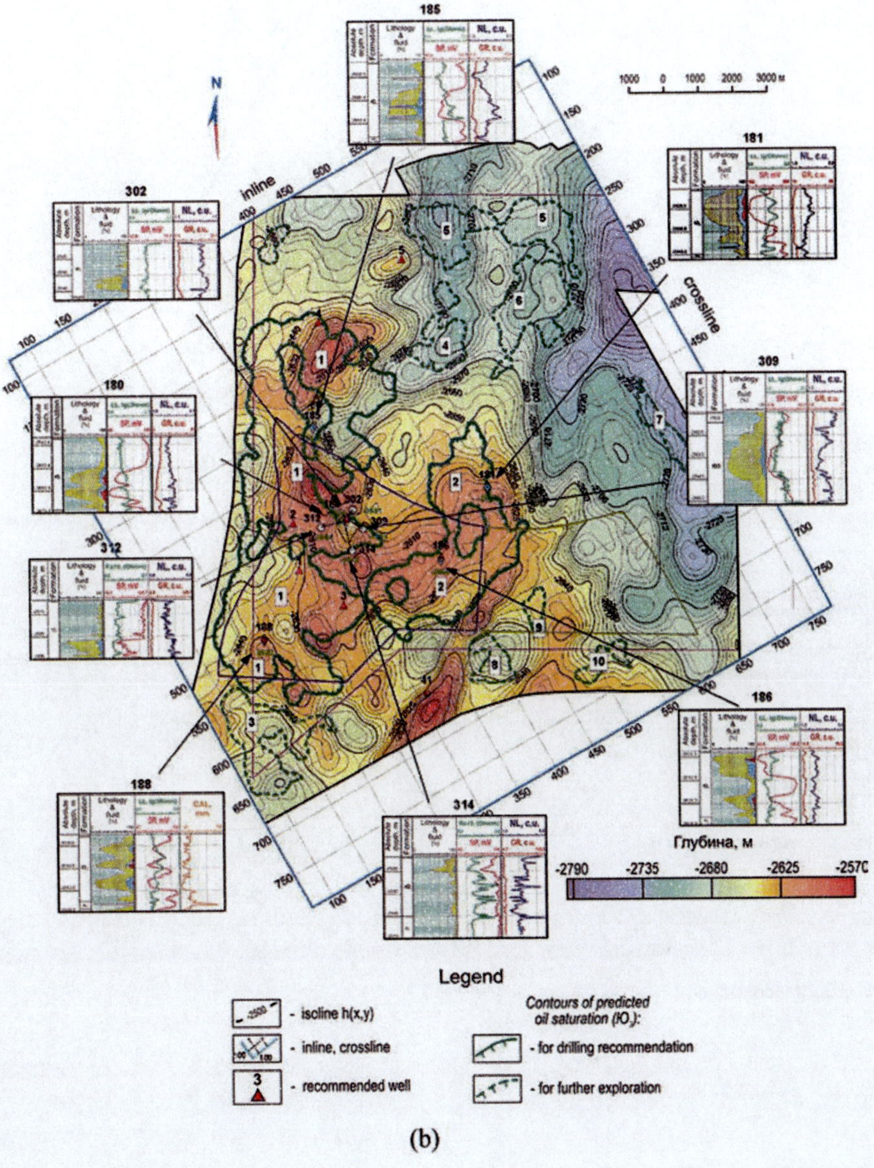


(b)

**Fig. 3.20** (continued)

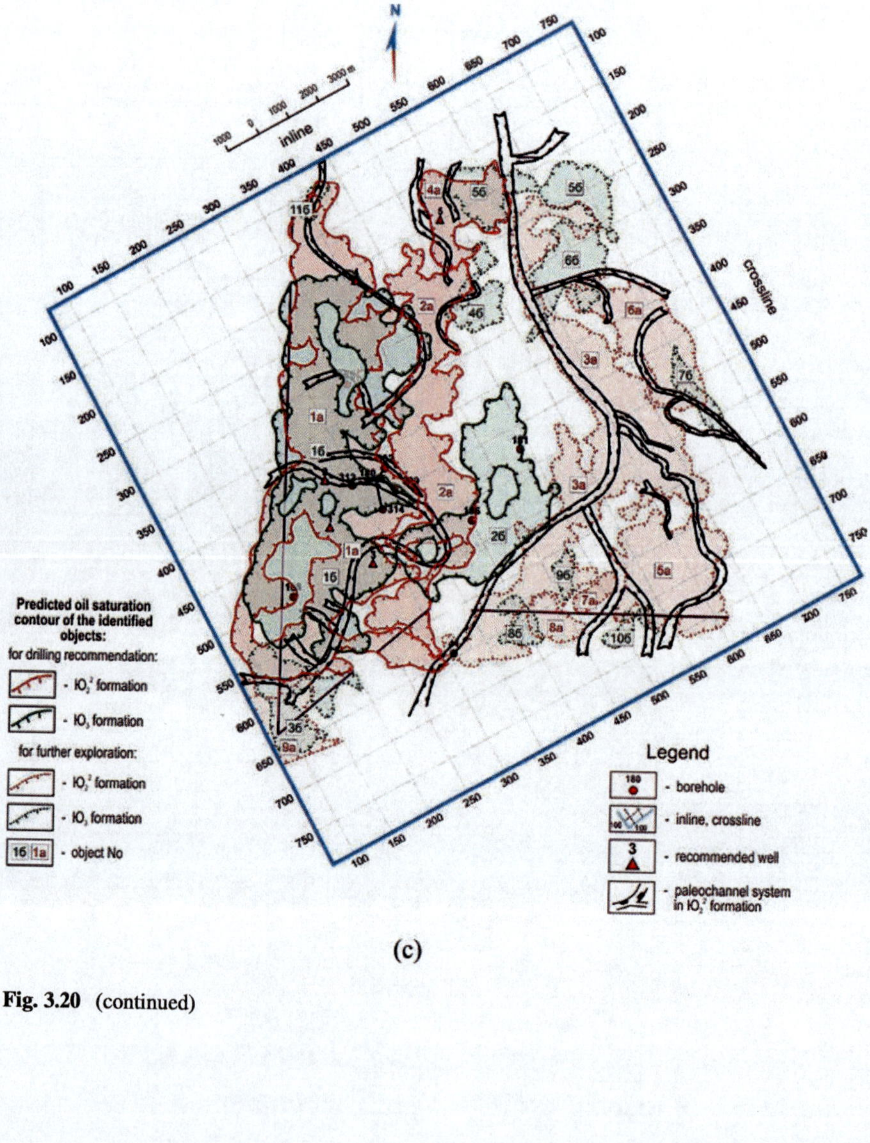


(c)

**Fig. 3.20** (continued)

### 3.3.8 *Facies Zoning of Productive and Promising Deposits*

Determination of deposits formation conditions (sedimentation) by clustering and forecasting of various facies groups in order to identify facies substitution zones, heterogeneity zones in target sediments is one of the directions for studying the detailed internal structure and lithological-facies variability of productive sediments.

Based on the use of cluster analysis methods, the problem of determining facies substitution sites on the studied area using high-resolution seismic data and digital models of various geological indicators in the form of corresponding distributions on 3D grids is solved. On this basis, facies groups in target sediments are investigated (including in thin-layer productive strata), cluster distribution maps are constructed by AI values or by specified sets of geological indicators. The identification of the facies belonging of the obtained classification results according to the studied deposits is carried out on the basis of their correlation with electrometric SP models (spontaneous polarization) and their corresponding sedimentation models of facies (reference analogues).

For example, in conditions of a small number of boreholes on the research area, methods of classifying traces by their record form were used. The results obtained are a sequence of color-classified model facies traces reflecting the heterogeneity of seismic data. On the basis of these data, the corresponding maps of seismic facies are formed, as shown in a number of examples (Figs. 3.21 and 3.22). Only the values of effective acoustic impedance and reflection coefficients were used as input data for the classification of seismic facies.

### 3.3.9 *Integrated Interpretation of Geological and Geophysical Data*

Using the method of high-resolution seismics, results are obtained that characterize the distribution of the lithological composition and reservoir properties of rocks in the studied section. Under favorable seismogeological conditions, it is possible to determine the nature and degree of fluid saturation of reservoir rocks. The very process of extracting the necessary information about geological indicators from seismic data acts as a process of integrated geological interpretation of geophysical data and substantiation of the geological structure of specifically studied oil-prospective objects. This process includes both the analysis of various geological and geophysical data, and a number of automated procedures performed in a certain sequence. The task of complex interpretation is the analysis of all information about the environment, in other words, the construction of a set of interpretive models of the environment, reflecting all those features of the geological structure and geological history that collectively determine the oil and gas prospects.

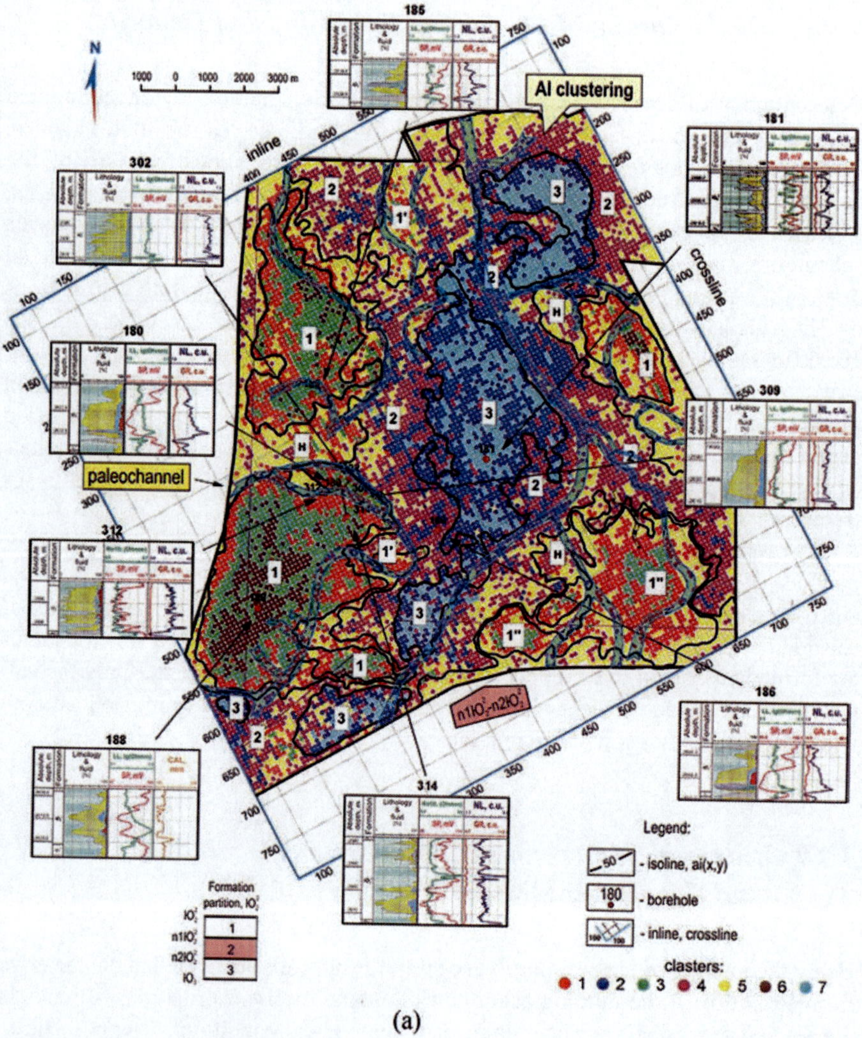


**Fig. 3.21** Clusters distribution maps of average acoustic impedance: (**a**) the $Ю_2^2$ horizon between the boundaries of $n1Ю_2^2$– $n2Ю_2^2$; (**b**) distribution of clusters on a 3D image of the $Ю_2^2$ reservoir surface with a system of paleochannels; (**c**) clusters of interval time (dt), porosity (Kp), coherence (coh), acoustic impedance (ai) of the $Ю_2^2$ horizon between the boundaries of $n1Ю_2^2$– $n2Ю_2^2$; (**d**) the $Ю_0$ horizon between the boundaries of $n1Ю_0$–$Ю_1$; (**e**) the $Ю_3$ horizon between the boundaries of $n1Ю_3$–$n3Ю_3$. KhMAO

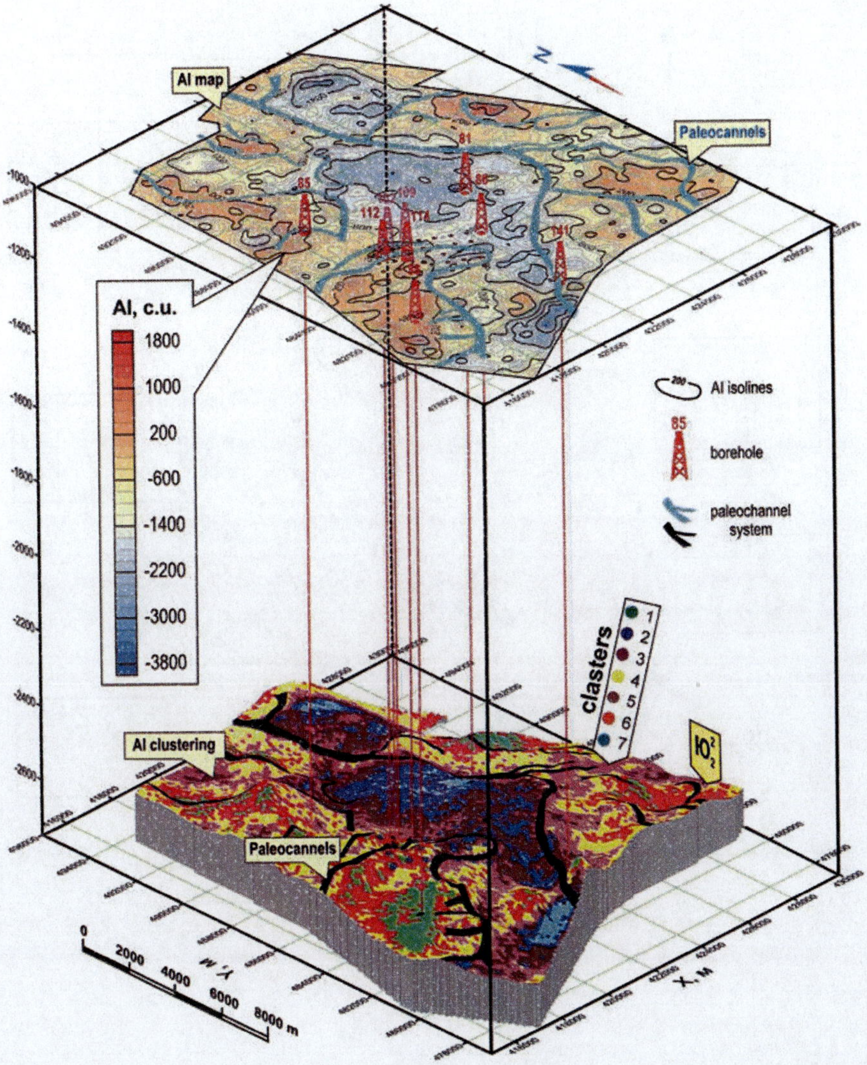


(b)

**Fig. 3.21** (continued)

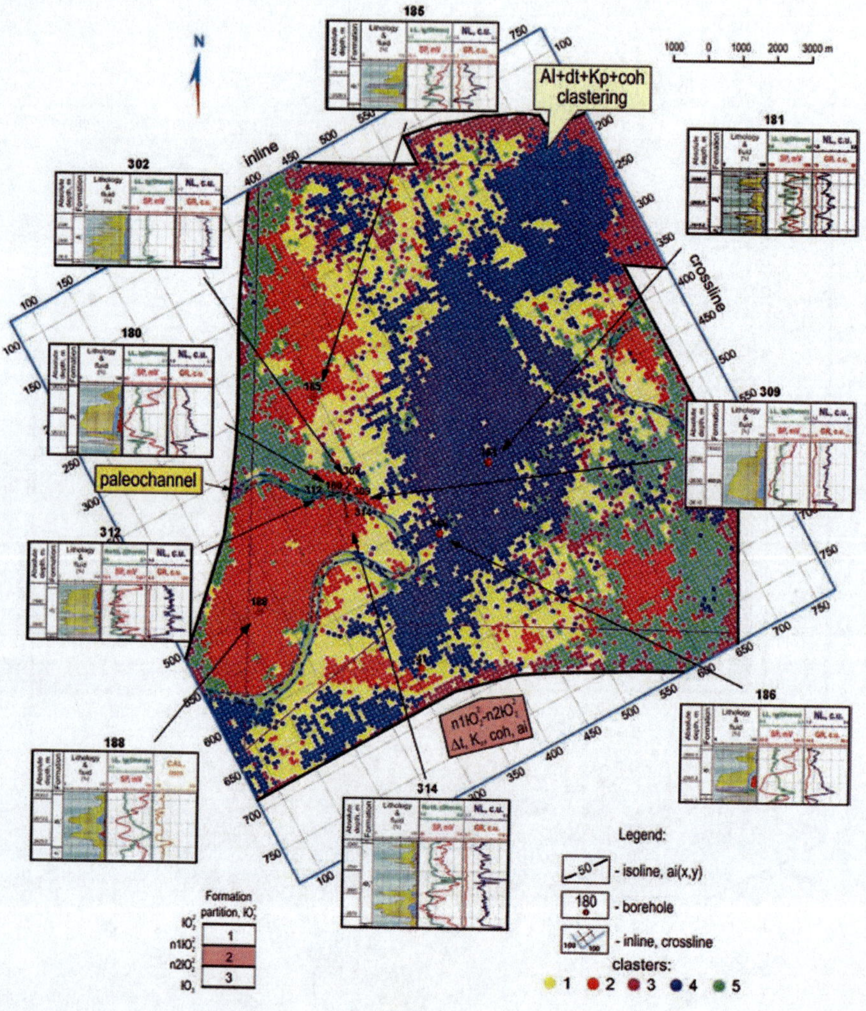


(c)

Fig. 3.21 (continued)

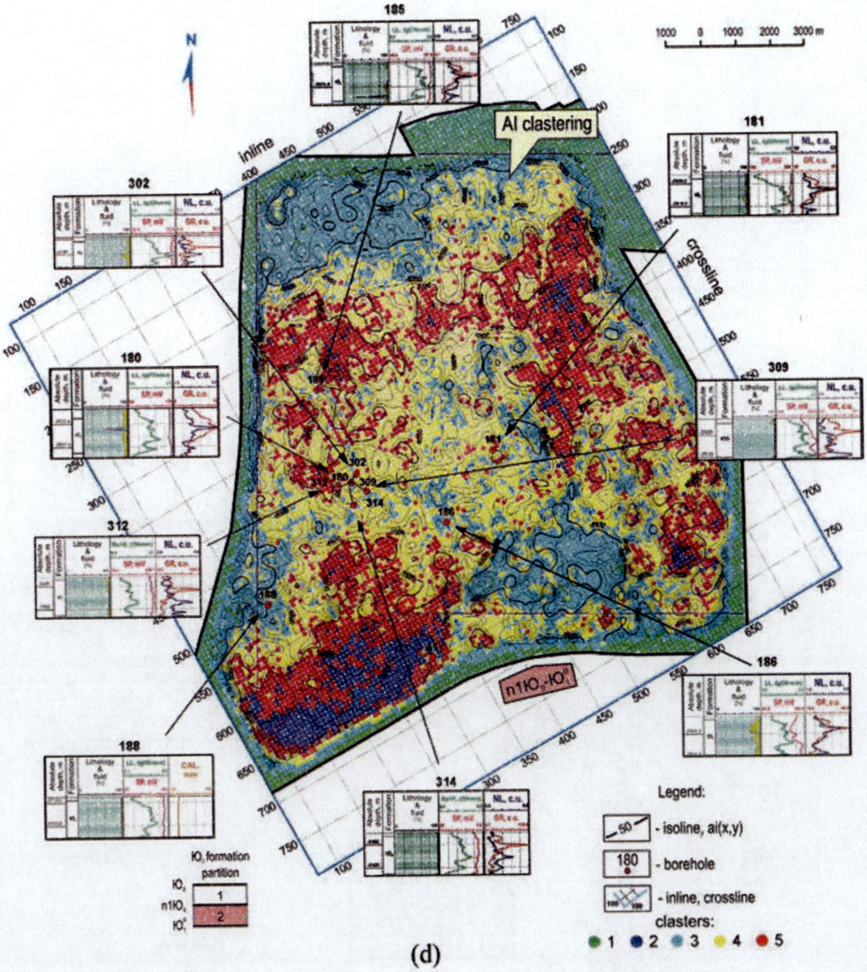


(d)

**Fig. 3.21** (continued)

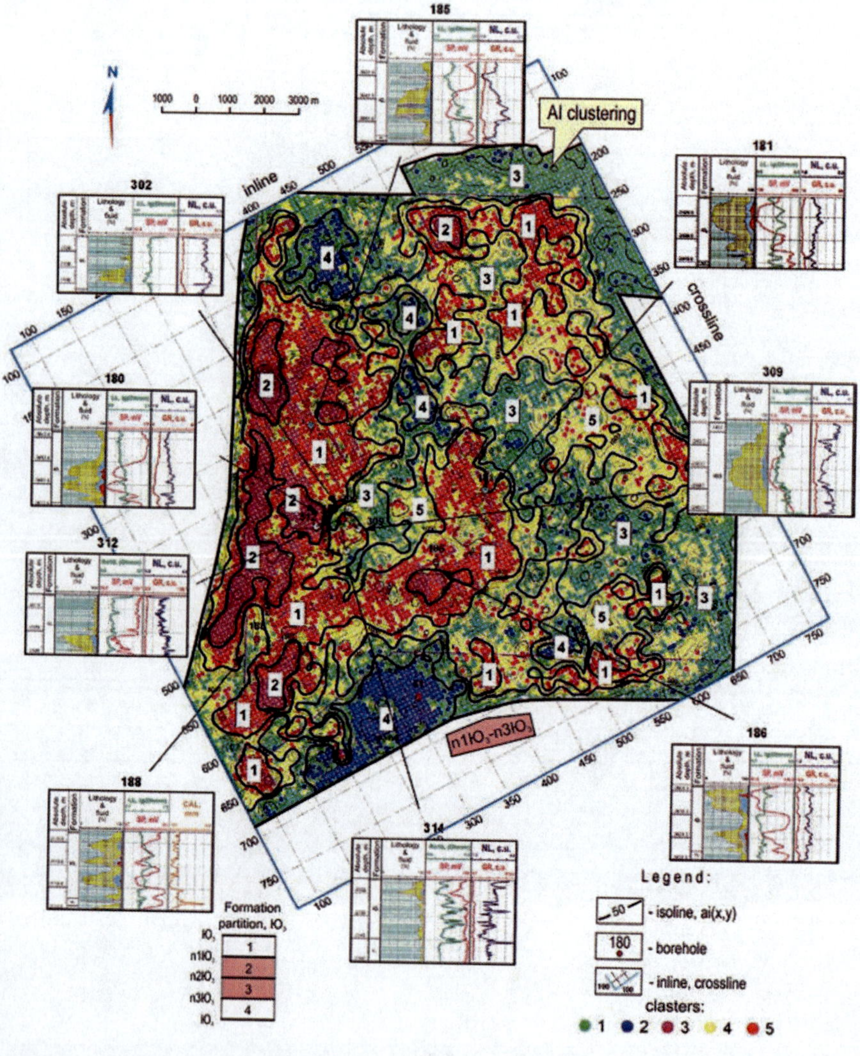


(e)

Fig. 3.21 (continued)

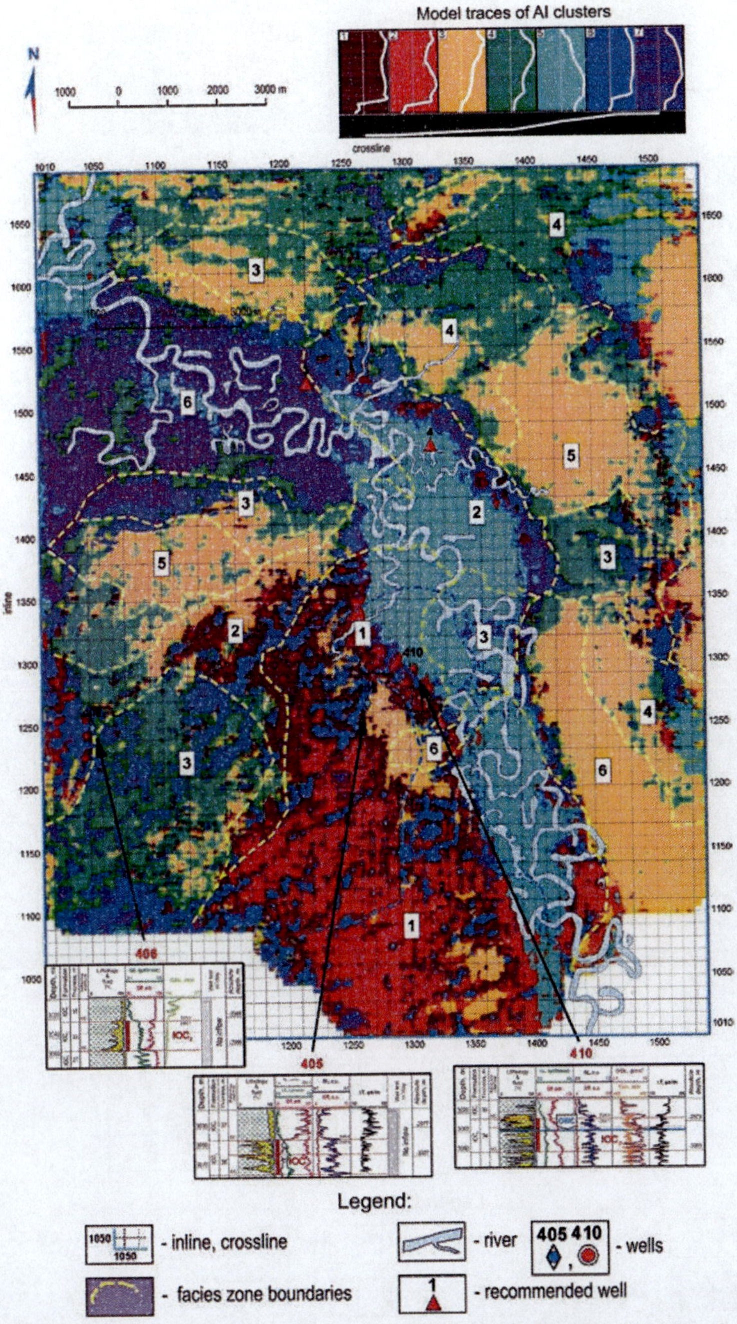


**Fig. 3.22** Facies zones for the deposits of the ЮС$_2$ formation, obtained on the basis of the classification of the RC (**b**), superimposed on the result of the classification by the values of AI (**a**); facies zone based on the classification of the RC (**b**); facies distribution map by AI in the interval of the ЮС$_6$ formation deposits (**c**). KhMAO

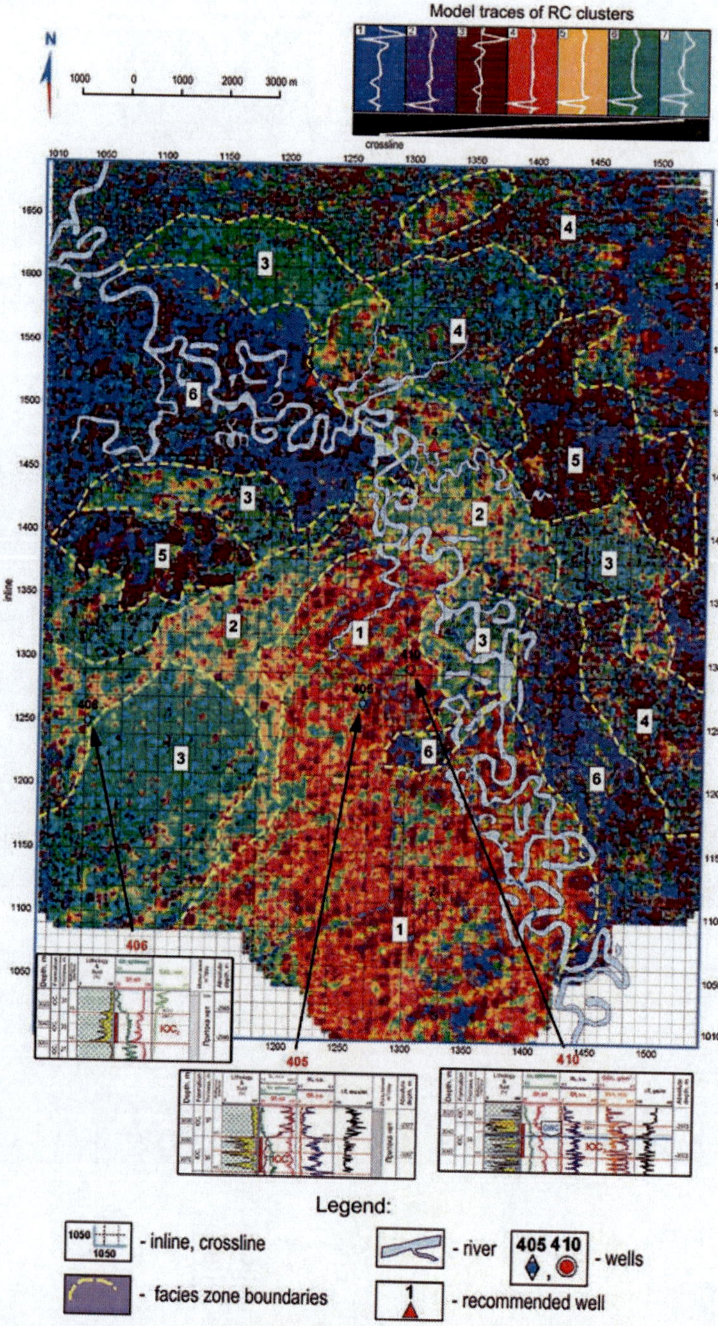


(b)

**Fig. 3.22** (continued)

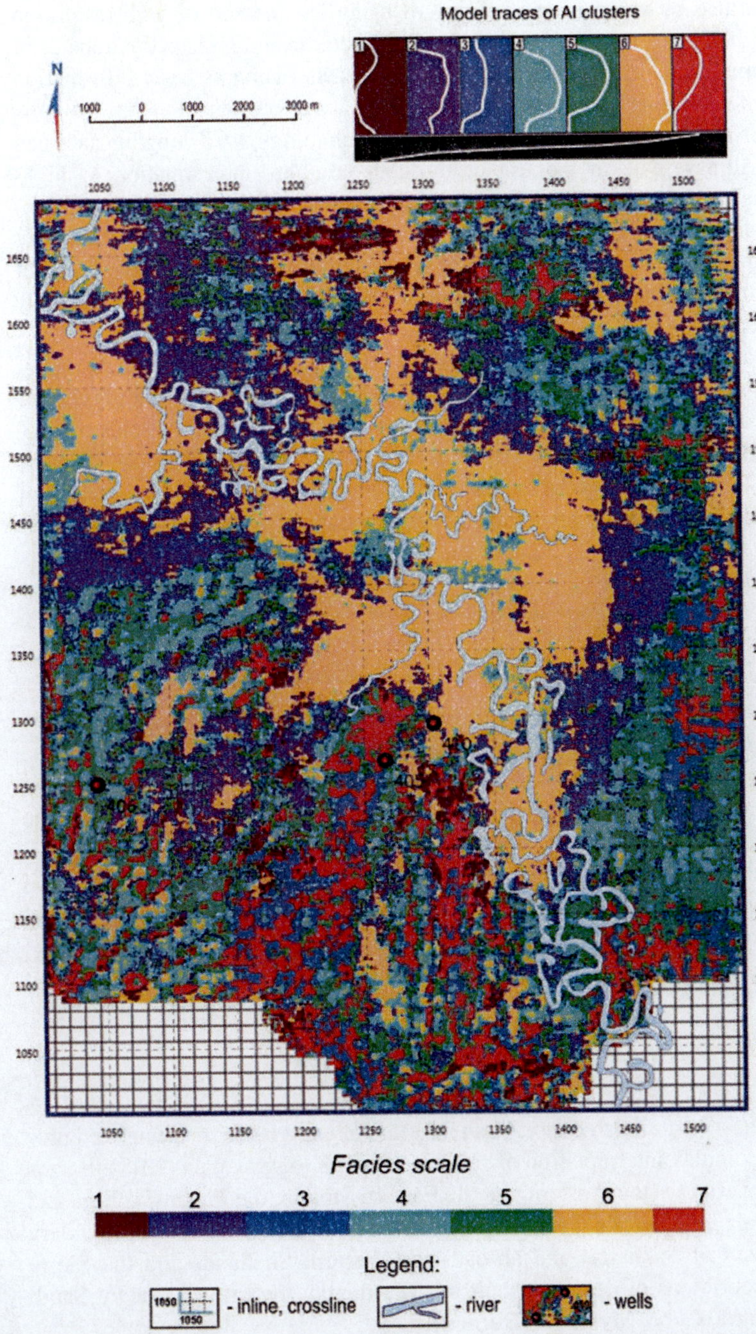


(c)

**Fig. 3.22** (continued)

It should also be noted the peculiarity of using the method of high-resolution seismics in the search for oil deposits of various genesis and sizes is the transfer of unconventional seismic methods to a geological basis (using as basic information the results of extracting identification information about geological matter from the dynamics of a seismic record using HRS-Geo technology, well logging data and borehole drilling). Due to the use of in-depth complex interpretation of high-resolution seismic data, well logging data and the results of facies analysis of high-resolution data, a number of new promising areas in reservoir formations are identified, the contours of oil content are significantly refined. The main types of traps in reservoir formations are often traps of lithologically and tectonically shielded type.

As an example, Figs. 3.23, 3.24, 3.25, 3.26, 3.27 and 3.28 presents the results of the study of oil-prospective facilities for a number of areas located in different territories. In particular, the initial data for them were the materials of seismic observations obtained on the territories of the Western Siberia (Umytin, S-Yutym, Z-Tugrov areas), the Timan-Pechora (Savinobor area) and on the northwestern section of Block E (Shvi Man Taung search site), Myanmar.

### 3.3.10 *Assessment of the Component Composition of Geological Indicators and Hydrocarbon Resources*

When determining from seismic data the volumes of recoverable hydrocarbon resources from deposits of the studied horizons and formations, both constant and variable calculation parameters are used (which are pre–determined and used in the examination of the main calculation parameters for categories C1–C3).

Recoverable oil reserves and resources for each of the specifically considered objects are calculated by the volumetric method under standard conditions using known calculation formulas.

### 3.3.11 *The Technology Application Experience*

The authors have accumulated more than 30 years of experience in specialized processing and complex interpretation of seismic and well logging data (GBS) in various regions of Russia (the Western and the Eastern Siberia, the Timan-Pechora and the Volga-Ural provinces, the South Dagestan shelf of the Caspian Sea, the Boundary Cenozoic basin of Sakhalin) and abroad (the Western Shetlands and the Sierra Leone – in conditions of deep-sea shelf of the Atlantic, the Pripyat trough, Saudi Arabia, Bulgaria, Cuba, Myanmar (Burma)).

Practically for all studied oil-promising objects, using the software tools of the HRS-Geo technology, positive results were obtained, further confirmed by the results of drilling deep wells.

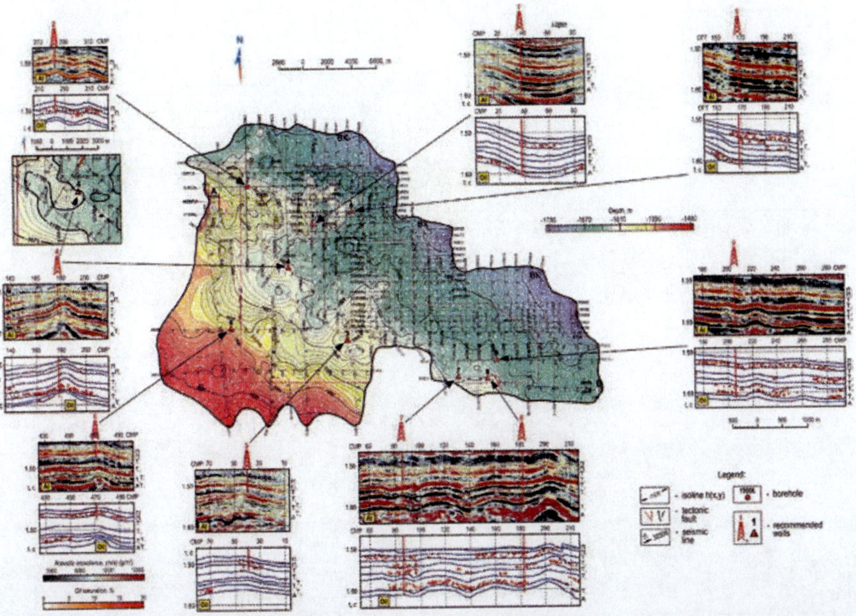

**Fig. 3.23** Recommended wells for optimal opening up of predicted oil-saturated objects. Shaim OGR, KhMAO

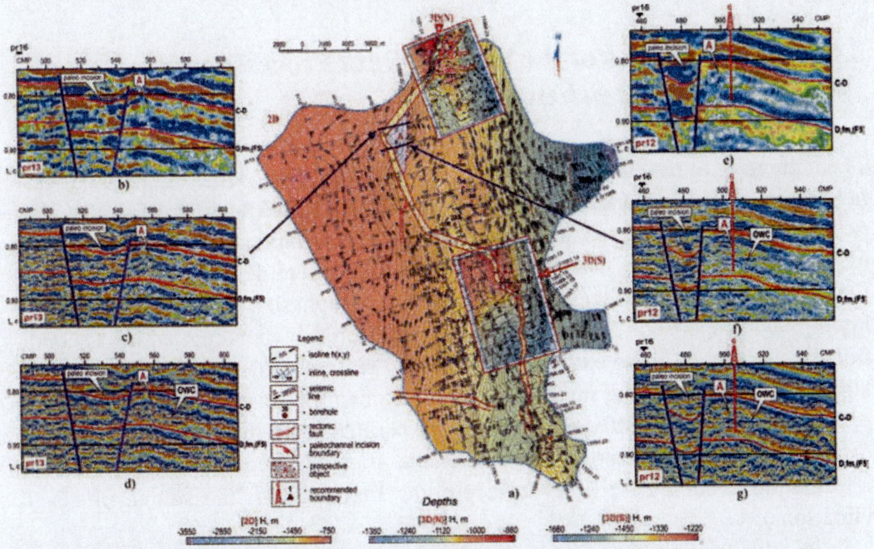

**Fig. 3.24** Structural map along the top of the deposits of the Nyumylgsky ($D_3$nm) upper substage of the Frasnian stage – horizon C-D and the manifestation of a promising object A on 2D sections of acoustic impedance and reflection coefficients. Timan-Pechora province

**Fig. 3.25** The prediction hydrocarbon distribution density map along the $Ю_3$ horizon of the Tyumen formation. KhMAO

## 3.4 Main Features of the Seismic Data Processing Technological Scheme

The authors have developed and considered in detail the scheme of reflected wave fields forming in the works [9, 10]. On this basis, the well-known processes of elastic seismic oscillations propagation in the real geoenvironment and the mechanisms of transformation (processing) of these oscillations were determined. At the same time, a certain novelty of the presented scheme is emphasized, which consists in the highlighted property of the wave field – its interference character, which is conditionally attributed by the authors to the block "Noises". The main "efforts" of the algorithm for solving the inverse dynamic problem of seismics should be aimed primarily at separating information about the parameters of the medium (in the form of corresponding distributions of the reflectivities and acoustic impedance) and information about the seismic pulse probing the medium (the source of elastic vibrations).

It is noted the parameters of the reflected wave field depend directly on the distribution of acoustic impedance and reflection coefficients in the explored section. Consequently, the heterogeneities of the geological medium, manifested in changes in acoustic impedance and reflection coefficients and related changes occurring in

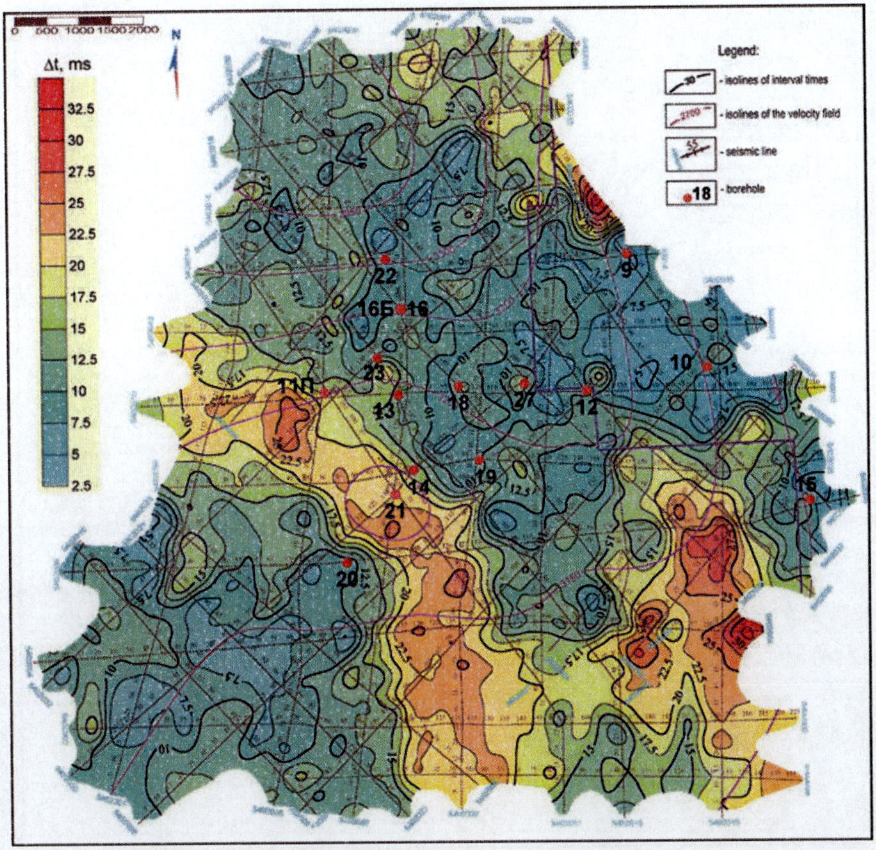


**Fig. 3.26** The map of interval times between the reflecting horizons of Ю$_{7-9}$ (Tyumen suite) and rd (Radom pack). Shaim OGR, KhMAO

lithology, porosity and saturation, are reflected in the characteristics of the wave field. Such changes should be adequate to the geological parameters of the environment under study. Therefore, the features of the wave field that are adequate to the corresponding changes in AI and RC must have true amplitude ratios.

These features, which manifest themselves in the structure of the wave field, can be considered from two positions: preserving the ratio of reflection energies and preserving the waveform according to changes in reflection coefficients for the sequence of layers occurrence of the studied real medium. Preserving the energy ratios of reflected waves during the entire data processing cycle allows us to correctly extract information about the acoustic impedance of the medium from the wave field, and preserving the phase spectrum of reflections provides adequate resolution (as shown above in Chaps. 1 and 2) and possible subsequent correlation of the boundaries of the layers on the reconstructed sections of the reflection coefficients. This approach to the processing and interpretation of seismic data ultimately makes

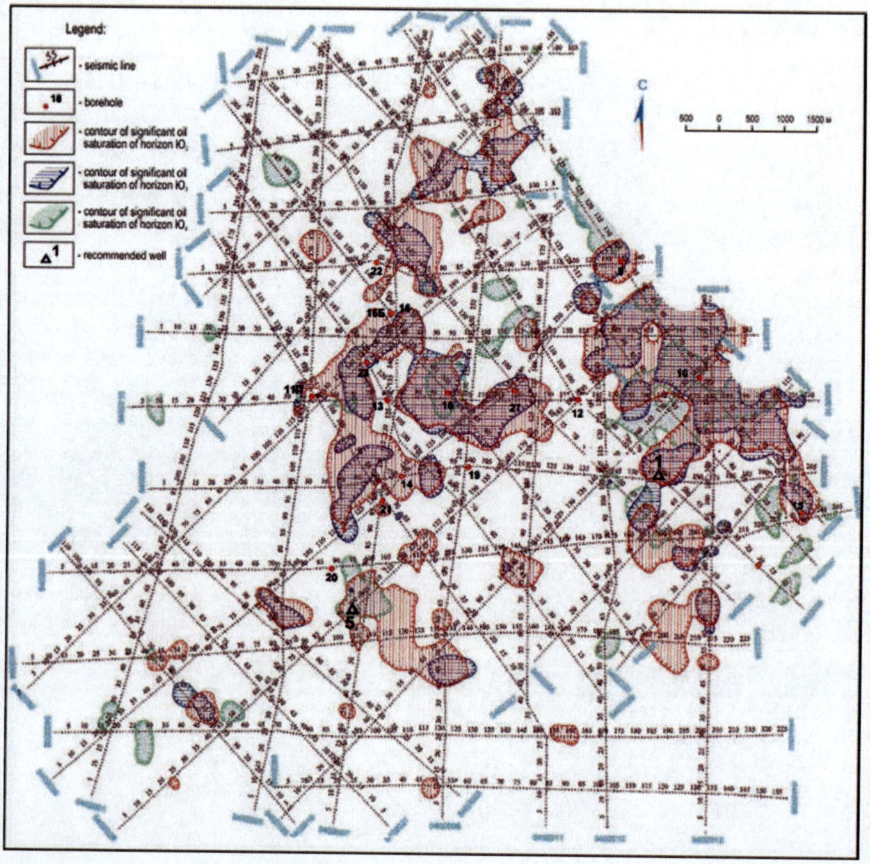


**Fig. 3.27** Map of significant oil and gas saturation in reservoir layers of the $Ю_2$, $Ю_3$ and $Ю_4$ horizons deposits of the Tyumen formation. Shaim OGR, KhMAO

it possible to accurately describe the thin-layer model of the real environment (as shown below in Chap. 5).

From the above and the results of the studies presented in [9, 10], an important goal of preserving the contributions of geological and geophysical parameters of the medium and the simplest amplitude ratios in the distribution of reflection coefficients in the wave field. At the same time, the problem also arises to achieve maximum spatial detail in the presence of both random and regular interference waves. Here, the contradictory nature of processing between the amplitude, vertical and lateral resolutions and the preservation of a significant influence of the target predicted geological and geophysical parameters of the section under study is also manifested. Preliminary processing of seismic observation materials, therefore, is a kind of optimization work. Namely, with the maximum attenuation of various types of noise, it is necessary to maintain the greatest information content on the predicted geological indicators and the best spatial detail. This important feature of seismic

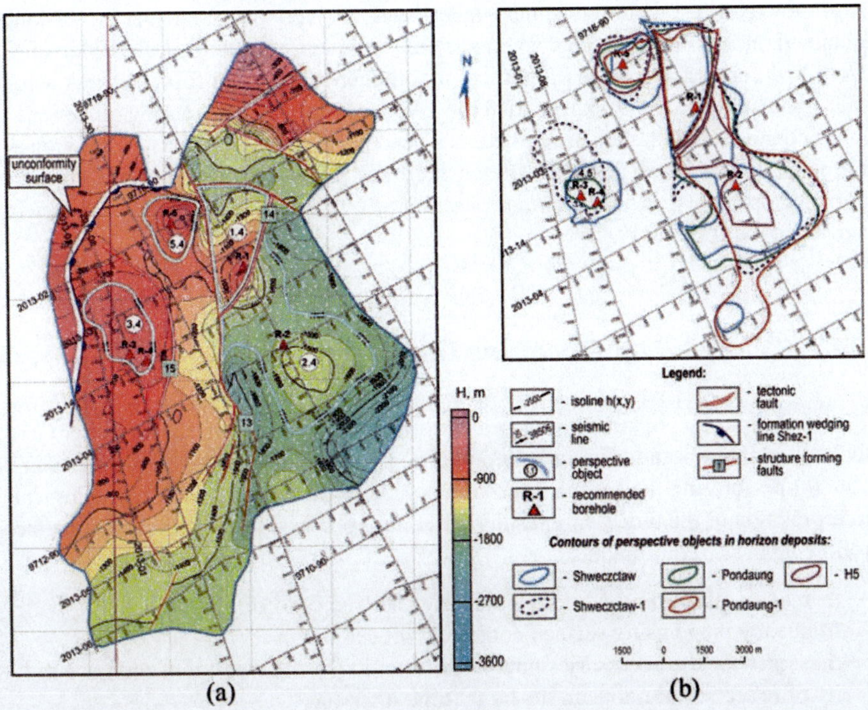


**Fig. 3.28** Structural map along the top of the Shwezetaw formation deposits (Lower Oligocene) – the Shwez horizon (**a**). Contour map of the identified promising structures on the deposits of horizons H5 (Paleocene), Pondaung (Middle Eocene), Pondaung-1, Shwez (Lower Oligocene), Shwez-1 and Red (Middle Oligocene) (**b**). Myanmar

data processing is largely related to the problem of identifying so-called "weak seismic signals" (see Sect. 3.2 above).

It is also worth noting in real terrigenous environments, unevenly overlapping strata of sand-siltstone and clay formations by acoustic properties represent a diverse alternation of layers with increased and decreased acoustic impedance, characterized by their various combinations and values of time thicknesses. Differences in the type of lithology, porosity and saturation of the reservoir also represent in the context of low-contrast acoustic characteristics of the medium. Such a geological situation leads to the formation of low–energy reflected waves – weak seismic signals (local responses from low-contrast inhomogeneities), which in a thin-layer medium are under the influence of strong interference of primary, as well as the background of multiple waves (usually manifested at a wavelength of $\lambda = 100$–$250$ m, probing the real medium). Often in such a situation, significant complications of the wave pattern are observed due to the fact the acoustic properties of reservoir rocks and impermeable rocks (cap rocks) overlap, creating ambiguity in tracing the phases of the reflecting boundaries. In this case, when processing seismic data, it is easy to lose local responses – contributions of reflected waves from low-contrast heterogeneities under study. When processing seismic data obtained under similar geological conditions, it is necessary to preserve reflections from low-contrast (acoustic properties) inhomogeneities in the form of weak seismic signals against the background of the influence of sufficiently strong reflections.

In connection with the above, there is a need for a special approach in the preparation of the original materials of the field seismic survey to solve the inverse dynamic problem – *special digital processing of seismic survey data by the reflected wave method (CMPM).*

### 3.4.1 Special Data Processing Graph

The main purpose of processing seismic observation materials in the HRS-Geo technology is to obtain non-migrated and migrated records of seismic traces suitable for performing the seismic inversion procedure and further detailed geological interpretation of the processing results. The entire processing process must meet the following basic requirements:

- it must be performed while maintaining the true ratio of amplitudes in a wide frequency range using surface-consistent procedures;
- must provide the necessary signal-to-noise ratio (≥1.0), resolution and traceability of reflecting horizons in the target time interval.

The first distinguishing feature of the applied technological processing scheme from the known standard graphs is the exclusion from the data processing of the procedures of trace-by-trace deconvolution, multi-chanal FX deconvolution by shot-receive distance and corrective filtration (somehow distorting the dynamics of the seismic record, making it not recoverable in relation to the thin-layer acoustic heterogeneities of the real environment).

The second distinctive feature of the technological processing scheme is the inclusion of a seismic inversion procedure in its composition to restore a model of a thin-layer real geological environment and the possibility of solving the inverse dynamic problem of seismics on this basis. Moreover, this procedure is widely used both for original seismograms processing and for final migrated seismic records processing. From a formal standpoint, the seismic inversion procedure is used instead of the deconvolution procedure [9, 10].

Primary seismic materials were directly processed according to a single special processing graph. The geophysical procedures were applied in the following sequence:

- description, application and control of observation geometry;
- editing of seismograms and traces;
- calculation for geometric divergence;

- seismic inversion – inversion of the original seismograms into traces of effective reflection coefficients;
- trace normalization;
- the first automatic editing of static corrections;
- preliminary velocity analysis;
- the second edit of static corrections;
- computing for the low-frequency component of static corrections;
- preliminary stacking velocity analysis;
- third edit of static corrections;
- attenuation of multiple waves;
- final correction of kinematics;
- correction of residual phase shifts;
- CMP seismogram stacking;
- migration of initial seismograms and final time sections;
- processing of stack traces;
- solving the inverse dynamic problem in order to obtain sections and cubes of reflection coefficients and acoustic impedances.

At the final stage of data transformation, pre- and post-migration processing of sections or a cube of seismic records is applied using a number of additional seismic record regularization procedures (spectral whitening, filtering in various domains). The objectives of this stage are to reduce the background of multiple waves, suppress random noise, correct residual phase shifts due to under- or overestimation of static and kinematic corrections at previous stages of the processing graph, etc.

The obtained processing and inversion results in the form of sections and cubes of the wave field, acoustic impedance and reflection coefficients, together with the well logging processing data and additional a priori digital information, are sent to the "interpretation system" block (see above in Sect. 3.3).

The use of such a processing graph ensured, first of all, the preservation of adequate recoverable dynamics of the seismic record (adequate in relation to the subsequent restoration of detailed seismoacoustic inhomogeneities of the real environment – effective reflection coefficients), as well as the saving of identification information in the seismic data about the most important geological indicators for prospecting and industrial exploration.

The main condition for the effective use of seismic data was the selection and correlation on the studied sections of the 3D-cube of seismic data of the reflected wave dynamics variability (i.e., rather fine dynamic features of the seismic record) from the target productive horizons, depending on the subtle features of the studied geological section.

### 3.4.2 Sets of Additional Procedures for Seismic Record Regularization

In addition to the above features of the non-standard approach to processing seismic data, at its final stage (after the final procedures for correcting static and kinematic corrections), pre- and post-migration processing of the 3D seismic data cube was applied using a whole set of additional seismic record regularization procedures. These procedures, on the one hand, somewhat improved the regularity of the seismic record, and on the other hand, did not significantly distort its dynamic features. In particular, the following procedures were chosen: MIGR is migration of seismic records; RADON is suppression of multiples by modeling reflected waves on a set of parabolas and subtracting the noise model from the traces of the original seismograms (the procedure for selecting waves in the τ-p transformation domain [11]); COH is the correction of residual phase shifts for the under- and overestimations of static and kinematic corrections; RNA is attenuation of irregular noise of the seismic record (FX-deconvolution); RHU is a single-channel filter for open channel records with irregular component cut-off on the right and left along the amplitude spectrum; τ-p (Filter-50%) – filtering in the τ-p-region with attenuation of the residual background of irregular noise (including feedback – the percentage (50%) of the original data added to the filtered data in the τ-p-domain), as well as various combinations of these procedures. At the same time, in addition to the cube of standard processing, 8 variants of such processing were sorted out. The obtained results were compared in advance in the form of time cubes of CMP seismic records, as well as cubes of effective acoustic impedance (AI) and reflection coefficients (RC).

In particular, such options (versions) were:

1. standard processing cube (version: std);
2. cube with waveform correction (version: rh);
3. cube: original seismograms + MIGR (version: hrs);
4. cube: RADON+COH + MIGR+*τ-p* (Filter-50%) (version: hrs1);
5. cube: RADON+COH + MIGR+*τ-p* (Filter-50%) + RHU (version: hrs2);
6. cube with RNA: RADON+COH + MIGR+RNA (100) (version: hrs3);
7. cube with RNA: RADON+COH + MIGR+*τ-p* (Filter-50%) + RHU + RNA (100) (version: hrs4);
8. sum of RNA cubes (50%) before and without migration: RADON+COH + RNA (50%) + MIGR (version: hrs5);
9. cube with RNA: RADON+COH + RNA (100) + MIGR+RNA (100) (version: hrs6).

It should be noted the processing before the MIGR procedure is performed along the original seismograms, and after this procedure, for the 3D-cubes of the seismic record. According to the results of such processing, based on the application of the seismic inversion procedure, the corresponding 3D-cubes of AI and RC are ultimately obtained. Thus, each of the versions marked above becomes characterized by three types of 3D cube results: seismic record, effective acoustic impedance (AI) and reflection coefficient (RC).

From the set of the results obtained in the form of 3D-cubes of seismic records, 3D-cubes of AI and 3D-cubes of RC, one of the results closest to the optimal variant was chosen. It was accepted as the main one. According to the chosen version of the final result, the necessary structures are further construced and the prediction of the target geological and geophysical parameters of the explored real environment is implemented in full using high-resolution seismic data.

In particular, from the listed versions of results from their enumeration, that with the following sequential chain of processing procedures was most often chosen: RADON+COH + RNA (100) + MIGR+RNA (100) (version: hrs6). At the same time, the rest of the results obtained from other (not the main) variants of such processing also participated to a greater or lesser extent in the general process of processing and interpreting seismic data as some additional material. The need to use such additional material almost always arises when a very complex structure of the real environment appears in the section for various geological reasons ("muddy" scattering section, non-specular reflecting boundaries, weak seismic signals, etc.).

A special processing graph used at the stage of seismic data pre-processing makes it possible to preserve the dynamics of the seismic record to the maximum extent. In this processing graph, only those procedures were used that provide a "reconstructed" seismic record dynamics that is adequate to the actual distribution of acoustic heterogeneities in the section under study. In its composition, the procedure of seismic inversion based on the original seismograms is widely used. This procedure makes it possible to expand the spectrum of the seismic record to the maximum, significantly increase its resolution, and regularize the dynamic parameters of the record. In this case, the effect of interference is excluded from the traces of the seismic record due to the use of the seismic inversion procedure. This provides the maximum vertical resolution of seismic data and, ultimately, makes it relatively easy to identify the features of the detailed internal structure of the productive parts of the section and their host deposits. In general, this circumstance was very relevant when predicting the most important geological indicators for prospecting and industrial exploration, such as lithological composition, reservoir properties, the nature and degree of fluid saturation of reservoir rocks, based on seismic data.

## 3.5 Summary

The materials of the complex geological interpretation of high-resolution acoustic characteristics (HRS) together with the geological indicators of the real environment extracted from these characteristics imply the presence of a reasonable geological basis. It is quite obvious this foundation should be based on fairly well-developed and largely well-known theoretical foundations for predicting the oil and gas potential of the subsurface. These foundations and the directions of hydrocarbon prospecting and exploration based on them should be focused primarily on the system of geostructural, lithological, stratigraphic principles and criteria, as well as on the regularities of the placement of oil and gas accumulations and on various reasonable criteria for the forming and predicting of the oil and gas potential of the studied subsurface.

1. This section analyzes in brief the well-known criteria for predicting the oil and gas content of productive deposits from the standpoint of the principles of system analysis, used to prognosis a group of criteria with a set of indicators inherent in each of them, which make it possible to solve the target problem with sufficient reliability, provided that the indicators meet certain requirements. This, in turn, implies taking into account the issue of forecasting, the degree of the object study, the nature of the information received, the conditions of necessity and sufficiency, comparative simplicity, economic efficiency and objectively measurable information, scientifically based relationships and features of their changes in space and time.
2. The interpretation specificity of the results obtained using high-resolution seismic (HRS) methods is analyzed. The combinations of the found geological indicators make it possible to use them to study the features of the reservoir formation, the conditions of occurrence, the structure and regularities of the sedimentary complexes alternation, the establishment of oil and gas saturation and signs of oil shows, the ranking of oil-saturated rock according to filtration and capacitance parameters, the clarification of paleogeographic conditions for the sediments accumulation and the systematization of materials for calculating predictive hydrocarbon resources.
3. It is shown on specific examples the selection of weak seismic signals in the original wave field against the background of a high level of noise and reflected waves from strong interfaces is one of the important tasks of the complex interpretation of seismic data.
4. The features of complex data interpretation based on the use of HRS-Geo technology, production well geophysics and borehole drilling are demonstrated. The peculiarity that one has to face in the conditions of detailed interpretation of seismic data using the developed technology, and their use in the process of prospecting and exploration of oil and gas deposits are consistently analyzed. In particular, the features of solving the main seismogeological problems using the procedures of the HRS-Geo technology are briefly considered. Almost each of these tasks is illustrated by numerous examples confirming the effectiveness of their solution. The main directions of complex interpretation of HRS data on the use of HRS-Geo technology are presented by the authors on the pages of the site: *www.hrseismic-tech.com*

One of the main features of the technological scheme for processing seismic data using the HRS-Geo technology, as noted above (in Chaps. 1 and 2), was the development and application of a special (optimal) processing graph and a set of seismic record regularization procedures. Based on the use of such a processing graph and record regularization procedures, the maximum vertical resolution of seismic data is ensured, which, in turn, makes it possible to identify the features of the detailed internal structure of the productive parts of the section and their host deposits.

# References


1. Aksenov, A. A., Zaydel'son M. I., & Vaynbaum, S. Ya. (1986). *Kriterii i metodika prognozirovaniya neftegazonosnosti karbonatnykh otlozheniy* (Criteria and methods for predicting the oil and gas content of carbonate deposits) (p. 136). M., Nedra.
2. Amelin, I. D., Bad'yanov, V. A., Vendel'shteyn, B. Y., Gamzikov, V. K., Gutman, I. S., Kanalin, V. G., Movmyga, G. T., Stasenkov, V. V., Chalovskiy, V. I., & Yaroshenko, A. A. (1989). *Podschet zapasov nefti, gaza, kondensata i soderzhashchikhsya v nikh komponentov* (Calculation of reserves of oil, gas, condensate and components contained in them). Spravochnik. Pod red. Stasenkova V.V., Gutmana I.S. (Handbook: edited by Stasenkova V.V., Gutmana I.S.) (p. 270). M., Nedra.
3. Bakirov, A. A., & Bakirov E. A. (1986). *Sistemnyye issledovaniya pri prognozirovanii neftegazonosnosti nedr* (System research in predicting the oil and gas content of the subsoil) (p. 204). M., Nedra.
4. Bakirov, A. A., Bakirov, E. A., Gabrielyants, G. A., Kerimov, V. Y., Mstislavskaya, L. P. (2012). Teoreticheskiye osnovy poiskov i razvedki nefti i gaza (Theoretical foundations of prospecting and exploration of oil and gas). In *Kniga 1. Teoreticheskiye osnovy prognozirovaniya neftegazonosnosti nedr (podrazdel 9.3)* (Book 1. Theoretical foundations for pridicting the oil and gas potential of the subsoil (subsection 9.3)) (p. 412). M., Nedra.
5. Brodov, L. Y., Gogonenkov, G. N., & Tal'virskiy, D. B.. (1979). Effektivnost' seysmorazvedki pri pryamykh poiskakh zalezhey nefti i gaza (Efficiency of seismic exploration in direct searches for oil and gas deposits). In *Sovershenstvovaniye metodov podgotovki ploshchadey dlya poiskov nefti i gaza*. Vyp. 214 (Improvement of methods for preparing areas for oil and gas exploration. Issue. 214). M., VNIGNI.
6. Zalyayev, N. Z. (1990). *Metodika avtomatizirovannoy interpretatsii geofizicheskikh issledovaniy skvazhin* (Technique of automated interpretation of well logging) (p. 142). Minsk University Edition.
7. Larochkina, I. A., & Bazarevskaya, V. G. (1998). Otsenka perspektiv slaboizuchennykh lokal'no-neftenosnykh gorizontov Romashkinskogo mestorozhdeniya: Opyt razvedki i razrabotki Romashkinskogo i drugikh krupnykh neftyanykh mestorozhdeniy Volgo-Kamskogo regiona (Evaluation of the prospects of poorly studied locally oil-bearing horizons of the Romashkinskoye field: Experience in exploration and development of the Romashkinskoye and other large oil fields of the Volga-Kama region). In *Trudy nauch.-prakt. konfer., posvyashchennoy 50-letiyu otkrytiya devonskoy nefti Romashkinskogo mestorozhdeniya* (Proceedings of scientific-practical. conference dedicated to the 50th anniversary of the discovery of the Devonian oil of the Romashkinskoye field) (pp. 136–140). Novoye Znaniye (New Knowledge).
8. Mushin, I. A., Brodov, L. Yu., Kozlov, Ye. A., & Khat'yanov, F. I. (1990). *Strukturno-formatsionnaya interpretatsiya seysmicheskikh dannykh* (Structural formational interpretation of seismic data) (p. 299). M., Nedra.
9. Trofimov, V. L., Khaziyev, F. F., & Trofimova, A. V. (2019). *Tekhnologiya vysokorazreshayushchey seysmiki pri poiskakh i razvedke neftegazovykh zalezhey* (High-resolution seismic technology in prospecting and exploration of oil and gas deposits) (p. 343). M., EAGE Geomodel LLC.
10. Trofimov, V. L., Khaziev, F. F., & Trofimova, A. V. (2022). *Oil and gas reservoir prospecting and exploration. High-resolution seismic (HRS) techniques and technology*. Springer. https://doi.org/10.1007/978-3-030-84389-2
11. Bel'fer, I. K., Goloshubin, G. M., & Shcherbich, A. V. (1987). Ispol'zovaniye v seysmorazvedke t-p preobrazovaniya. Razvedochnaya geofizika (Use in seismic exploration of $\tau$-p transformations). In *Obzor. inform. / VNII ekonomiki miner. syr'ya i geol.-razved. rabot.* (A review information: VNII economics mineral raw materials and geological exploration works) (p. 54). M., VIEMS.

# Chapter 4
# Wells Data Geophysical Research Interpretation Features

**Abstract** As part of the HRS-Geo technology, an interpretation system was created, which is based on the principles of continuous functional transformations along the well logging data for the study of borehole sections. The results of such transformations are necessary for optimal matching of the obtained physical parameters with the corresponding high-resolution seismic (HRS) ones. The features of the development and improvement of the methodology for processing and interpreting data from standard well logging methods according to the technique of integrating geophysical data by means of their functional transformations into information systems are considered. The results of logging data interpretation were obtained through the wide application of the well logs normalization method and their functional transformations, elimination of distortions in the original log curves, data integration when identifying complex reservoirs, assessing the porosity and density of various lithologies rocks, finding factors affecting geophysical parameters, determining influence bound water on resistivity in assessing the saturation of clay reservoirs and the effect of fluid composition on readings of normalized porosity and elastic wave velocity curves. The physical essence and principles of the normalized diagrams interpretation lie in the fact, with the help of the neutron gamma-ray logging curve, electrical resistivity variations caused by changes in the porosity of rocks along the section are excluded to consideration, and thereby areas of increased rock resistance due to their oil saturation are visually represented.

As noted in Sect. 3.3, a complex determination of the composition and properties of rocks of oil-bearing strata according to well logging data is necessary for the of a detailed geoacoustic model formation [1]. This model in the studied interval of the section, as is known, contains information about the distribution of acoustic velocity, density, lithology, reservoir properties and oil saturation of the reservoir rocks. Such a section model is formed at the stage of combining the results of solving the inverse dynamic seismic problem and well logging data processing due to adequate



V. L. Trofimov et al., *Theory and Practice of High Resolution Seismic HRS-Geo Technology Application*, https://doi.org/10.1007/978-3-031-41590-6_4

"convergence" of high-resolution seismic data and well logging data (the results of their automated interpretation).

At the same time, it should be noted one of the most effective ways to interpret well logging data obtained in wells is the method of functional transformations of geophysical parameters that are integrated into information systems [1]. The algorithms of the geophysical parameters functional transformation system are based on the technique developed by N.Z. Zalyaev and his colleagues [1–5]. As noted by these authors, the expediency of continuous data processing follows from the uninterrupted nature of the sedimentation itself and the structure of geological sections.

In addition to the continuity of processing and analysis of well logging data, this method provides a quantitative interconnection of all geophysical parameters based on their direct relationships, as well as the integrated use of data and the results of solving particular problems. The physical prerequisite for solving such problems is the different sensitivity of geophysical parameters to variations in lithological, reservoir and fluid properties. Using the feature of different effect of well logging methods to the same parameters, it is possible to build normalized measurement curves. When the indications of certain methods are normalized in rocks of one type according to the lithology and the nature of the fluid, a characteristic discrepancy of indications for rocks of another type is observed. The effect of a particular property is revealed by sequential compensation of distorting factors by comparative analysis of normalized diagrams of various methods in certain combinations, after which the result is calibrated to the value of the desired parameter, corrections are introduced and quantified in the appropriate dimension.

When integrating data from a significant number of well logging methods, a whole range of rock characteristics is determined. According to the set of parameters used, located in a certain system, the material composition, porosity, content of bound water and useful capacity are determined, the oil and gas saturation of reservoirs is estimated, a number of very important physical characteristics of geological sections are calculated. There are extensive publications on the method of functional transformations based on the integration of geophysical data [1–4].

For the joint integrated use of borehole data and high-resolution seismic data, an interpretive system Well Processing System (WPS) was created in the HRS-Geo technology, which is based on the principles of continuous functional data transformations for studying well sections. This system has a software-technological block, on the basis of which the physical parameters of the section, extracted from well logging data, are optimally matched with the corresponding high-resolution seismic parameters (obtained using HRS-Geo technology).

The processing and automated interpretation of well logging data and their combined integrated use with high-resolution seismic data (HRS-Geo technology) are described in detail by the authors [40, 41]. In this subsection, the authors briefly touch upon the main features of solving the problems of processing and automated interpretation of well logging data, which were developed and successfully applied directly by Zalyaev N.Z. and his colleagues in the well field geophysics department of BelNIGRI. They were dealing with GBS data obtained in sections with a complex lithological composition. These developments, as is known, served as the basis for the creation of an automated interpretation system – the INGEF system [1, 3, 4, 6].

In addition, we note the numerous studies (by domestic and foreign authors – companies) have been devoted to modern technologies for processing and interpreting well logging data, of which it is worth highlighting the most well-known [7–12]. The results of these studies were also used to a certain extent by the authors [1, 3, 4, 6] in the creation, development and improvement INGEF systems.

The authors [40, 41], together with employees of the Department of Field Geophysics of BelNIGRI, carried out work on the integration of seismic exploration and logging data on a number of materials. In addition, the authors were interested in such works because, in terms of physical, geological and informational meaning, they covered almost the main spectrum of geophysical and geological characteristics of the parameters extracted from well logging data and from the dynamic parameters of high-resolution seismic data.

In this section, we will briefly focus on the main methodological features of the creation and development of methods for processing and interpreting well logging materials, which were performed by the authors [1, 3, 4, 6]. In general, it should be noted the lithology of rocks is a multicomponent system, and geophysical parameters are functions of many variables, depending on many factors. Under these conditions, for the correct and reliable geological identification of the properties of the section, it is very important to consistently perform operations to determine the material composition of rocks, porosity, taking into account the lithological heterogeneity of rocks, including differences in the structure of the pore space, determination of bound water and effective porosity by normalization methods, integration of the results of transformations and assessment of the complex reservoirs saturation nature and, in general, the influence of hydrocarbon formation on geophysical parameters.

In order to effectively address these issues, we will in short focus below on the main features of the development and improvement of the processing methodology and interpretation of the standard well logging methods, which are aimed at significantly speeding up the data processing. At the same time, the main features of the logging data transformation aimed at quantitative and qualitative study of complex lithological composition sections are considered. In particular, the sequential solution of the following tasks and operations is implemented.

## 4.1 Application of the Logging Diagrams Normalization Method

The authors [1, 3–6, 14–16], while improving the methods of integrating geophysical parameters, developing algorithms and programs of the INGEF system, effectively develop and use the "normalization method" of logging diagrams measured in wells. This method is widely used both for measuring acoustic data and for normalizing neutron and electrical logging curves.

For a more complete presentation of the normalization method, we will outline the main logging methods in brief: sonic, neutron and electrical logging, the parameters of which are included in the basic formulas for accounting for measurement

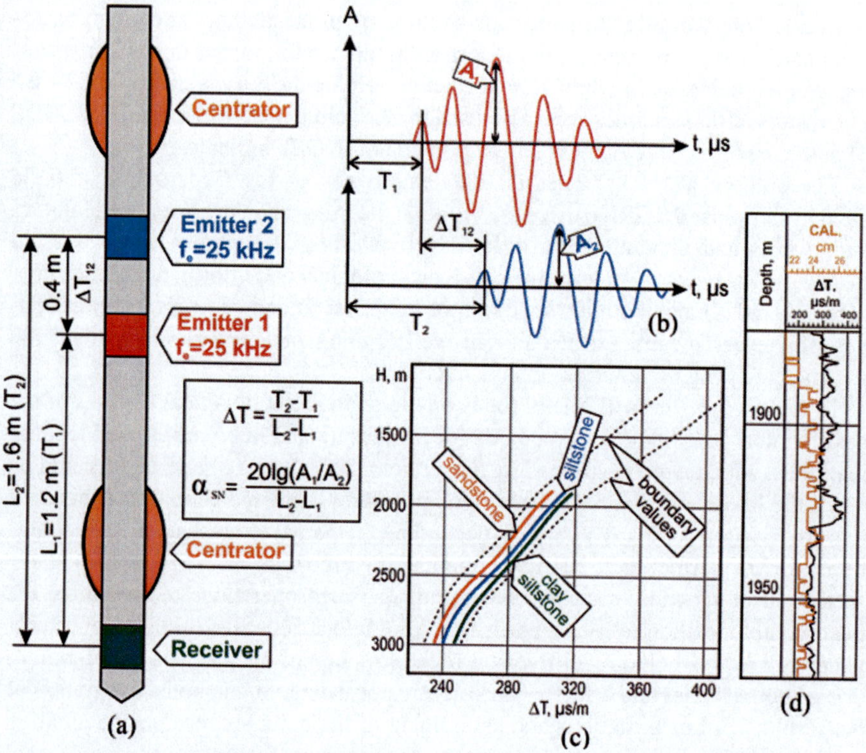


**Fig. 4.1** The method of borehole acoustic logging: (**a**) the scheme of the device SPAK-6; (**b**) the parameters of the sonde signals; (**c**) the dependence of the acoustic waves interval time vs the depth for polymictic reservoirs of Cretaceous age of the West Siberian plain; (**d**) an example of recording of SN and CAL. (According to Refs. [40–42])

corrections and functional transformations of diagrams and using materials from [40–42].

Sonic logging (SN) is based on measuring the kinematic and dynamic parameters of elastic waves that propagate in the borehole rocks. Figure 4.1 schematically shows an acoustic sonde (*a*) with geometric and recorded parameters, pulses with their characteristics (*b*), an example of the average interval time dependence vs depth for polymictic reservoirs of Cretaceous age of the West Siberian Plain (*c*) and an example of record SN measurements (*d*). An acoustic three-element probe consists of two emitters and a receiver. The emitters are located at distances $L_1$ and $L_2$ from the pulse receiver (Fig. 4.1a). When the tool moving along the borehole, the following parameters of the signals emitted by the sources and received by the detector are continuously recorded: $T_1$ is the travel time of the head wave from the emitter located at a distance of $L_1$ (the time of the first break); $T_2$ is the travel time of the head wave from the emitter located at a distance of $L_2$; $A_1$, $A_2$ are the amplitudes of the first wave arrivals from the sources 1 and 2, respectively. According to these parameters, the following kinematic and dynamic characteristics are calculated:

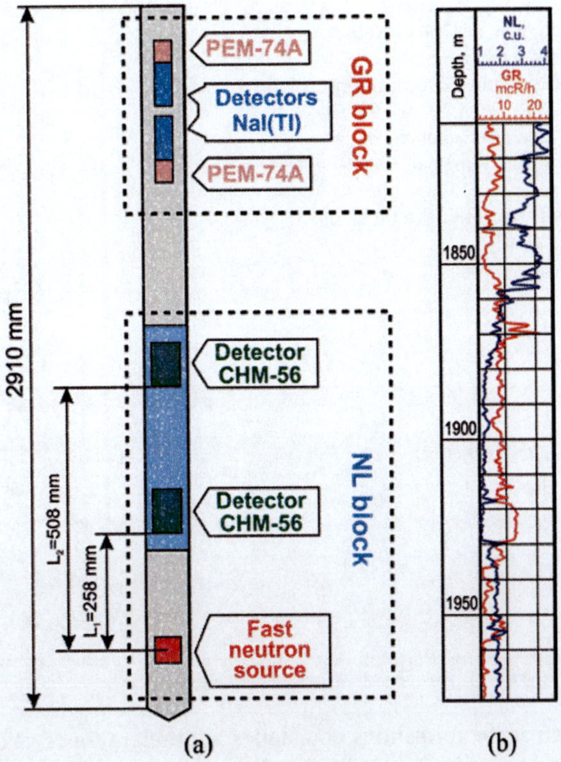


**Fig. 4.2** The neutron and gamma logging method of wells: (**a**) the scheme of device installations of the CRK-01 device for recording the readings of GR and NL; (**b**) an example of the NL and GR measure. (According to Refs. [40–42])

$$\Delta T_{12} = T_2 - T_1, \quad \Delta T = \frac{\Delta T_{12}}{L_2 - L_1}, \quad a_{SN} = \frac{20 \cdot \log\left(A_1 / A_2\right)}{L_2 - L_1} \tag{4.1}$$

where $\Delta T_{12}$ is the difference in travel times from sources 1 and 2 in µs; ΔT is the specific travel time in µs/m; $\alpha_{SN}$ is the apparent absorption coefficient of the longitudinal wave. Figure 4.1b schematically shows the formation and structure of signals. Figure 4.1c gives one example of the kinematic parameters use in the interpretation of the Cretaceous deposits deep structure. On Fig. 4.1d a diagram of the records of SN and CAL are presented.

The presented combined probe in Fig. 4.2 is intended for conducting studies using the method of two-probe neutron-neutron logging on thermal and epithermal neutrons (NLT and NLET), neutron gamma ray logging (NL) and gamma ray logging (GR). The device consists of a neutron source and two detectors located at a distance $L_1 = 258$ mm and $L_2 = 508$ mm from the source (NL block) (Fig. 4.2a). In the upper part of the sonde in the GR unit, detectors with photomultipliers are installed to detect natural electromagnetic gamma radiation. According to the measurements of the described probe, the hydrogen content (water-saturated porosity) and the intensity of natural radiation of rocks are determined. Due to the finite size of the probe (the presence of distances $L_1$ and $L_2$), as in the case of using the SN method, when recording readings, distortions occur when the radiation detectors

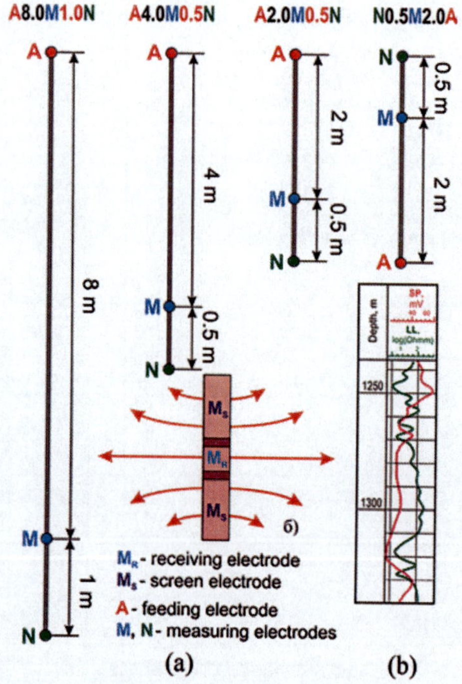


**Fig. 4.3** The lateral logging method of wells: (**a**) the scheme of sonde installations for recording of LL and LLS; (**b**) the scheme of shielding at LL (**c**) an example of recording LL and SP. (According to Refs. [40–42])

cross the formations boundaries at different times, and which should be corrected. There is also a contribution from the natural radiation of rocks in the NL measurements, they also need to be excluded, for example, using data obtained from the GR block (Fig. 4.2a). Figure 4.2b shows an example of registering GR and NL readings.

Figure 4.3 shows an example of the electrical logging devices for the LL and LLS methods. There is also a set of tools with different bases between the feeding and measuring electrodes, and the need to take into consideration the probe length factor also remains (Fig. 4.3a). A feature in the design of the LL probe is the use of shielding electrodes for more efficient measurement of rock resistance (Fig. 4.3b). Figure 4.3c is an graph example of LL records.

A. When evaluating the main geological and geophysical parameters of the section by the normalization method, the definition of this operation is given [11]. In this case, normalization is the combination of two logging curves, one of which is previously converted into units of measurement of the other one using a regression equation so that the studied layers are highlighted by significant discrepancies of the combined curves. Normalization is effective if the following conditions are met: (a) the normalized curves should depend on the same lithological-petrophysical parameter (property A); (b) another property B that needs to be identified as a result of normalization should affect both curves in different ways. Then it is possible to construct a regression equation of the relationship between the curves over the layers with a zero value of property B. This will ensure the curves diverge after normalization in the section where the values of the property B are different from zero.

Normalization of the following logging curves is applied [11]:

(a) to isolate reservoir layers – GR and any porosity logging (NL, SN, GGL); here property A is the clay content, B is the porosity, the regression equation is based on dense layers and clays;
(b) to highlight oil and gas–bearing (or fractured) intervals – LL (or IL) and SN (or NL), while property A is granular porosity, property B is oil and gas saturation (or fracturing); the regression equation is based on dense layers, clays and aquifer granular reservoirs;
(c) to distinguish intervals with different lithological characteristics (or gas–bearing intervals) – NL, GGL, etc.; here A is the porosity of layers of known lithology (for example, limestones), B is a different lithological composition of the rock matrix (for example, anhydrites, dolomites, salt); the regression equation is based on layers with known lithology.

The specified operation for normalization of logging curves is used by the authors [1, 3–6, 14–16, etc.] quite widely in the complex interpretation of geophysical parameters by the method of functional transformations.

B. When using sonic logging, due to the non-simultaneous crossing of the well diameter change boundaries by the logged sonde, the measurement components differ and z-shaped distortions occur [1]. To eliminate z-distortions to the already registered diagram $\Delta T$, according to $T_1$ and $T_2$ data, a calculated diagram $\Delta T_{calc}$ corresponding to the tool with the reverse arrangement of the device elements is additionally compiled. To eliminate distortions, the authors [1] use a universal method, the essence of which is to take into account the borehole component by normalizing the $T_1$ and $T_2$ readings (Fig. 4.4). When dividing the values of $T_1$ and $T_2$ by the corresponding lengths of the probes $L_1$ and $L_2$, the obtained values of the components from the rocks are aligned:

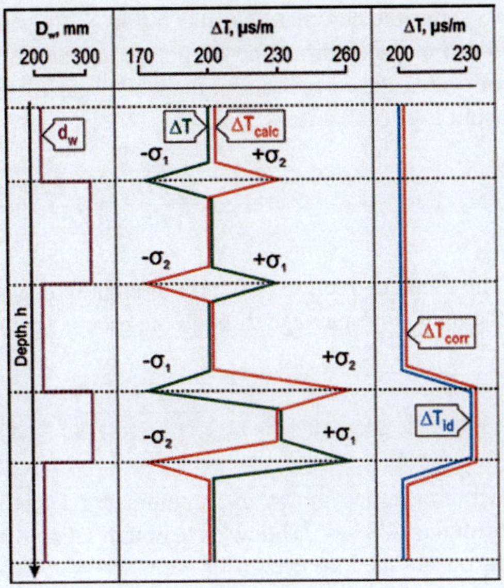


**Fig. 4.4** Z-distortion compensation scheme. $D_w$ is the diameter of the well; ΔT is the interval time. (According to Ref. [1])

$$T_1 / L_1 = T_r / L_1 + T_w = T_r + T_w / L_1 \tag{4.2}$$

$$T_2 / L_2 = T_r / L_2 + T_w = T_r + T_w / L_2 \tag{4.3}$$

where the readings $T_1$ and $T_2$ are expressed as:

$$T_1 = T_w + T_r \cdot L_1, \tag{4.4}$$

$$T_2 = T_w + T_r \cdot L_2 \tag{4.5}$$

$T_w$ is the time acoustic passage waves through the borehole medium with a constant diameter of the well at both measurements. It is almost the same, but the time of passage through the rocks differs, which depends on the distance between the emitter and the receiver in both $L_1$ and $L_2$ probes (with the same time interval after 1 m of the $T_r$ rock). The readings are expressed in terms of the passage times of the borehole medium $T_w$, rock $T_r$ and distances $L_1$ and $L_2$ according to formulas (4.4) and (4.5).

The difference between these transformations of (4.2) and (4.3) parameters will mostly reflect the influence of the borehole component, since the influence of the velocity characteristics of the rocks is mutually compensated. In order to eliminate borderline discrepancies, curves $T_1$ and $T_2$ are normalized by the differentiation of the section with special filters that ensure the identity of these curves in terms of vertical resolution. As shown in [1], measurements $T_1$ and $T_2$ may have inaccuracies due to incorrect fixation of the first break and skipping cycles. Cycle gaps are established by comparative analysis of normalized $T_{1n}$ and $T_{2n}$ curves. These curves are combined in places of normal recording, therefore, cycle skips are marked by a "rebound" of the distorted curve in the direction of an increase in time of at least 20 microseconds. Such distorted readings on one of the curves are replaced by a normal record on the other.

After the complete normalization of the curves for sensitivity to rocks and the detail of the stratification of the section, as well as taking into account the skipping of cycles, the correction for the well is calculated according to the formula obtained from Eqs. (4.2) and (4.3):

$$T_w = \frac{T_2 / L_2 - T_1 / L_1}{1 / L_2 - 1 / L_1} = \frac{T_{2n} - T_{1n}}{1 / L_2 - 1 / L_1}. \tag{4.6}$$

The found correction (4.6) is entered into the values $T_1$ (4.2) and the integral time of the only wave passage through the rock is found:

$$T_r = \frac{T_1 - T_w}{L_1} = \frac{1}{L_1}\left(T_1 - \frac{T_{2n} - T_{1n}}{1 / L_2 - 1 / L_1}\right). \tag{4.7}$$

Experience has shown the recommended methods of correcting distortions using formulas (4.2)–(4.7) allow us to obtain reliable results on SN, in complex sand-clay-carbonate sections even (Fig. 4.5).

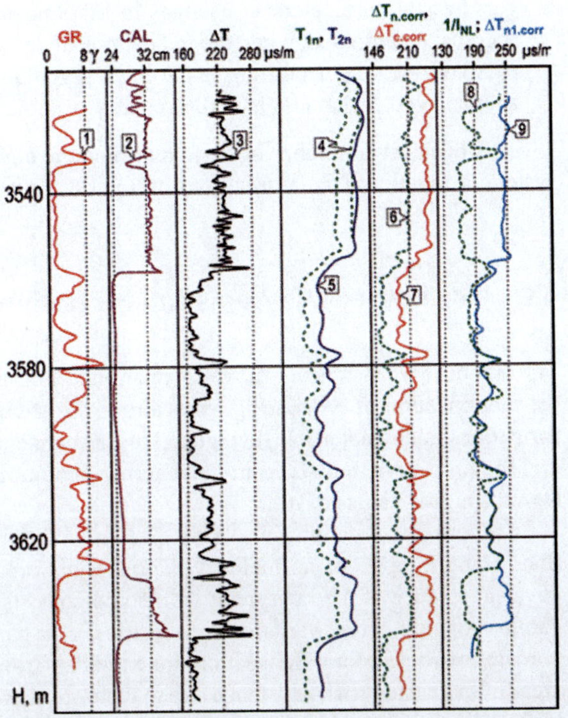


**Fig. 4.5** Comparison of corrected curves ΔT with these of other geophysical measurements: 1-GR; 2-CAL; 3-ΔT; 4-$T_1$; 5-$T_2$; 6-$\Delta T_{n.corr}$; 7-$\Delta T_{c.corr}$; 8–1/$I_{NL}$; 9-$\Delta T_{n1.corr}$. (According to Ref. [1])

Along with SN diagram, the gamma-gamma logging (GGL) method is also used. In the diagrams of the GGL, due to the non-simultaneous intersection of the interlayers with large and small sondes, z-shaped distortions occur in the GGL device (as in sonic logging). Depending on the properties of the media against the large and small tools, either overstated or underestimated GGL readings are obtained compared to the active property of the reservoir under study. Practice has shown distortions in the GGL diagrams are more common than in the SN diagrams, since they arise not only from cavities, but also from the heterogeneity of the rocks themselves. Since the distortions of GGL and SN are of the same nature, they can be eliminated, in principle, by the same techniques as for SN. At the same time, the most accurate method is that of compensation for z–distortions, which is based on summation of the registered GGL – $F_8$ diagram with the calculated curve for the case of a reverse probe according to the readings of the large and small GGL sondes [1].

C. In accordance with [1, 3–6, 13–16], when extracting information about one particular property of rocks, many unknowns have to be encountered. To ensure multifactorial analysis of geophysical data and mutual correction of the results, the well logging data processing methodology is based on the integration of various geophysical parameters into a single information system. The consistency of the parameters, as noted earlier, can be achieved by functional transformations of logging diagrams based on their direct relationships, which makes it possible to normalize the scale of readings of various methods. The physical prerequisite for solving the problems is the different sensitivity of various geophysical parameters to changes in lithological and reservoir properties. In accordance with this, when the indications of certain methods are normalized in rocks of one type according to the lithology and the nature of the rock fluid of another type, a characteristic discrepancy of indications is observed.

Let's consider a number of basic and fundamental features implemented in the system of functional transformations of logging diagrams.

## 4.2 Normalized Diagrams Interpretation Features

The authors [17] also noted the interpretation of normalized diagrams is reduced to the identification of oil-bearing reservoirs in the studied section. For oil–saturated formations, a characteristic divergence of normalized curves is observed – the lateral logging curve departs from the neutron gamma-ray logging curve to the right, towards higher values of $\rho_a$.

In accordance with this, it is worth dwelling a little more on the physical foundations and principles of normalized diagrams interpretation. At the same time, the physical essence of the interpretation method consists in the fact using the neutron gamma-logging curve, variations in electrical resistance caused by changes in the porosity of rocks along the section are excluded from consideration, and thereby areas of increasing rock resistance due to their oil saturation are clearly represented. If only water-saturated rocks are present in the studied section interval, the readings of both lateral and neutron gamma logging will chiefly depend on their porosity and the normalized curves of these methods approximately coincide with each other. Since when combining the normalized lateral and neutron gamma-logging curves in the aquifers, some equivalence is achieved, the neutron gamma-logging curve can be considered as a resistance curve characterizing the change in the electrical resistance section at their full water saturation, because oil content practically does not affect the indications of this method.

At the same time, the appearance of oil in the rocks significantly increases their electrical resistance, which is reflected in the lateral logging readings accordingly. The presence of oil in the rocks leads to a noticeable divergence of the normalized curves, while the magnitude of the divergence reflects the degree of oil saturation of the rocks and, in its physical essence, corresponds to the resistance increase coefficient or the saturation parameter $P_s$. Indeed, in this case, the lateral logging curve characterizes the resistivity of an oil-saturated reservoir ($\rho_r$), and the neutron gamma-ray curve taken on a resistivity scale (resistivity of the same reservoir), but at 100% water saturation ($\rho_{wr}$). The difference between these values on a logarithmic scale corresponds to their ratio on a linear scale, and the ratio $\rho_r/\rho_{wr}$ is $P_s$.

However, according to the normalized diagrams, only some fictitious value of the parameter $P_s$ (i.e. $P_{sf}$) is obtained, which does not always coincide with the true one. Coincidence of $P_{sf}$ with $P_s$ could be expected if there was a functional dependence between $I_{NL}$ and $lg\ \rho_r$. In fact, the relationship between these parameters is only correlative. In addition, to build normalized diagrams, the observed curves of lateral and neutron gamma ray logging are used without making corrections for the influence of various factors, which can also be the cause of a possible discrepancy between $P_{sf}$ and $P_s$.

The $P_{sf}$ value, like $P_s$, can be used to assess the commercial significance of the identified oil-saturated intervals of the section. Thus, the experience of interpreting normalized diagrams, for example, under the conditions of the Pripyat depression, shows at $P_{sf}$ values below 3, no commercial oil inflows are obtained. In this regard, $P_{sf} = 3$ can be taken as the critical (limit) value of this parameter, above which, in most cases, one can expect commercial oil inflows from the studied reservoirs.

The value $P_{sf} > 3$ is a necessary, but still insufficient criterion for identifying oil-bearing reservoirs in a section and assessing their commercial significance. In certain cases, the discrepancy between the normalized lateral and neutron logging curves may exceed the indicated $P_{sf}$ value and even significantly, but the rocks may not contain commercial oil or turn out to be non-oil-bearing.

The presence of crystallization and hygroscopic water in gypsum and clays leads to a false overestimation of the porosity of clayey and gypsum-bearing varieties of carbonate rocks when assessed according to the NL data. For these rocks, the same divergences of the normalized curves can be observed, as in the case of oil-bearing rocks. If gamma-ray logging data is successfully used to estimate clay content, then it is practically impossible to identify areas of gypsum content of rocks from logging data. However, according to the available materials, it should be noted it is extremely rare for rocks taken for oil-bearing to be gypsum-bearing. It is also obvious gypsum rocks are not widely distributed.

Changes in the structure of the pore space along the section, the development of cavernosity and fracturing in carbonate reservoirs significantly affect the normalizability of the lateral and neutron gamma-ray logging curves and may cause their divergence. This is due to the fact at the same values of the total porosity of water-saturated rocks with different structures differ in resistivity. Thus, fracturing leads to an underestimation of the $\rho_{wr}$ in comparison with these in granular reservoirs, cavernosity leads to an overestimation of the $\rho_{wr}$.

The influence of the pore space structure to the resistivity of rocks can be assessed using a graph constructed from calculated data using the formula [18]:

$$P_p = \frac{\rho_{wr}}{\rho_w} = \frac{3\cdot(1-k_{cav})}{(1+2\cdot k_{cav})\cdot(2\cdot k_{fr}+3\cdot k_{gr})}, \tag{4.8}$$

where $k_{gr}$, $k_{cav}$, $k_{fr}$ are the coefficients of granular, cavernous and fractured porosity, respectively. Formula (4.8) is obtained under the assumption the cracks (fractures) are oriented along planes in three mutually perpendicular directions, the cavities are in the form of random spherical inclusions of various sizes, and the blocks between them are represented by rocks with granular porosity, obeying the dependence:

$$P_p = \frac{1}{k_{gr}^2}. \tag{4.9}$$

According to these data, resistivity dependence curves on the total porosity are obtained at different ratios of $k_{gr}$, $k_{cav}$ and $k_{fr}$.

## 4.3 Integration of Well Logging Data in the Identification of Complex Reservoirs

As shown in [19, 20], with the combined use of these resistivity and neutron (NL) methods, it is possible to more confidently determine the porosity and oil saturation coefficients of carbonate reservoirs than using individual methods. This possibility is also preserved in those cases when the determination of the specified parameters from the materials of separate methods is difficult or even impossible.

The essence of such a complex method lies in the calibration of NL readings on the scale of porosity according to electrical logging data. In obviously water-saturated intervals of the section with an assumed close relationship between the electrical resistance and porosity of rocks in terms of resistivity, using its known dependence on porosity [8], the values of rock porosity and, thereby, the scale of porosity, which can later be extended to the NL curve, are established. Using this scale, it is possible to determine the porosity of rocks in the oil-bearing intervals of the section, where the resistivity method has the limitations noted above.

Based on this principle, when determining the porosity of carbonate rocks, as noted above, normalized electrical and neutron gamma-ray logs are used. When the normalized curves are combined in the water saturated part of the section, a certain equivalence of the curves is achieved. As a result, the porosity scale determined from the resistance curve can be extended to the NL curve, which can be used to estimate the porosity of carbonate rocks, except for clay and gypsum areas, when they are saturated with both formation water and oil.

With the same mineralization of the drilling mud and formation waters, according to the lateral logging data, it is possible to simply and fairly exactly estimate the resistivity of water-bearing carbonate formations, which is of no small importance for the correct determination of their porosity. When estimating porosity using a complex of electric and neutron gamma-ray logging, it is of great importance to identify intervals where it is assumed there is a quite close relationship between the porosity coefficient $k_p$ and the porosity parameter $P_p$ and knowledge of the nature of this relationship. One of the main advantages of using normalized diagrams lies in the fact that it is easier to identify such intervals: the presence and tightness of the relation can be evaluated by the degree of normalizability (coincidence of the shape and amplitudes) of the resistance curves and NL. The good normalizability of the curves obviously indicates, first, the presence of a close relationship and, second, indirectly, the predominance of granular rock varieties in the section, since with a sharp predominance of caverns or fractures, one would expect a violation of the relationship between $P_p$ and $k_p$, and, consequently, and between $\rho_r$ and $I_{NL}$.

In this regard, with good normalizability of the lateral and neutron gamma-ray logs, it may be possible to use the dependence $P_p = 1/k_p^2$ most characteristic for carbonate reservoirs with granular porosity to establish the porosity scale [38]. In this case, it is advisable to use the refined dependence established as a result of cores measurements from the studied intervals of the section, but it is not always available and often absent.

Graduation of neutron gamma-ray logging diagrams on the scale of porosity is carried out specifically for each well by the resistivity method. In this case, for a large number of points, the probability of errors due to inaccurate measurements of neutron parameters is largely eliminated. As a result, the reliability of the porosity estimate is improved. The advantages also include the simplicity of data processing, the possibility of determining porosity in the entire studied interval, and the clarity of the results.

In the presence of granular reservoirs in the carbonate section, the errors due to inaccurate estimation of $\rho_r$ and the inconsistency of the used dependence $\rho_r = f(k_o)$ for this type of rocks, obviously, will not exceed those values that are possible when estimating $k_p$ according to the porosity parameter $P_p$ in reservoirs of this type (according to the available data, the average errors in this case are 20–25%). Since the calibration of the porosity-scale neutron gamma-ray curve is based on the actual materials obtained in each specific well, the influence of well conditions on the porosity results appears to be minimized.

When establishing the scale of porosity, the dependence $P_p = \rho_{wr}/\rho_w = 1/k_p^2$ was used. Formation water resistance $\rho_w$ under well conditions is assumed to be 0.03 Ohm•m. The resistivity of water saturated rocks was determined from the lateral logging curve, the scale of which was refined according to the LLS.

The discrepancy of the normalized curves of the lateral and neutron gamma-ray logs against the water-bearing part of the section, represented by non-clay carbonate rocks, generally does not exceed 25% on the scale of porosity. Therefore, such an error can also be allowed when determining $k_p$ due to a possible change in the structure of the pore space, rock lithology and well conditions, if the dependence $P_p = f(k_p)$ in the reference intervals of the section is close to the accepted one.

To determine the oil saturation factor, the value of the saturation parameter $P_s$ is usually used, which is the ratio of the resistance of oil-bearing rocks $\rho_r$ to their resistance at full saturation with formation water $\rho_{wr}$. In the conditions of a carbonate section, when drilling wells with saline mud, in most cases it is possible to confidently estimate the resistance of oil-saturated formations from lateral logging data, but using direct measurements in wells, it is impossible to obtain data on the resistance of these rocks at their full water saturation. Determining $\rho_{wr}$ of oil-bearing formations by calculation is not always feasible due to the lack of reliable data on $k_p$. In this regard, to estimate the oil saturation factor, it is recommended to use the value of the fictitious saturation parameter $P_{sf}$, obtained from the normalized resistivity and neutron gamma log (NL) diagrams.

Since, as mentioned above, with good normalizability of the resistivity and NL curves in the aquifers of the section, a certain equivalence of these curves is achieved, it is legitimate, together with the porosity scale, to extend the NL curve and the resistivity scale of the electrical logging charts. With this assumption, the NL curve, taken on a resistivity scale, in the oil-bearing part of the section characterizes the resistivity of rocks at their full water saturation, i.e. $\rho_{wr}$ of oil-bearing reservoirs, and the ratio of the readings of the resistivity curves and the oil-bearing reservoirs and NL in these layers, obtained from the normalized logarithmic resistivity scale, is the fictitious saturation parameter $P_{sf}$. The degree of conformity of $P_{sf}$ with the true value of $P_s$ depends on the accuracy of determining the resistivity of an oil-bearing formation by the method of electric logging and on how correctly the NL curve reflects $\rho_{wr}$ of this formation. In the absence of deep penetration of the filtrate along the lateral logging diagram, it is possible to determine $\rho_r$ of oil-saturated carbonate formations with the required accuracy. In doubtful cases, the obtained values of $\rho_r$ can be refined according to the LLS data.

A change in the structure of the pore space has a relatively small effect on the resistance of oil-bearing reservoirs, since the electrical conductivity of such reservoirs is mainly determined by the presence of formation water contained in small pores, which have approximately the same structure in reservoirs of any type. Large pores, cavities and cracks are filled with oil. Therefore, with the same volumetric water content of oil-bearing carbonate reservoirs, they should have approximately the same resistance values, regardless of the nature of the structure of large pores filled with oil.

The resistance of the reservoir at its full water saturation of the $\rho_{wr}$, obtained from the NL curve taken at the resistance scale, will correspond to the true value of this parameter only in the case of the development of predominantly granular porosity in carbonate reservoirs. But the method of normalizing the diagram just provides for the achievement of equivalence of resistance curves and NL in aquifers of this type. Therefore, although in the case of a cavernous or fractured oil reservoir, the values of the $\rho_{wr}$ obtained according to the specified NL will be correspondingly lower or higher than the actual values of the $\rho_{wr}$. However, they will be close to the resistance of water saturated granular reservoirs with the same values of total porosity as that of the reservoir under study. This is explained by the fact the NL curve does not reflect a change in the structure of the reservoir pore space, but only the total porosity.

Thus, according to the normalized curves of lateral and neutron gamma logging, the values of $\rho_r$ and $\rho_{wr}$ for oil-bearing rocks are obtained, the value of which does not depend much on the nature of the structure of the pore space. In accordance with this, it can be assumed the values of the $P_{sf}$ parameter determined by these diagrams will be more closely related to the amount of oil saturation of rocks ($k_o$) than the actual $P_s$ values. This conclusion is of great practical importance, because it becomes possible to assess the $k_p$ of oil-bearing carbonate reservoirs with a complex structure of the pore space, which until recently was considered an unsolved problem.

In accordance with the above, the $P_{sf}$ is likely to correspond to the $P_s$ parameter for granular reservoirs with the same oil saturation as the studied reservoir with any type of porosity (cavernous, fractured, mixed). On this basis, during the transition from $P_{sf}$ to $k_o$ with a known approximation, the $P_s = 1/(1-k_o)^2$ dependence characteristic of carbonate granular reservoirs can be used. At the same time, of course, one should take into account the possible degree of approximation of the obtained values of $k_o$ to their true values. Therefore, in the intervals where the normalized diagrams best match, it is desirable to clarify the nature of this dependence by measurements on cores.

The reliability of the $k_o$ assessment depends on the accuracy of the determination of the $P_{sf}$. Therefore, for these purposes, the diagrams should be normalized especially carefully and it is desirable to combine them along clearly granular reservoirs or along sections where they best repeat each other, and these sections should be sufficiently long (at least 20–30 m).

The physical preconditions and the obtaining of experimental data convince the use of $P_{sf}$ instead of $P_s$ to assess the $k_s$ of carbonate reservoirs is not worse, and in many cases (fracturing, cavernosity) may be even better and will give more accurate values of $k_o$ [21]. If we take into consideration the value of the $P_{sf}$ parameter is fairly accurate and easy to obtain using normalized lateral and neutron gamma logging diagrams, and the $P_s$ of carbonate reservoirs in many cases cannot be determined, this method is of great practical importance.

Normalization of the LL and NL porosity diagrams consists in rearranging the LL curve in the logarithmic scale of resistance, which is set in accordance with the deviations on the NL curve, and in combining these diagrams in water-saturated reference formations. When there are no reference layers, the “geophysical parameter vs porosity” technique is used [21].

The accuracy of determining $P_{sf}$ can be judged based on the degree of formalizability of the lateral and neutron gamma-ray logging curves in the aquifer part of the section. The available factual materials indicate the discrepancies of the normalized curves in the resistance scale rarely exceed 30–40% (on average 10–15%). Corresponding errors can also be expected when estimating $P_{sf}$. They turn out to be quite acceptable [8, 20, 21].

Thus, the use of porosity-normalized resistance and NL diagrams makes it possible to more confidently assess the oil saturation of carbonate reservoirs than by other methods. Moreover, this possibility persists when the determination of $k_o$ from the materials of individual logging methods is difficult or even impossible, for example, in the case of a complex structure of the pore space of rocks (cavernosity, fracturing).

## 4.4 Porosity Estimation Using the Normalization Method

In accordance with [22], it is noted the neutron gamma-ray logging (NL) method is most widely used to assess porosity in oilfield geophysics. However, according to NL alone, reliable results can be obtained only in pure carbonate deposits and porosity for clay rocks is not accurately determined. The NL readings reflect porosity through the total hydrogen content, and clays contain a significant volume of bound water (up to 40%) and hydrogen, respectively. This water has the same effect on the NL readings as any pore liquid. Therefore, there is a need to account for bound water. For this purpose, it is possible to use, as is known [23], GR data. In this case, the exclusion of the influence of bound water is carried out according to the formula:

$$K_p = K_{p.NL} - C_{cl} \cdot W_{bn} \tag{4.10}$$

where $K_{p.NL}$ is the porosity obtained from NL, $C_{cl}$ is the volume content of clay material, determined by the GR diagrams, $W_{bn}$ is the content of bound water per unit volume of clay material (determined by laboratory method).

This method (4.10), despite its fundamental correctness, does not find wide practical application. The main reason is that the low accuracy of the definition of the $C_{cl}$ and the establishment of the $W_{bn}$ is associated with great technical difficulties. All this affects the final results and discredits the principle of attracting GR data for the purpose of porosity refinement.

In [22], a method is proposed for taking into account the influence of bound water according to GR data, but only using the GR readings directly, bypassing the transition from them to clay and without the $W_{bn}$ data. This is done by comparing the GR curve with the NL curve after normalizing them in sensitivity to the content of bound water $K_{p.bn}$.

The physical meaning of such normalization is in rocks which contain only bound water (impermeable clay and dense rocks), the readings of both GR and NL depend in general on the volume of this water. Therefore, by the appropriate functional transformation of the curves by the specified method, it is possible to achieve their coincidence. For the conversion, such a function is selected so that a linear dependence is obtained between the readings of the GR and NL.

The readings of NL linearly depend on the logarithm of porosity, If you normalize the curve of GR with NL, then it must be transformed so as to obtain a logarithmic scale for $K_{p.bn}$. It is known there is a linear relationship between the readings of GR ($J_{GR}$) and clay content ($C_{cl}$) [24]. It is obvious the dependence of $J_{GR}$ with $K_{p.bn}$ will be close to linear. It follows to normalize the GR and NL, it is enough to transform the GR curve from a linear to a logarithmic scale. In practice, the logarithmic scale module is chosen in such a way as to obtain the same deviation amplitudes of both curves in impermeable clay and dense rocks.

To interpret the transformation, the GR is inverted and superimposed on the NL curve, while achieving a combination of curves in dense and clay rocks. This is a graphical subtraction from the total porosity, characterized by the readings of NL, the volume content of bound water $K_{p.bn}$. The normalized GR curve is like a conditional zero line, and the departure of the NL curve from it towards high $K_p$ values means the presence of a container filled with a mobile liquid, i.e. effective porosity. The amount of porosity can also be quantified. It will be equal to the difference in porosity values taken respectively from the GR diagrams expressed in the same $K_p$ scale. On NL diagrams, it is usually available or it can be installed.

The paper [22] provides an example of determining the effective porosity of one of the wells (subsalt deposits, interval 3250–3265 m, 7 Ostashkovich well, Pripyat trough) (Fig. 4.6). Here it is 3%, while the total porosity for NL is 5%. At the same time, the effective porosity (3%) is occupied by a mobile fluid, in this case oil. This determination is confirmed by comparing the results obtained with the oil saturation of rocks determined by the normalized charts of NL and GR. Oil saturation, defined as the difference between the total porosity of NL and the volume of bound water by LL, coincides with the value of effective porosity found by NL-LL. At the same

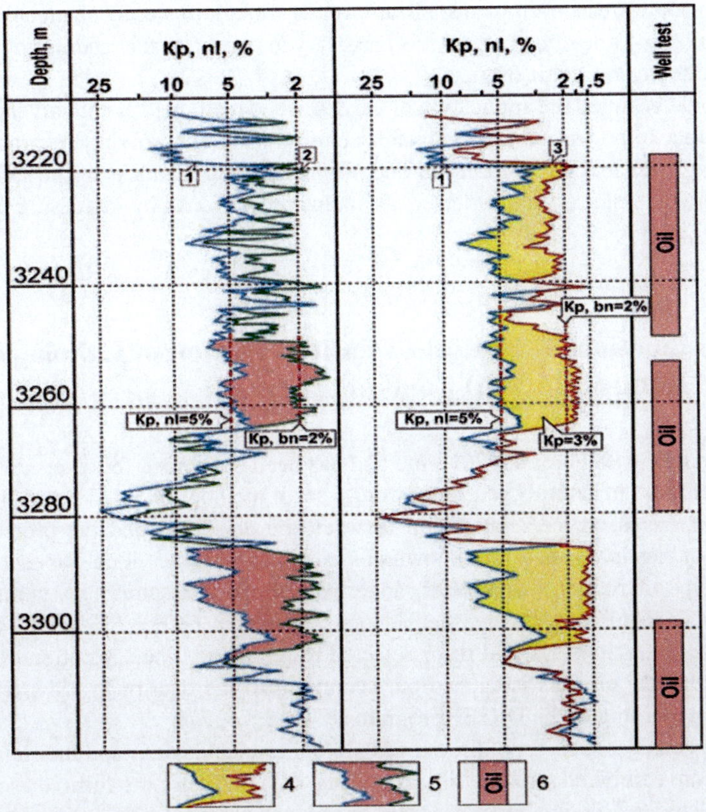


**Fig. 4.6** An example of evaluating effective porosity by normalizing GR and NL diagrams: 1-NL curve; 2-LL curve; 3-GR curve; 4-intervals containing effective porosity; 5-oil-saturated intervals; 6-test result. (According to Ref. [22])

time, it can be seen the porosity determined by the complex of normalized diagrams of NL and GR is close to effective.

Cases of good coincidence of similar results are not a single. They are especially characteristic of high clayey carbonate rocks and sandstones. Such conditions should be considered favorable for the successful application of the method. However, it is not always possible to obtain effective porosity, for example, in chalk-like limestones, where the amount of bound water is controlled not only by the content of clays, but also by the fine pores of the bedrock.

The limitations of the method are due to the fact that it does not allow for the consideration of bound water in gypsum rocks. There may be difficulties in using it in the absence of reference clay or dense layers. But this difficulty is surmountable. In such cases, the curves can be normalized based on the data of calibration and standardization of the chart record scales.

When using the described method, you should first study the nature of the relationship between the indications of GR and NL in each specific region, as they may differ. In some areas, during normalization, it is enough to simply change the scale of the recording linearly, in others it is necessary to logarithm the readings of the NL or write the inverse value of the $J_{GR}$.

It should be noted the application of the described method [22] not only increases the accuracy of porosity estimation, but also proves useful in solving the problem of separating formations into reservoir and non-reservoir, regardless of the saturation of rocks with water or oil, to detect salt inclusions in rocks by increasing the values of NL.

## 4.5 Estimation of the Rocks Density of Different Lithologies in Natural (In Situ) Conditions

In accordance with [1–3, 25, 26] a method has been developed for determining the density of rocks in natural occurrence using a set of methods of SN, GR and NL. The method is based on the relationship between the density σ and the propagation velocity of elastic waves ν. It is known this relationship is individual for each lithological type of rocks. Nevertheless, adhering to certain assumptions, taking into account the required accuracy, these dependencies can be combined into several generalized functions ΔT and σ for selected rock groups. The separation of rocks into lithological clusters is carried out according to the data of the SN, NL, GR complex according to the INGEF programs [2–4, 6, 25, 27].

In the considered method, limestones, dolomites, sandstones and their clay differences are combined into the first group (Fig. 4.7). In their pure form, these rocks are characterized by an almost general dependence provides a transition from one parameter to another with an accuracy of 0.03 g/cm³ according to the formula [5, 26]:

$$\sigma = 3.425 - 4.31 \cdot \Delta T \tag{4.11}$$

This line turns at a certain angle with the vertex at the matrix values of the parameters, depending on the coefficient of bound water saturation, during the clay formation of rocks. For $k_{bn} = 1$, it is described by the expression:

$$\sigma = 3.150 - 2.5 \cdot \Delta T \tag{4.12}$$

With this in mind, it is possible to write a general formula for calculating the density of both pure and clay sedimentary rocks:

$$\sigma = 2.75 - \left(4.3 - 1.8\,k_{bn}\right) \cdot \left(\Delta T - 160\right) \tag{4.13}$$

The calculation of the density based on the averaged dependence ensures the accuracy of its determination of about 0.06 g/cm³. To find the $k_{bn}$, data from the GR or SP method are used.

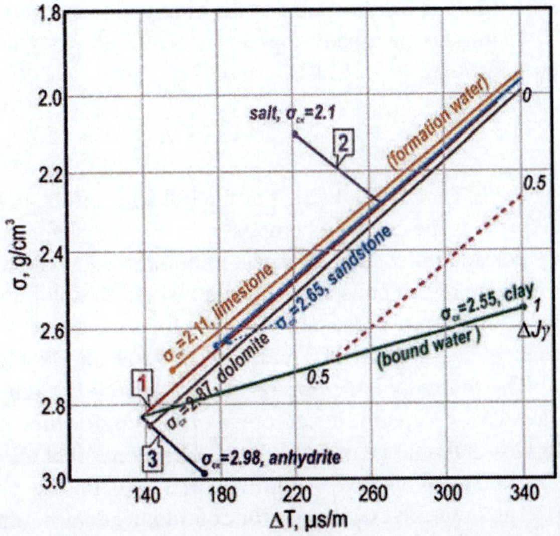


**Fig. 4.7** Chart for determining the rock density of different composition according to sonic log data. (According to Ref. [26])

The second group of rocks consists of salts and their clay differences, the density of which:

$$\sigma = 1.275 - 3.75 \cdot \Delta T \tag{4.14}$$

The third group of rocks are anhydrites, for which:

$$\sigma = 1.940 - 6.0 \cdot \Delta T \tag{4.15}$$

Thus, an important point in determining the density by the acoustic properties of rocks is the division of rocks into groups by material composition. This is carried out by the SN and NL complex. In order to reveal information about the lithology and to suppress the interfering influence of other factors, such as porosity, clay content, technical condition of the well, the method of functional transformations is used. According to this method, some functions of geophysical parameters are compared in pairs, selected in such a way as to bring them to a comparable scale for one or more excluded components. To determine the material composition, the values of $\Delta T$ and $T_{NL} = c \cdot \lg(I_{NL} - I_{NL0})$ are compared, where $I_{NL0}$ is the value level corresponding to 30–35% porosity. Such a transformation corresponds to the nature of NL and requires a minimum of additional information. It allows you to work with an arbitrary scale without reference data and without reference horizons. Experimentally, it was found in rocks homogeneous in material composition, there is a close to linear relationship between these two parameters. The angular coefficients of the relation lines in different rocks are close. Taking into account the influence of the change in the diameter of the well, the NL conversion formula has the form:

$$T_{NL} = 90 \cdot \lg\left(I_{NL} - I_{NL0}\right) + 1.26 \cdot \Delta d_w, \tag{4.16}$$

where $\Delta d_w$ is the deviation of the bore diameter from the nominal one.

Using this parameter together with $\Delta T$ allows you to determine the lithological composition:

$$LIT + A = \Delta T + \Delta T_{NL}, \tag{4.17}$$

where LIT is the total effect of the influence of the material composition on $\Delta T$ and $\Delta T_{NL}$; A is the constant component.

For dolomites, the LIT has a minimum value taken as 0 (thereby eliminating the influence of the constant A). With respect to dolomites, the LIT has the following values: for pure limestones 30–35, for sandstones – 60, for anhydrites – 50, for salts – 90–100 μs/m. For mixtures, LIT has intermediate values.

The lithology is determined by LIT and $\Delta T$. Carbonate rocks are considered to have LIT <35 μs/m, for dolomites LIT = 0, for limestones – 30–35 μs/m, for mixtures – in the range of 0–35 μs/m. Sandstones include rocks whose LIT varies from 50 to 80 μs/m, and $\Delta T$ – from 190 to 280 μs/m.

The described technique for continuous determining the physical and lithological characteristics is implemented in the DENSI program [16, 28]. The initial data are logging diagrams of SN, NL, GR and CAL, information on the scale of methods, well conditions data. The following parameters are computed: lithology, total porosity, bound water content, rock density, elastic wave propagation velocity, vertical hodograph, reflection coefficient and reservoir velocities at specified intervals.

Comparison of the density determination results by the described method with the densities of rocks along the core, despite the complex section, gives similar values.

## 4.6 The Main Factors Affecting the Geophysical Parameters

In continuation of the topic related to the determination of the rocks lithological characteristics – the material composition and their structural and textural features, it should be noted they are the main factors affecting the geophysical parameters [1, 3, 9, 29]. In this regard, it is not accidental carbonate deposits and sandy-clay sections are studied as independent objects by various methodological techniques [9–11]. However, in practice, even when separating objects (which is not always possible), there are heterogeneities within each of them, the failure to account for which leads to significant errors. For complex sections where both rocks occur, this approach is unacceptable, since the interpretation of geophysical materials is not reliable without a detailed systematic consideration of the lithological factor. For this reason, the study of sections should begin with the determination of the material composition of rocks based on geophysical data, and all other tasks should be solved taking into account and in accordance with the results obtained.

The main initial data for lithological determinations, taking into consideration the technical capabilities of geophysical studies of wells, were selected materials SN (interval time curve $\Delta T$) and NL [1, 3].

The physical basis of the lithological division is the difference in the influence of the different rocks matrix and individual minerals (the effect of lithology) on the indications of these methods. Other geological factors significantly complicate the detection of the lithology effect, for example, the influence of rock porosity, also affect of the SN and NL. To eliminate this interference, the curves $\Delta T$ (SN) and $I_{NL}$ should be brought to a comparable scale in porosity sensitivity and the difference between the transformed curves should be calculated. The porosity factor is mutually compensated, and the resulting difference gives the effect of lithological changes.

The readings of $\Delta T$ and $I_{NL}$ are related to porosity according to different laws. In order to ensure the same sensitivity of the SN and NL curves to changes in the porosity of rocks, it is necessary to functionally transform their indications and transform the curves into forms satisfying the equations:

$$F_{\mathrm{SN}} = K_{\mathrm{p}} + \mathrm{LIT}_{\mathrm{SN}} \tag{4.18}$$

$$F_{\mathrm{NL}} = K_{\mathrm{p}} + \mathrm{LIT}_{\mathrm{NL}} \tag{4.19}$$

where $F_{\mathrm{SN}}$ and $F_{\mathrm{NL}}$ are the porosity–normalized readings of SN and NL, respectively; $\mathrm{LIT}_{\mathrm{SN}}$ and $\mathrm{LIT}_{\mathrm{NL}}$ are lithological effects affecting the readings of SN and NL.

There is a well-known time average equation for estimating porosity by SN:

$$K_{\mathrm{p}} = \frac{\Delta T - \Delta T_{\mathrm{mx}}}{\Delta T_{\mathrm{f}} - \Delta T_{\mathrm{mx}}}, \tag{4.20}$$

where $\Delta T_{\mathrm{mx}}$ and $\Delta T_{\mathrm{f}}$ are the interval travel times of the rock matrix and the filler fluid, respectively.

However, experience shows, according to Eq. (4.20), overestimated porosity values are obtained in high-porous rocks. A good example is the near–surface unconsolidated deposits of rocks. Here, the formation velocities of elastic waves are 1800–2000 m/s. The porosity calculated according to Eq. (4.20) on the basis of such velocities are obtained equal to 70–100%. This incredibly high $K_{\mathrm{p}}$ value raises doubts about the validity of using Eq. (4.20). This issue was studied in detail by comparing the readings of ΔT with the porosity estimated from the data of density and neutron logging, as well as the electrical resistance of aquifers [1]. As a result, a new dependence has been established, on the basis of which it is possible to study both low-porous and high-porous sections, which is expressed by the equation:

$$K_{\mathrm{p}} = \mathrm{Q}\frac{V_{\mathrm{mx}} - V}{V_{\mathrm{mx}} - V_{\mathrm{f}}}, \tag{4.21}$$

or expressing in terms of specific times:

$$K_{\mathrm{p}} = \mathrm{Q}\frac{\left(1 - \Delta T_{\mathrm{mx}} / \Delta T\right)\Delta T_{\mathrm{f}}}{\Delta T_{\mathrm{f}} - \Delta T_{\mathrm{mx}}}, \tag{4.22}$$

where $V_{mx}$, $V_f$, $V$ are the velocity of elastic waves propagation ($V = 1/\Delta T$) in the rock matrix, the filling fluid and measured, respectively; $\Delta T_{mx}$, $\Delta T_f$, $\Delta T$ are the travel wave times in the matrix, fluid, and measured in the rock, respectively. Q is the coefficient of critical porosity of an acoustically rigid matrix.

The graphs of the obtained dependence and the time average equations are shown in (Fig. 4.8), from which their significant difference can be seen. Other empirical SN dependencies obtained for specific sediments and lithotypes of rocks are also known in the literature [12, 30–32]. It turned out most of them fit well into the new dependence as special cases combined by this dependence into one general pattern.

This dependence, as the most reasonable and universal, is used for functional transformations $\Delta T$ in order to obtain a normalized porosity curve $F_{SN}$.

For the NL method, an expression derived from theoretical data is known [8]:

$$I_{NL} = B \cdot e^{\mu L}, \tag{4.23}$$

where $B$ is the hardware constant; $\mu$ is the effective absorption coefficient of the neutron flux; $e$ is the base of the natural logarithm; $L$ is the size of the sonde.

The effective absorption coefficient is determined by the total content of elements with abnormal neutron properties and is proportional to the hydrogen content or the porosity of the $K_p$:

$$\mu \approx \alpha \cdot K_p, \tag{4.24}$$

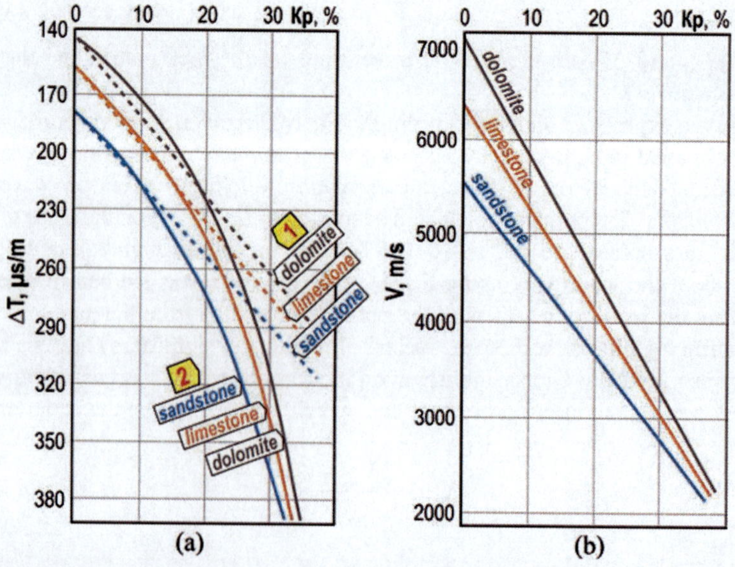


**Fig. 4.8** Dependences of ΔT and velocity V vs porosity: (**a**) according to the time average equation (4.1) according to the formula (4.19) and (4.2) according to the dependence (4.21); (**b**) according to the (4.20). (To Ref. [1])

Substituting equality (4.24) into Eq. (4.23) and taking logarithm, the dependence for NL is obtained:

$$\lg I_{\mathrm{NL}} = -\alpha \cdot K_{\mathrm{p}} + K, \tag{4.25}$$

where α is the proportionality factor; K is the total constant.

At the same time, it can be seen by taking the logarithm of the NL readings, one can obtain a curve reflecting porosity on a linear scale (4.25). Such a function can be used to construct a normalized NL curve:

$$F_{\mathrm{NL}} + C = -H \cdot \lg I_{\mathrm{NL}}, \tag{4.26}$$

where $F_{NL}$ is a function of NL readings equivalent to $F_{SN}$; H and C are scale factors.

The readings of NL have a constant component, which does not depend on the properties of the rocks, but on the direct impact of a radioactive source, so the values of $I_{NL}$ are counted from some conditional zero. Based on graph materials, as well as the comparison results of the normalized $F_{NL}$ and $F_{SN}$ curves, it was found the conditional zero line should be drawn at the level of readings of the NL, corresponding to a porosity value of 30–35%. In this case, it is important to accurately establish the level on which the linearity of the $F_{NL}$ scale depends. This level in practice is often carried out according to the minimum NL readings in clays. However, studies have shown that the hydrogen content of clays varies over a wide range [33]. For each specific well, it is desirable to calculate the conditional zero level $I_{NL0}$ from the electrical resistance $\rho_{cl}$ in clays using the formula:

$$I_{\mathrm{NL0}} = \left(4\sqrt{\frac{0,28}{\rho_{\mathrm{cl}}}} + 0,4\right) I_{\mathrm{NL\,cl}}, \tag{4.27}$$

where $\rho_{cl}$ and $I_{NL\,cl}$ are indications of LL and NL methods in the well under study.

When logarithm $I_{NL}$, in addition to obtaining a linear relationship between the parameters $F_{NL}$ and $F_{SN}$, standardization of the amplitudes of $F_{NL}$ is achieved, which is observed for any units of registration of NL diagrams (conventional units, millimeters, pulses, etc.), since according to the expressions:

$$\begin{gathered} \lg\left(n_1 I_{\mathrm{NL}}\right) = F_{\mathrm{NL}} + \lg n_1, \\ \lg\left(n_2 I_{\mathrm{NL}}\right) = F_{\mathrm{NL}} + \lg n_2 \end{gathered} \tag{4.28}$$

changing the scale of the NL in $n_1$ and $n_2$ leads only to a parallel shift of the transformed $F_{NL}$ curve and does not affect the angular coefficient of dependencies (4.28). Consequently, by logarithmic transformation, it is possible to bring the NL readings into a form equivalent to the $F_{SN}$ scale, even if there are no reference data or reference horizons for scaling the NL readings.

It has been established for pure dolomites, limestones (for each separately), the angular dependence coefficient lg $I_{NL}$ vs $K_p$ or $K_T$ (normalization coefficient H) remains almost the same regardless of the type of device, borehole conditions, and rocks depth. The normalization coefficient in this case is 25.

With a logarithmic representation of the NL readings, accounting for borehole conditions is also simplified, since a change in the diameter of the well leads to a parallel shift of dependencies. Accounting for changes in diameter is described by the expression:

$$F_{NL} + C = H \cdot \lg I_{NL} + D \cdot \Delta d_w, \tag{4.29}$$

where $D$ is a constant set according to graphs (for saline solutions D = 1.25); $\Delta d_w$ is deviation of the well diameter from the nominal in any direction in cm.

In order to identify the effect of lithology and determine the composition of rocks, the normalized curve $F_{NL}$ (4.18) is superimposed on the curve $F_{SN}$ (4.19) and combined across rocks of the same composition (Fig. 4.9a). If the curves coincide in configuration and the amplitudes change equally, then the composition of the rocks is the same, if the curves diverge, the lithological composition is changed.

To determine the composition of the studied rocks, it is necessary to link the curves at least in one interval with the known composition. To do this, it is enough to set a rock of any lithology. In practice, it is more convenient to tie the curves according to limestones, which are always present in carbonate sections. In addition, the same nature of the relation of $F_{NL}$ with $F_{SN}$, as in limestones, is observed in marls and mudstones, which can be used as rocks of the same class. This makes it possible to easily link, even if there are no pure limestones. When combining the curves $F_{NL}$ and $F_{SN}$ for the class of "limestones" in the intervals with other rocks, discrepancies are obtained. In the dolomites, the $F_{SN}$ departs from the $F_{NL}$ curve in the direction of low $K_p$ values, and in anhydrites, sandstones and salts – in the direction of high ones, while the velocity of the mineralogical matrix of these rocks changes. Analysis of the results obtained under various conditions has shown each rock-forming mineral is characterized by a certain amount of divergence and a constant that practically does not depend on the age and depth of the rocks. In this case, we are talking about a mineral matrix (rock without pores), the properties of which do not depend on these conditions, However, it is not true the same minerals with the same crystal lattice structure have different acoustic wave velocities in the matrix [8, 32].

When combining the normalized curves $F_{NL}$ and $F_{SN}$ in limestones, other lithological differences are characterized by the following divergence values: dolomites – 7%, anhydrite – 5%, sandstone – 6%, salt – 20% and mudstones – 0–2%. These values are called the lithological sign (indication) ΔLIT [1]. The values of ΔLIT have pure rock differences, and in lithologically mixed rocks, the value of ΔLIT will change in proportion to the change in composition. According to the deviation of ΔLIT from the values for pure rocks, a quantitative assessment of the composition is possible.

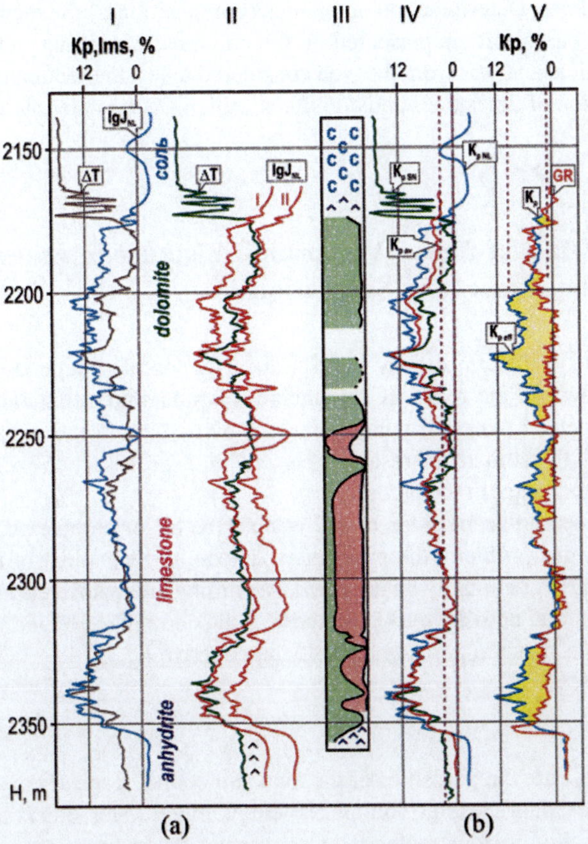


**Fig. 4.9** Determination of lithological composition (**a**) and porosity (**b**) and comparison of the obtained data from chemical analysis data. (According to Ref. [1])

The lithological sign is determined by the formula:

$$\Delta \text{LIT} + Y = F_{\text{SN}} + F_{\text{NL}}, \tag{4.30}$$

where Y is the shift constant for combining the normalized parameters $F_{\text{SN}}$ and $F_{\text{NL}}$ in limestones at ΔLIT equal to zero.

In order to construct a lithological quantitative determination of the rocks composition along the section, in addition to the main normalized curves, an auxiliary curve $F_{\text{NL}}$ is applied with a shift of −7% according to the lithological feature of pure dolomites (Fig. 4.9a) [1]. As a result, we get two lines along the $F_{\text{NL}}$: one of them $F_{\text{NLlm}}$ limits the possible position for $F_{\text{SN}}$ in pure limestones, and the other $F_{\text{NLdol}}$ in dolomites. With this construction, the curve $F_{\text{SN}}$ divides the distance between the bounding lines in proportion to the content of individual components in the rock – dolomite and limestone. A lithological column of a two-component system is obtained between the two bounding lines of the $F_{\text{NL}}$. The appearance of the third component in the composition of the rock will be marked by the exit of $\Delta T$ beyond the boundary lines. Determination of the rocks composition by the method of graphical layout of parameters is presented on the columns. According to the results of the core chemical analysis from the well considered in [1], the section is represented by an alternation of limestones and dolomites, which are very clearly recognized by this technique.

## 4.7 The Effect of Bound Water on Resistance in Assessing the Clay Reservoirs Saturation

The total porosity determined by the SN and NL complex depends on the total hydrogen content in the rocks. It also includes bound water contained in clay or other fine-dispersed material, which does not participate in the dynamics of the reservoir fluid. To estimate the reservoir capacity, it is necessary to exclude the proportion of bound water in the pores.

To determine and account for bound water, GR, SP or complex data are used. When solving this problem, the parameters reflecting the content of bound water (GR, SP, etc.) are functionally transformed. According to the difference between the readings of the total porosity curve and these of the normalized bound water saturation one, there is a useful porosity close to the effective $K_{p.eff}$:

$$K_{p.eff} = K_{p.tot} - K_{p.bn}, \tag{4.31}$$

where $K_{p.tot}$, $K_{p.bn}$ are the porosities of the total and bound, respectively.

Some original data show the volume content of bound water, others relative ones. Depending on this, various methods of converting the source data are proposed, requiring consideration of each separately.

The method of normalizing the GR readings is based on the dependence of gamma activity on the clay content and the content of bound water in it, but the methods of using this dependence differ from the generally accepted [8, 9, etc.], in which the GR method first determines the clay content, and then the moisture content of clays in the studied sections by the laboratory method. According to the data obtained, the volume of bound water is calculated. The whole process is unnecessarily time-consuming, and mistakes may be made in each operation, which in turn may affect the correctness of the results.

In the considered technique, the GR diagram is normalized and compared with the porosity curve of the $K_{p.tot}$ with their combination on dense and clay rocks. The porosity scale for the $K_p$ curve also extends to the GR curve and is used to report the bound water content on it.

The nature of the dependence of GR vs bound water was studied by numerous comparisons of GR readings with the total porosity of logging in sections composed only of impermeable rocks. Such rocks are deposits of the Voronezh and Evlanov horizons of the Pripyat trough [1]. These deposits are represented by mudstones, marls, dense limestones, anhydrites, from which inflow are not obtained during well tests, and therefore it can be assumed they contain only bound water. The relationships of gamma activity with insoluble residue and core clay content were also investigated. Studies show in carbonate sections, the dependence of gamma activity on the content of bound water is close to linear. Therefore, when normalizing the GR curve with the porosity curve, it is transformed linearly by the formula:

$$K_{\text{p.bn}} = \frac{K_{\text{p.cl}}}{I_{\text{GR cl}}} I_{\text{GR}} = \alpha I_{\text{GR}}, \tag{4.32}$$

where $K_{\text{p.bn}}$ is the content of bound water in units of porosity; $K_{\text{p.cl}}$ is the arithmetic mean value of the total porosity (humidity) of the clay interval in the section under study; $I_{\text{GR cl}}$, $I_{\text{GR}}$ are GR readings in that clay interval and the point under study (counting from the minimum level of GR readings); α is coefficient of conversion of GR values into the content of bound water.

However, the experience of using the GR method has shown the instability of the coefficient of conversion of GR into the volume content of bound water. In different cases, various values of α are obtained, and a significant difference is observed not only in diverse zones and areas, but also within the same stratum of individual wells.

The GR method is not always effective due to the increased radioactivity of some non-clay rocks (polymictic sandstones, bituminous rocks, etc.). For such cases, other methods are recommended, such as the method of complex conversion of SP measurements [1].

As is known, the SP potential curve reflects the diffusion-adsorption activity of rocks. The difference between the SP readings in rocks with different reservoir properties occurs as a result of a disturbance of ion diffusion between drilling mud and reservoir water, which occurs due to the difference in the concentration of their mineralization. This violation appears due to the selective adsorption of ions of the same sign by rock particles. Adsorbed ions in the case of micropore systems (clays, marls, etc.), occupying interrelated channels, prevent the passage of ions of another sign. As a rule, cations are adsorbed by rock particles and therefore only they can diffuse through finely dispersed systems. With free diffusion taking place in reservoirs with free water, $Cl^-$ – anions diffuse from more concentrated to less concentrated solutions ahead of cations [34].

Due to the existing diffusion of ions in rocks, the PS readings have two limiting values: one minimum (in the case of unobstructed diffusion, high porous reservoirs are close enough to such a state), the other maximum (in the case of the filling of all channels with adsorption-bound water, when only cations can be diffused). The latter case is typical for clay or when all the pores and channels are occupied by clay materials. Intermediate PS values are obtained when there are conditions in the rocks for the presence of both bound and formation (free) water. Therefore, the PS method by its physical nature is an indicator of the ratio of bound and free water in the pores.

The SP readings are most closely related to the so-called reduced absorption capacity $q$, which represents the ratio of the volume of adsorption-bound water to the total pore volume [34]. Adsorption water forms the main share of bound water in the rock. There is a direct relationship between the total volume of bound water and the adsorption content. Based on this, the relative amplitude of the SP between the maximum and minimum levels of its values will have a dependence:

$$U_{SP} = f\left(K_{bn}\right) \tag{4.33}$$

The dependence (4.33) is nonlinear in nature [9, 32] and does not have a strictly analytical expression. There is no data on the nature of dependence in the literature. In order to study the dependence (4.33), the SP readings were compared with the values of the $K_{p.bn}/K_{p.tot} = K_{bn}$ calculated from the results of measurements by other well logging methods. Also, the SP data were compared with the ones of the water saturation coefficient in strata where the rocks were represented only by oil-bearing or impermeable differences, i.e. when the rocks really contained only bound water. As a result of data comparisons, it was found the dependence (4.33) is linear in some cases, nonlinear in others, and its curve has both a concave and convex appearance. In the literature, this is explained by the difference in the adsorption activity of clays [8, 9] and the dependence for each specific object of study is recommended to be established empirically. Since it is not always possible to establish the dependence empirically, an attempt is made in [1] to find a more general solution to this issue that does not depend on geological and technical conditions. Thus, in the analysis of borehole materials, attention was paid to the dependence of the $U_{SP}$ values on the value of the ratio $\rho_{p.z}/\rho_r$:

$$1 - U_{SP} = K \cdot \lg \rho_{p.z} / \rho_r \tag{4.34}$$

Dependence (4.34) is observed at a constant ratio of the filtrate resistance $\rho_f$ of the solution and the formation water resistance $\rho_w$: $\rho_f/\rho_w$. Changes in the value of $\rho_{p.z}/\rho_r$ at a constant $\rho_f/\rho_w$ occur due to the presence and influence of surface-bound water:

$$\frac{\rho_{p.z}}{\rho_r} = \frac{\rho_{mx.p.z}}{\rho_{mx.r}} = \frac{K_{bn}\sigma_{bn} + \left(1 - K_{bn}\right)\sigma_w}{K_{bn}\sigma_{bn} + \left(1 - K_{bn}\right)\sigma_f}, \tag{4.35}$$

where $\sigma_w$ and $\sigma_{bn}$, $\sigma_f$ are the conductivity of formation and bound water, filtrate from the well fluid, respectively; $\rho_{mx.r}$ and $\rho_{mx.p.z}$ are the resistances of fluid mixtures inside the formation and in the zone of penetration of the mud filtrate.

Solving Eqs. (4.34) and (4.35), we obtain the general equation for SP readings:

$$K_{bn} = \frac{P_{s.f.c}\left(A^{U_{SP}} - 1\right)}{\left(A - A^{U_{SP}}\right) + P_{s.f.c}\left(A^{U_{SP}} - 1\right)} \tag{4.36}$$

The parameters required to solve Eq. (4.36) are found from well logging data:

$A = \rho_f/\rho_w$ is calculated from the maximum $U_{SP}$ amplitudes in this section [8]; $P_{s.f.c} = \rho_{bn}/\rho_w$ is calculated from the readings of the normalized resistance diagrams and $K_p$ in the clays. After substituting into Eq. (4.36) the most characteristic values of the formation water resistance $\rho_w$ and filtrate $\rho_f$, the dependences of the SP on the bound water saturation for various conditions are calculated (the bound water resistance is assumed to be constant 0.3 Ohm•m and its graphs are plotted [1]).

These calculations confirmed dependence (4.33) can be both linear and non-linear, and its curves are convex or concave. These graphs are in good agreement with the graphs of empirical dependences of SP on relative clay content (proportional to $K_{bn}$), established from the data of cores laboratory studies [32]. The definitions of bound water saturation are confirmed by the results obtained by other well logging methods ($\rho_r + K_p$).

When determining the effective porosity, the obtained bound water saturation $K_{bn}$ is recalculated into the volumetric content $K_{p.bn}$: $K_{p.bn} = K_{bn} \cdot K_{p.tot}$ and is substituted into formula (4.31).

The SP method is widely used in practice to determine porosity, clay content, lithological division, etc. as independent method [8, 9, 34]. The incorrectness of such use of SP can be shown by comparing it with the results of complex interpretation of SP data on actual well logging data [1]. In the absence of effective porosity in argillaceous-carbonate rocks, the SP curve has maximum readings typical for clays, although the clay content in them is low, and the total porosity varies significantly. These rocks are impermeable, and all the water in the pores is in a bound state. The SP curve behaves this way because it reflects the bound water saturation ratio and not the shale ratio, which in this case is 100%. Only after multiplying the values of $K_{bn}$ by the total porosity can one obtain the volumetric content of bound water $K_{p.bn}$, which is proportional to the clay content. Then the $K_{p.bn}$ curve is in good agreement with the GR diagram and reflects the presence of clay content. Thus, neither the clay content nor the porosity of rocks can be determined directly from the SP diagram without transformations.

The recommended model of interpretation of SP data gives the most objective results under a variety of geological and technical conditions in wells. Such a complex use of SP and NL data for clay assessment is well known [34], however, the method discussed above differs in that instead of NL readings, $K_{p.bn}$ values corrected for lithology are involved and $K_{bn}$ is calculated according to the SP data, taking into account the specific conditions of the wells. This makes it possible to unify the quantitative interpretation of the SP data and obtain good results when studying lithologically heterogeneous sections.

One of the papers presents data showing the manifestation of surface conductivity when samples are saturated with solutions of different concentrations, depending on their clay content. The effect of surface conductivity is visible with a fresh solution of $\rho_s = 3$ Ohm•m, but practically not noticeable with a solution of $\rho_s = 0.27$ Ohm•m [35]. This indicates solution has a resistance close to that of a stationary liquid, i.e. bound water.

The stability of the clay resistance is explained by the fact the concentration of bound water, the distribution and behavior of ions in it is controlled by the electrostatic field of the surface charges of the solid itself. When the mineralization of the external solution changes, the total volume charge, concentration and mobility of ions in the double layer, while preserving the substance, remains almost constant.

Therefore, the resistance of this layer, and, accordingly, the surface–bound water, should not change.

Based on the analysis of logging materials from different regions, as well as the results of experimental studies, it was found the resistance of bound water is an almost constant value, in the first approximation it can be assumed to be equal to 0.28 Ohm•m. It is obvious if the reservoir water has a resistance greater than this value, then in the presence of bound water in the rock it is underestimated and surface conductivity occurs, if the resistance is lower, then bound water affects its overestimation according to the change in the total conductivity of fluids $\sigma_{mx}$:

$$\sigma_{mx} = K_{bn}\sigma_{bn} + (1 - K_{bn})\sigma_w \approx 4K_{bn} + (1 - K_{bn})\sigma_w, \tag{4.37}$$

where $K_{bn}$ is the coefficient of bound water saturation; $\sigma_{bn} = 1/\rho_{bn}$, $\sigma_w = 1/\rho_w$, $\sigma_{mx} = 1/\rho_{mx}$.

The results of studying the electrical properties of bound water are of methodological importance in developing a method for interpreting the electrical resistance of rocks in assessing their oil content. In sections composed of water-saturated rocks, they can be used to identify aquifers and evaluate bound water.

To highlight the sign of bound water in rocks, the resistance and porosity curves are normalized to a logarithmic scale in accordance with the well-known Archie equation:

$$\lg \rho_r = -2 \lg K_r + \lg \rho_w \tag{4.38}$$

The normalized $f_{Kp}$ curve coincides with the $f_{\rho r}$ curve in one position along water-bearing sandstones, and in the other, when shifted parallel to the $f_{Kp}$ curve, along clays. The two $f_{Kp}$ curves serve as possible boundary lines for rock resistivity; in the first position, for the case of filling pores with free water; in the second position, for the bound state of water. Fluctuations in the resistance curve between such limiting lines occur in accordance with the content of one or another water in the rocks.

According to the position of such a dividing line, it is possible to quantify the value of $K_{bn}$:

$$K_{bn} = \frac{\rho_w / \rho_{mx} - 1}{\rho_w / \rho_{bn} - 1} = \frac{\dfrac{1}{\rho_{mx} / \rho_w} - 1}{\dfrac{1}{\rho_{bn} / \rho_w} - 1}, \tag{4.39}$$

where $\rho_{bn}/\rho_w = P$s.f.p is the deviation of the resistance in clays from the level of the $f_{Kp}$ lines for sandstones; $\rho_{mx}/\rho_w$ is the deviation of the resistance in the studied reservoir from the same line.

To confirm the results, in [1] a comparison is made of the bound water saturation curve $K_{bn}$, obtained from $\rho_r + K_p$ (indicator of sandstones), with the SP curve, which is a known indicator of bound water saturation. Calculations of the values of $K_{bn}$ are identical with the readings of the SP curve. The results of two independent methods are the same.

As a result, it should be noted the laws of the influence of bound water on geophysical parameters put forward a series of new methods for determining its content and create opportunities for a more reliable estimate of effective porosity. This is extremely necessary for the identification of clayed oil and gas reservoirs and the correct dissection of complex media. The results of determining the effective porosity were also checked against other data. In particular, the values of the determined "useful" capacity are in good agreement with the materials of nuclear magnetic logging (NML), showing directly the content of the mobile fluid in the reservoir.

## 4.8 Accounting for Porosity When Interpreting Electrical Resistivity

When taking into account porosity, one should first of all determine which porosity should be used in the calculations. In the current practice, to compute the resistance of water-saturated rock $\rho_{w.r}$, as a rule, the value of open porosity or porosity, from which bound water was excluded, is used. However, if it is not included in the calculations, the porosity in clays turns out to be close to zero and the calculated resistance value will tend to infinity. In reality, the value of clay resistance is close to the minimum value in geological sections of 2–10 Ohm•m. This means the electrical conductivity of rocks is determined by both free and bound water. Bound water, although immobile, is also electrically conductive. If it is not included in the calculations, then it follows bound water is classified by us as a non-conductive component of the rock, and this is a fundamental mistake. In the method [1], when analyzing the electrical resistance of rocks, it is proposed to take into consideration the total amount of water in rocks, including bound water, but with the allocation of its special characteristics. A comparison of the proposed polyfluid and traditional models of rock electrical resistivity is shown in (Fig. 4.10).

To fully account for the rock-saturating waters, the resistance curve is normalized with the total porosity by appropriate functional transformations. The normalized curves of $\rho_r$ and $K_{p.tot}$ are generally combined in a water-saturated interval and their difference is established. If in other layers the saturating liquid is represented by the same water, the normalized resistance curve $F_{ps}$ and its calculated porosity equivalent $F_{Kp}$ coincide in shape and amplitudes. If oil or gas turns out to be in the pores instead of water, an additional increment is observed on the resistance. As a result, oil and gas saturated reservoirs are marked by a significant discrepancy between the normalized resistance and porosity curves, and the magnitude of the positive discrepancy reflects the amount of oil and gas in the reservoir.

To obtain the scale equivalence of the curves $\rho_r$ and $K_p$, different transformation functions can be used. One of them may be to represent the resistance curve on a hyperbolic scale, in which it becomes proportional to the porosity curve:

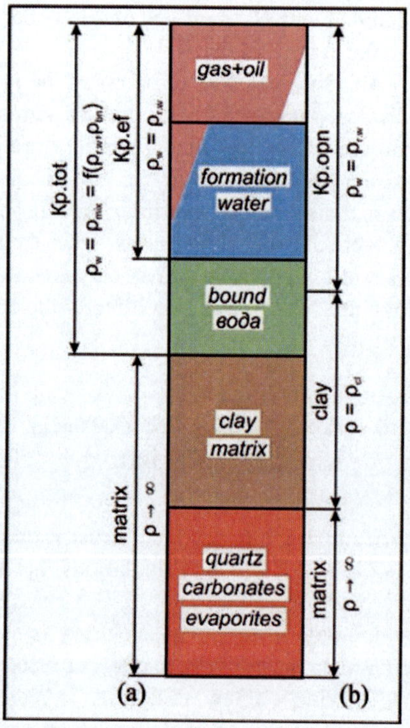


**Fig. 4.10** Models of electrical resistance: (**a**), (**b**) – for the INGEF method and traditional methods, respectively. (According to Ref. [1])

$$K_{\mathrm{p}} = \frac{a\sqrt[m]{\rho_{\mathrm{w}}}}{\sqrt{\rho_{\mathrm{r}}}}, \tag{4.40}$$

where *a* and *m* are constants for the object under study.

Since the solution of Eq. (4.40) requires additional data on the formation water resistivity $\rho_w$ and the value of *a*, and the interpreter often does not have them, the authors of [1] proposed a more convenient method – using the properties of the logarithmic function. Both compared parameters – electrical resistance and porosity are converted to a logarithmic scale according to Eq. (4.40):

$$\lg \rho_{\mathrm{r}} = -m \cdot \lg K_{\mathrm{p}} + \left(\lg \rho_{\mathrm{w}} + \lg a\right). \tag{4.41}$$

Equation (4.41) shows in order to normalize the curves $\rho_r$ and $K_p$, it is sufficient to logarithm them with modules in the ratio *1/m*, while the values *a* and $\rho_w$ do not affect the proportionality coefficient of the logarithms $\rho_r$ and $K_p$. In this expression, they represent an additive component. Since the value of $\rho_w$ practically does not change during the interval study of the section, its influence is reduced to a parallel shift of the curves. The deviation amplitudes of the normalized curves with stable formation water resistivity and a value will be the same. In this case, to take into account the formation water resistance, it is sufficient to link the parameters $\rho_r$ and $K_p$ for one layer that does not contain oil and clay material. This allows us to solve the problem of quantitative comparison of the readings of the normalized curves $F_{\rho r}$ and $F_p$ without involving data on $\rho_w$. In addition, errors that may arise due to inaccuracies in the scale factors of resistance and porosity are excluded, since these are also additive errors, which greatly simplifies processing.

Many years of research experience [1] showed for pure rock differences, the best normalization is achieved at $m = 2$. The same result was obtained for clays.

## 4.9 Influence of Fluid Composition on the Readings of Normalized Porosity and Elastic Wave Velocity Curves

The solution to the problem of direct predicting involves the identification of oil and gas deposits [1, 5]. Therefore, it would be most valuable to obtain identification information directly about the content of oil or gas in rocks. The possibility of seismic exploration to solve this important but complex task has a certain physical basis. It consists in the difference between the acoustic properties of hydrocarbons and water in reservoir conditions [12]. However, according to the results of mass studies of wells by acoustic logging in different conditions and areas, the possibility of differentiating aquifers and oil and gas reservoirs by acoustic parameters did not turn out to be as obvious as expected, which cast doubt on the prospects of seismic exploration in the field of predicting [5]. In this regard, the authors [1, 5] conducted a detailed analysis of logging materials taking into account the geological and technical conditions for their production. The analysis, in particular, showed the absence of the expected effect is due to the limitations of the acoustic logging equipment used to obtain data on the true velocity of oil and gas reservoirs. The acoustic signal from the emitter to the recorder passes along the shortest path – along the downhole zone. Due to the high frequency of waves (25 kHz), the method has a small radius of study. The recorded signal is thus determined by the penetration zone from which oil and gas are washed and carried deep into the reservoir. In this case, oil and gas reservoirs will be characterized in the same way as aquifers. According to the materials obtained immediately after the opening of the layers and in the absence of penetration (as evidenced by the data of micrologging sondes, micro-lateral logging, cavernometer), there is a noticeable effect of gas and oil on acoustic parameters. This was found by the same method of comparing functionally transformed NL and SL curves in lithologically homogeneous sections.

Of the fluids, gas has the most significant influence on these parameters. Due to the lower hydrogen content in the gas and its low density in comparison with liquid fluids, gas saturation leads to an underestimation of porosity by NL and an increase in the interval time $\Delta T$ by SN. As a result, when comparing curves, gas-bearing reservoirs without penetration are marked by their divergence (Fig. 4.11a). To assess the effect of gas on each method, the behavior of curves in aquifers and gas reservoirs was compared separately. For these purposes, sections were specially selected where the reservoir rocks have sufficiently stable porosity. As can be seen from (Fig. 4.11b), in the case of water saturation of rocks, the readings of SN and NL throughout the stratum remain unchanged without significant variations. Therefore, in the same section along the adjacent gas well, it is possible to extrapolate the level of the porosity from the aquifer part of the section to the gas-bearing one of the section (Fig. 4.11a). The deviation from this level of the curves ΔT and NL reflects the effects of the gas influence on them, which turns out to be significant. In sandstones, which in this case have a porosity of 20%, the presence of gas leads to an overestimation of ΔT compared to the aquiferous case by 50–60 μs/m, which is equivalent to an increase in porosity up to 15%.

In the literature, the most sensitive parameter to the gas content is considered [8] to be the gamma-gamma ray logging (GGL) method, which is based on a lower density of gas compared to liquid fluids. The factor of gas influence on the GGL readings was also estimated by comparing the normalized ones of $F_{GGL}$ with $F_{NL}$ in terrigenous sections with stable porosity. It has been established gas saturation affects the values of $F_{GGL}$ in almost the same way as $F_{SN}$, but in the opposite direction with respect to $F_{NL}$.

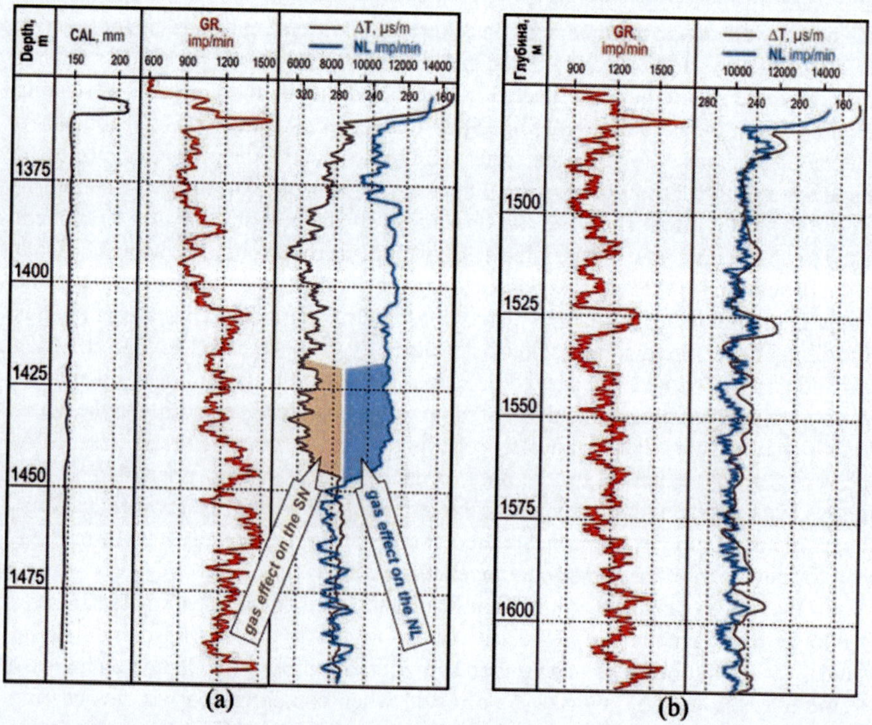


**Fig. 4.11** Effect of reservoir gas saturation on geophysical parameters: **(a)** section with gas-bearing sandstones; **(b)** water-saturated section. (According to Ref. [4, 5])

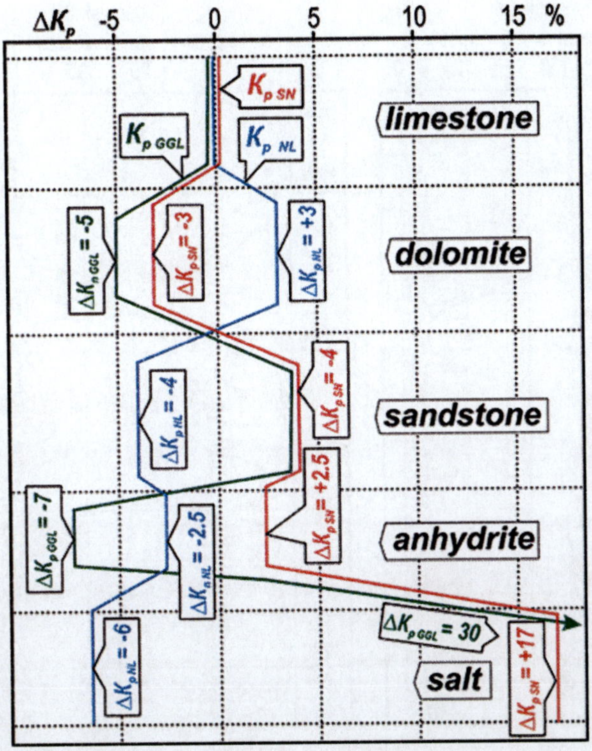


**Fig. 4.12** Generalized scheme of the lithology effect on the readings of NL, SN, GR for various rocks. (According to Ref. [1])

A generalized scheme of the influence of pure lithological differences and gas separately on the readings of $F_{NL}$, $F_{SN}$ and $F_{GGL}$, compiled based on the results of the studies, is shown in (Fig. 4.12) [1].

The effect of oil saturation of rocks was studied by logging materials in subsalt carbonate deposits, in which logging is carried out after opening, since this is the lower productive object, as a result of which well mud penetration is usually not observed. This is indicated by the maximum readings of micro lateral logging and other small-depth sondes.

As shown by the results of comparing the normalized curves ΔT(SN) and NL for two oil wells (Fig. 4.13), in oil-bearing intervals, the SN curve noticeably deviates from the NL curve towards a decrease in velocity. The influence of lithology is excluded here, since according to the core and adjacent wells, the Semiluk deposits are represented only by dolomites. In quantitative terms, the effect of oil saturation in this case is 20–30 μs/m. With an average porosity of these reservoirs $K_p = 8\%$, the observed effect leads to a false overestimation of the SN porosity by about 1.8 times. An attempt was made to determine the fluid ΔT in productive reservoirs using the average time equation. In this case, the values $\Delta T_f$ = 770–740 μs/m ($V_f$ = 1300–1350 m/s) were obtained. These values are in good agreement with

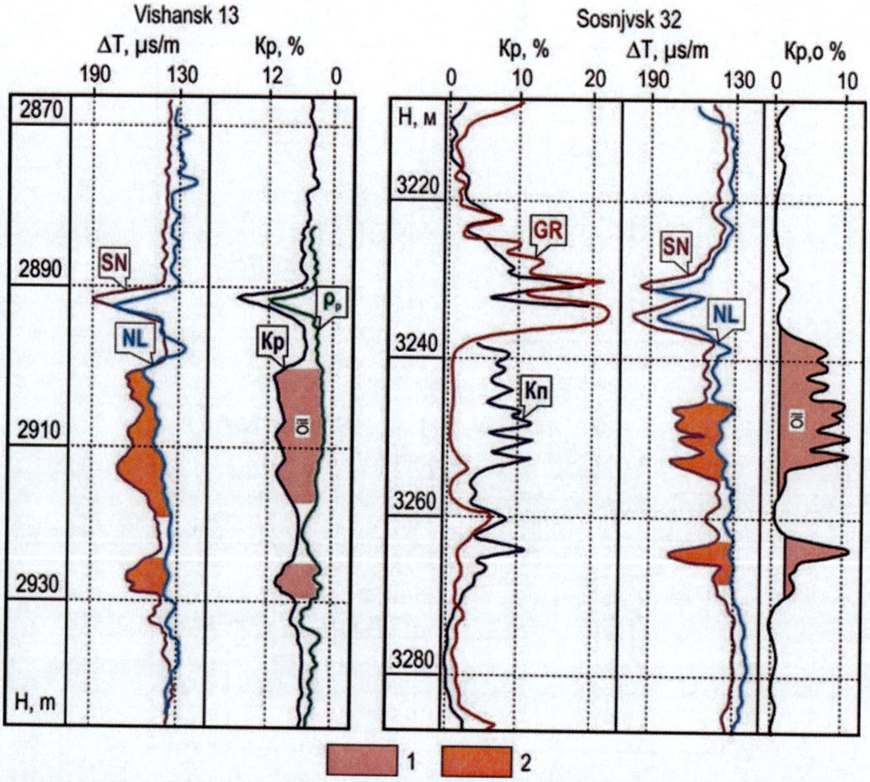


**Fig. 4.13** The effect of oil saturation on the velocity of elastic waves (SN): 1 – oil saturation according to well logging survey; 2 – the effect of oil on the SN curve. (According to Ref. [5])

experimental measurements on oil samples under reservoir thermodynamic conditions (T = 70 °C, P = 300 atm) [5].

Positive results on actual logging materials are experimental evidence of the difference between oil and gas-bearing and water-bearing reservoirs (in terms of acoustic properties). They confirm the possibility of predicting from seismic data of gas and oil deposits under appropriate favorable conditions. However, due to the influence of many factors on the velocity characteristics, the use of one such parameter seems unreliable. In such a situation, the use of high-resolution dynamic parameters in the form of RC and AI seems to be one of the main directions for improving and further developing the seismic research method [40, 41]. In support of this, we consider below (in Chap. 5) specific examples of the provisions and principles application set out in the above Chaps. 1–4 and the referenced authors' monograph.

As some results of the complex interpretation of logging data by the method of functional transformations and their integration into the information system, performed by the authors [1–6, 14, 17, 19, 22, 27, 28, 36–39] for complex clayey carbonate and terrigenous sections are given below as separate examples (Figs. 4.14, 4.15, 4.16 and 4.17).

As has been repeatedly noted, the authors of [40, 41] and this monograph widely use well logging data, namely the results of functional transformations of

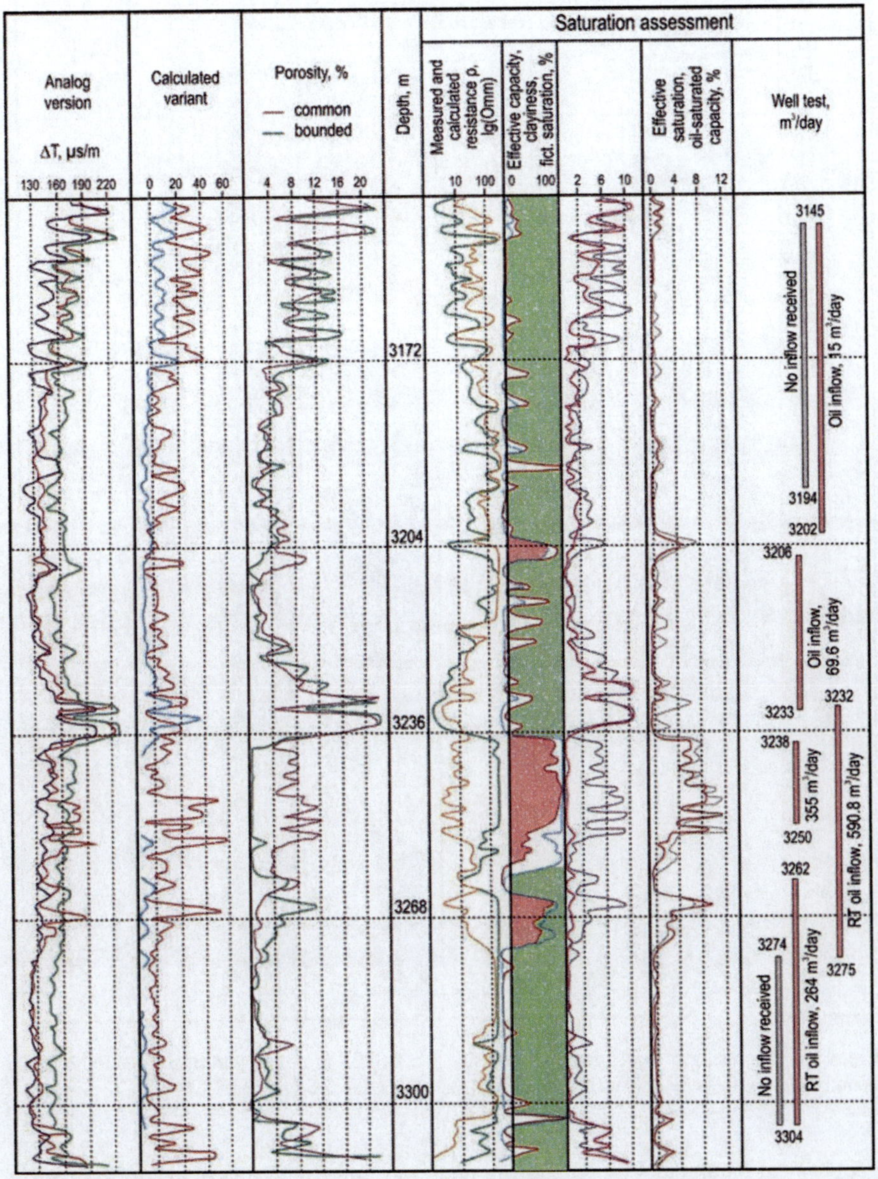


**Fig. 4.14** An example of a study using the INGEF system of clay-carbonate sections (subsalt strata of the Pripyat Trough). (According to Ref. [4])

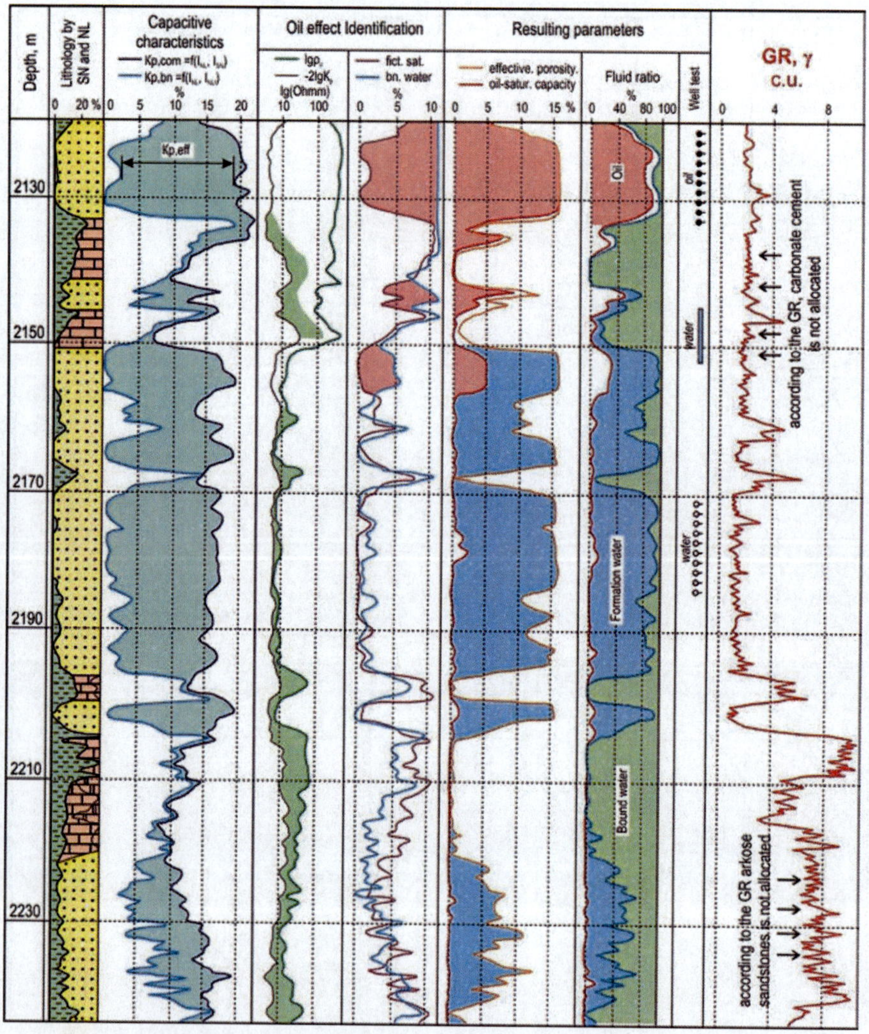


Fig. 4.15 Processing of well logging data using the INDEF program with determination of the bound water in rocks by SN and NL complex (Elsk 26). (According to Ref. [4])

geophysical parameters by integrating them into information systems, in the complex interpretation of high resolution seismic (HRS) data. This ensures the continuity of data processing and analysis, the quantitative linkage of all parameters used on the basis of their genetic relationships. Based on a set of these parameters located in a certain system, the material composition, porosity, the content of bound water and the useful capacity of the reservoir are established, the oil and gas saturation of the reservoirs is estimated, and a number of physical characteristics of the

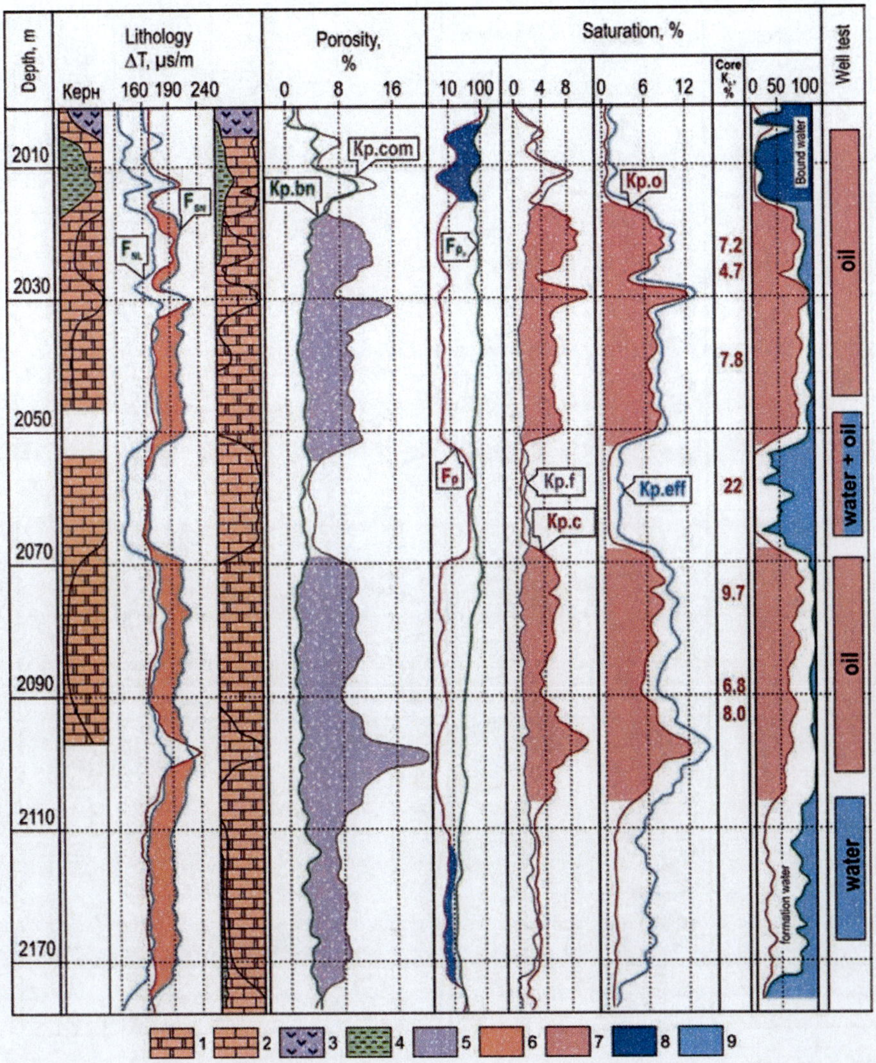


**Fig. 4.16** Determination of lithology, porosity, oil saturation of carbonate sections by the method of geophysical parameters integration in the INGEF system: 1-dolomite, 2-limestone, 3-anhydride, 4-clay, 5-reservoir, 6-lithological factor, 7-oil, 8-bound water, 9-formation water. (According to Ref. [1])

geological section are calculated. On this basis, the construction of lithological-stratigraphic columns and detailed geoacoustic models in vertical sections of the well section is carried out. At the same time, the authors, according to the HRS seismic data, often build predictive lithological-stratigraphic columns in the

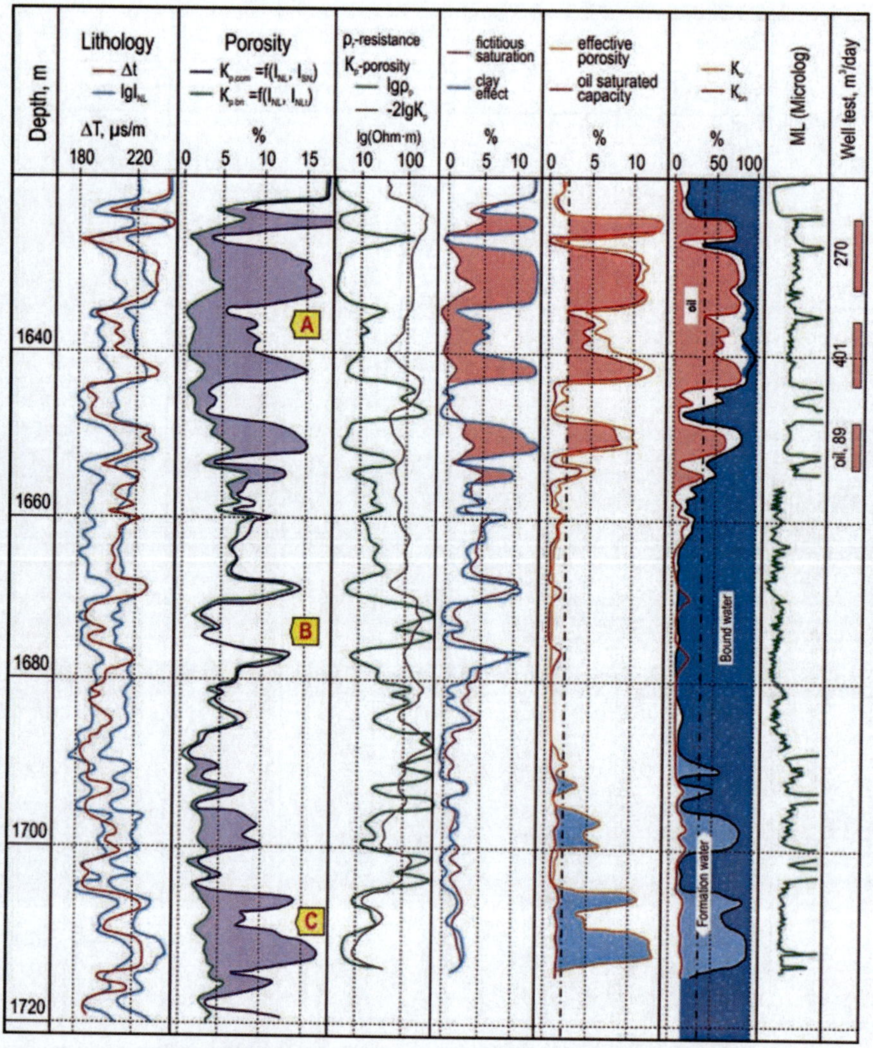


**Fig. 4.17** An example of interpretation of the NL, SN, SP, IL data set using the INGEF system of sections containing aquifers (C), oil-bearing reservoirs (A) and dense rocks (B) (Timan-Pechora province). (According to Ref. [5])

inter-well space with a discretization step of seismic data, using exclusively HRS data with a relatively high prediction probability.

Below, in the next Chap. 5, examples of the application of the HRS-Geo technology on materials obtained on the territory of the Timan-Pechora province are considered.

## 4.10 Summary

For the joint use of high-resolution seismic data obtained using the HRS-Geo technology and well logging data, an interpretive system was created as part of the technology, which is based on the principles of continuous functional data transformations for studying well sections [40, 41]. This system has a software-technological block, on the basis of which the physical parameters, extracted from well logging data, are optimally matched with the corresponding high-resolution seismic parameters obtained using the HRS-Geo technology.

1. Chapter 4 discusses the main features of the development and improvement of the methodology for processing and interpreting data from standard well logging methods according to the method of integrating geophysical data by functionally transforming them into information systems. These features of log data transformation are aimed at quantitative and qualitative study of well sections with complex lithological composition. At the same time, sequential execution of operations is carried out to determine the material composition of rocks, porosity, taking into account their lithological heterogeneity, differences in the structure of the pore space, determination of bound water and effective porosity by means of normalization, integration of the results of transformations, assessment of the complex reservoirs saturation nature and the influence of reservoir hydrocarbons on geophysical parameters.
2. Positive results of automated interpretation of well logging data were obtained due to the wide application of the method of normalizing logs and their functional transformations, eliminating distortions in the original log curves, integrating data when identifying complex reservoirs, assessing the porosity and density of various lithologies rocks, determining factors affecting geophysical parameters, the effect of bound water on resistivity in assessing the saturation of clay reservoirs, accounting for porosity in interpreting electrical resistivity, and the effect of fluid composition on readings of normalized porosity and elastic wave velocity curves.
3. The physical essence and principles of interpretation of normalized diagrams in this case lie in the fact that with the help of the neutron gamma ray logging curve, variations in electrical resistance caused by changes in the porosity of rocks along the section are excluded from consideration, and thus areas of increasing rock resistance due to their oil saturation are visually represented. Since when the normalized curves of lateral and neutron gamma-ray logging in water-bearing intervals are combined, some equivalence is achieved, the neutron gamma-ray logging curve can be considered as a resistance curve characterizing the change in electrical resistance along the section at their full water saturation, because the oil content on the readings of this method is practically not affects.
4. In the complex interpretation of high-resolution seismic (HRS) data, well logging data, namely the results of functional transformations of geophysical parameters, provide, as already noted, the continuity of data processing and analysis, the quantitative linkage of all used parameters based on their genetic

relationships, etc. On this basis, the construction of lithologic stratigraphic columns and detailed geoacoustic models in vertical sections of the well section is carried out. At the same time, the authors, according to the HRS seismic data, also build predictive lithological-stratigraphic columns in the inter-well space with a discretization step of seismic data, using exclusively HRS data with a sufficiently high prediction probability.

## References


1. Zalyayev, N. Z. (1990). *Metodika avtomatizirovannoy interpretatsii geofizicheskikh issledovaniy skvazhin* (Technique of automated interpretation of well logging) (p. 142). Minsk: University Edition.
2. Zalyayev, N. Z., Kaufman, R. M., & Mashara, L. P. (1978). *Sistema obrabotki i shirokikh geofizicheskikh parametrov metoda funktsional'nykh preobrazovaniy (INGEF)* (System of processing and integration of geophysical parameters by the method of functional transformations (INGEF)). V sb.: Problemy strukturnogo analiza v geofizike (Collection: Problems of structural analysis in geophysics) (pp. 100–112). Minsk: BelNIGRI.
3. Zalyayev, N. Z. (1981). *Izucheniye razrezov slozhnogo litologicheskogo sostava po dannym geofizicheskikh issledovaniy v skvazhinakh* (Study of complex lithological composition according to geophysical surveys in wells). Region., razved. i promysl. geofizika. VIEMS Review., M, p. 51.
4. Zalyayev, N. Z. (1981). *Kompleksnaya interpretatsiya geofizicheskikh parametrov funktsional'nymi preobrazovaniyami s pomoshch'yu EVM* (Complex interpretation of geophysical parameters by functional transformations with the help of a computer) (p. 150). Minsk: BelNIGRI.
5. Zalyayev, N. Z. (1984). Vliyaniye litologii i plastovykh uglevodorodov na geofizicheskiye parametry (Influence of lithology and reservoir hydrocarbons on geophysical parameters). Book: *Metodika i rezul'taty geologo-geofizicheskikh issledovaniy v Pripyatskom progibe* (Methodology and results of geological and geophysical research in the Pripyat Trough) (pp. 61–74). Minsk. Science and Technology.
6. Zalyayev, N. Z., Kaufman, R. M., & Mashara, L. P. (1981). *O funktsiyakh, ispol'zuyemykh pri obrabotke promyslovo-geofizicheskikh materialov metodom funktsional'nykh preobrazovaniy (Sistema INGEF)* (On the functions used in the processing of field geophysical materials by the method of functional transformations (INGEF System)). V sb.: Metody geofizicheskikh issledovaniy skvazhin (Collection: Methods of geophysical research of wells) (pp. 16–24). Minsk: BelNIGRI.
7. Debrand, R. (1972). *Teoriya i interpretatsiya rezul'tatov geofizicheskikh metodov issledovaniy skvazhin* (Theory and interpretation of the results of geophysical methods of well research) (p. 288). M., Nedra.
8. Dakhnov, V. N. (1985). *Geofizicheskiye metody opredeleniya kollektorskikh svoystv i neftegazonasyshcheniya gornykh porod* (Geophysical methods for determining reservoir properties) (p. 310). M., Nedra.
9. Vendel'shteyn, B. Yu., & Rezvanov, R. A. (1978). *Geofizicheskiye metody opredeleniya parametrov neftegazovykh kollektorov (pri podschete zapasov i proyektirovanii razrabotki mestorozhdeniy)* (Geophysical methods for determining the parameters of oil and gas reservoirs (when calculating reserves and designing field development)) (p. 318). M., Nedra.
10. Itenberg, S. S., & Shnurman, G. A. (1984). *Interpretatsiya rezul'tatov karotazha slozhnykh kollektorov* (Interpretation of logging results for complex reservoirs) (p. 256). M., Nedra.

11. Sokhranov, N. N., Aksel'rod, S. M., Zundelevich, I. M., & Churinova, I. M. (1989). *Obrabotka i interpretatsiya dannykh promyslovykh geofizicheskikh issledovaniy na EVM* (Processing and interpretation of field geophysical research data on a computer) (p. 240). Spravochnik/Pod red. Sokhranova N.N. / (Handbook: edited by Sokhranov N.N.). M., Nedra.
12. Ivakin, B. N., Karus, Ye. V., Kuznetsov, O. L. (1978). *Akusticheskiy metod issledovaniya skvazhin* (Acoustic method of well research) (p. 320). M., Nedra.
13. Masyukov, V. V., & Kaufman, R. M. (1981). *Avtomatizirovannaya interpretatsiya karotazha skvazhin v Pripyatskom progibe* (Automated interpretation of well logging in the Pripyat trough). V sb.: Metody geofizicheskikh issledovaniy skvazhin (Collection: Methods of geophysical research of wells) (pp. 3–16). Minsk: BelNIGRI.
14. Zalyayev, N. Z., Kaufman, R. M., Mashara, L. P. (1982). *Kompleksnaya interpretatsiya dannykh karotazha v peschano-glinistykh razrezakh* (Integrated interpretation of logging data in sandy-clay sections). V sb.: Metodika i rezul'taty interpretatsii karotazha neftyanykh skvazhin (Collection: Methodology and results of oil well logging interpretation) (pp. 16–25). Minsk: BelNIGRI.
15. Kaufman, R. M. (1982). *Issledovaniye ustoychivosti resheniya zadachi opredeleniya kollektorskikh svoystv porod razreza po programme INGEF* (Investigation of the solution stability of the problem for determining the rocks reservoir properties according to the INGEF program). V sb.: Metodika i rezul'taty interpretatsii karotazha neftyanykh skvazhin (pp. 25–30). Minsk: BelNIGRI (Collection: Methodology and results of oil well logging interpretation).
16. Zalyayev, N. Z., Kaufman, R. M., Mashara, L. P., Astapenko, V. A., & Gabrielyan, E. Kh. (1982). *Printsipy postroyeniya i organizatsii sistemy obrabotki rezul'tatov tsifrovoy zapisi karotazha* (Principles of construction and organization of a system for processing the results of digital logging recording). V sb.: Metodika i rezul'taty interpretatsii karotazha neftyanykh skvazhin (Collection: Methodology and results of oil well logging interpretation) (pp. 53–57). Minsk: BelNIGRI.
17. Zalyayev, N. Z., & Masyukov, V. V. (1970). *Otsenka neftenosnosti karbonatnykh otlozheniy Pripyatskoy vpadiny po promyslovo-geofizicheskim dannym* (Oil-bearing capacity estimation of the Pripyat depression carbonate deposits according to field geophysical data). V kn.: Geofizicheskiye metody poiskov i razvedki nefti i gaza v Pripyatskoy vpadine (Collection: Geophysical methods of prospecting and exploration of oil and gas in the Pripyat depression) (pp. 80–91). Minsk: Izd. Nauka i tekhnika (Ed. Science and Technology).
18. Basin, Ya. N., Berman, L. B., & Neyman, V. S. (1964). *O vozmozhnosti vydeleniya kavernoznykh i treshchinovatykh izvestnyakov metodami promyslovoy geofiziki* (On the possibility of separating cavernous and fractured limestones using field geophysics methods). Prikl. geofizika, vyp. 39 (Appl. geophysics, Vol. 39). M., Nedra.
19. Zalyayev, N. Z., Masyukov, V. V., & Orlov L. I. (1974). *Opredeleniye neftenasyshchennosti karbonatnykh kollektorov po dannym promyslovoy geofiziki* (Determination of oil saturation in carbonate reservoirs according to field geophysics data). V sb.: Pripyatskaya vpadina. Geologicheskiye rezul'taty, metodika i tsifrovaya obrabotka geofizicheskikh issledovaniy (Collection: Pripyat depression. Geological results, methodology and digital processing of geophysical surveys) (pp. 157–163). Minsk: BelNIGRI.
20. Zalyayev, N. Z. (1970). *Opredeleniye poristosti i neftenasyshchennosti karbonatnykh porod po dannym elektricheskogo i radioaktivnogo karotazha* (Determination of porosity and oil saturation of carbonate rocks according to electrical and radioactive logging data). V kn.: Geofizicheskiye metody poiskov i razvedki nefti i gaza v Pripyatskoy vpadine (Collection: Geophysical methods of prospecting and exploration for oil and gas in the Pripyat depression) (pp. 92–97). Izd. Nauka i tekhnika (Ed. Science and Technology).
21. Masyukov, V. V., & Kuznyanyy, V. A. i dr. (1971). *Metodicheskiye rekomendatsii po provedeniyu geofizicheskikh issledovaniy skvazhin Pripyatskogo progiba i interpretatsii poluchennykh rezul'tataov* (Methodological recommendations for conducting geophysical surveys of wells of the Pripyat trough and interpretation of the results obtained). Minsk: Izd. Nauka i tekhnika (Ed. Science and Technology).

22. Zalyayev, N. Z. (1974). *Otsenka effektivnoy poristosti glinistykh porod sposobom normalizatsii diagramm radioaktivnogo karotazha* (Evaluation of the clay rocks effective porosity by the method of radioactive logging diagrams normalization). V sb.: Pripyatskaya vpadina. Geologicheskiye rezul'taty, metodika i tsifrovaya obrabotka geofizicheskikh issledovaniy (Collection: Pripyat Depression. Geological results, methodology and digital processing of geophysical surveys) (pp. 216–220). Minsk: BelNIGRI.
23. Larionov, V. V. (1969). *Radiometriya skvazhin* (Well radiometry). M., Nedra.
24. Malinin, V. F., & Ruchkin, A. V. (1967). Ispol'zovaniye dannykh gamma-karotazha dlya litologicheskogo raschleneniya otlozheniy devona v vostochnoy chasti Pripyatskoy vpadiny (Use of gamma-ray data for lithological dissection of Devonian deposits in the eastern part of the Pripyat depression). V kn.: *Metodika i geologicheskiye rezul'taty geofizicheskikh issledovaniy Pripyatskoy vpadiny* (In the book: Methodology and geological results of geophysical studies of the Pripyat depression), Minsk, Nauka i tekhnika (Science and Technology).
25. Varlamov, A. S., & Filatov, V. G. (1983). *Opredeleniye plotnosti gornykh porod i geologicheskikh ob"yektov* (Determination of the density of rocks and geological objects) (p. 216). M., Nedra.
26. Zalyayev, N. Z., & Bondarenko, B. V. (1977). *Algoritm opredeleniya plotnosti porod kompleksu geofizicheskikh issledovaniy skvazhin* (Algorithm for determining the density of rocks for a complex of geophysical well surveys). V sb.: Pripyatskaya vpadina. Izucheniye slozhnopostroyennykh struktur geofizicheskimi metodami (Collection: Pripyat Depression. The study of complex structures by geophysical methods) (pp. 156–161). Minsk: BelNIGRI.
27. Zalyayev, N. Z., Kaufman, R. M., & Mashara, L. P. (1985). *Vydeleniye polimiktovykh peschanikov po dannym standartnogo karotazha (programma TerINGEF)* (Identification of polymictic sandstones according to standard logging data (TerINGEF program)). V sb.: Interpretatsiya materialov karotazha v slozhnykh razrezakh nefteperspektivnykh tolshch (Collection: Interpretation of logging data in complex sections of oil-promising strata) (pp. 13–21). Minsk: BelNIGRI.
28. Zalyayev, N. Z., Mogilevich, M. V., & Astapenko, V. A. (1984). *Izucheniye fizicheskikh svoystv porod terrigennogo razreza po materialam karotazha skvazhin* (Physical properties study of the terrigenous rocks based on well logging data). V sb.: Interpretatsiya materialov karotazha v slozhnykh razrezakh nefteperspektivnykh tolshch (Collection: Interpretation of logging data in complex sections of oil-promising strata) (pp. 40–49). Minsk: BelNIGRI.
29. Zalyayev, N. Z. (1984). *Vliyaniye litologii i plastovykh uglevodorodov ne geofizicheskiye parametry* (The influence of lithology and reservoir hydrocarbons are not geophysical parameters). V kn.: Metodika i rezul'taty geologo-geofizicheskikh neftepoiskovykh issledovaniy v Pripyatskom progibe (Collection: Methodology and results of geological and geophysical oil exploration in the Pripyat trough) (pp. 61–74). Minsk: Izd. Nauka i tekhnika.
30. Ellanskiy, M. M. (1978). *Petrofizicheskiye svyazi i kompleksnaya interpretatsiya dannykh promyslovoy geofiziki* (Petrophysical relations and complex interpretation of production geophysics data) (p. 215). M., Nedra.
31. Ellanskiy, M. M., & Yenikeyev, B. N. (1991). *Ispol'zovaniye mnogomernykh svyazey v neftegazovoy geologii* (The use of multidimensional relationships in oil and gas geology) (p. 205). M., Nedra.
32. Bulatova, M. M., Volkova, Ye. A., & Dubov, Ye. F. (1970). *Akusticheskiy karotazh* (Acoustic logging) (L., p. 264).
33. Zalyayev, N. Z., Kononov, A. I., & Drobysheva, V. G. (1985). *Otsenka paleotermobaricheskikh usloviy po rezul'tatam GIS. Geologiya nefti i gaza* (Evaluation of paleothermobaric conditions based on logging results) (№ 8, p. 13).
34. Vendel'shteyn, B. Yu. (1966). *Issledovaniye razrezov neftyanykh i gazovykh skvazhin metodom sobstvennykh potentsialov* (Investigation of oil and gas wells sections by the method of self-potentials) (p. 206). M., Nedra.
35. Vendel'shteyn, B. Yu. (1962). Ob uchete glinistosti pri opredelenii poristosti po dannym metoda soprotivleniy (On accounting for clay content in determining porosity using the data of the resistance method). *Prikladnaya geofizika*. Vyp. 32 (Applied geophysics. Issue. 32), M., p. 8.

36. Kaufman, R. M., & Mashara, L. P. (1985). *Sposoby opredeleniya soderzhaniya svyazannoy vody v porodakh po programmam INGEF* (Methods for determining the content of bound water in rocks according to the INGEF programs). V sb.: Interpretatsiya materialov karotazha v slozhnykh razrezakh nefteperspektivnykh tolshch (Collection: Interpretation of logging data in complex sections of oil-promising strata) (pp. 58–66). Minsk: BelNIGRI.
37. Zalyayev, N. Z., & Masyukov, V. V. (1985). *Problemy vydeleniya i otsenki peschanykh kollektorov slozhnogo sostava v mezhsolevykh otlozheniyakh po karotazhu i puti povysheniya effektivnosti promyslovo-geofizicheskikh issledovaniy* (Problems of identifying and evaluating sandy reservoirs of complex composition in intersalt deposits by logging and ways to improve the efficiency of field geophysical surveys). V sb.: Tipy zalezhey nefti Pripyatskogo progiba i ikh prognoz (Collection: Types of oil deposits in the Pripyat trough and their prediction) (pp. 142–155). Minsk: BelNIGRI.
38. Masyukov, V. V., & Zalyayev, N. Z. (1989). *Vyyavleniye bituminoznosti glinistykh obrazovaniy po materialam karotazha skvazhin* (Collection: Peculiarities of formation and placement of oil deposits in the Pripyat trough). V sb.: Osobennosti formirovaniya i razmeshcheniya zalezhey nefti v Pripyatskom progibe (pp. 110–116). Minsk: BelNIGRI.
39. Zalyayev, N. Z. (1989). *Rol' svyazannoy vody v formirovanii zalezhey nefti* (The role of bound water in the formation of oil deposits). V sb.: Osobennosti formirovaniya i razmeshcheniya zalezhey nefti v Pripyatskom progibe (Collection: "Peculiarities of formation and placement of oil deposits in the Pripyat trough") (pp. 117–128). Minsk: BelNIGRI.
40. Trofimov, V. L., Khaziyev, F. F., & Trofimova, A. V. (2019). *Tekhnologiya vysokorazreshayushchey seysmiki pri poiskakh i razvedke neftegazovykh zalezhey* (High-resolution seismic technology in prospecting and exploration of oil and gas deposits) (p. 343). M., EAGE Geomodel LLC.
41. Trofimov, V. L., Khaziev, F. F, & Trofimova A. V. (2022). *Oil and gas reservoir prospecting and exploration. High-Resolution Seismic (HRS) techniques and technology*. Springer (Hardcover Book, eBook). https://doi.org/10.1007/978-3-030-84389-2
42. Geofizicheskiye metody issledovaniya skvazhin. (2022). (Geophysical methods of well research). https://studwood.net/1919624/matematika_himiya_fizika/geofizicheskie_metody_issledovaniya_neftyanyh_i_gazovyh_skvazhin

# Chapter 5
# Detail Interpretation of High Resolution Seismic Data. Timan-Pechora Province

**Abstract** The results of the HRS-Geo technology application on the materials obtained in seismogeological conditions in the exploration areas of the Timan-Pechora Province (TPP) are presented. At the first of these sites, located in the southern part of the Mechayu-Pashnin swell and the Tebuk stage of the TPP, the geological structure of local uplifts was refined based on the results of 3D-data acoustic impedance (AI) and reflection coefficient (RC) cubes. The results of structural constructions reinterpretation are analyzed considering drilling data. Using the forms and values of the reference curves of seismic facies–clusters, oil-promising objects of structural, tectonically and lithologically screened types were identified in the deposits of the upper part of the Famennian stage. In the junction zone of the northeastern part of the Kolvin megaswell and the northwestern part of the Khoreyver depression, an assessment of the most important geological indicators of the real environment, the material composition and reservoir properties in the sediments of the Devonian and the Upper Silurian strata is made. Oil-perspective objects of a structural type in the target horizons were identified. In the central part of the Kolvin megaswell of the Pechora-Kolvin aulacogen and in the junction zone of the large Vozeya and Kharyagin swells, using 3D cubes and 2D sections AI and RC, the nature of the layering of sediments along productive and perspective horizons was studied. Detailed structural constructions for these horizons were made, sedimentation environments were considered by seismic facies classification in order to identify zones of facies substitution, zones of the geological heterogeneities in the process of studying the detailed internal structure and lithofacies variability of target deposits.

Theoretical and experimental estimates of the dynamic characteristics of the wave field by analyzing the information content and sensitivity of its spectra with respect to the resolution of the seismic record (see Chaps. 1 and 2), geological foundations oriented to the system of geostructural, lithological, stratigraphic, facies principles and criteria (see Chap. 3), and well logging data (see Chap. 4) for their application

 
V. L. Trofimov et al., *Theory and Practice of High Resolution Seismic HRS-Geo Technology Application*, https://doi.org/10.1007/978-3-031-41590-6_5

together with HRS data, were used within the framework of the HRS-Geo technology [13, 14] in the complex interpretation of seismic observation materials obtained in various regions of Russia and abroad. The technology was used primarily to find stable values of acoustic and geological indicators of the real geological environment. Practically for all studied oil prospects, positive results were obtained using HRS-Geo technology programs, which were later confirmed by the results of drilling deep wells.

In this chapter, when evaluating the effectiveness of perfomed studies, only a small part of the results obtained for one of the polygons, which is characterized by a rather complex geological structure, is presented. This test site was the Timan-Pechora oil and gas province, located in the northeastern outskirts of the Russian platform. It is characterized by a number of significant features, primarily due to its high tectonic activity, which also determines the specifics of its oil and gas potential [2–4, 6, 10–12, 15, 16, etc.].

Serious scientific work on this region has been carried out by many researchers and scientists. The bibliography of works of these researchers is quite extensive and it is almost impossible to list them (*the bibliography of the researchers we mentioned above contains a list of works, although it is very small in size*). At the same time, we note the authors of this monograph did not seek to reveal the geological knowledge of the region in all its aspects and peculiarities. The information presented below on the features of the geological structure of the region under consideration is presented by the authors in a brief form and only to the extent they were used as some justification and explanatory geological basis for the results of high-resolution seismic.

In each of the subsections below, the authors tried to follow a certain sequence, i.e. schemes for demonstrating the results of complex interpretation:

1. The geological setting in which the desired oil-prospective objects are (structural-tectonic and geodynamic features of the structure, distinctive geological patterns of the location of objects, conditions of sedimentogenesis (sedimentation) of the studied deposits, the formation of pore space and the oil and gas content of thin-layer reservoirs).
2. The geophysical material used is 2D or 3D seismic and well logging data for the wells available for the period of the survey.
3. Goals and objectives of the research.
4. The main results obtained on the basis of the use of software tools of the HRS-Geo technology, which were the material for a detailed complex interpretation:
   - 2D sections and 3D cubes of effective acoustic impedance (AI) and reflectance (RC) to study the detailed internal structure of the real geological environment with a vertical resolution equal to the seismic time sampling step;
   - the results of automated GBS data processing in the form of lithological and stratigraphic columns of wells necessary for the interpretation of high-resolution seismic data;
   - lithofacies replacements of deposits, zones of heterogeneities in productive and potentially productive horizons (thin layers), conditions for the formation of sediments (sedimentation) by clustering and predicting various groups of facies in target deposits;

- for a number of cases – predictive, most probable distributions of geological indicators in the studied areas – in the form of distributions of effective values of AI and RC, porosity, clay content, oil saturation, density of hydrocarbon distribution in productive and prospective reservoirs, allowing to establish the location regularities of reservoir rocks and oil and gas saturation, and others features of the geological structure.

5. Conclusions and prospects for setting up further exploration work.

Before considering the specific geological and geophysical conditions for studying the areas, let us dwell on some general aspects of the formation of the basin and the lithological and stratigraphic features of the sedimentary cover of the TPP.

## 5.1 Some Aspects of the Formation and Development of the Sedimentary Basin of the Timan-Pechora Province

Let us emphasize the main features and peculiarities of the geological structure of the Timan-Pechora oil and gas province (TP OGP) [1–12, 15, 16].

1. By its nature, the Timan-Pechora sedimentation basin (TP SB) is heterogeneous both in terms of structure and composition of its constituent formations. In the larger area of the western part of the basin, dislocations are predominantly of the platform and the inversion aulacogen types with a predominance of carbonate formations. In the east, from the inner zone of the marginal trough to the Central zone of the Ural and Paikhoi orogens, fold-overthrust deformations are already widespread, and among the formations, a significant portion belongs to molasses and formations of filling uncompensated depressions.
2. In the continental part of the Timan-Pechora Plate, the section of the cover begins with the Lower-Middle Ordovician terrigenous formation, replaced by Upper Ordovician-Lower Devonian carbonates; they constitute the Caledonian cycle of sedimentation. The Hercynian cycle includes terrigenous deposits of the Middle Devonian – Lower Frasnian, terrigenous-carbonate deposits – Middle Frasnian – Tounaisian with reefs, coal-bearing terrigenous deposits of the Lower-Middle Visean, and above there is a thick stratum of carbonates of the rest of the Carboniferous. This cycle ends with carbonates, anhydrites, Permian reds, and in the Triassic the plate becomes land. The Mesozoic-Cenozoic cycle is represented by the Middle Jurassic-Cretaceous marine and Neogene-Quaternary continental (except for the Upper Pliocene) terrigenous formations, separated over most of the plate by a sedimentation break in the Late Cretaceous-Paleogene.
3. The main structure elements of the Timan-Pechora plate within the land are from west to east: the Kanin-Timan ridge, consisting of a number of echelon-like blocks located and limited from the west by reverse-thrust blocks, and in general thrust over the edge of the Russian plate (Riphean on the Upper Paleozoic); Izhma-Pechora depression; Malozemel-Kolguev monocline; Pechoro-Kolvin aulacogen; Khoreyver depression superimposed on the Bolshezemelsky buried

arch; The Varandey-Adzvinsky block and further the pre-Ural-pre-Paikhoi system of troughs extends. All these structural elements are complicated by local uplifts grouped into swells. The Pechora-Kolvin aulacogen with axial trough and near-fault inversion swells, as well as the Varandey-Adzvin block, has the most stressed internal structure. The origin of the Pechora-Kolvin aulacogen belongs to the Silurian, but the most active development – to the Devonian – Early Carboniferous, and inversion deformations – to the Late Paleozoic ages. In general, the formation of oil and gas reservoirs due to chemical leaching is associated with ancient continental breaks in sedimentation on the territory of the TPP. Reservoir horizons traced for a considerable distance are confined to the Upper Ordovician, Lower Silurian, and Lower Devonian. The Lower Silurian rocks lying near the surface of the pre-Devonian nondepositional hiatus have the best reservoir properties.

4. The marine part of the Timan-Pechora Plate is separated from the continental zone by sinistral strike-slip faults extending along the coast. The structural relief of the basement surface to the north of this zone is smoothed out and individual structures fade out; the structures of the most eastern, Varandey-Adzvin zone, are the furthest traced. The thickness of the cover increases, especially due to the Devonian, Permian, and Triassic sediments. The northern limit of the Timan-Pechora plate, separating it from the Svalbarsk one, is another latitudinal zone, the southern branch of which extends in the east to the southern tip of Novaya Zemlya.
5. Specialists-geologists and scientists on the territory of the Timan-Pechora OGP carried out extensive lithofacies studies, numerous structural constructions, paleostructural analysis and analysis of oil and gas potential. In particular:

   (a) The evolution of sedimentation environments was traced throughout the entire period of the sedimentary cover formation of the TP SB. At the same time, the full bathymetric series, reflecting the features of the basin development, covers areas from continental high and low plains, to marine, coastal-marine and shelf settings, and to the continental slope (Figs. 5.1 and 5.2).
   (b) Based on the complex application of formational, historical-geological analyzes and stratigraphy of sequences, sedimentary rhythms/sequences of different orders are distinguished, which are the result of the interaction of global eustatics, regional tectonics, sediment supply and climate.
   (c) The formation regularities of accumulating and sealing oil source strata are outlined in relation with the staged development of the basin.
   (d) The relationship between the location of the main reservoir horizons and the features of the development of the main paleotectonic elements and them confined to the junction zones of large tectonic structures or zones of long-lived faults, to the arches and slopes of large stable paleo-elevations, and within the graben troughs and depressions on the shelf – to consedimentary uplifts within them is traced.
   (e) For all oil and gas complexes, the main genetic types of rock bodies are outlined – potential reservoirs of hydrocarbon reservoirs.

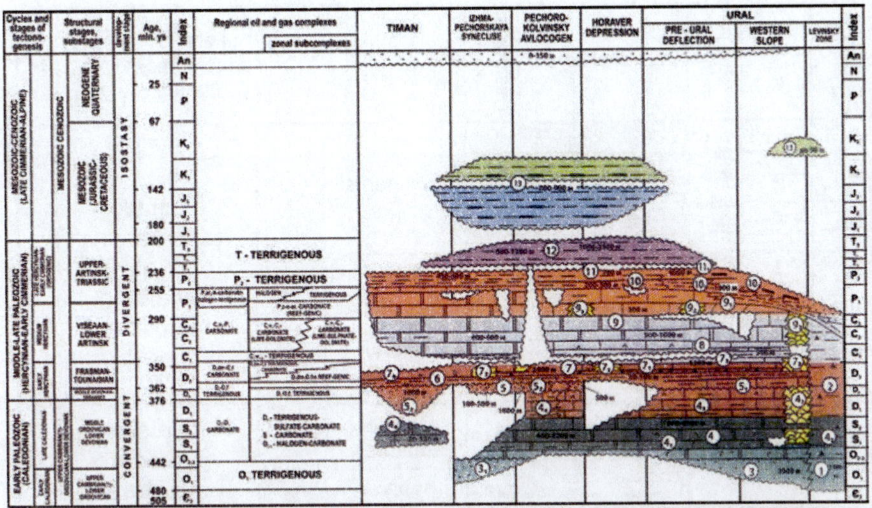


Formations and subformations: 1 is coarse-grained volcanogenic-sedimentary rift; 2 is limestone-siliceous-terrigenous; 3 is red-gray-colored terrigenous, $3_1$ is red-colored-oligomictic; 4 is carbonate, $4_1$ is halogen-terrigenous-carbonate, $4_2$ is dipsogenic, $4_3$ is sulfate-carbonate, $4_4$ is terrigenous-carbonate; 5 is gray-colored oligomictic terrigenous allochthonous, $5_1$ is carbonate-terrigenous autochthonous, $5_2$ is oligomictic graben; 6 is effusive-terrigenous; 7 is terrigenous-carbonate, $7_1$ is domanicoid, $7_2$ is riphogenic, $7_3$ is carbonate-terrigenous wedge-shaped filling; 8 is carbonaceous-terrigenous; 9 is carbonate, $9_1$ is marly (sezymskaya), $9_2$ is riphogenic; 10 is gray-colored lower molasse, $10_1$ is terrigenous-carbonate-halogen, $10_2$ is coal-bearing-terrigenous; 11 is gray-coastal-marine and lagoon-continental, $11_1$ is coal-bearing; 12 is red-colored continental; 13 is terrigenous and coal-bearing-terrigenous.

**Fig. 5.1** Structural and formational affiliation of oil and gas complexes of the Timan-Pechora sedimentation basin (according to [12])

(f) An important practical result was the creation of a unified catalog of the stratigraphic subdivision of wells in the Khoreyver, Varandey-Adzvin and S-Predural OGR. The result of lithofacies studies and structural constructions was the preparation of materials with a set of small-scale lithofacies, paleogeological and structural maps, facies and geological-geophysical profiles.

6. The geological history of the Timan-Pechora province development is characterized by the predominance of the marine sedimentation regime, which contributed to the formation of several oil and gas source strata and regional seals in the sedimentary cover. The main oil and gas producing strata in the province are predominantly terrigenous deposits of the Devonian (Lower Devonian, Upper Eifelian, Givetian and Pashian), carbonate-siliceous and terrigenous-carbonate rocks of the Frasnian and Famennian stages of the Upper Devonian, terrigenous and terrigenous-carbonate deposits of the Tournaisian age, coal-bearing formations of the Visean stage and terrigenous -carbonate deposits of the Lower Permian. In addition, as studies of recent years have shown, carbonate-argillaceous and carbonate deposits of the Ordovician – Lower Silurian are oil source deposits.
7. The basis for the allocation of OGC in the Timan-Pechora OGP is based on the features of the composition of geological formations and sub-formations, taking into account lateral and vertical zonality and those geological criteria that will determine the frames of structural stages and sub-stages. Paleogeographic, paleotectonic and lithological-facies prerequisites have determined the composition

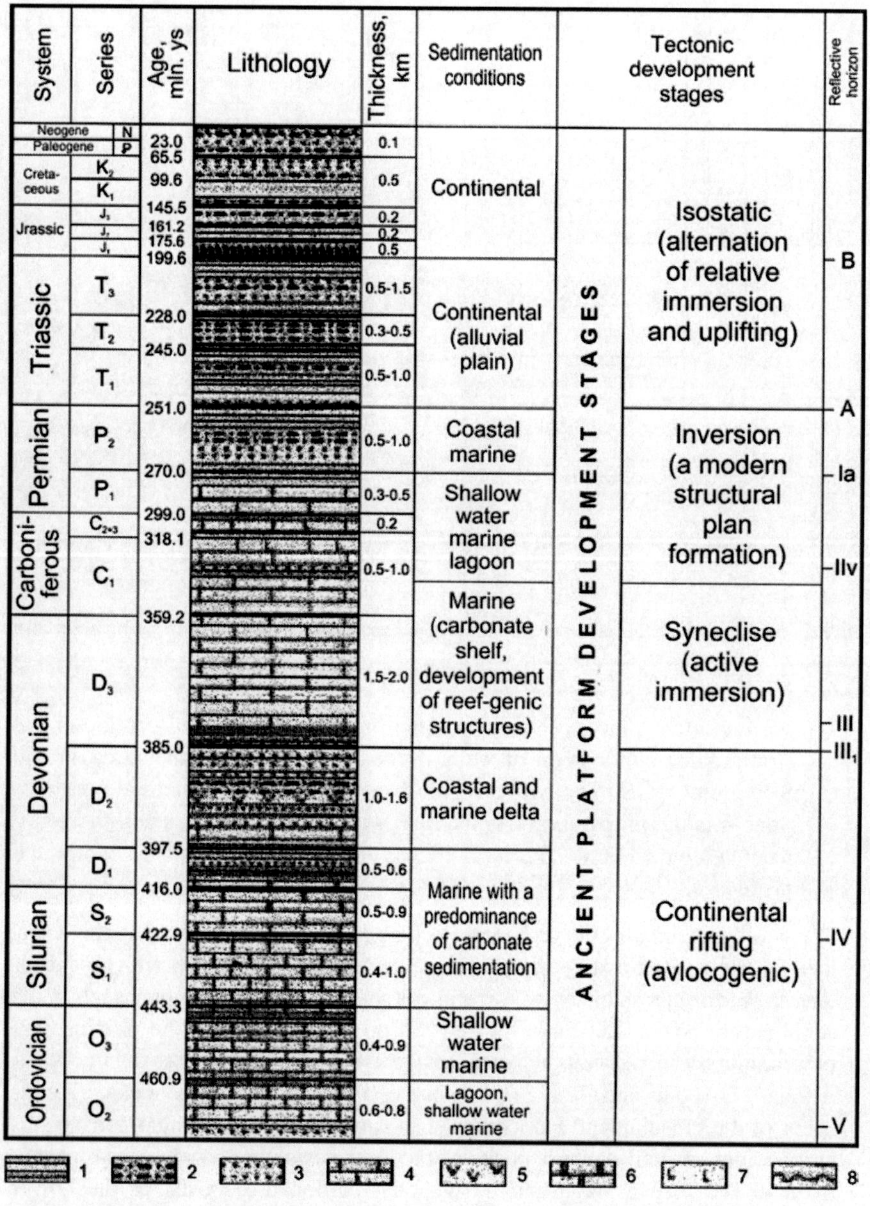


**Fig. 5.2** Development stages of the northeastern part of the East European platform (age column uses data from Episodes, 2001, Vol. 24, No. 2, pp. 102–114.) (according to [12])

of geological formations and their relationships in profile and area, which largely controls the distribution of reservoirs and seals and associated oil and gas accumulation zones. Since an individual tectonic regime is required for the accumulation of a certain formation or the paragenesis of subformations, the formation or their combination clearly fixes the development of individual paleotectonic

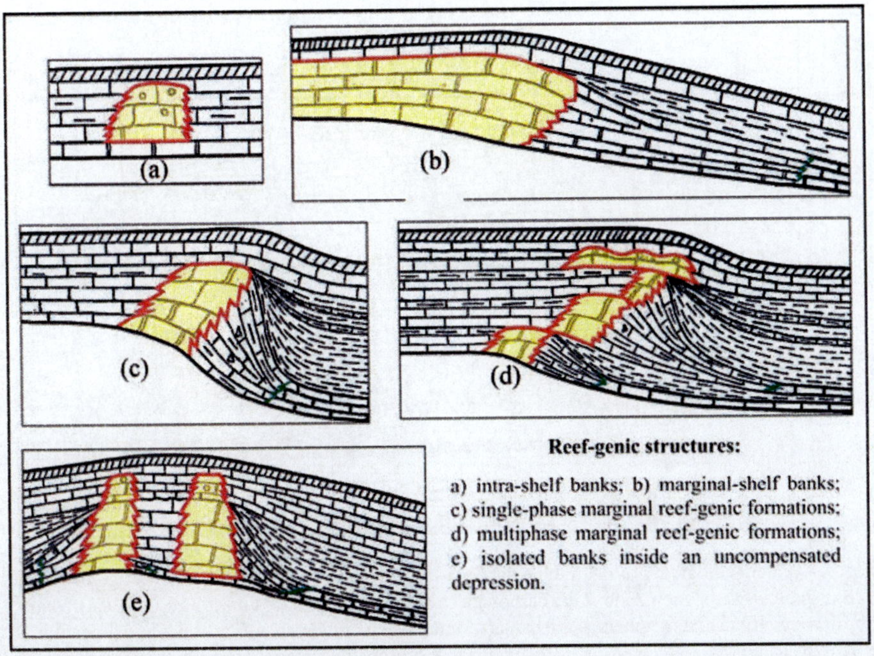


**Fig. 5.3** Types of reef-genic formations (according to [6])

structures and reflects all significant boundaries of the geological history of the region corresponding to the stages of tectogenesis. The stages of tectogenesis are grouped into cycles of geotectonic development, during which structural stages and sub-stages were formed. The subordination of cycles and phases of tectogenesis, structural stages and sub-stages, formations and sub-formations and oil and gas complexes of the TPP is shown in Table 5.1 and in Fig. 5.1.

8. In general, the Pechora deep-water basin arose due to short-term pre-sedimentary subsidence that occurred at the turn of the Middle and Late Frasnian (Figs. 5.3 and 5.4). Figure 5.3 shows the differences between marginal banks and marginal reefs that form marginal reef zones. Marginal reefs, compared with marginal banks, are characterized by higher and steeper slopes facing the adjacent deep-water basin. They indirectly differ in size, as well as in the thickness gradient of the fill strata adjacent to the marginal reef zones. Reefs are characterized by larger values than banks. The frontal – facing the deep-sea basin – the boundary of reefs and banks is determined without much difficulty both by lithological and seismic-geological features due to sharp changes in the thickness of pure carbonate strata, their replacement by clayey-carbonate slope deposits or adjoining clayey strata of filling to them. It is more difficult to determine the rear boundary of reef bodies, since the transition to back-reef facies is often gradual. Conventionally, it is carried out by a significant reduction in the thickness of the carbonate body and the replacement of pure carbonates with clayey ones.

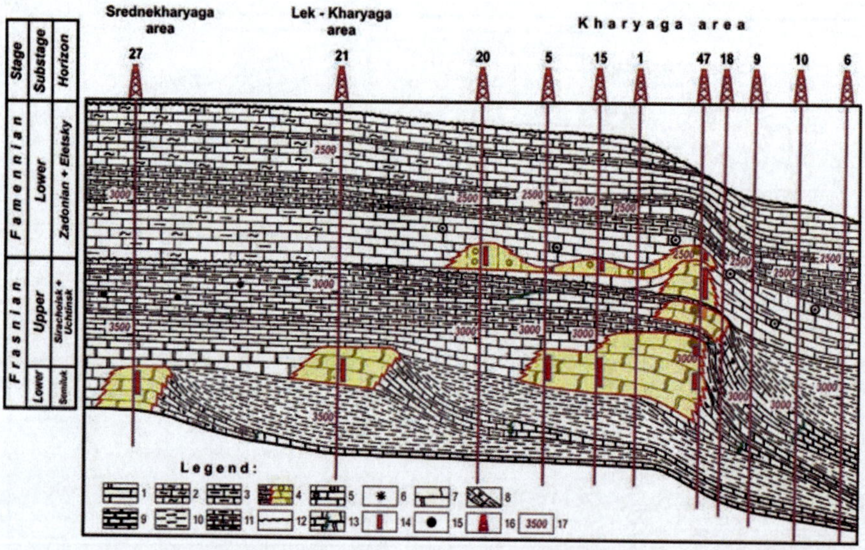


Fig. 5.4 Paleogeological profile of the Kharyaga section of the Frasnian marginal reef-genic zone. 1–5 is shallow-water shelf zone: 1 are carbonate strata, 2 are "nodular" limestones, 3 are carbonate-argillaceous strata and members, 4 are banks of interaclastic limestones, 5 are oncolitic limestones; 6 is redness color; 7 are reef-genic formations; 8 are pre-reef slope deposits; 9 are deposits of depression facies; 10, 11 are filling strata: 10 is clayey, 11 is carbonate-clayey; 12 are surfaces of erosion and stratigraphic breaks; 13 are zones of sharp lithofacies replacements; 14 are reservoir intervals; 15 are oil deposits; 16 are well sections; 17 is well depth, m (according to [6])

In areas where there are three to five large reef bodies located one above the other in sections of the marginal reef zone, it can be difficult to distinguish between reef formations of the Sirachoi and Ukhta age [6]. The Sirachoy bodies lie on clayey strata of the fill, usually have large thicknesses (100–300 m), have steep frontal slopes (width 0.5–1.5 km), and in back-reef sections they correspond to rather extended members of weakly clayey carbonates.

The Ukhta reef formations are replaced by red-colored and variegated carbonate-argillaceous deposits towards the reef back. When compared with the Sirachoy ones, they differed in greater variability in thickness (20–150 m), steepness of slopes. The distances at which young reef bodies are displaced relative to ancient those within the marginal zone also vary considerably. For example, in the Pashorsk area, the displacement of the Ukhta bodies relative to the Domanik ones is 3–4 km, in the sections of the Kharyaga swell – up to 20–25 km, but the most powerful the Sirachoi and Ukhta reef bodies are compactly located at the frontal bound of the entire zone (Fig. 5.4).

It should be noted, however, the uplifts that arise in the form of isolated banks and single reefs, then overlain by shallow-water-shelf carbonate layers, are often too gentle for the formation of productive traps, so that large-amplitude productive traps also formed here only when superimposed on banks and single reefs of tectonic structures. It is obvious the main significance of the reef zones and the oil and gas content of the Upper Devonian-Tournaisian subcomplex was not in the formation of structures and traps, but in the fact permeable reef bodies and zones were probable routes for active vertical, and sometimes lateral migration of hydrocarbons, both from the underlying more ancient deposits, and from the enclosing strata of the subcomplex. Above the powerful, extended reef areas, oil and gas accumulation zones were formed, with which most of the deposits discovered today are associated.

9. Quite significant materials on the typification and description of the natural reservoirs structure that have various genetic relation of the hydrocarbon complexes of the Timan-Pechora oil and gas field, including the distribution patterns of oil and gas deposits, are given in the monograph [12]. Here, the identified regularities in the placement of various morphogenetic types of natural reservoirs and the confinement of hydrocarbon accumulations to them are considered based on the use of a complex of formational, lithological-facies, historical-geological analyzes, taking into account the sedimentation features of reservoir-rocks properties and their post-sedimentary changes. In particular, they make it possible to reasonably search for and explore hydrocarbon deposits, which also ensures the discovery of a significant number of deposits in the TPP. Here, in particular, natural reservoirs of oil and gas bearing complexes (OGCs) have been identified: Lower Ordovician terrigenous; Middle Ordovician-Lower Devonian carbonate; Middle Devonian-Franzian terrigenous; Domanik-Tournaisian carbonate; Lower-Upper Visean terrigenous; Upper Visean-Lower Permian carbonate; Lower Permian carbonate-halogen-terrigenous; Upper Permian terrigenous; Triassic terrigenous.
10. The distribution of hydrocarbon deposits in the sediments of various systems and in various structural zones of the TP OGP is not the same. Industrial hydrocarbon potential and oil and gas appearance have been established in almost all formations that make up the structural stages and substages of the sedimentary cover, except for the Lower Ordovician and Jurassic-Cretaceous terrigenous deposits, although residual oil saturation was noted in the Jurassic sands and Ordovician sandstones according to the core study. A wide variety of lithological and paleotectonic conditions of oil and gas content of the Middle Devonian-Lower Frasnian terrigenous complex was revealed, associated with the primary lithofacies variability of deposits, the manifestation of stratigraphic breaks, erosion, the three-layer structure of reservoirs, and the features of catagenic transformation of rocks in various structural zones: aulacogen, foredeep, on stable blocks [6]. It should be noted the dependence of oil and gas potential with marginal reef zones and, to a lesser extent, with isolated banks as the most important lithological criterion for the Upper Devonian-Tournaisian predominantly carbonate subcomplex. The deposits are located in layers of shallow-water shelf carbonates above reef formations and under zonal and local seals in shallow-water shelf strata. The zonal Upper Frasnian (Ukhta) and Famennian (Elets-Ust-Pechora) seals control the placement of the vast majority of deposits in the subcomplex. The largest zone of oil and gas accumulation in the Upper Devonian-Tournaisian complex in the north of the province coincides with the band of adjacent Frasnian marginal reef zones and banks, which is oriented sharply discordant with the strike of large structures of the modern structural

plan. An analysis of the productive traps showed contrary to popular belief, they are all structural, often combined (structural-lithological, structural-stratigraphic), with structural-reef traps being a minority.

11. The location of oil and gas fields in the Timan-Pechora province is due to a number of factors, among which tectonic plays an important role [10].

Most of the accumulations of hydrocarbons are confined to systems of disjunctive disturbances. This dependence is also confirmed by the close relation of the sedimentary cover dislocations, within which the deposits are discovered, with the block structure of the basement. This is especially evident in tectonically mobile zones, where deposits are associated with near- and supra-fault structures. In more stable basins, this trend is less expressed, but here, too, part of the deposits is controlled by deep faults. The most important features of a regional fault are their structure-controlling role, age, and mode of manifestation at the platform stage of development. The nature of their impact on the processes of oil accumulation largely depends on this. As an example, the process of oil and gas accumulation during the formation of this zone at the junction of the Kolvin megaswell and the Khoreyver depression is considered. This zone is characterized by a block, sharply differentiated structure, asynchrony of tectonic movements in time, which predetermined the mismatch of structural plans, and in some cases their inverse relationships along different marker horizons [10]. The formation of traps in such zones is due to a system of intersecting longitudinal and transverse faults, which form a mosaic structure of horst and monoclinal blocks, with which individual deposits are associated. The modern structural plan often does not reflect the paleotectonic conditions for the formation of deposits. In this regard, the main attention in the near-fault zones should be given to the identification and search for paleotraps, which could later turn into epigenetic (sealed) or reformat. The asynchronous movements of the blocks lead to a change in their hypsometric position in the section of the sedimentary cover. As a result, the formation (complete or partial) of the deposits in the lowered blocks and the vertical migration of hydrocarbons into the uplifted ones occur. Moreover, in the presence of a conductive plane, a flow from younger deposits to ancient ones is possible (for example, from the Middle Devonian to the Silurian).

Thus, due to the high activity of tectonic development in the northeast of the Russian Platform, inversion movements were more pronounced, which led to structural changes, the emergence of important productive strata and erosion. A reflection of the greater tectonic activity of the Timan-Pechora province is also the large values of the minimum amplitudes of structures in productive traps.

In these complicated lithological and tectonic conditions, in order to increase the efficiency of geological exploration, further in-depth development of complex criteria for the zonal and local forecast of the oil and gas content of the Paleozoic strata of the Timan-Pechora province is necessary.

## 5.2 Detailing the Structure of Productive and Promising Horizons of the Subdomanik Sediment Complex (Southern Part of the Michayu-Pashnin Swell of the Izhma-Pechora Depression)

The subsection discusses the features of the formation and geological structure of the most important oil-prospective strata of the sedimentary cover in the study area – within one of the areas of the southern part of the Michayu-Pashin swell and the Tebuk step, which complicate the Omra-Lyzha saddle of the Izhma-Pechora structure. The purpose of the research is to interpret the materials of geological and geophysical works in order to refine the geological and geophysical model of the Savinobor field using high-resolution seismic data. Accounting the velocity heterogeneities in the upper part of the section, preprocessing and reinterpretation of the previously obtained CDP 3D survey data was performed. A processing graph has been selected to obtain high-resolution time sections in the studied interval of the wave field. Using a system of functional transformations, a complex interpretation of logging data in the terrigenous-carbonate profile of wells 3 and 301 of the field, located in the southern and central parts of the area, was performed.

Using the high-resolution seismic method, 3D cubes of effective acoustic impedance (AI) and reflection coefficient (RC) were obtained – models of a real geoenvironment, which depict a detailed internal structure of the studied interval. All the necessary maps were built: isochrones, structural, isopach (interval times), maps of seismic facies zoning for the productive strata of the Frasnian subdomanik complex and propective horizons: Ik ($P_1k$), ($P_1ar$), II-III (C-D), (F5 $D_3fm_2$), $D_3zd$, $D_3dm$, $D_3dzr$, B-3к + B-2п, $D_2st(k)$. In addition, the results of reinterpretation of structural constructions are analyzed considering drilling data from wells 3 and 301. A classification of seismic facies is given based on the use of shapes and values (reference curves – clusters) of 3D cubes of AI and RC in deposits of productive and promising horizons – Ik, ($P_1ar$), II-III (C-D), $D_3zd$, B-3к + B-2п and $D_2st(k)$. Separate oil-prospective objects of structural, tectonically and lithologically screened types were identified in the deposits of the upper part of the Famennian horizon (II-III (C-D)), the upper part of the Zadon horizon ($D_3zd$) and in the reservoir beds of the B-3 k + B-2p units of the Dzhyer-Yaran deposits horizon ($D_3dzr$ + jr).

The obtained results of seismic data reinterpretation served as the basis for further research – a reasonable revision (reassessment) of reserves and resources of hydrocarbons in the investigated exploration area, primarily for deposits of the productive formation B-3к + B-2п ($D_3dzr$ + jr).

### 5.2.1 *Geological Structure Features of the Study Area*

In the current structural plan, the Michayu-Pashninsky swell is a chain of echelon-shaped medium-sized (6÷12 × 3÷7 km) brachyanticlinal folds of north-northwest strike with a total length of 160 km, limiting the Izhma-Pechora depression of the

Pechora syneclise from the Verkhnepechora depression of the east pre-Ural foredeep. The southern part of the swell is characterized by a sharply asymmetric structure: the western wings of the structures pass into a large flexure (possibly a flexure-fault zone) in the sedimentary cover, confined to the zone of the Ilych-Chikshin deep fault, the eastern ones are much younger.

In the direction from north to south, the dome of structures subsides and their amplitudes increase. The absolute depths of the structure arches along the base of the Domanik suite correspond to the following values: Michayu −2040 m, S-Savinobor −2140 m, Savinobor −2385 m, V-Savinobor −2400 m, Pashnin −2300 m. The highest values of absolute depths are along the top of the Artinskian carbonates: Michayu −785 m, S-Savinobor −749 m, Savinobor −870 m, V-Savinobor −868 m, Pashnin −822 m. In both cases, the lowest levels are arches of the V-Savinobor and Savinobor structures.

Based on the seismic exploration work carried out here earlier in 1990–1991, the basement surface, broken by tectonic faults of various orientations into separate blocks, gradually plunges to the east (*Nosov A.P., Kozitsina N.P.* et al., *2007*). The hypsometric position of the arched part is −4600 m. The age of the foundation is Riphean-Vendian. According to the Silurian deposits and the basement surface (horizons $IV_2$, VI), the Savinobor structure is presumably complicated by intrusion penetrating into the Lower Silurian deposits. According to the deposits of the Upper Silurian (horizon $IV_2$), the structure is presented in the form of a structural nose, complicated from the north by tectonic faults. The dimensions of the structure in the isohypse contour −3125 m are 4.6 × 1.0 km, the amplitude is 65 m. A low-amplitude structure was revealed in the Upper Devonian terrigenous subdomanik deposits (horizon $IIIf_1$). It is an anticline fold of submeridional strike, 6.0 × 1.7 km in size in the isohypse contour −2500 m, amplitude 50 m. From the west and east, the structure is limited by low-amplitude tectonic faults.

In the area under study, it is of particular interest to consider the conditions of sedimentation, namely, the characteristics of the main conditions of terrigenous sedimentation of subdomanik deposits and the determination of geological models of hydrocarbon reservoirs for this part of the section. In this regard, the development of delta formations in the Middle Devonian strata of the Timan-Pechora province, which was previously pointed out by a number of authors is of the greatest interest (*V.A. Grossgeim (1971), G.F. Rozhkov (1984), N.S. Oknova (1983), A.M. Sklovsky (1980), etc.*) [12]. These authors considered the Middle Devonian strata of Timan as continental-river, and in the south of the Izhma-Pechora syneclise, along the Yarega Dzhyer – Pashnya line, the environments of the paleo-delta of a fan-shaped configuration, strongly protruding into the open sea, were reconstructed (Fig. 5.5). As noted in [12], these constructions were based on the analysis of a large amount of data on fractional sieve granulometry and the azimuthal orientation of flow-type textures. However, the "Starooskol" horizon, for which the most detailed analysis was carried out, at that time included strata of different ages (actually Starooskol, Yaran and Dzhyer), divided over most of the territory by a break. However, the overall picture illustrating the influence of the relatively large pra-Yarega flow is quite clear (Fig. 5.5). This pra-Yarega flow also spread in the Lower Frasnian time to the Savinobor prospecting area.

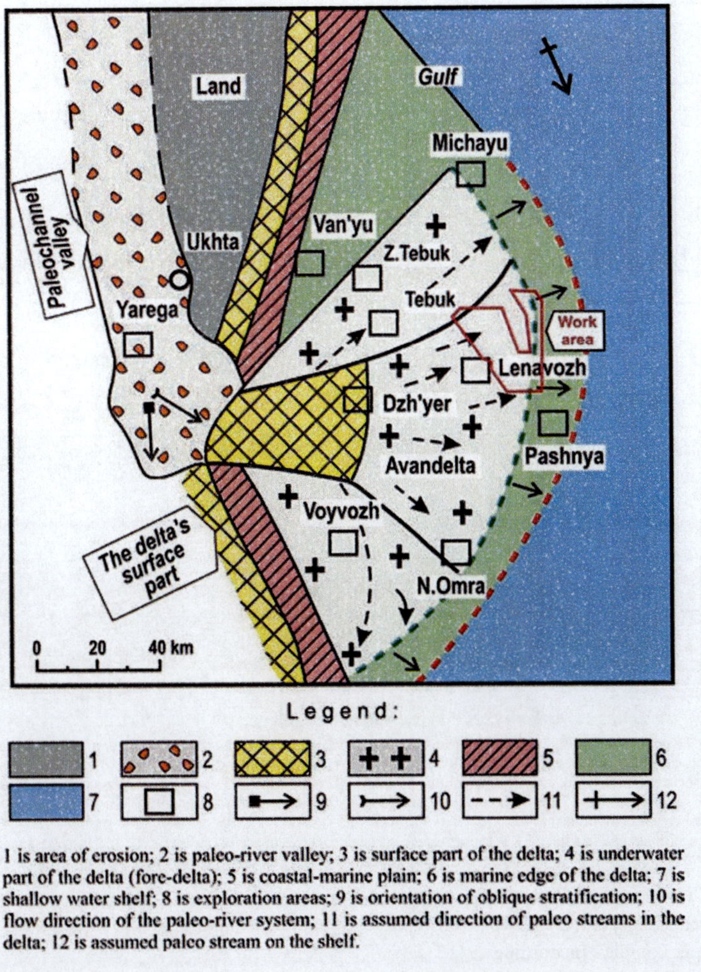


**Fig. 5.5** Paleodelta reconstruction of a relatively large pre-Yareg stream (according to [12])

As an example of massive reservoirs in the sediments of the Yaran and Dzhyer horizons, formed in shallow sandy-silty sediments (composing these reservoirs) under the influence of one of the largest main flows from the north-northwest along the western side of the Pechora-Kolvin aulacogen, Fig. 5.6 shows sections of the structure of the Lower Frasnian deposits, which are built on a significant amount of well data. On this Fig. 5.6 clearly shows the complex geometry of the Lower Frasnian natural reservoirs of the type the “filling nested valleys” and coastal-marine genesis [12]. The presented sedimentation environments as a whole determined the irregularity of the stratigraphic volumes of the reservoirs in the study area, the seaward nature of the reservoir, the coarsening of the granulometric composition of sandstone layers (especially the Dzhyer, Yaran and Starooskol), as well as their relationship with the underlying and overlying deposits.

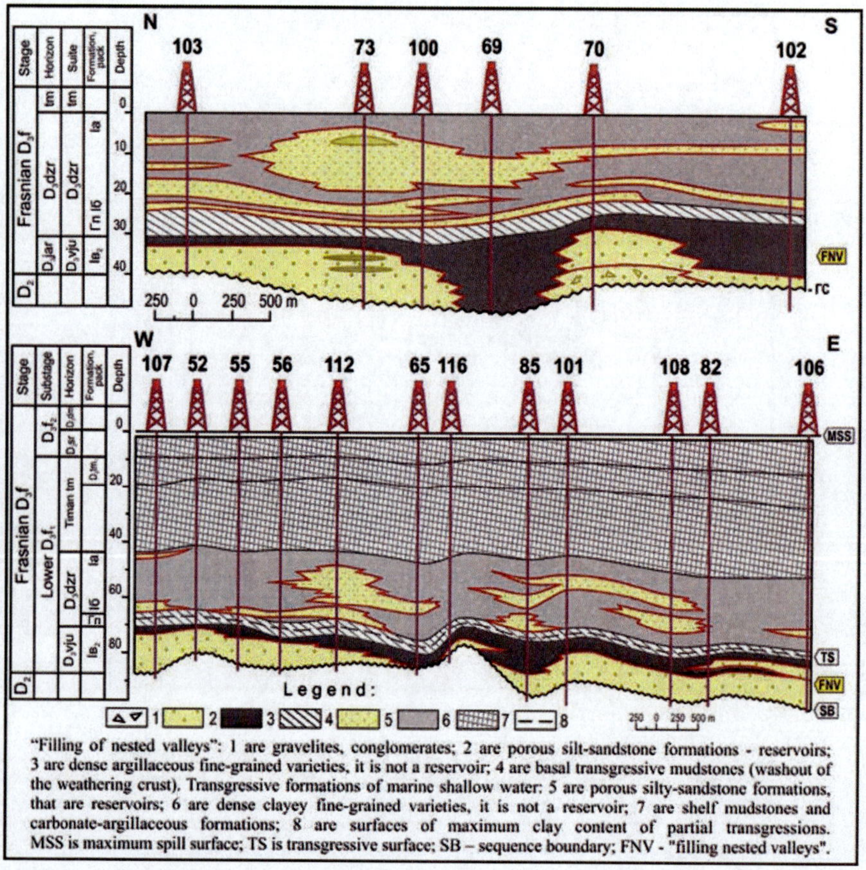


**Fig. 5.6** The structure of the Lower Frasnian natural sub-reservoirs of the "filling nested valleys" type (sequence $Iv_2$) and coastal-marine genesis (layers Ia-б) and their position within the Frasnian subdomanik complex (according to [12])

The results of structural constructions for the most productive deposits – the top of the B-3 formation of the Yaran horizon, made on the basis of the use of 2D seismic data, are shown in Fig. 5.7. It is clearly seen here the complex reclinal of the Savinobor anticline structure is distinctly traced along the deposits under consideration along an extended tectonic fault, recorded throughout the prospecting area from northwest to southeast. The structure, to which productive well 3 Savinobor is confined, is separated from the northwestern uplifted area (from the location of well 25) by a sublatitudinal tectonic fault, which is probably one of the reasons for the absence of hydrocarbons in well No. 25 (Fig. 5.7). Naturally, the structural constructions made on the basis of CDP 2D materials turn out to be not sufficiently accurate and detailed. But these constructions give a good first idea about the features of the geological structure of the B-3 formation at the work site.

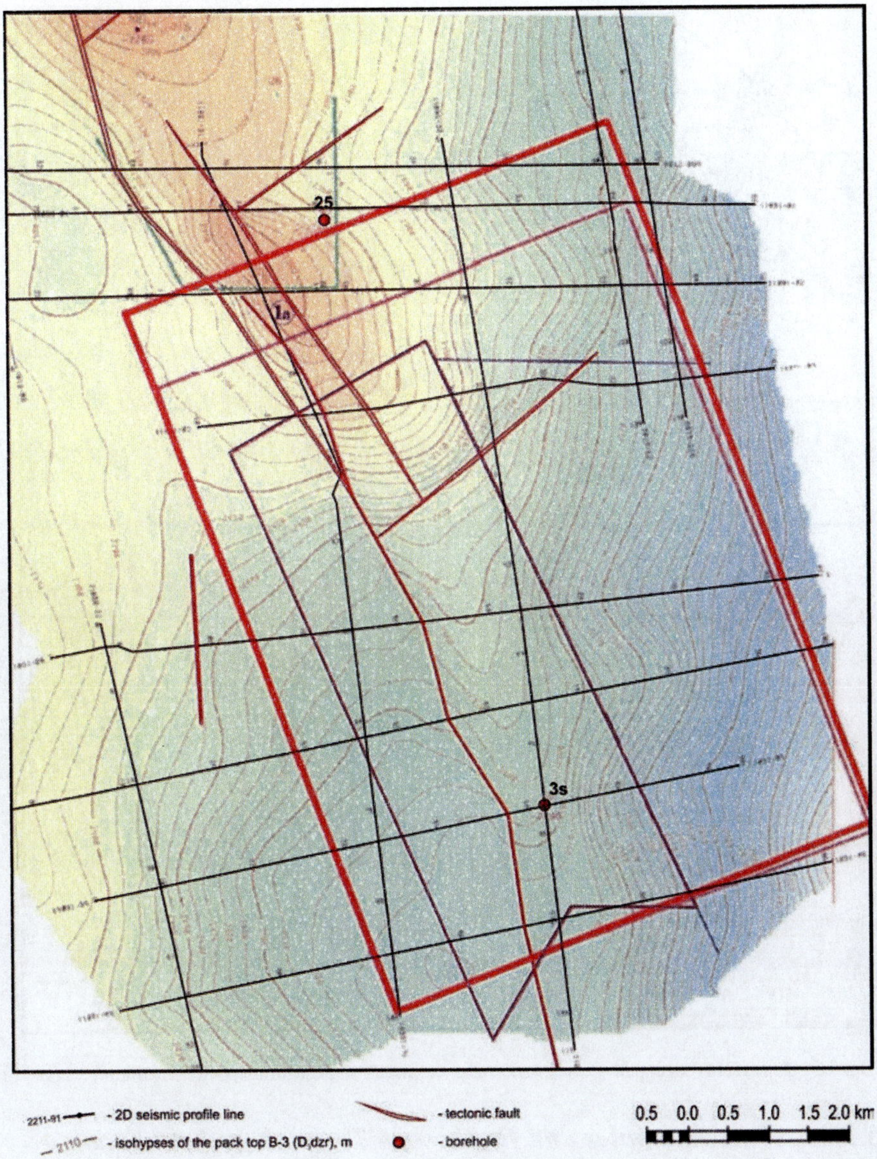


**Fig. 5.7** The 3D work site within the Savinobor field at object B-3 ($D_3jr$) (according to JSC "Komineft", 2000)

In addition to the construction of the foundation surface, made according to seismic data, similar constructions were also made in the area of work according to gravity and magnetic survey data (Fig. 5.8) (*Karpyuk T.A., Shabalin S.V., Osmanova N.S. and others, 2005*). On Fig. 5.8 it is clearly seen within the Savinobor prospecting area in the east direction there is a gradual subsidence in the form of a large-inclined monocline, stepwise plunging into the Upper Pechora depression.

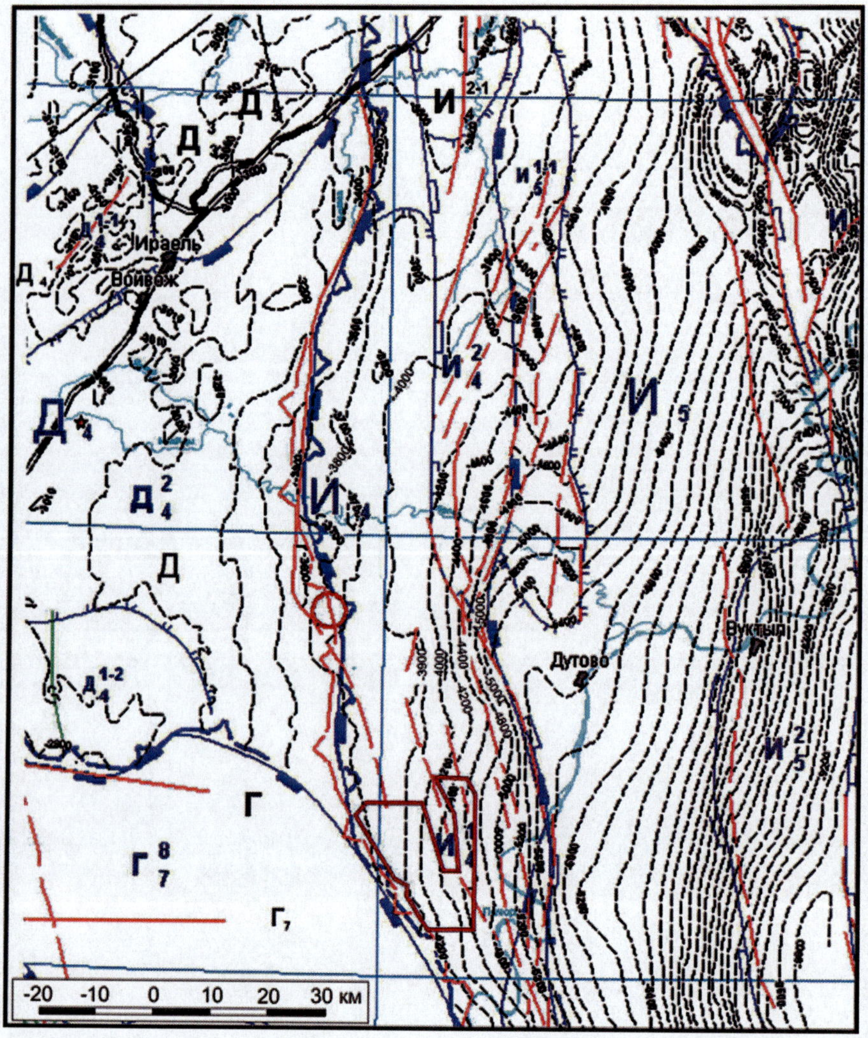


**Fig. 5.8** Copy from the map of the basment surface of the TPP (according to JSC "Komineft", 2001)

## 5.2.2 *Some Examples and Features of Technology Application*

The obtained materials of seismic 2D and 3D CDP-surveys and geological and geophysical works of well logging in the area under consideration were preliminarily processed and interpreted according to the methods and technologies (using processing and interpreting complexes and computer equipment) described in some detail by us above (in Chap. 3) and in the work [13, 14]. It should be noted on the Savinobor area, as a whole, areal detailing work was carried out – seismic survey of two 3D cubes with a volume of 56.39 km$^2$ (3D(N) – northern cube) and 74.5 km$^2$ (3D(S) – southern cube), as well as prospecting work – 2D seismic survey with a volume of 262 km. Additionally, in the complex interpretation of seismic data, 2D seismic materials of previous years were used in the amount of 407.6 km. The main processing of seismic data was carried out using the Western Geco OMEGA2 processing system. Integrated interpretation of seismic data was performed using software tools – the HRS-Geo technology developed by the authors and the interpretation system GeoGraphix R2004.1 v.8.5 of the Landmark company.

On the sections of the cubes of effective acoustic impedance (AI) and reflectivities (RC), in the process of interpretation, the main reflecting boundaries were previously identified and correlated, corresponding to the horizons of productive and promising deposits from the Kazanian stage of the Upper Permian to the deposits of the Wenlockian stage of the Lower Silurian. In particular, the following main reflecting boundaries were correlated, which were assigned indices in accordance with their lithological and stratigraphic confinement:

1. **I$P_2$kz** is the top of the Kazanian stage ($P_2$kz) deposits of the upper Permian, identified with the eroded surface of the carbonate layer. The positive boundary is a positive jump in AI values associated with the transition (from top to bottom) of the marl layer of the Tartarian stage into carbonates.
2. **Ik** is the bottom of the Kungurian ($P_1$k) deposits of the Lower Permian. Positive reflector along the AI, associated with the boundary of the transition from terrigenous deposits of the Lower Permian to carbonate deposits. Stratigraphically, the reference horizon Ik is assigned to the base of the terrigenous Kungur deposits. There is a certain amount of conventionality in linking the reflection to the bottom of the kungur, because the reflection is formed from the lithological-stratigraphic benchmark – the top of "pure" carbonates, which in this case is of Kungur age (the top of the Filippov ($P_1$fl) horizon) and stands out 15–20 m above the base of the Kungur deposits. This reflective border best suits the bottom of the Kungur reflections.
3. **II-III(C-D)** is the top of the deposits of the Nyumylga horizon ($D_3$nm) of the upper substage of the Famennian stage of the Devonian. The positive interface along the AI corresponding to the top of the Upper Devonian carbonate deposits in contact with the overlying clayey member of the Tournaisian ($C_1$t) stage. The horizon controls the picture of the angular unconformity of the Famennian strata exposed under erosion with the overlying Carboniferous deposits. The amplitude and quality of the boundary depends on the thickness of the Tournaisian clayey patch overlying the eroded surface of the Upper Devonian deposits.
4. **III$fm_2$ (Ф5 $D_3fm_2$)** is top of layer Ф5 of the middle substage of the Famennian stage of the Upper Devonian. Positive reflecting boundary along the AI, confined to the top of the Ф5 carbonate layer, identified in the lower part of the Middle Famenian.
5. **III$fm_1$ (Ф1 $D_3fm_1$)** is top of layer Ф1 of the lower substage of the Famennian stage of the upper Devonian. The positive reflector along the AI, confined to the top of the Ф1 carbonate layer, identified in the lower part of the Yelets horizon.
6. **$D_3$zd** is the top of the deposits of the Zadon horizon of the lower substage ($D_3fm_1$) of the Famennian stage of the upper Devonian. The positive interface

along the AI corresponding to the top of the carbonate deposits in contact with the overlying clay member.

7. **IIId ($D_3dm$)** is the bottom of the Domanik horizon of the Middle substage ($D_3f_2$) of the Frasnian stage of the Upper Devonian deposits (*the top of the Sargay horizon of the Middle substage of the Frasnian stage*). Negative reflective interface along the AI, formed near the base of the Domanik deposits. The reflector is confined to the boundary of high-velocity deposits of the Domanik and lower-velocity deposits of the Sargay horizon.
8. **IIItm ($D_3tm$)** is top of the deposits of the Timan horizon of the Lower substage ($D_3f_1$) of the Frasnian stage of the Upper Devonian. Positive reflective boundary along the AI, traced in the study area and confined to the top of deposits composed of dark gray clays with subordinate interlayers of limestone and volcanic sand, which are overlain from above by bituminous shales, clayey with interlayers of limestones, marls and clays of the Sargay horizon ($D_3sr$).
9. **IIItm1 ($D_3tm1$)** is reflecting interface within the deposits of the Timan horizon of the Lower substage ($D_3f_1$) of the Frasnian stage of the Upper Devonian. Predominantly negative in AI reflecting boundary (in some parts of the area turning into a positive one), confined to the top of deposits of high clayey limestones, clays and siltstones and volcanic sand.
10. **$D_3$dzr** is top of the deposits of the Dzhyer horizon of the lower substage ($D_3f_1$) of the Frasnian stage of the upper Devonian. Positive AI reflector confined to the top of deposits composed of two potentially productive strata I-b, I-a, which are represented by dark gray siltstones, quartz, micaceous and dark gray mudstones, platy, with interlayers of dark gray clays. The boundary is also distinguished on well logging curves (in particular, on the PZ curve) in the form of a high-resistance logging benchmark at the boundary of the Dzhyer and Timan deposits.
11. **B-4** is top of deposits of layer B-4 of the Yaran horizon ($D_3jr$) of the lower substage ($D_3f_1$) of the Frasnian stage of the upper Devonian. Reflecting boundary positive in AI, confined to the top of productive deposits composed of fine-grained, gray-brown quartz sandstones and dark gray, quartz, micaceous siltstones.
12. **B-3** is top of deposits of layer B-3 of the Yaran horizon of the Lower substage ($D_3f_1$) of the Frasnian stage of the Upper Devonian. Positive AI reflector confined to the top of productive deposits composed of fine-grained, gray-brown quartz sandstones and dark gray, quartz, micaceous siltstones, separated by clayey and dense rocks.
13. **B-2** is top of deposits of layer B-2 of the Yaran horizon of the Lower substage ($D_3f_1$) of the Frasnian stage of the Upper Devonian. Positive AI reflective interface confined to the top of productive deposits composed of sandstones, siltstones and clays separated by low-resistivity clay interlayers.
14. **B-1** is top of deposits of layer B-1 of the Yaran horizon of the lower substage ($D_3f_1$) of the Frasnian stage of the upper Devonian. A positive AI reflecting boundary confined to the top of potentially productive deposits, composed in the same way as the B-2 reservoir, of sandstones, siltstones and clays separated by low-resistivity clay interlayers.

15. **IIst(k)** is the top of the deposits of the Starooskol suprahorizon ($D_2$st(k)) of the Givetian stage of the Middle Devonian (*simultaneously the bottom of the Yaran horizon of the Frasnian stage*). The reflecting interface, positive in AI, is the top of the stratum, represented by sand–siltstone deposits, which is indexed as the "main stratum" and characterized by negative SP anomalies and minimal GR readings. The sandstones are gray, light gray, multi-grained, well sorted. Siltstones quartz and chlorite-quartz, gray and gray-green, with frequent siderite concretions.
16. **IIst(p)** is the base of the Starooskol deposits ($D_2$st(p)) of the Givetian Stage of the Middle Devonian. A positive AI reflector formed near the erosion boundary of the underlying Eifelian deposits ($D_2$ef) of the Middle Devonian.
17. **IIIkd** is the top of the Eifelian stage ($D_2$ef) Kedrov horizon ($D_2$kd) of the Middle Devonian deposits, composed essentially of dark gray and gray-brown clays with slip mirrors. A positive AI reflector formed at the top of the carbonate layer II (erosion boundary), represented by gray, clayey, cryptocrystalline limestones.
18. **III-IV $S_2$p** is the top of the Przhidolian stage ($S_2$p) Greben horizon ($S_2$gr) of the Upper series ($S_2^2$) of the Silurian deposits. It is a positive AI reflective horizon that forms at the boundary of the transition (from top to bottom) of clays into dense clayey carbonates. The horizon is composed of limestones, marls and mudstones.
19. **IV $S_2$ld** is top of the Ludlovian stage ($S_2$ld) Gerdyu horizon ($S_2$gj) of the Upper series ($S_2^1$) of the Silurian deposits. The positive boundary is a positive jump in AI values associated with the transition (from top to bottom) of the clay interlayer into carbonates. The horizon is characterized by a layered structure, represented by clayey limestones, secondary dolomites with rare detritus, their clayey varieties and thin layers of mudstones.
20. **IV $S_1$v** is the top of the deposits of the Sed'elian horizon ($S_1$sd) of the Wenlockian stage ($S_1$v) of the lower series ($S_1^2$) of the Silurian. The positive boundary is a positive jump in the AI values, is associated with the transition (from top to bottom) of the interlayers of marl-anhydrite-marl into pure carbonates. The horizon, as is known, is represented by limestones and dolomites, the fractures are filled with clays and calcite.

Based on the use of seismic data from AI and RC for 2D and 3D surveys in the study area, structural constructions were made for all these horizons. Let us give only some examples were used in the joint complex interpretation of HRS and well log data.

So, Fig. 5.9 shows the structural forms of the Nyumylga horizon ($D_3$nm) deposits top – horizon II-III (C-D) of the Upper substage of the Famennian stage of the Devonian in the form of a structural map h(x, y), which reflect the structure of possible perspective objects in carbonate deposits of the Upper Devonian. These objects are in direct contact with the overlying clay member occurring in the lowest part of the Tournaisian stage ($C_1$t) Chernyshin suprahorizon ($C_1$čn).

General patterns of tectonic structure on such maps are consistent with the Lower Permian deposits. There is a monoclinal submersion of the sediments of the horizon in question in the east and partially in the southeast directions. At the same time, the

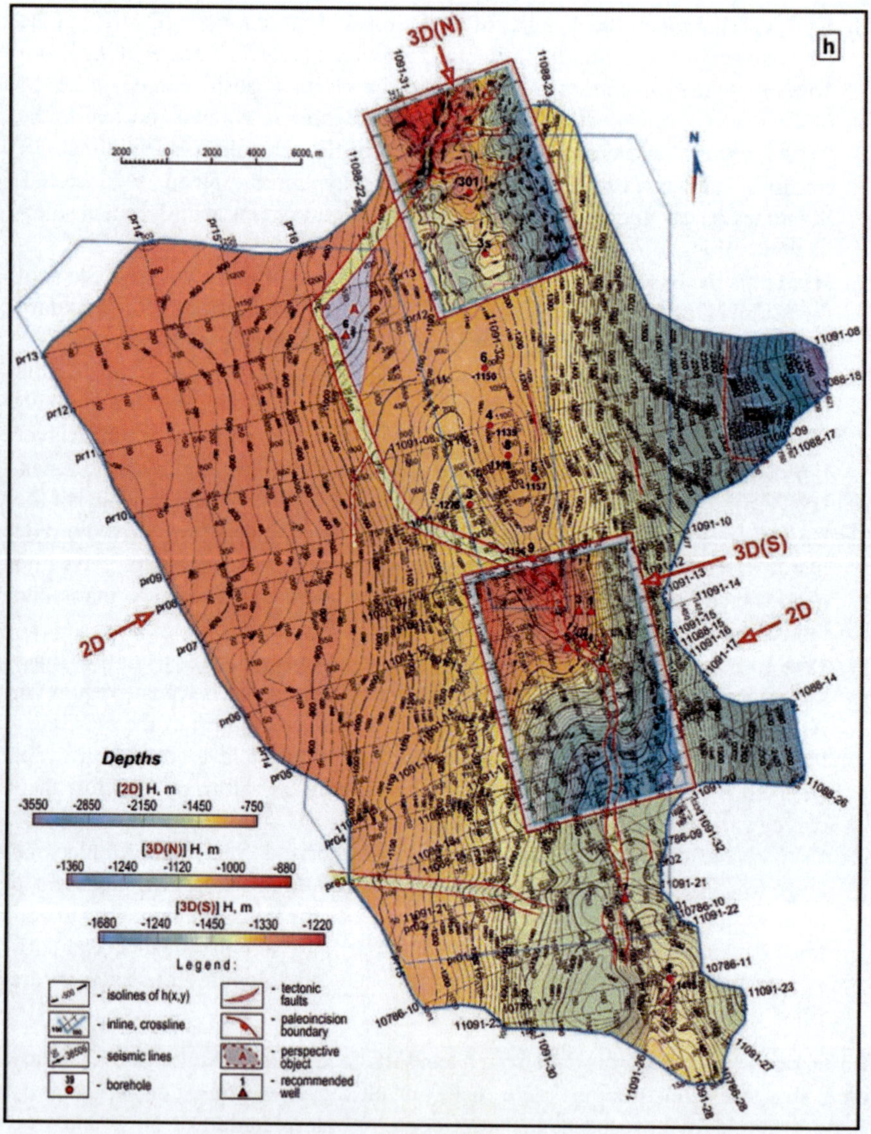


**Fig. 5.9** Structural map along the top of the deposits of the upper part of the Nyumylyg horizon ($D_3$nm). Upper substage of the Famennian stage – horizon II-III (C-D)

morphology of the surface of the deposits of horizon II-III (C-D) also retains the inherited character of development with the underlying horizons of the Devonian deposits.

A distinctive feature of the geological structure of the sediments under consideration within the Savinobor area is their complication with a paleoincision formed by a rather dynamic paleo-river flow, which manifested itself in the Upper Famennian time. On Fig. 5.9 the contour of the channel of this paleo-river is marked with red lines. The sediments of the upper substage of the Famennian stage (horizon II-III(C-D)) are indented by the channel of this paleo-river and its tributaries in different parts of the area of 2D and 3D surveys. It seems that one of the most interesting in terms of oil and gas potential is the site located in the northern part of the area (immediately adjacent to the northeastern bend of the paleo-river channel contour, designated by us as a perspective object "A"). A tectonically screened trap is quite confidently fixed on this bend of the paleochannel. At the same time, on fragments of sections AI and RC along 2D profiles pr13 and pr12, promising object A confidently appears in the deposits under consideration, the contour of the paleo-river channel is clearly visible together with the tectonic faults accompanying these contours (see above in Fig. 3.24).

It should be noted the structural forms of the upper Devonian Famennian Stage Ф5 layer ($D_3fm_2$ (Ф5 $D_3fm_2$)) largely repeat the geological structure features of the dolomite and calcareous deposits of this section part. In these deposits, as noted above, oil sweatings are often observed (in well 3 Savinobor, rare fractures with accumulations of liquid oil were found in the considered interval of the section). Here, the largest anticline structure (with its southern periclinal part) is confined to the northern part of the 3D seismic data cube (3D(N) – northern cube). A distinctive feature of this anticline structure is its complication by an incision formed by a rather dynamic paleo-channel flow, which manifested itself in the Upper Famennian time. On Fig. 5.10 the channel of this paleo-river is marked with red contour lines. These contours of the paleo-river turn out to be less highlighted, but the main paleo-channel manifested itself quite clearly. To the south of the presented contour of the paleo-river, in the 3D-survey area, several local uplifts, small in their linear dimensions and amplitude, are also distinguished.

On the coherence map of the wave field amplitude cube along the paleo isochronous surface of horizon II-III (C-D) (Fig. 5.11), the irregularity of the seismic record (in the form of background dimming of increased intensity) clearly marked the above noted contour of the paleo-river channel and the "sleeves" of smaller ones adjacent to it paleoincisions. At the same time, the dimming lines of the dynamics of the seismic record significantly refine the contours of both banks of the paleo-river (Fig. 5.11). It should be emphasized in the rest of the study area, the coherence of the wave field amplitudes manifests itself in a somewhat different way from that part of the geological environment associated with the contour of the paleo-river channel (the areas of irregularity in the dynamics of seismic recording in the area outside the paleo-river have an isometric shape).

For the upper part of the deposits of the Nyumylga horizon ($D_3nm$ + 15 ms), seismic facies were classified for approximately the same purpose using the neural network technology (NNT) of clustering (Stratimagic software package) according to the shape of the reference curves and their values for the cube of acoustic impedance (AI), wave field (WF) and reflectivity (RC) (Fig. 5.12a–c). The classification results obtained in this way for almost all three forms of seismic data (AI, WF, and RC) confidently show the contour system of the paleochannel and its paleotributaries. In addition, zones with increased AI and a special form of reference curve shape distribution are also confidently distinguished over the entire area of the area (Fig. 5.12a). These zones appear in the form of a brown-red-yellow palette

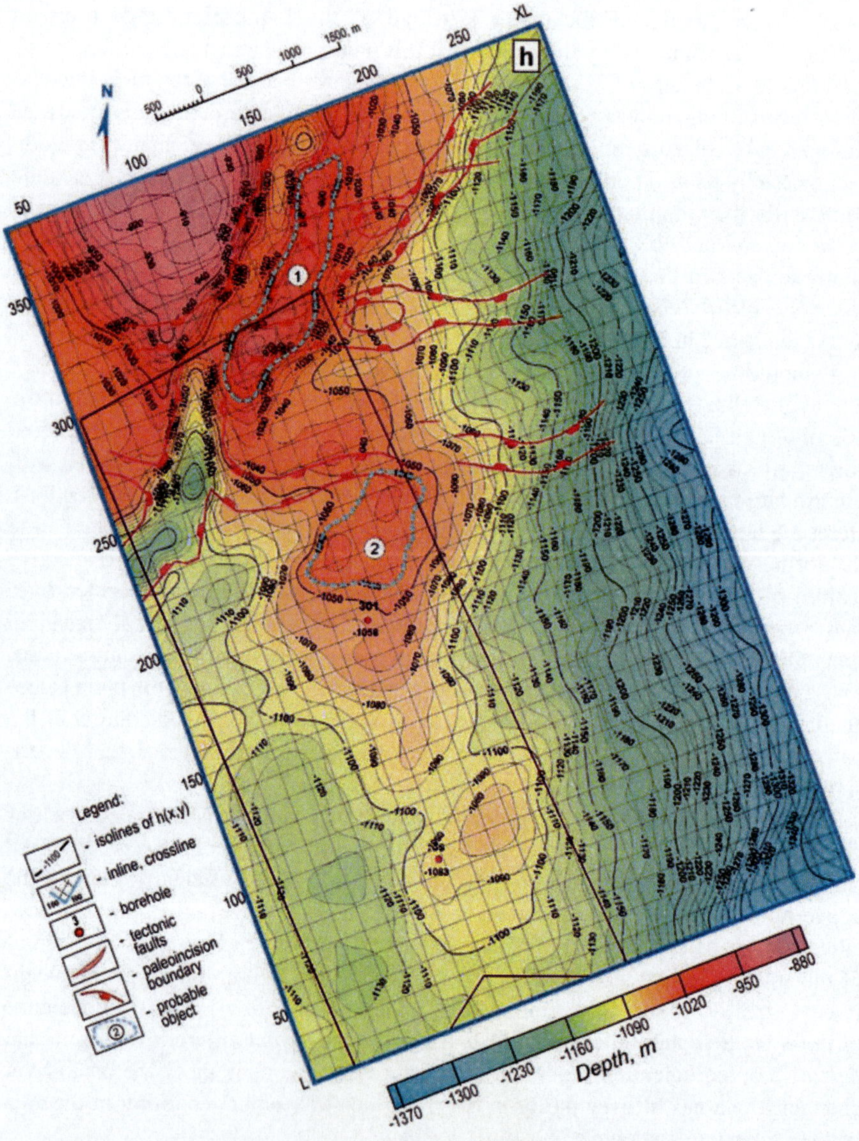


**Fig. 5.10** Structural map along the top of the deposits of the upper part of the Nyumylyg horizon ($D_3$nm). Upper substage of the Famennian stage – horizon II-III (C-D)

(dolomitic-calcareous carbonate varieties are associated with a high probability with this part of the section).

Zones with slightly lower AI values are confidently manifested both within the contour of the main paleochannel, and in a number of parts of the area adjacent to it, and in the central part of the area, located just to the east of wells 3 and 301 Savinobor. All these areas are characterized by a blue-green palette. These zones are most likely associated with calcareous-argillaceous formations of carbonates.

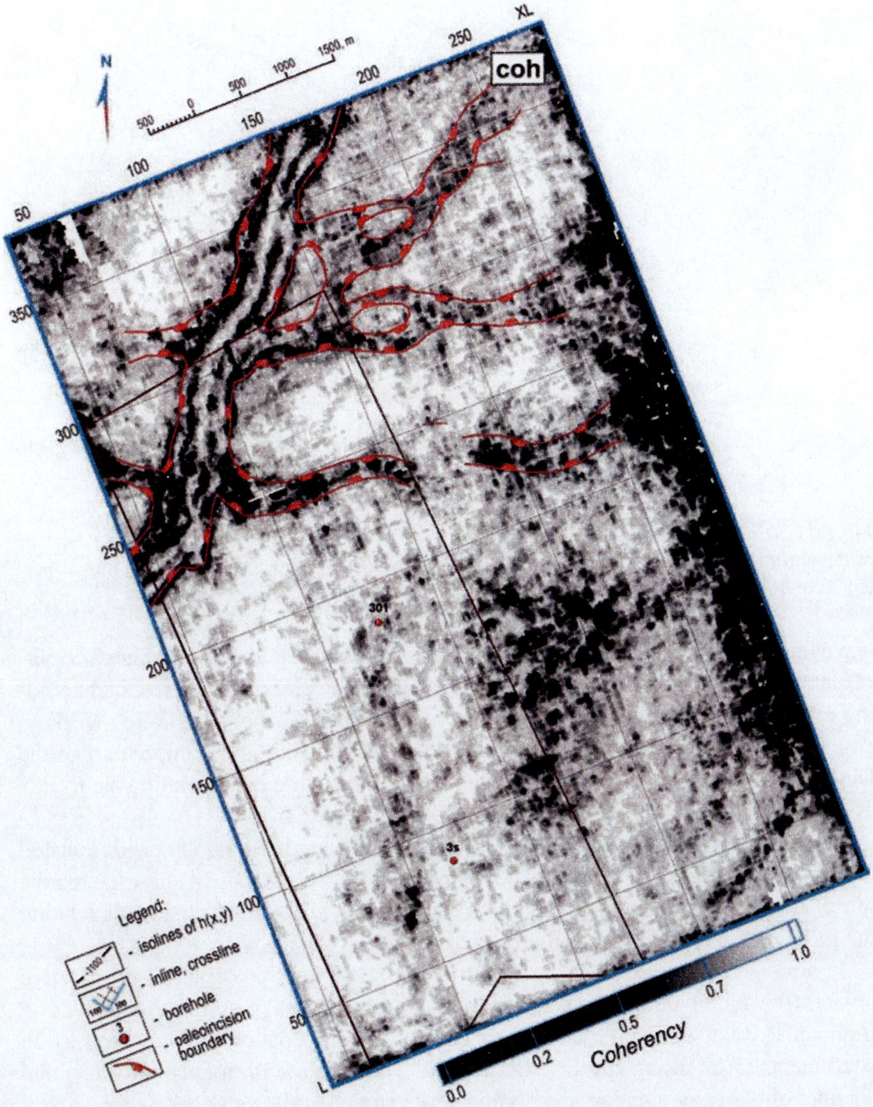


**Fig. 5.11** Coherence map of the wave field amplitude cube along the paleoisochronous surface of horizon II-III (C-D)

It is also worth noting the main priority of clustering results with a high probability should be given to those obtained on the basis of using acoustic impedance (AI) as initial data. The next priority (ranking) are the results obtained using reflection coefficients (RC). These results somehow provide a certain detail on the results of classification according to AI. The results of classification according to WF data, of course, are some starting point (but nevertheless a distorted result of the seismic

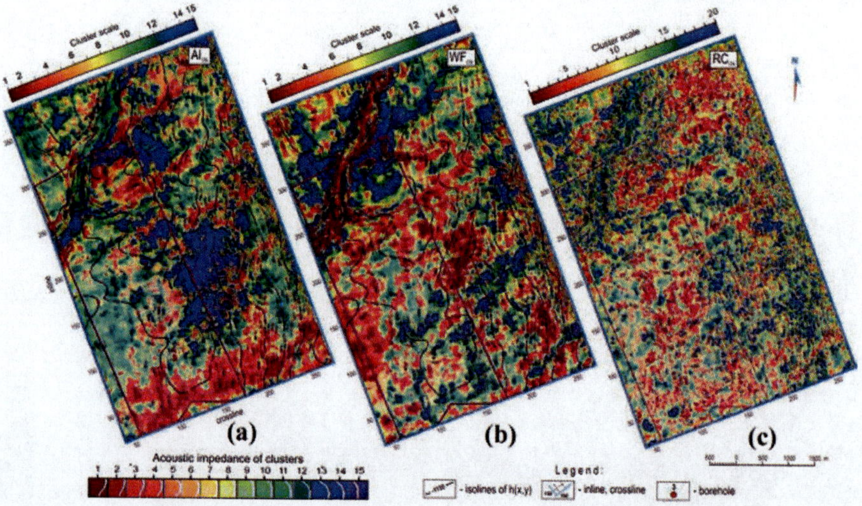


**Fig. 5.12** Distribution maps of cube clusters of the upper part Nyumylga horizon ($D_3$nm) deposits of the upper substage of the Famennian stage – horizon II-III (C-D) for: (**a**) acoustic impedance; (**b**) wave field; (**c**) reflection coefficients

wave interference phenomenon), with which one can compare, i.e. evaluate the corresponding increase in the information content of the classification results according to the data of AI and RC. Naturally, when planning further exploration work on structural and lithological objects associated with deposits of the upper part of the Devonian Famennian stage, it will be necessary to take into consideration the results of the carbonates classification obtained above (Fig. 5.12).

To obtain a more or less complete picture of the identified paleoincision located in the northwestern part of the prospecting area, Figs. 5.13 and 5.14 show the results of its "scanning" in the form of sections that pass the 3D cube of seismic data along the most characteristic directions selected.

Figures 5.13a–c and 5.14a–c, in particular, show how possible oil-prospective objects (the main paleoincision and adjacent coastal ledges) manifest themselves on fragments: wave field (WF), sections of acoustic impedance (AI) and reflection coefficients (RC) along the line inline 350. These inline directions cut the main channel of the paleoincision at a certain acute angle. At the same time, it is clearly visible how the target object itself is confidently traced on the presented sections (paleoincision on the AI sections) and its internal filling with sediments (on the RC sections), as well as the object with the assigned conditional number 1. According to this object 1, a possible level of OWC is revealed in the RC section. On the structure map (Fig. 5.10), this object is represented with the same number 1. It is outlined by a line of equal level having an absolute mark (a.m.) of −1000 m.

In addition to the above, it should be noted that the formal transfer to wave fields (to time sections) of reflecting boundaries (along the horizons Ik, II-III (C-D) and $D_3fm_2$ (Ф5)) shows that the selection of the above objects (which are quite

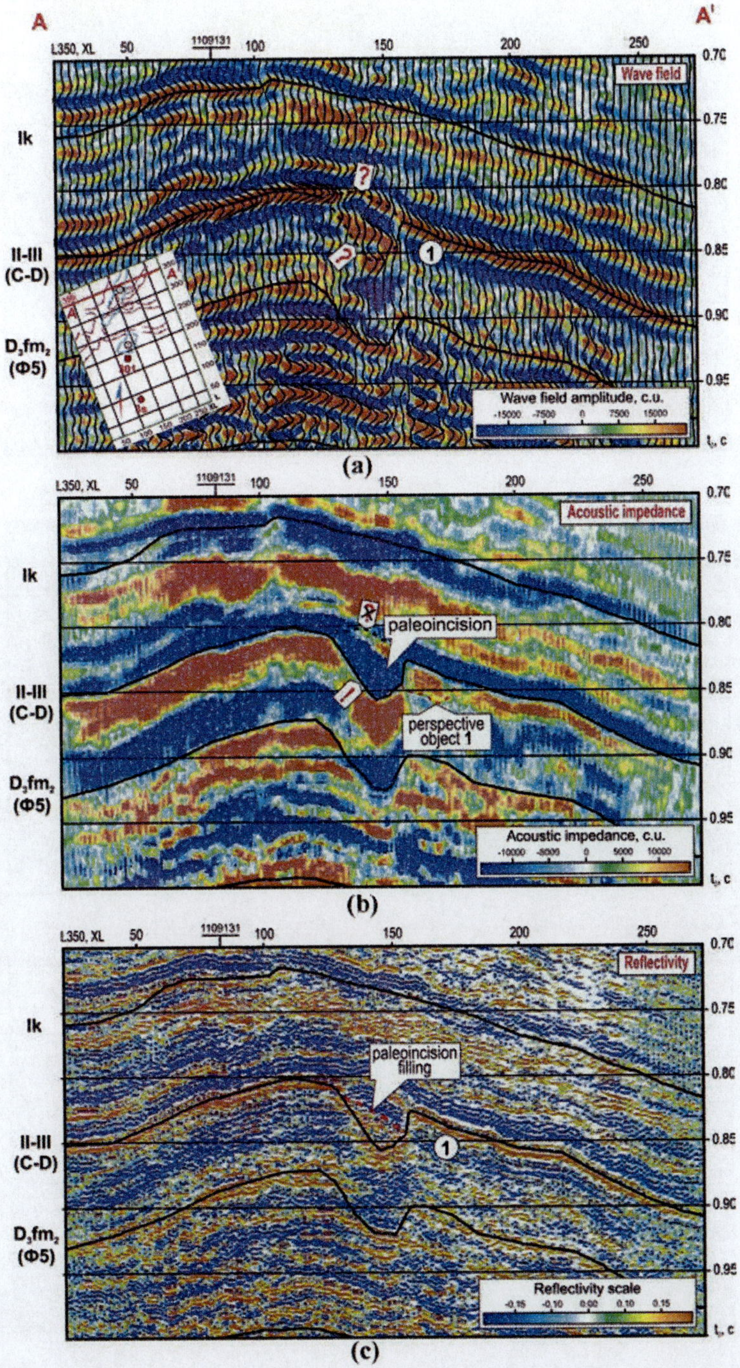


**Fig. 5.13** Objects that differ in kinematic and dynamic features of reflected waves from the surrounding real medium in deposits of horizons II-III(C-D) and $D_3fm_2$ (Φ5) on fragments of sections along the inline 350 (A-A$^I$): **(a)** wave field; **(b)** acoustic impedance; **(c)** reflection coefficients

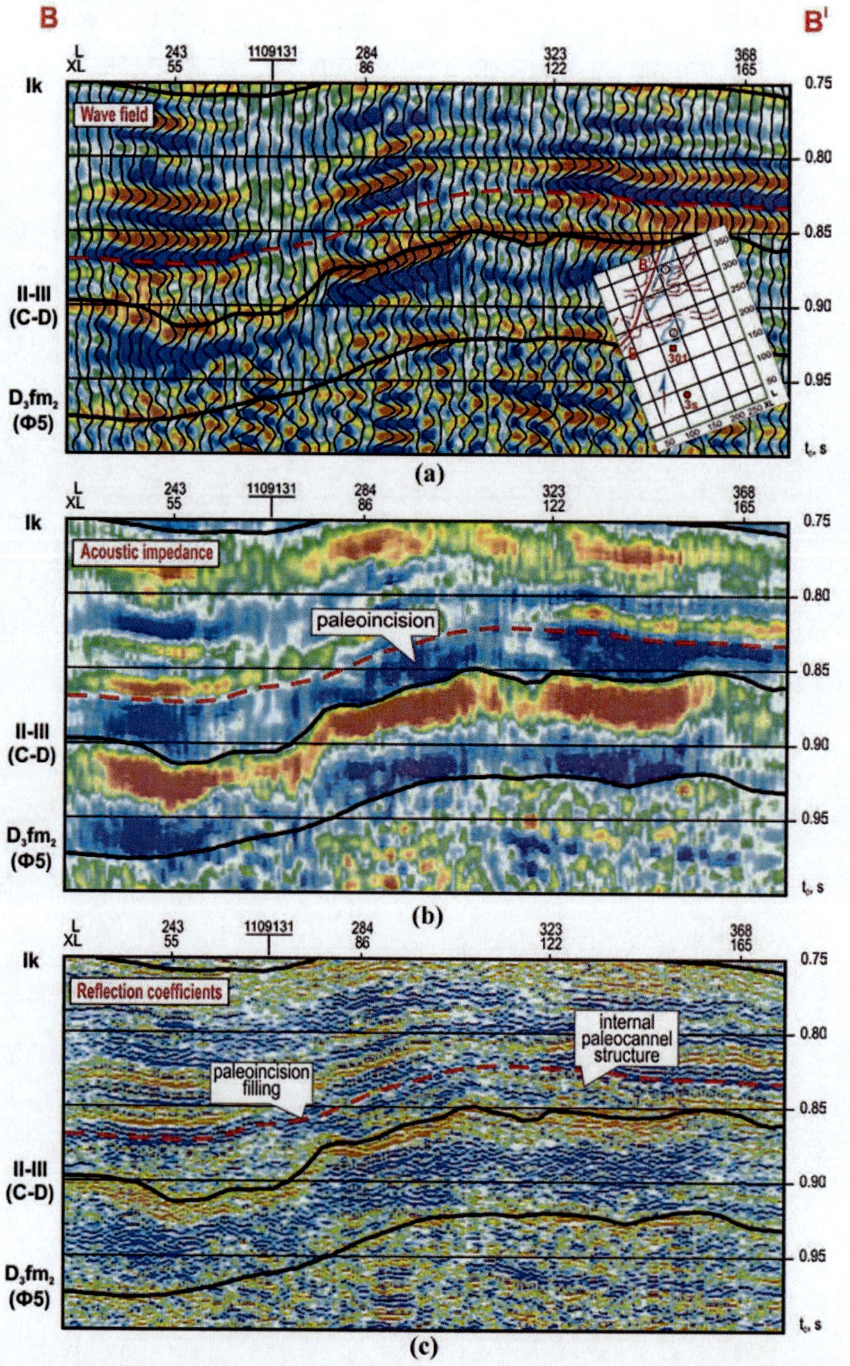


**Fig. 5.14** Objects that differ in kinematic and dynamic features of reflected waves from the host real medium in deposits of horizons II-III(C-D) and $D_3fm_2(\Phi5)$ on fragments of sections along the B-B$^I$ line (along the central axis of the paleoincision): (**a**) wave field; (**b**) acoustic impedance; (**c**) reflection coefficients

confidently recorded on the sections AI and RC) on the CDP time sections is possible only with a large degree of uncertainty (Fig. 5.13a). At the same time, the channel of the paleoincision on the CDP time sections can be easily confused with objects of the type “intrashelf banks” or “single-phase reef formations” (not to mention the impossibility of identifying the proposed OWC on the time sections).

On fragments of sections AI and RC (Fig. 5.14b, c), along the profile formed along the central (fairway) axis of the paleo-incision (along the line B-B), the profile relief of the paleo-river bottom is clearly visible and changing (according to the values of RC and the internal structure) sedimentation of the bed of the paleo-incision. The fact of ambiguity in the display of these indicated diagnostic features (paleoincision and its filling) on the interference structure of the wave field is also obvious – on a fragment of the CDP time section (Fig. 5.14a).

### 5.2.3 *Processing and Interpretation of Well Logging Data*

The method of processing and automated interpretation of well logging data, based on the integration of geophysical parameters into a single geological information system through functional transformations of well logs, was developed by N.Z. Zalyaev and quite similarly described by us above (in Chap. 4) and in [13, 14]. In this subsection, we will consider only the results of processing and interpretation of well logging data for two wells 3 and 301, located in the southern and central parts of the exploration area (3D(N) – area of the Savinobor field) (Fig. 5.9). The results obtained in this case are one of the important components of the geological indicators forecast using seismic sections and a 3D cube of seismic records.

For well 3 Savinobor, which opened up the Upper Silurian deposits, the available logging data (*recorded in the appropriate LAS-formats*) are represented by a set of standard logging methods. These are data from radioactive logging methods (GR, NL), electrometry (RZ, SP, IL, LL), caliper logging (CAL) and acoustic (SN ($\Delta$T)) logging. For productive intervals of the section in well 3, test results are also available.

In well 301, drilled to a depth of 2770 m, in the sediments, including the productive deposits of the Frasnian subdomanik complex, electrometry studies of standard (PS, LL, IL) and microlateral (MLL) logging, VIKIZ, radioactive logging (GR, NLT), caliper logging were carried out (CAL), density logging (GGL), thermometry, moisture logging and inclinometry. For productive intervals of the section in well 301, tests were also carried out.

The geological section of **well 3 Savinobor** (Fig. 5.15), located in the southern half of the Savinobor prospecting area, was studied in detail in the interval of occurrence of productive and promising deposits from the Ufimian stage ($P_2u$) of the Upper Permian ($P_2$) to the Przhidolian stage ($S_2p$) of the Upper Silurian ($S_2$). It is clearly seen from the figure, in general, the studied section is composed of alternating carbonate and terrigenous facies, which are heterogeneous in structure. Terrigenous deposits are significantly represented by intercalation of sandstones,

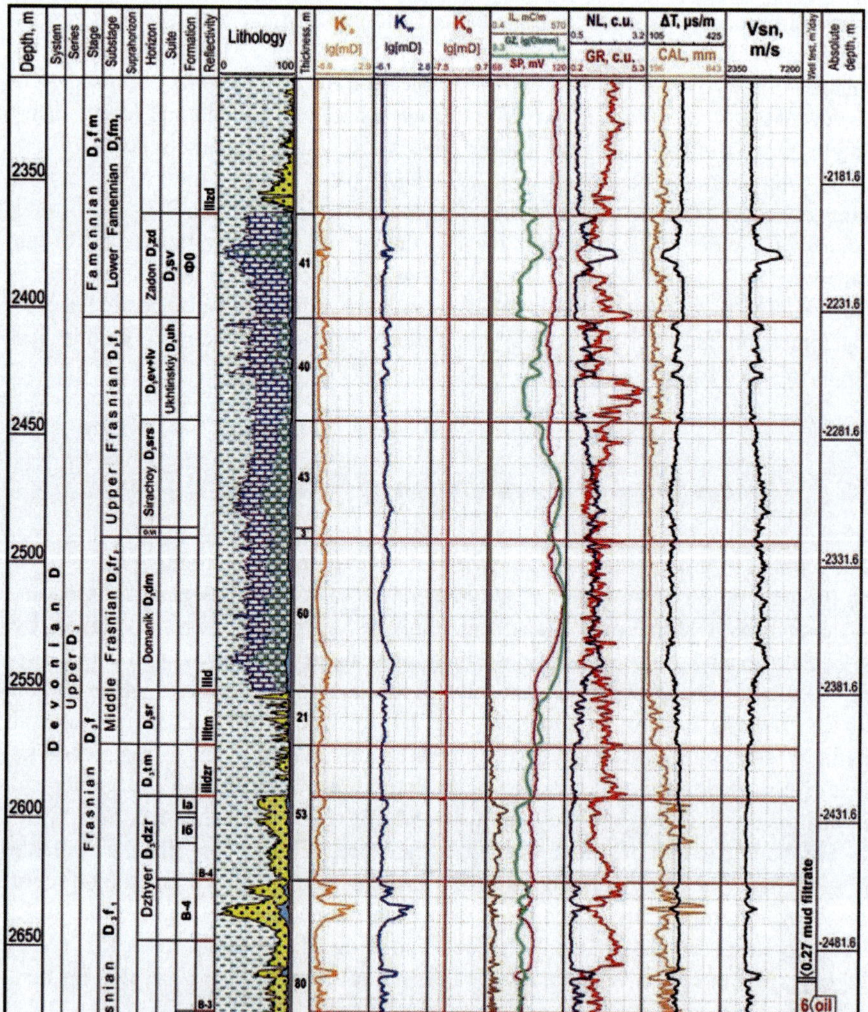

**Fig. 5.15** Lithological-stratigraphic column with the distribution of the saturation nature of reservoir rocks in well 3 in the interval of the Lower Famennian and Frasnian deposits

siltstones, and clays with rather rare and relatively thin layers of carbonates. Carbonate deposits are mainly represented by a different combination (volume ratio) of limestones, dolomites and clays with very rare interlayers of terrigenous material.

Considering the section sequentially from top to bottom, it can be noted individual permeable interlayers 2–7 m thick stand out in the lower part of the deposits of the Filippov horizon ($P_1fl$) of the Kungurian stage ($P_1k$) of the Lower Permian in the depth interval of 1011–1033 m. These interlayers are water-saturated, have porosity $K_p \approx 11–15\%$. In general, the deposits of this horizon are represented by both carbonate (limestone, dolomite) and terrigenous formations (argillite, siltstone, sand, sandstone). Clay intervals of the section turn out to be comparable in thickness to sandy-siltstone interlayers. In these clay intervals, the content of clay material is $C_{cl.} \approx 80–95\%$.

The permeable intervals of the Sakmara ($P_1s$), Artinsk ($P_1ar$) stages of the Lower Permian, Bashkirian and Moscovian ($C_2b + m$) stages of the Middle Carboniferous, represented by carbonate deposits (primarily limestones with a small admixture of dolomites) and occurring at depths of 1038–1141 m, are characterized by the porosity – $K_p \approx 7–13\%$. A distinctive feature of these intervals is their confinedness to those parts of the section that are characterized by relatively high dolomitization.

The carbonate deposits of the Serpukhovian ($C_1s$), Visean ($C_1v$) and Tournaisian ($C_1t$) stages of the Lower Carboniferous (in the depth interval 1141–1234 m) contain heterogeneous reservoir properties and lithology (according to the ratio of limestones and dolomites), predominantly water-saturated intervals with the porosity $K_p \approx 5–10\%$.

In the sediments of the Nyumylga ($D_3nm$) horizon of the Upper Famennian ($D_3fm_3$) stage of the Upper Devonian, terrigenous deposits are fixed (at depths of 1242–1252 m), composed originally of greenish-gray clays. There are practically no reservoirs here.

Further, at a fairly significant interval of the section (at depths of 1252–2306 m), carbonate formations are heterogeneous accoding to lithology (mainly in terms of the ratio of limestones, clayey limestones and dolomites) and their reservoir properties in deposits of the Upper ($D_3fm_3$)-, Middle ($D_3fm_1$)-, and Lower Famennian ($D_3fm_1$) substages of the Upper Devonian ($D_3$). In a number of intervals, associated first of all with the presence of dolomitic, porous, cavernous limestones, permeable water-saturated interlayers appear. The porosity in them is characterized by the values $K_p \approx 5–15\%$.

The deposits confined to the lowest part of the Yelets ($D_3el$) horizon of the Lower Famennian ($D_3fm_1$) Upper Devonian substage ($D_3$) are characterized (in the depth interval of 2305–2363 m) as a heavily clayey section with their gradual replacement in the lower half of the considered interval by sandstones and siltstones. There are practically no reservoirs in this part of the section.

In the interval, starting from the top of the Zadon ($D_3zd$) horizon of the Lower Famennian ($D_3fm_1$) substage and ending with the base of the Domanik ($D_3dm$) horizon of the Middle Frasnian ($D_3f_2$) substage of the Upper Devonian ($D_3$) (2363–2550 m), an approximately similar pattern is observed, which manifests itself and in the above-mentioned depth interval of 1252–2306 m. Here, according to the lithology of carbonate rocks (by the ratio of limestones, clayey limestones and dolomites), according to reservoir properties (by the location of permeable interlayers in the section and in the characteristics of porosity values), approximately the same features of the geological structure are manifested. The porosity of permeable interlayers here varies in the range $K_p \approx 5–7\%$.

The deposits of the Sargay horizon ($D_3sr$) of the Middle Frasnian ($D_3f_2$) substage and the Timan horizon ($D_3tm$) of the Lower Frasnian ($D_3f_1$) substage (in the depth interval of 2550–2624 m) are characterized by a rather strong claying of the section (as noted above, these deposits are represented by dark gray clays with subordinate interlayers of limestone and volcanic sand). Clay content for this part of the section as a whole varies within $C_{cl.} \approx 65–100\%$. There are practically no reservoirs in this part of the section. In this regard, the deposits of the Sargay and Timan horizons in the study area serve as a regional screen-seal.

For productive and potentially productive deposits occurring in the lower part of the Lower Frasnian ($D_3f_1$) substage – the Yaran ($D_3jr$) and Dzhyer ($D_3dzr$) horizons (in the depth interval 2624–2778 m) and composed of sandstones, fine-grained quartz sandstones, quartz, micaceous siltstones, separated among themselves clayey and dense rocks, dark gray argillites, platy, with interlayers of dark gray clays, permeable water and oil saturated interlayers. Two oil-saturated interlayers, separated by a relatively thin clay bridge, are confined to the deposits of the lower half of the B-3 productive formation (at depths of 2682.5–2691.8 m). In accordance with the results of well logging data processing by the method of functional transformations of well logs (Fig. 5.15), porosity for two oil-saturated interlayers is characterized by $K_p \approx 12.6–18.6\%$, oil saturation – $K_o \approx 48.7–76.9\%$, clay content – $C_{cl.} \approx 10.8–47.6\%$. For the same set of considered permeable interlayers, the reservoir properties are characterized by the following values: absolute permeability (graph $k_a$) $k_a \approx 0.04–5.62$ mD, relative water permeability (graph $k_w$) $k_w \approx 0.01–0.09$ mD and relative oil permeability (graph $k_o$) $k_o \approx 0.01–4.86$ mD (Fig. 5.15).

When testing the depth interval of 2687–2692 m, performed here by the tester of the KII-2 M-146 formations in an open hole, an influx of highly degassed oil with a density of 0.85 g/cm$^2$ in a volume of 0.5 m$^3$ was obtained for 30 min of standing on the inflow. At the same time, the oil production rate, calculated from the pressure change, amounted to Qn = 25.2 m$^3$/day. Further testing in the cased hole (intervals 2680.3–2688.3; 2675.1–2678.1 m) also confirmed the oil saturation of the B-3 formation.

The other permeable interlayers of the considered interval (2624–2778 m) both in the deposits of the Dzhyer ($D_3dzr$) horizon (in potentially productive units Ia and Ib) and in the deposits of the Yaran ($D_3jr$) horizon (in units B-4, B-2 and B-1), turn out to be water-saturated.

Sediments of the Starooskol suprahorizon ($D_2st$) of the Givetian ($D_2zv$) stage, as well as the underlying deposits of the Kolvin ($D_2kv$), Omrin ($D_2om$), partly Kedrov ($D_2kd$) Upper Eifelian ($D_2ef_2$) substage, as well as the Lower Eifelian ($D_2ef1$) and Upper Silurian ($S_2p$) substages, are characterized as terrigenous sandy-clayey formations. Almost all permeable intervals of the section (most of which are located in the sediments of the Starooskol suprahorizon) manifest themselves as water-saturated interlayers. Heterogeneous sandstone formations are characterized by porosity $K_p \approx 10–22\%$. Genetically (often also hydrodynamically) interconnected permeable layers of these horizons are composed of well-sorted gray, light gray sandstones and quartz and chlorite-quartz, gray and gray-green siltstones, with frequent siderite nodules. The impermeable intervals of the section are mostly high clayed rocks. Clay content for this part of the section $C_{cl.} \approx 80–100\%$. The largest volume of clay material is confined to the deposits of the Omrin ($D_2$om) and partially Kedrov ($D_2$kd) horizons. The upper part of the Lower Eifelian ($D_2ef_1$) substage is also strongly clayed (Fig. 5.15). Here, practically in all permeable intervals of the section there is no oil saturation, except for individual interlayers, in which only signs of oil saturation "break through".

The geological section of well 301 (Fig. 5.16), located in the central part of the Savinobor exploration area, has been studied in most detail in the interval of occurrence of productive and promising formations of the Lower Famennian ($D_3$fm) and Frasnian ($D_3$fr) deposits. As in the section of the previous well, the oil prospects here are somehow associated with the terrigenous deposits of the Lower Frasnian, which include the facies of the Yaran ($D_3$jr) and Dzhyer ($D_3$dzr) horizons, i.e. layers B-1÷B-4, Iб and Ia.

In the interval, starting from the top of the Zadon ($D_3$zd) horizon of the Lower Famennian ($D_3fm_1$) substage and ending with the base of the Domanik ($D_3$dm) horizon of the Middle Frasnian ($D_3f_2$) substage of the Upper Devonian ($D_3$) (2346–2543 m), carbonate rocks occur (with different ratios limestones, argillaceous limestones and dolomites). According to the reservoir properties (to the location of permeable interlayers in the section and in the characteristics of porosity), an interval of permeable interlayers is noted (at depths of 2452–2534 m), in which a change in porosity is observed in the range $K_p \approx 5–11\%$.

The deposits of the Sargay horizon ($D_3$sr) of the Middle Frasnian ($D_3f_2$) substage and the Timan horizon ($D_3$tm) of the Lower Frasnian ($D_3f_1$) substage (in the depth interval of 2543–2624 m) are characterized by a rather strong claying of the section (the deposits here are represented by dark gray clays with subordinate interlayers of limestone and volcanic sand). Clay content for this part of the section as a whole varies within $C_{cl.} \approx 35–95\%$. Reservoirs in this part of the section are absent with rare exceptions. In this regard, the deposits of the Sargay and Timan horizons in the study area serve as a regional screen-seal.

For productive and potentially productive deposits occurring in the lower part of the Lower Frasnian ($D_3f_1$) substage – the Yaran ($D_3$jr) and Dzhyer ($D_3$dzr) horizons (in the depth interval 2624–2768 m) and composed of sandstones, fine-grained quartz sandstones, quartz, micaceous siltstones, separated among themselves clayey and dense rocks, dark gray, platy mudstones, with interlayers of dark gray clays, water-saturated permeable interlayers with signs of oil saturation according to the processing of logging materials (such signs are noted in the deposits of the B-3, B-2 and B-1 formations). At the same time, there is a fairly frequent alternation of significantly flooded thin-layered interlayers, separated from each other by relatively thin clay bridges (at depths of 2660–2766 m). In accordance with the results of well

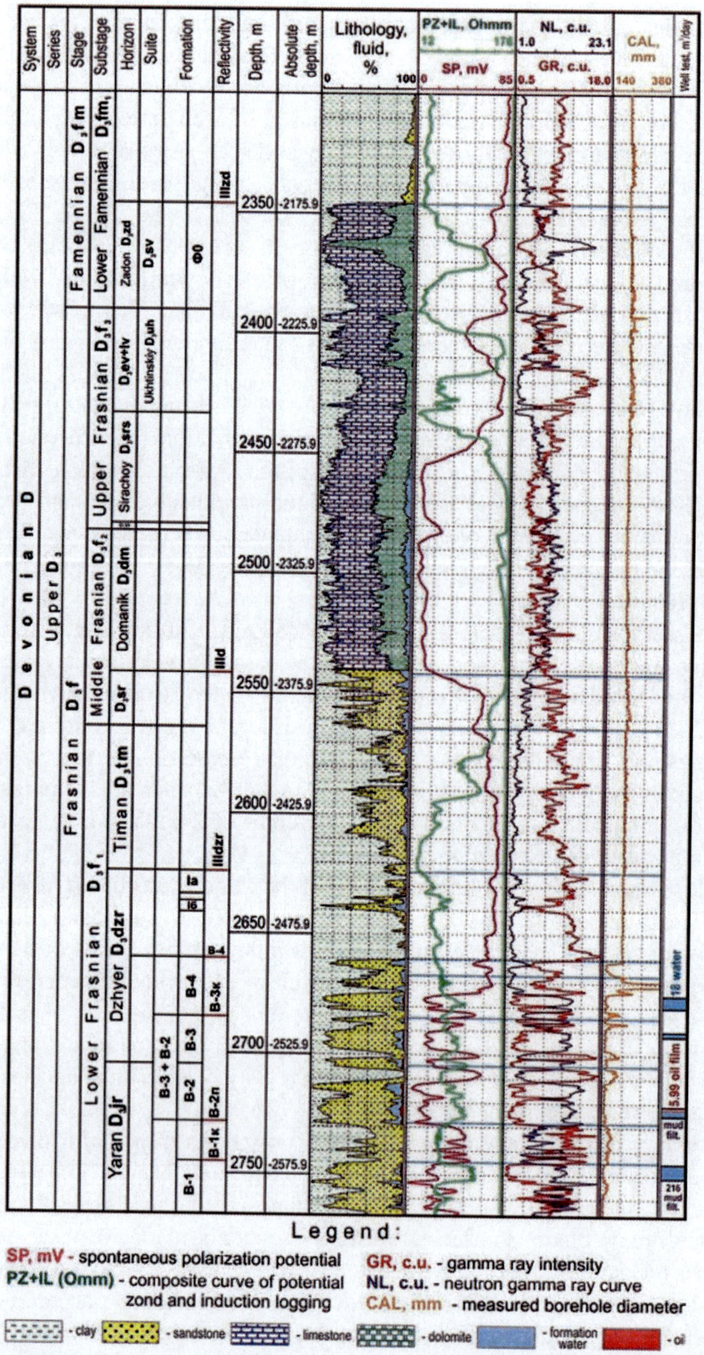


**Fig. 5.16** Lithological-stratigraphic column with the distribution of the saturation nature of reservoir rocks in well 301 in the interval of the Lower Famennian and Frasnian deposits

logging data processing by the method of functional transformations of well logs (Fig. 5.16), the porosity of these water- and partially oil-saturated interlayers is characterized by $K_p \approx 10–21\%$, oil saturation $K_o$ is not more than 3–5%, clay content is $C_{cl.} \approx 15–75\%$.

When testing different depth intervals using test equipment of the KII-95 type and in the cased well, the layers B-3 + B-2 in well 301 turned out to be unproductive. So, in the depth interval of 2663–2712.5 m, a drilling fluid with formation water with a volume of 1600 liters was obtained at formation pressure $P_{fm} = 275.08$ atm and depression 122.7 atm. An approximately similar test result was obtained in the depth interval of 2663–2700.5 m: a drilling mud was obtained with a reservoir fluid with a volume of 781.2 liters at formation pressure $P_{fm} = 278.4$ atm and depression 83.55 atm. In the interval of 2727–2757.2 m, a drilling mud with formation salt water with a volume of 3500 liters was obtained at formation pressure $P_{fm} = 290.31$ atm and depression 102.72 atm. In the interval 2721–2725 m, a salt water filtrate was obtained with an oil film and density of this 1.08 g/cm$^3$ in a volume of 0.158 m$^3$ at a reservoir pressure of $P_{fm} = 281.91$ atm. and depression 136.77 atm. In the depth interval of 2745–2751 m, saline water with a density of 1.12 g/cm$^3$ in a volume of 2.8 m$^3$ was obtained at a pressure of $P_{fm} = 285.76$ atm and depression 108.49 atm.

The results of swabbing the section intervals 2675–2680 and 2690–2692 m showed: in the first cycle, 14.65 m$^3$ of water inflow was taken; in cycle 2, 4.83 m$^3$ was recieved.

According to the results of petrographic analysis, which includes macro- and micro-descriptions of rocks, rocks with a brownish tint are identified here, as probably containing OM. At the same time, according to the chemical composition of water extracts of rocks – pore solutions, the most probable "initial" nature of saturation of the studied rock interval is formation water.

Deposits of the Starooskol suprahorizon ($D_2$st) of the Givetian ($D_2$zv) stage were not opened up by borehole 301 (in this regard, well logging data are not available for this part of the section).

Figure 5.17 shows a correlation scheme for wells 3 and 301. According to such correlation schemes of lithostratigraphic units ("packs", horizons) built from the original well logs or from the corresponding results of their processing and interpretation, changes in the features of the geological structure of the target horizons are usually estimated. In particular, we can talk about the nature of changes in the thicknesses of productive and potentially productive intervals, oil-saturated thicknesses, lithofacies replacements, pinchouts, etc. In addition, they help evaluate the capabilities of the seismic method in identifying target reflectors and evaluating the consistency of structural mapping results (and in some cases, the results of dynamic seismic data processing) with well data. Typically, such horizon correlation schemes are built for a large number of wells and for different directions of the studied area. The used lithological-stratigraphic columns of wells (Fig. 5.17) are tuned according to their hypsometric levels, that is, in accordance with their absolute marks – geological breakdowns of lithostratigraphic units of the studied section. For each of the wells in the correlation scheme, individual wellbore elongations were taken into account, i.e. well inclinometry data.

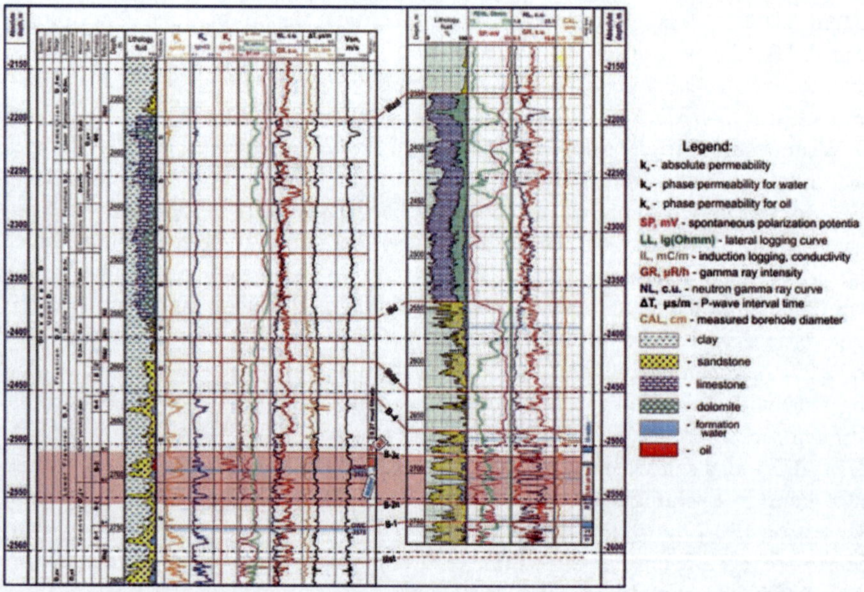


**Fig. 5.17** The correlation scheme for wells 3, 301

Despite the absolute marks of wells 3 and 301 on the tops and bottoms of the productive horizon B-3 are very close and with a relatively small distance from each other, the presence of a sufficient volume of sand reservoirs in them, in well 301 horizon B-3 is not productive (as the test results showed). The reasons and results of this will be discussed below.

In general, the obtained information on the intervals of processed wells is directly used in the complex interpretation of drilling, well logging and high-resolution seismic data.

### 5.2.4 *Target Deposits Structure Features According to Seismic Profiles*

Analysis of seismic data along lines processed using a special graph and passing through vertical sections of wells 3 and 301 with elements of the interpretation performed, in particular, shows the following.

The profile passing along the inline 90 of the sublatitudinal direction in the alignment with the vertical section of the well 3, the features of the geological structure in the form of distributions of the effective acoustic impedances and reflection

**Fig. 5.18** Identification of target reflecting interfaces on sections of effective acoustic impedances (**a**) and reflection coefficients (**b**) along the inline 90 (3D-cube) in the vicinity of the well location 3s

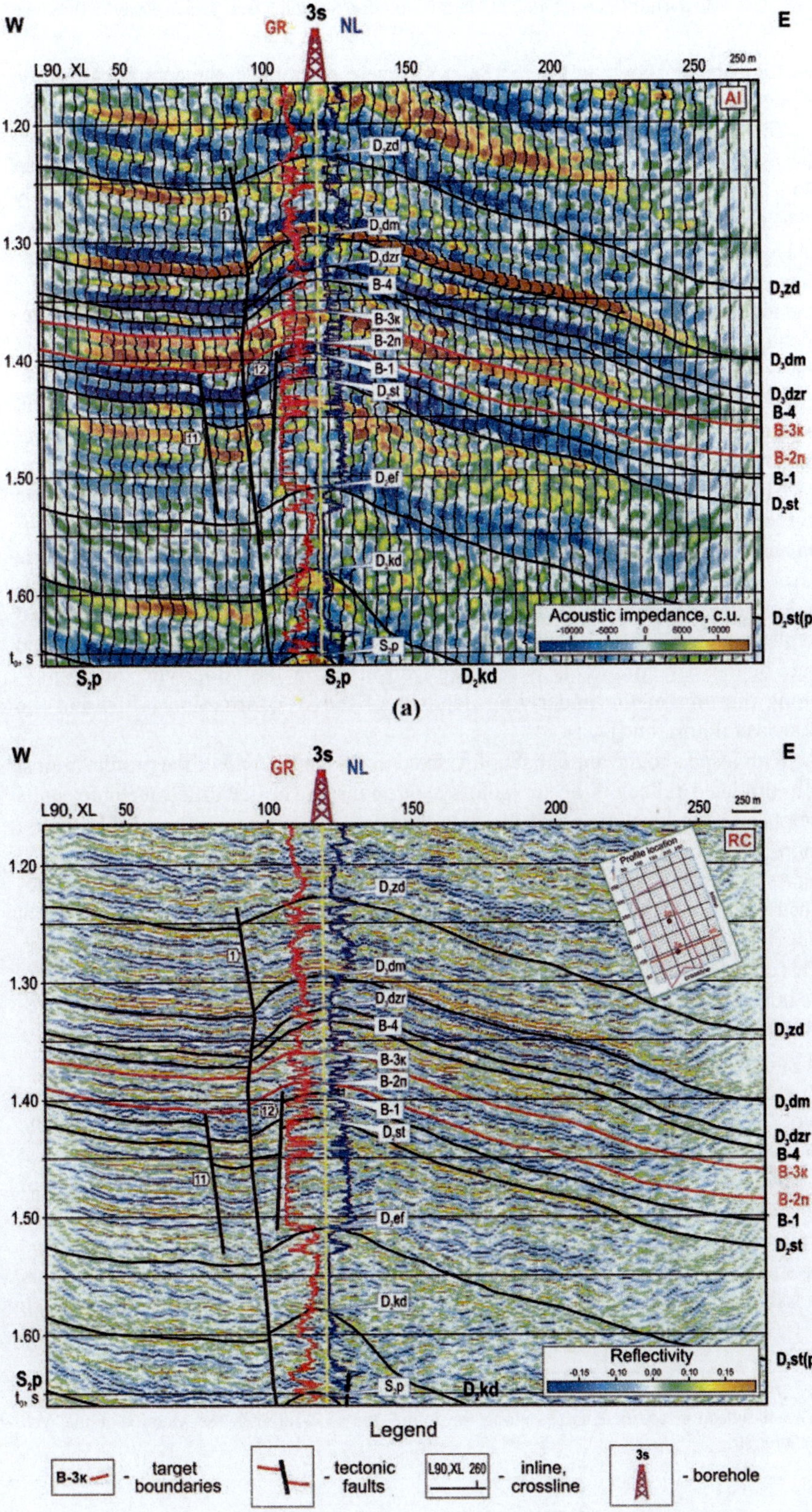

W
E
3s
GR
NL
L90, XL
50
100
150
200
250
250 m
AI
1.20
1.30
1.40
1.50
1.60
t0, s
D3zd
D3dm
D3dzr
B-4
B-3к
B-2п
B-1
D3st
D3ef
D3kd
D2st(p)
S2p
D2kd
Acoustic impedance, c.u.
-10000
-5000
0
5000
10000
(a)
RC
Profile location
Reflectivity
-0,15
-0,10
0,00
0,10
0,15
Legend
B-3к
- target boundaries
- tectonic faults
L90,XL 260
- inline, crossline
3s
- borehole

coefficients are shown in Fig. 5.18a, b. On these sections, in the central part of the inline, the anticlinal uplift is clearly manifested along almost all correlated reflectors.

The most complex geological structure is found in the subdomanik deposits, which are associated with the productive and potentially productive layers of the Yaran ($D_3$jr), Dzhyer ($D_3$dzr), and Starooskol ($D_2$st) horizons. This complexity is primarily reflected in the extremely complicated distribution of effective acoustic impedance (AI) values (Fig. 5.18a) and reflection coefficients (RC) (Fig. 5.18b) in productive deposits. In some cases, it is not possible to unambiguously correlate target reflectors and low-amplitude faults in a thin-layer section. Nevertheless, the main structural elements of the section are clearly distinguished. As noted above (in Sect. 5.1), here, in the conditions of the development of avandelta formations in the Middle Devonian time (in a fan-shaped paleodelta setting), in the deposits of the Yaran and Dzhyer horizons, formed in shallow sandy-silty deposits with frequent breaks in sedimentation, massive reservoirs were formed (tectonically and lithologically shielded traps).

Along the profile passing along the inline 185 of the sublatitudinal direction in the alignment with the vertical section of the well section 301, as in the previous case along the inline 90 profile, the features of the geological structure in the form of the distribution of the effective acoustic impedances and reflection coefficients are shown in Fig. 5.19a, b. In productive formation B-3 + B-2, the drilled exploratory well 301 is located approximately 350 m eastward from the arched part of the structure. This is also well observed in the underlying horizons – along the tops of the underlying deposits of the B-1 formation ($D_3$jr) and the Starooskol horizon ($D_2$st).

With respect to the anticlinal uplift, fixed in the central part of the profile, almost all correlated reflectors occur quite synchronously. The identified tectonic faults emphasize the localization of the anticlinal uplift along the entire profile with a more gentle dip of the traced layers in the east direction.

As in the previous case, the subdomanik deposits, which are associated with the productive and potentially productive layers of the Yaran ($D_3$jr), Dzhyer ($D_3$dzr) and Starooskol ($D_2$st) horizons, are distinguished by the greatest complexity of the geological structure. This complexity is primarily reflected in the extremely complicated distribution of acoustic impedance (AI) (Fig. 5.19a) and reflection coefficients (RC) (Fig. 5.19b) in these productive deposits. In some cases, this also does not allow one to unambiguously correlate target reflectors and low-amplitude faults in a thin-layer section.

However, the correlation of reflecting boundaries along the target layers, performed with such conditions of the geological structure, ultimately made it possible to make structural constructions and, on this basis, to identify the main structural elements of the section. Here, as noted above (in Sect. 5.1), in the conditions of development of avandelta formations in the Middle Devonian time (in the setting of a fan-shaped paleo-delta), in the deposits of the Yaran and Dzhyer horizons, formed in shallow sandy-aleurite deposits with frequent interruptions in sedimentation, massive-stratal reservoirs (tectonically and lithologically shielded traps). Such

**Fig. 5.19** Identification of target reflecting interfaces on sections of effective acoustic impedances (**a**) and reflection coefficients (**b**) along the inline 185 (3D-cube) in the vicinity of the well location 301

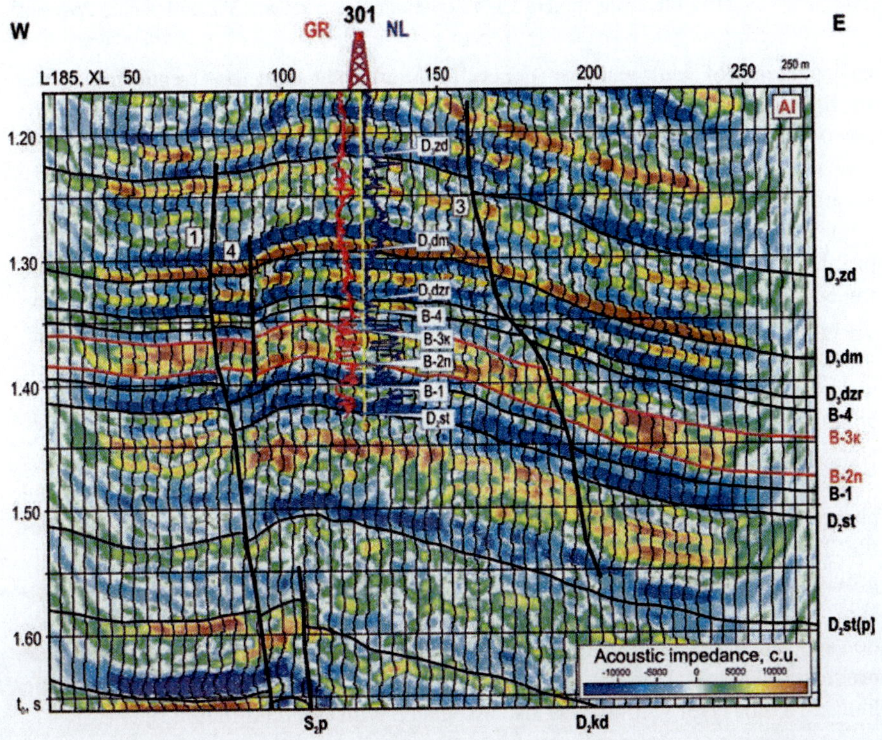
W
GR
301
NL
E
L185, XL 50
100
150
200
250
250 m
AI
1.20
1.30
1.40
1.50
1.60
$t_0$, s
$D_3zd$
$D_3dm$
$D_3dzr$
B-4
B-3к
B-2n
B-1
$D_2st$
$D_2st(p)$
1
3
4
$S_2p$
$D_2kd$
Acoustic impedance, c.u.
-10000
-5000
0
5000
10000

(a)

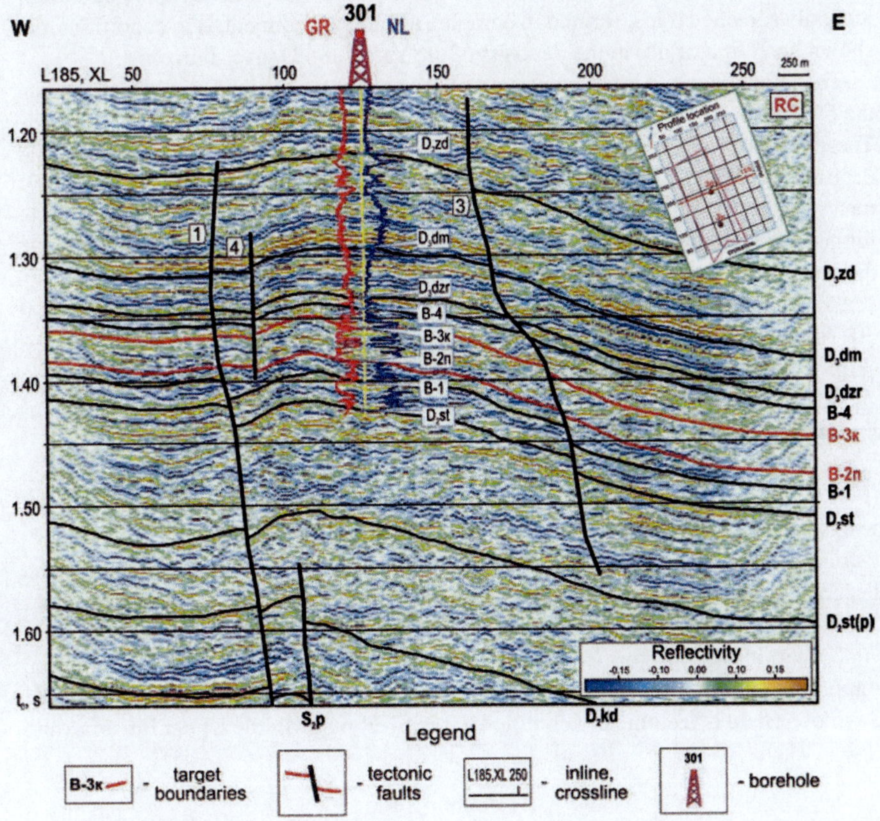
W
GR
301
NL
E
L185, XL 50
100
150
200
250
250 m
RC
Profile location
1.20
1.30
1.40
1.50
1.60
$t_0$, s
$D_3zd$
$D_3dm$
$D_3dzr$
B-4
B-3к
B-2n
B-1
$D_2st$
$D_2st(p)$
1
3
4
$S_2p$
$D_2kd$
Reflectivity
-0.15
-0.10
0.00
0.10
0.15
Legend
B-3к - target boundaries
- tectonic faults
L185,XL 250 - inline, crossline
301 - borehole

environments of sedimentation generally determined here the irregularity of the stratigraphic volumes of the reservoirs in the study area, the seaward nature of the reservoir, the coarsening of the granulometric composition of sandstone layers (especially the Dzhyer, Yaran and Starooskol), as well as their relationship with the underlying and overlying deposits [12].

*A composite profile passing in the alignment of wells 3s and 301*. According to this profile of the submeridional direction, the features of the distributions of the acoustic impedances (Fig. 5.20a), reflection coefficients (Fig. 5.20b) and the wave pattern on the migrated time section (Fig. 5.20c). Here, in all the considered sections, in general, monoclinal subsidence of the target horizons from the northwest to the southeast is observed. At the same time, an anticlinal uplift of small amplitude is recorded in the area of well 3 Savinobor. For well 301, it should be noted its location of relatively weak structural anticlinal forms (possible traps) is not quite emphasized.

The complexity of the geological structure in the subdomanik deposits, noted above along the inlines 90 and 185, is also fully expressed here on the composite profile. This complexity of the section primarily refers to the productive and potentially productive formations of the Yaran ($D_3$jr), Dzhyer ($D_3$dzr) and Starooskol ($D_2$st) horizons. The heterogeneity of the structure is most highlighted in the form of the distribution acoustic impedances (AI) (Fig. 5.20a) and reflectivites (RC) (Fig. 5.20b) in these productive deposits. Correlation of target reflectors, as well as low-amplitude tectonic faults in a thin-layer section, was also associated with certain difficulties here.

As a result of the performed structural constructions, however, it is possible to trace the main structural elements of the objects under study. Tectonically and lithologically screened traps, formed in complex avandelta sedimentation conditions, are shown here, first of all, in the deposits of the Yaran and Dzhyer horizons.

The reflecting boundaries correlated on the sections AI and RC are also placed on the CDP time section along the line of the considered composite profile (Fig. 5.20c). These reflectors, which are directly confined to the terrigenous Frasnian subdomanik deposits, generally characterize the possibility of tracing reflectors using only elements of the wave field structure. It can be seen these possibilities are significantly limited. On the time section (Fig. 5.20c) using the elements of the wave field structure, the identification of individual reflecting boundaries is not unambiguous.

The composite profile clearly shows that approximately from the location of well 301 there is a gradual uplift (uplift) of sediments in the north-west direction (Fig. 5.20c), i.e. in the direction of the S-Savinobor structural zone, which corresponds to the general picture of the Michayu-Pashninskiy zone of oil and gas accumulation of a structural class associated with a swell-like uplift [6, 12].

Judging by the behavior of the wings of the anticline fold relative to its arch (Figs. 5.18 and 5.19), it is worth noting its asymmetric structure: the steeper dip of the wings occurs to the west (in the direction of a regionally traced tectonic fault of submeridional strike and, in general, towards the Izhma-Pechora depression of the Pechora syneclise), as well as a more gentle occurrence of the wings of the structures in the east direction (towards the Upper Pechora depression of the pre-Ural trough).

Concerning the performed structural constructions, one more, in our opinion, important feature of the geological structure should be noted. The work area of the Savinobor field is tectonically confined to the overlying Middle Upper Frasnian and

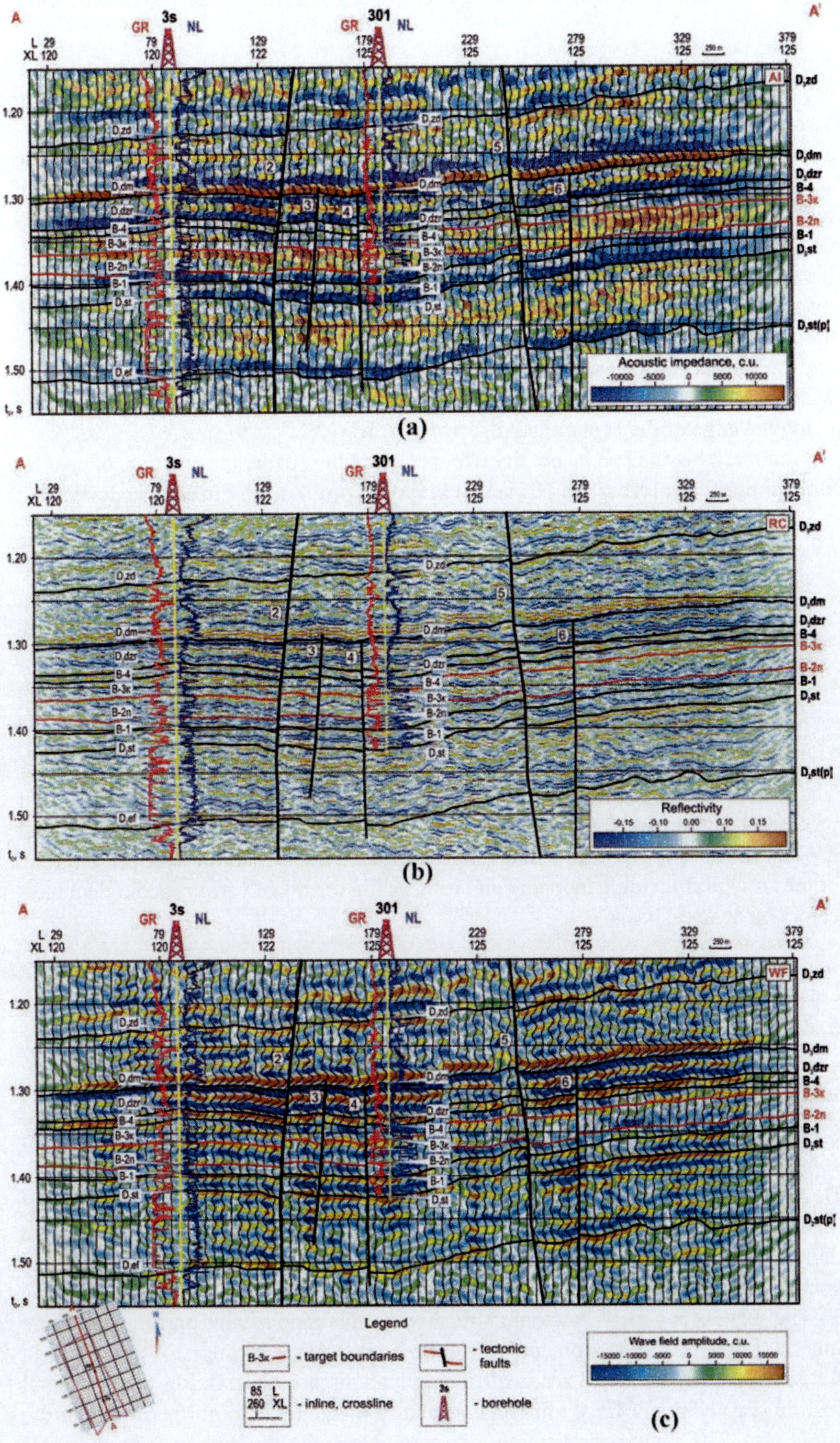


**Fig. 5.20** Seismic sections along a composite profile passing through wells 3 s and 301 in the interval of correlation the Frasnian subdomanik complex: (**a**) acoustic impedance; (**b**) reflection coefficients; (**c**) wave field

Lower Famennian deposits (namely, those lie higher relative to the reflecting boundary IIId associated with the base of the Domanik horizon) is practically confined to a complex lithofacies zone. Such a location of the prospecting area here is associated with the transition from the open shelf zone with mixed terrigenous-carbonate sedimentation (western part of the study area) through a relatively narrow zone of reef facies development (approximately the central part of the area) to a moderately deep-water zone with uncompensated clay-siliceous-bituminous-carbonate sedimentation (eastern part of work site). This allows us to hope for the possibility of discovering barrier reefs (reef formations, marginal banks) here in these deposits (in the Middle Upper Frasnian and Lower Famennian). Here, as is known, carbonate and carbonate-argillaceous deposits accumulated under the conditions of a shallow water shelf, and a marginal reef zone with Upper Devonian reef formations formed along the edge of the accumulative terrace [6, 10–12].

Summarizing the results obtained for seismic lines (using the example of profiles along inline 90 and crossline 185 and a composite profile), we can note the following:

(a) on seismic sections of high resolution and according to the effective acoustic impedances (AI) and reflection coefficients (RC) in the section plane, the heterogeneity of the studied deposits structure the is most reliably traced, including the features of the manifestation of natural hydrocarbon reservoirs in these heterogeneities;
(b) based on the distributions of increased and decreased AI values, it is possible to trace zones of facies substitutions, sand formations that occur in isolation, and areas of formation of possible non-anticlinal hydrocarbon traps.

The results of the profile interpretation of seismic data over the entire volume of processed sections-traverses of the 3D seismic cube data, of course, are the input data for the next stage of processing and automated interpretation – the presentation of the structural-tectonic mapping and forecasting results of various geological indicators on an areal basis.

### 5.2.5 *Some Geological Structure Features of the Zadon Horizon*

Structural forms of the top of the deposits of the Zadon horizon ($D_3zd$) the Lower substage ($D_3fm_1$), occurring in the lowest part of the Famennian stage of the Upper Devonian, in the form of a structural map h(x, y) (Fig. 5.21) reflect the structure of possible promising objects in the carbonate deposits of the Upper Devonian. Zadon dark gray clays occur in these deposits, dense, strong limestones, bituminous in some areas, with interlayers of clay shales.

The regularities of the tectonic structure on this map retain, on the whole, the inherited nature of development with underlying and overlying deposits. Here, in the sediments of the $D_3zd$ reservoir, complications are observed in the periclinal region of a relatively large anticlinal structure located in the northern part of the study area. The same complications are associated with the dome of the anticlinal fold, which extends from the indicated pericline into the southern half of the area. Complications also persist on the monoclinal wings of this fold. At the same time, the surface morphology of the $D_3zd$ horizon deposits generally repeats that of the underlying horizons of the Upper Devonian deposits – the horizons of the Frasnian stage.

The structural forms and tectonic features of the identified local structures along the surface h(x,y) of the $D_3zd$ horizon (designated as the Ф0 marker layer) of the Upper Devonian Lower Famennian deposits largely repeat those the overlying and underlying deposits (Fig. 5.21). It should only be marked that in the sediments

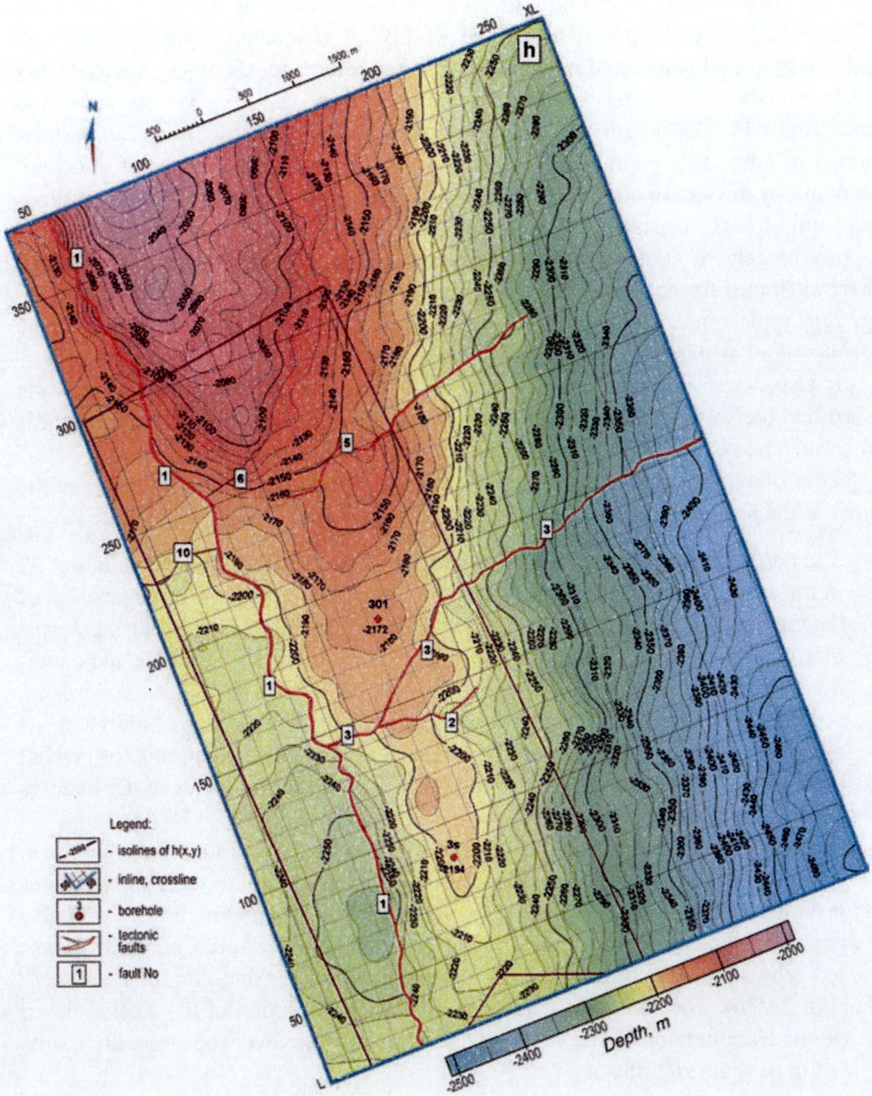


Fig. 5.21 Structural map along the top of the Ф0 formation – deposits of the Zadon horizon ($D_3zd$)

under consideration ($D_3zd$) individual tectonic faults begin to be traced (not always extended and in their full manifestation), which "originate" generally in deeper structural levels.

For the upper part of the deposits of the Zadon horizon ($D_3zd$ + 15 ms), seismic facies were classified using the neural network technology (NNT) clustering (implemented in the Stratimagic software package) according to the shape of the curves and their values for the cube of acoustic impedance (AI), wave field (WF) (Fig. 5.22a, b). At the same time, the obtained classification results for both forms of seismic data (AI and WF) show zoning of the section inhomogeneity in the shape of the envelope reference curves and their values. To the greatest extent, it manifests itself on the map of the distribution of clusters in the cube of acoustic impedance (Fig. 5.22a). In particular, here, east of well 3, an anomalous zone is traced with relatively low AI values and with the shape of the reference AI curves characterizing their relative decrease (approximately within inline = 90–180 and crossline = 150–240) with a distinct blue palette (surrounded by brown-red-yellow-green palette). This anomaly, 2.5 × 2.5 km in size, occupies an area of ≈ 6 $km^2$. Judging by the nature of the distribution of AI values (on the corresponding inlines and crosslines), this is most likely a reef formation of an intrashelf bank type, approximately of siliceous-clayey-limestone-dolomite composition [12]. Usually here such reef formations are adjacent to the marginal reef zones, with which the gradual replacement of dolomites by limestones from the shallow shelf towards the deep-water basin is most often associated.

It also does not exclude the possibility of the presence of cavernosity in the specified reef bank (reservoir with pore-cavernous porosity), as well as filling this inhomogeneity with appropriate fluids, including hydrocarbons.

The following features of the geological structure of the studied section in the area of the proposed reef formation should also be noted (Figs. 5.22 and 5.23):

- The northern edge of the revealed anomaly in terms of the shape and values of AI in the sediments of the Zadon horizon practically coincides with the position of the tectonic fault conditionally designated No. 3 (Figs. 5.22a, 5.23a, b). It seems that, according to the geological and tectonic features of the structure and conditions for the formation of oil-promising objects in the terrigenous subdomain sediment complex, this tectonic fault separates the V-Savinobor field (with its own structural features and a significant number of drilled productive wells) from the S-Savinobor one (also distinguished by its geological tectonic features of the structure).
- Despite the interference nature of the initial seismic wave field and the corresponding seismic facies classification results (Fig. 5.22b), in the zone of manifestation of a negative anomaly in the form and AI values (Fig. 5.22a), it is also possible to trace the location of this area – it manifests itself a relatively dense red-brown palette in comparison with the areas surrounding this anomaly (Fig. 5.22b). The zoning of heterogeneities in the structure of the section based on the classification of the wave field cube as a whole over the entire area turns out to be quite significant.

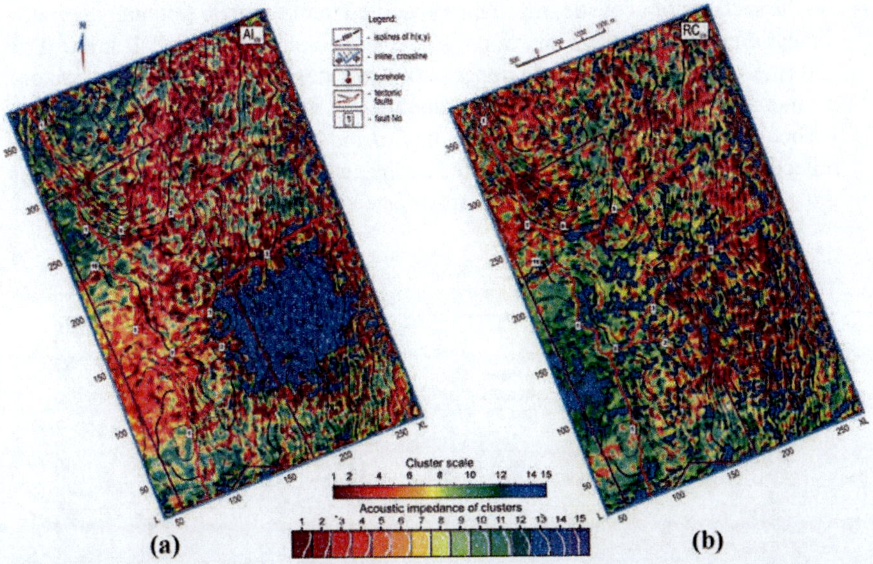


**Fig. 5.22** Distribution map of cube clusters: acoustic impedance (**a**), reflection coefficient (**b**) for the upper part of the Zadon horizon ($D_3zd$) – relative to the reflecting horizon $D_3zd$

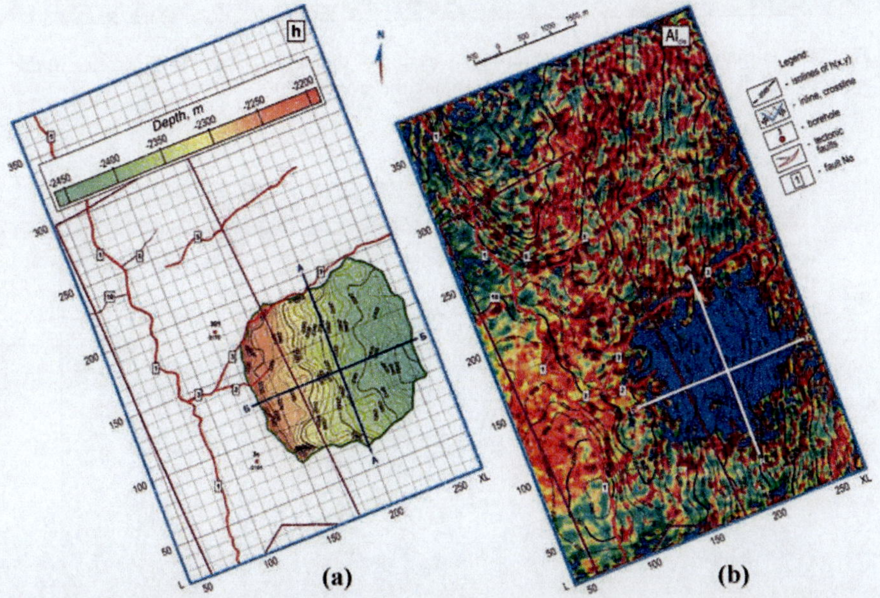


**Fig. 5.23** Display of the presumed bioherm structure in the sediments of the upper part of the Zadon horizon: (**a**) structural map along the base of the bioherm structure ($D_3zd$); (**b**) the distribution of facies according to the clustering of sediments in the upper part of the Zadon horizon

- On the sections of acoustic impedance specified in the control (mutually perpendicular directions – fragments of the AI sections along the lines B-B inline 125 and A-A crossline 200), in the upper part of the deposits of the Zadon horizon, an anomaly of low values of the AI is confidently traced (Figs. 5.24 and 5.25).
- It should be especially noted on the map of the distribution of clusters of the reflection coefficients cube constructed by the authors (the authors do not present this map here), the anomaly, which manifested itself in the results of classifica-

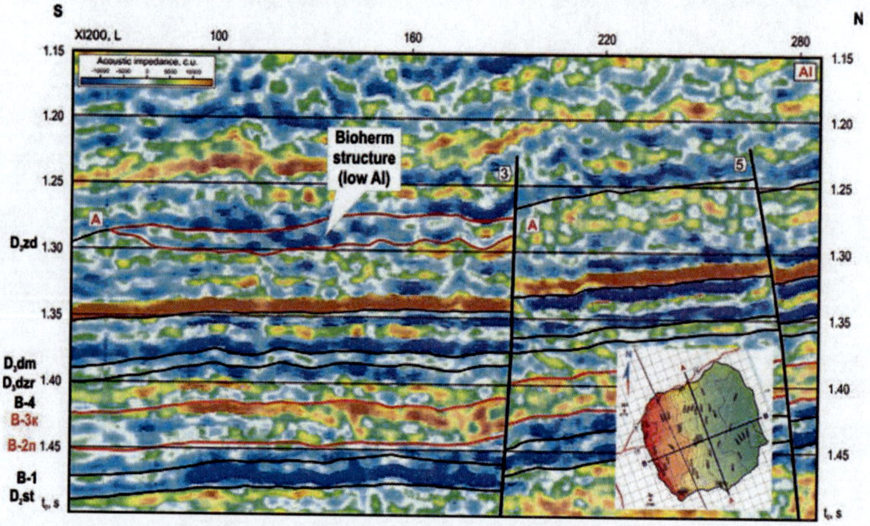


**Fig. 5.25** Display of a bioherm structure in the section of the field of acoustic impedance in the upper part of the Zadon horizon $D_3zd$ deposits along the crossine 200

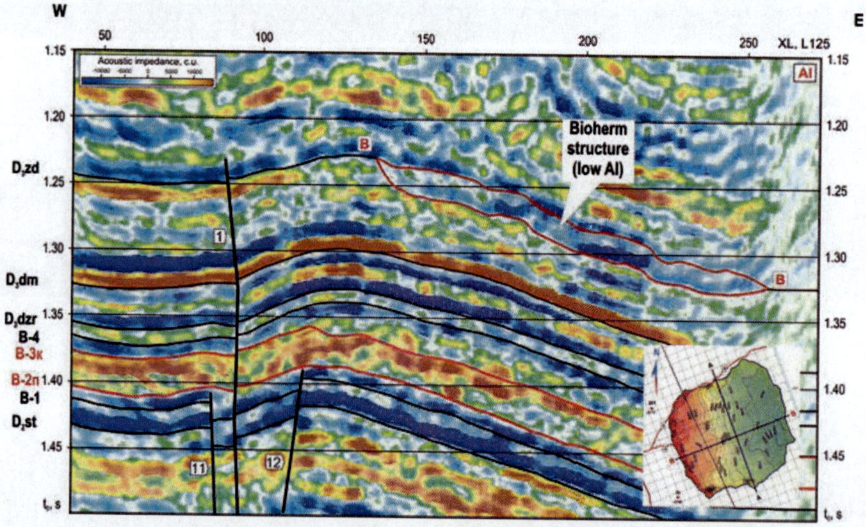


**Fig. 5.24** Display of a bioherm structure in the section of the field of acoustic impedance in the upper part of the Zadon horizon $D_3zd$ deposits along the inline 125

tion by the form and AI values (Fig. 5.22a), is practically not showed. This seems to indicate that the studied section – the upper part of the Zadon horizon ($D_3$zd + 15 ms) is purely carbonate in lithological composition (with possible relatively small variations in the proportions of limestones and dolomites in one direction or another), and a more significant the identification of the AI anomaly in the form of a seismic facies with a bright blue palette (Figs. 5.22a, and 5.23b) can be largely due to the appearance of porosity due to the strong cavernosity of the intrashelf bank and fluid saturation.

### 5.2.6 *Geological Structure Features of the Reservoir B-3к + B-2п*

Patterns of tectonic system and structural forms by maps h(x,y) for the top and bottom of the combined productive layer B-3к + B-2п ($D_3$dzr + jr) of the lower substage of the Upper Devonian Frasnian stage, presented respectively in Figs. 5.26a, b as a whole remain almost similar to the horizons considered above. The morphology of the horizons (layers) B-3 and B-2 (which are part of the combined layer B-3к + B-2п) deposits surface practically retains the inherited character of development with the underlying horizons of the Middle Devonian sediments. Beds B-3 and B-2 are composed of fine-grained, gray-brown quartz sandstones and dark gray, quartz, micaceous siltstones, which are separated by clayey and dense rocks. The deposits of these units, as noted above (in Sect. 5.1), belong to the main Middle Devonian-Frasnian terrigenous oil and gas complex (OGC). The beds B-3 and B-2 testing in well 3 of the Savinobor field, in particular, confirmed their oil saturation. Based on well testing and core analysis, productive intervals within the Upper Devonian section of well 301 were not identified.

The features and regularities of the tectonic structure of the deposits under consideration – the B-3к + B-2п layer of the Dzhyer-Yaran horizon ($D_3$dzr + jr) are largely similar to those for the overlying deposits. A comparatively large anticlinal structure, traced over the entire area of the prospecting area and increasing in its linear dimensions and hypsometry in the northern part of the 3D seismic data cube, turns out to be a fragmented series of tectonic faults of submeridional and sublatitudinal directions. At the same time, tectonic faults of sublatitudinal strike are as a rule characterized as low-amplitude.

The local brachyanticlinal structure, located in the southern half of the study area, in that of which well 3 Savinobor was drilled, is quite confidently localized along the top B-3к and the base B-2п of the considered formation B-3к + B-2п (Fig. 5.26a, b). This local structure, as already noted above, is complicated in its northern part by an insignificant, low-amplitude disturbance.

The structural forms of the deposits surfaces h(x,y) along the top of the B-3к and the base of the B-2п horizon ($D_3$dzr + jr) are almost similar to the above-mentioned forms. The main structural elements characterizing the tectonic structure in the

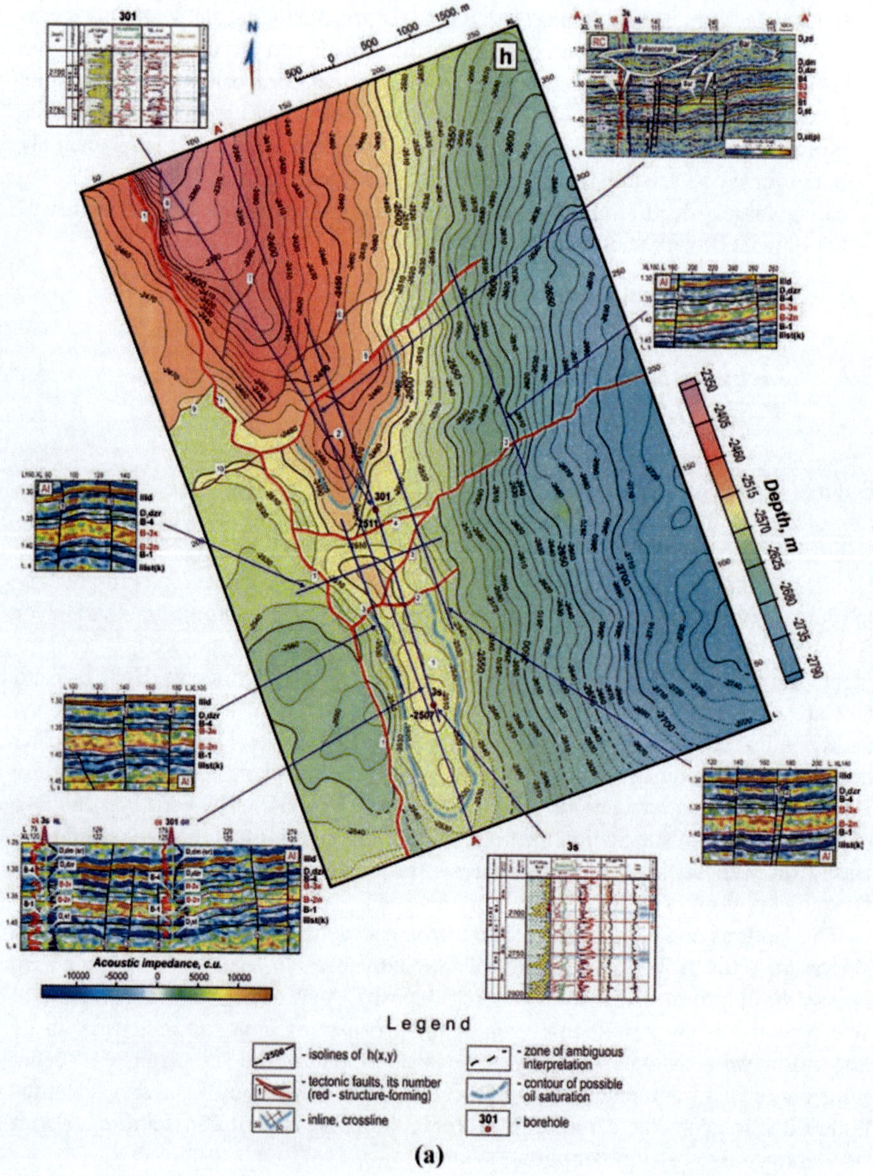


**Fig. 5.26** Structural maps: (**a**) along the top of reservoir B-3к of formation deposits B-3 ($D_3$dzr); (**b**) along the bottom of the reservoir B-2п deposits of the formation B-2 ($D_3$jr); (**c**) the isopach map between reflectors B-2п and B-3к

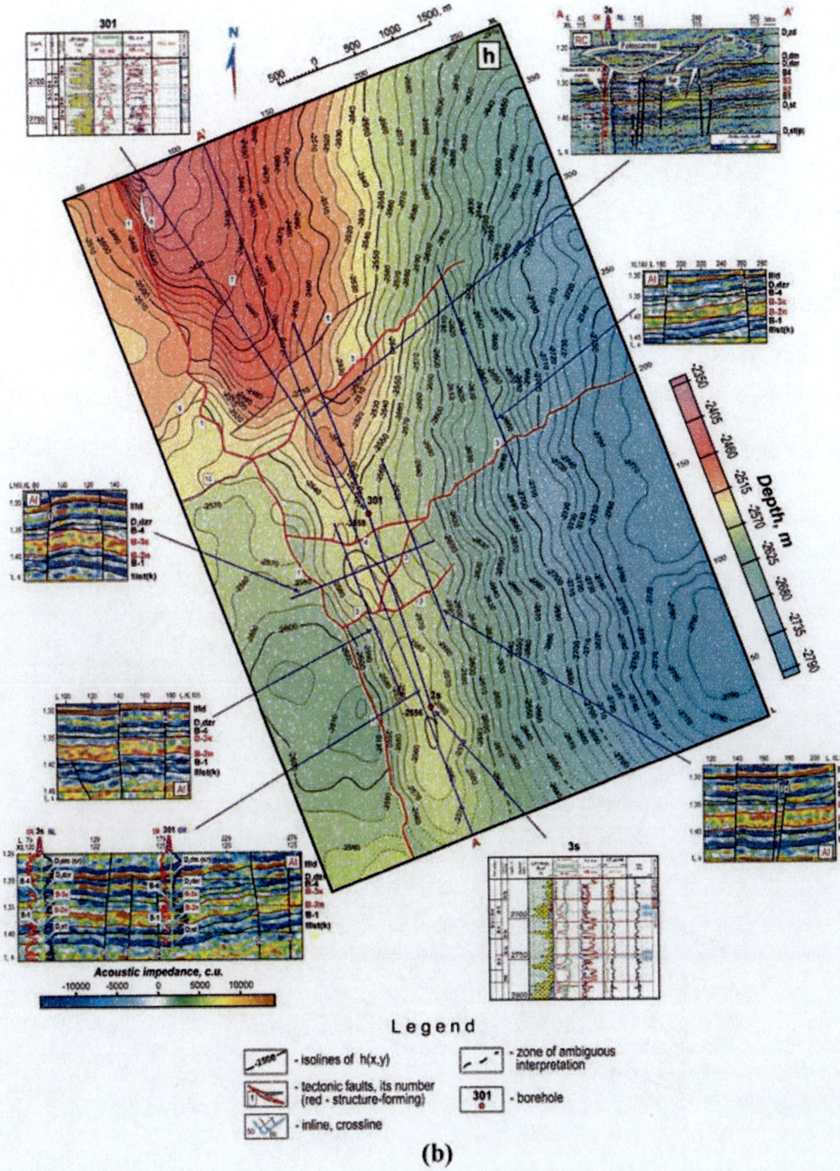


(b)

**Fig. 5.26** (continued)

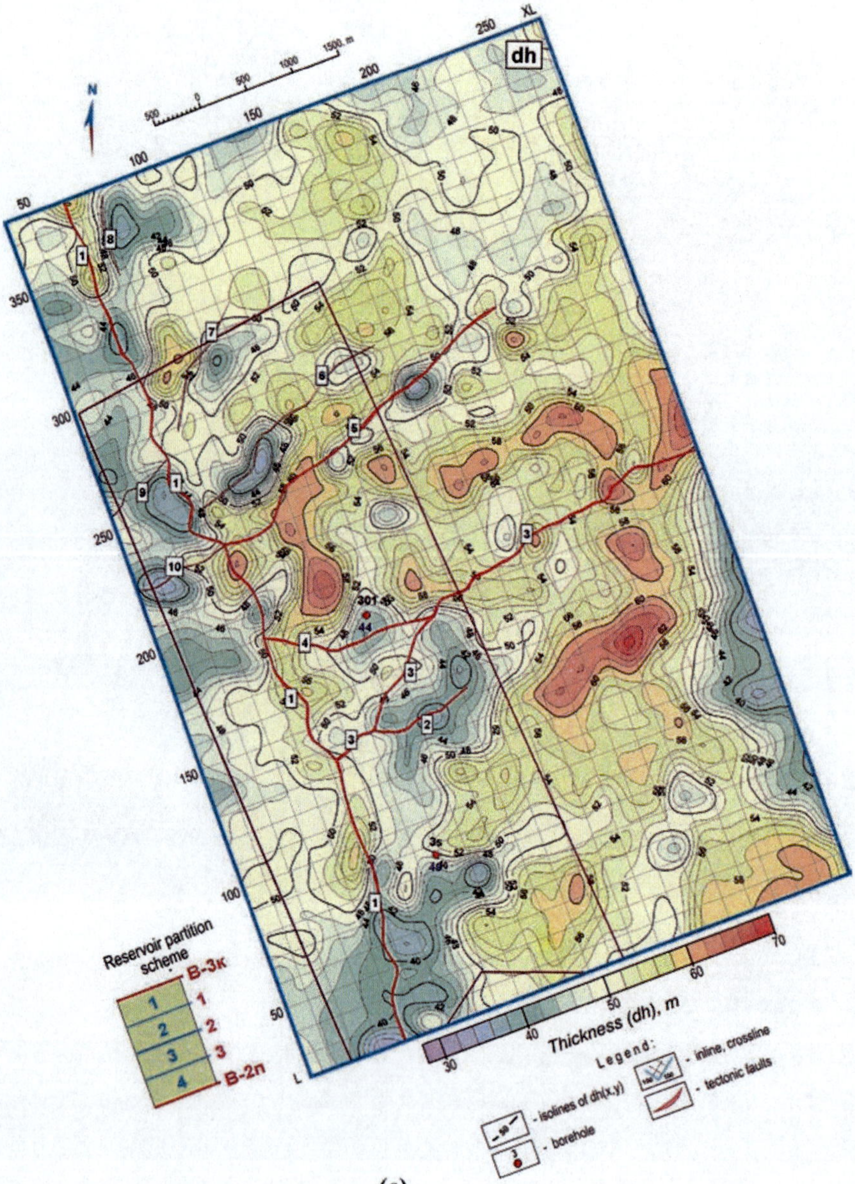


(c)

**Fig. 5.26** (continued)

considered structural zone according to the studied horizons are as a whole consistent with the general tectonic structure of the region under study.

To determine the thickness of the deposits of the B-3к + B-2п horizon ($D_3$dzr + jr), the constructed structural maps were used (Fig. 5.26a, b) and an isopach map – thicknesses $\Delta h_{(B\text{-}2п\text{–}B\text{-}3к)}$ was built, which is presented in Fig. 5.26c. In this figure, in particular, it can be seen the distribution over the area of increased thicknesses $\Delta h_{(B\text{-}2п\text{–}B\text{-}3к)}$ corresponds to a certain regularity, resembling some general annular anomalous zone with increased thicknesses. Moreover, this anomalous zone is localized ("breaks") into separate, somewhat disconnected, relatively short sections. This anomalous zone is confined in general to the central part of the study area. The thickness values here generally vary in the range of 49 ± 19 m (it can be seen the increased thickness values here are in the range of ≈ 54–68 m) (Fig. 5.26c). To this, it is only worth noting the drilled wells 3 and 301 do not fall into the zone of increased thicknesses $\Delta h_{(B\text{-}2п\text{–}B\text{-}3к)}$, which generally indicates a certain "reserve" in terms of the prospects for the output volume of the productive formation B-3к + B-2п horizon ($D_3$dzr + jr).

Of particular interest in terms of the detection of oil-promising objects may be structural forms identified in the sections (cubes) of reflection coefficients (RC), as shown in Fig. 5.27a, b. In these figures, in the arched part of the anticlinal structure (bounded in the north by tectonic disturbance No. 5, and in the south by tectonic fault No. 4) in the productive reservoir B-3к + B-2п ($D_3$dzr + jr), the proposed bar structure is allocated. This edifice could have been formed here in the process of sedimentation by an intense paleoflow of sedimentary material with the simultaneous participation in this process of contour limiters in the form of the above-mentioned tectonic faults. The areal distribution of such a bar structure is shown in Fig. 5.26a, b.

A somewhat different form of an oil-promising object can be established in the RC section (according to the features of its structural forms) in the form of an assumed paleochannel, as shown in Fig. 5.27a. Here, on the RC section along the control crossline 115, the internal structure of such a paleochannel is relatively well seen in the sediments of the Dzhyer horizon $D_3$dzr (directly in packs Ia, Ib and bed B-4). Such objects, as is known, are characterized, at least, by increased reservoir properties.

With regard to the results of the classification of seismic facies using the neural network technology (NNT) (based on the application of the *Stratimagic* software package) according to the shape and the values of these curves for the cubes of acoustic impedance (AI), wave field (WF) and reflection coefficient (RC), we note following.

Prior to seismic facies classification procedure using a special developed program, the time interval of 3D cubes AI, RC and WF, occupied by the B-3к + B-2п productive formation (with the corresponding top and bottom of this formation throughout the license area), was divided into four equal time thicknesses. On this basis, the layer B-3к + B-2п was divided into 4 uniform ones, i.e. interlayers of equal time thickness. Further, using these 4 interlayers (in accordance with such equivalent thicknesses), seismic facies were classified for a different combination of

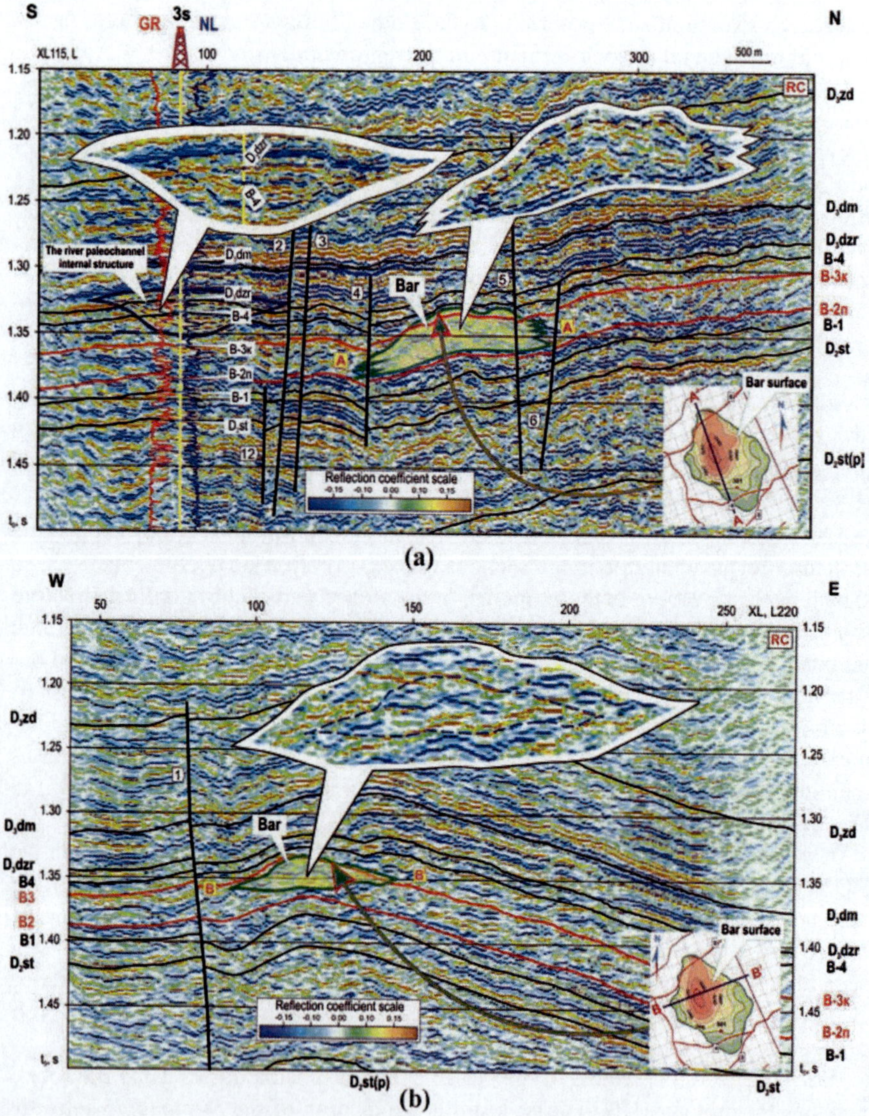


Fig. 5.27 Display of the internal structure of the river paleochannel and the bar on the section of the reflection coefficient field in the interval of the Frasnian subdomanik complex along: (**a**) the crossline 115; (**b**) the inline 220

these 4 interlayers. Such a differentiation of the total thickness of the B-3к + B-2п formation, in particular, shows the first two upper interlayers (2/4 of the total thickness) characterize in most cases the thickness of the B-3 formation ($D_3$dzr), as such, and the lower two interlayer (thickness 2/4 of the total one) characterizes mainly the thickness of the B-2 layer ($D_3$jr).

Based on the results of such differentiation of the B-3к + B-2п formation using the neural network technology (NNT) of clustering, the results of seismic facies classification were obtained (in accordance with the selected different combinations of 4 successively occurring interlayers), which are shown in Figs. 5.28, 5.29 and 5.30.

In particular, Fig. 5.28 gives a map of the clusters distribution for the shapes of the curves and their values of the AI and WF cubes as a whole throughout the entire thickness of the B-3к + B-2п formation. As a result of this classification, zoning is clearly highlighted in the cubes of AI, WF. Particular emphasis should be placed on the region of low AI values (and the shape of the corresponding reference curves, which are covered by the green-blue palette), traced from the southwestern corner of the work area to its northeastern corner with some expansion of the sector of such "coverage" in this indicated direction (Fig. 5.28a). It is worth noting approximately the same zonality is formed on the 3D WF cube (despite the significant manifestation of seismic wave interference in these seismic data) (Fig. 5.28b).

However, the spanned thickness of the combined layer B-3к + B-2п, shown in Fig. 5.27, in our opinion, cannot be considered quite acceptable in terms of optimizing the possible manifestation of changing seismic facies in different parts of this reservoir. In this regard, the classification results were considered according to the shape of the curves and their values for the upper half (1/2) of the reservoir (Fig. 5.29a–c), for the lower half (1/2) of the reservoir (Fig. 5.30a, b). Without dwelling on all the details and the classification subtleties of seismic facies with

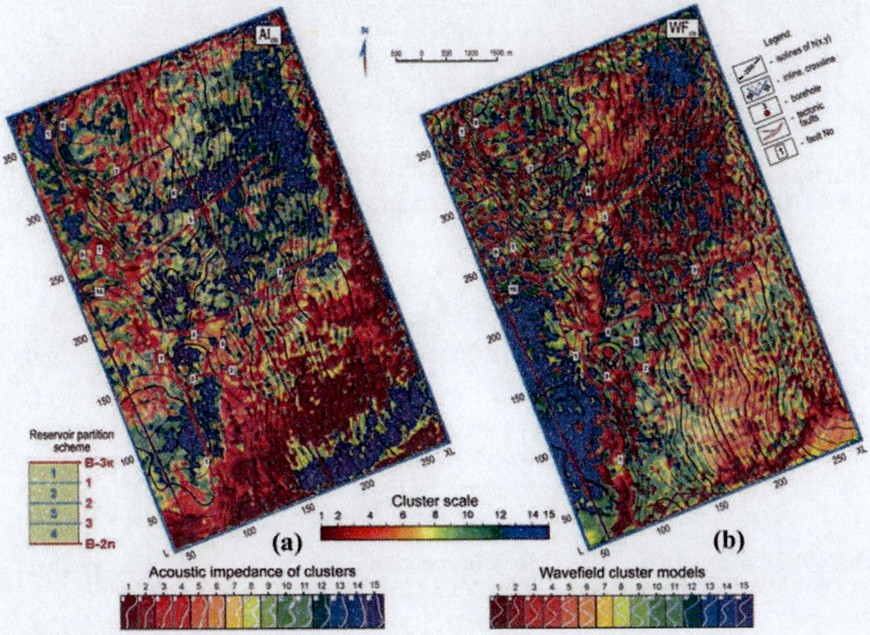


**Fig. 5.28** Distribution map of cube clusters: (**a**) acoustic impedance; (**b**) wave field for deposits of reservoir B-3к + B-2п ($D_3dzr + jr$)

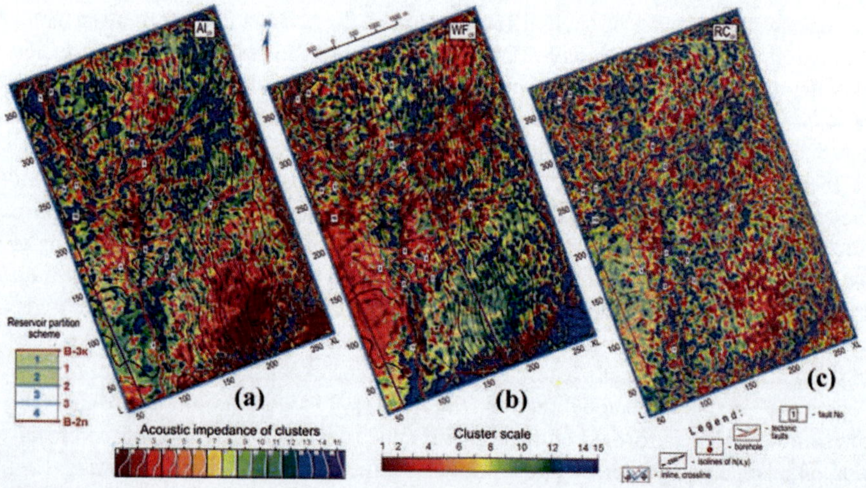


**Fig. 5.29** Distribution map of cube clusters: (**a**) acoustic impedance; (**b**) wave field; (**c**) reflection coefficient for deposits of the upper half of reservoir B-3к + B-2п ($D_3$dzr + jr)

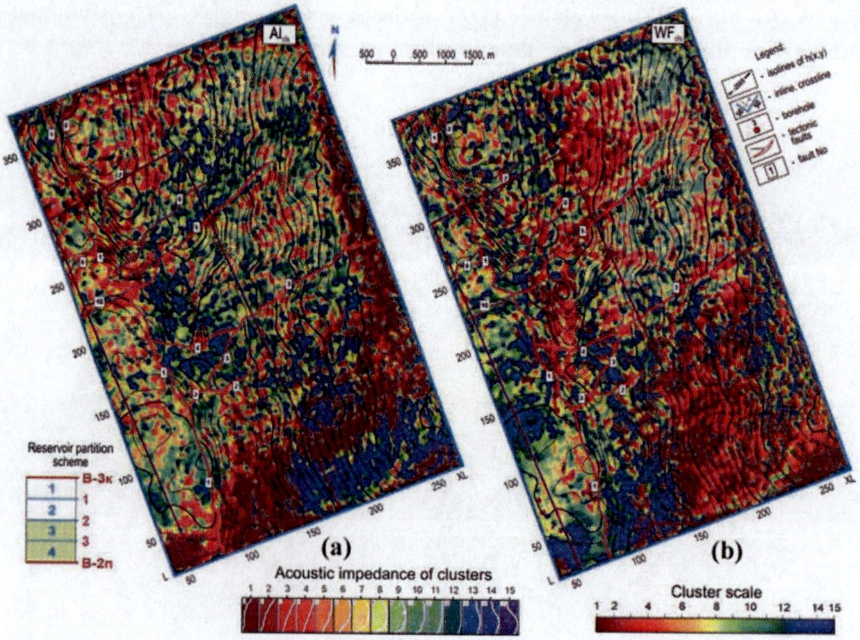


**Fig. 5.30** Distribution map of cube clusters: (**a**) acoustic impedance; (**b**) wave field for deposits of the lower half of the reservoir B-3к + B-2п ($D_3$dzr + jr)

different versions for the thicknesses of the given interlayers (for the combined productive formation B-3к + B-2п), we note only the main features of the results obtained. So, for almost the entire enumeration of results, it is worth noting the well-manifested zoning of seismic facies as a whole. It is almost in one way or another similar to the zoning shown in Fig. 5.28.

At the same time, only one of the zonations is most characteristic of the upper half of the B-3п + B-2п formation can be especially distinguished (Fig. 5.29a). This zoning is most responsible for the distribution of seismic facies, primarily only for the B-3 productive layer. Here, the greatest interest in the distribution of reservoirs and possible oil prospects can be associated with increased values of AI (and the corresponding shape of the reference curves on the cluster scale) in the tectonically shielded (according to fault No. 5) anticline fold located in the northwest relative to the wellhead 301 Savinobor (Fig. 5.29a). At the same time, the distribution of clusters over the study area by the shape of the curves and the AI values is largely consistent, and is well detailed by the shape and RC values (Fig. 5.29c).

In addition to the above, it can be noted under the conditions of avandelta sedimentation of deposits, the structure-forming submeridional tectonic fault No. 1, most likely, was some kind of barrier inside the fan of the paleoavandelta on the path of the water flow moving terrigenous sedimentary material. Along the line of this barrier, the water flow rate sharply reduced its intensity, as a result of which the flow rate decreased outside it, which ultimately led to a partial increase in the thickness of the terrigenous formations of the Dzhyer and Yaran horizons.

The combination of packs B-3 and B-2 into a single reservoir B-3к + B-2п made it possible, on the one hand, to facilitate the task of correlating reflectors throughout the entire seismic data cube (due to the use of the so-called "strong" reflecting boundaries), and on the other hand, it made it possible to divide this reservoir into more or less uniform interlayers (in particular, into four interlayers) using the example of seismic facies classification (according to the shape of the reference curves and their values).

Approximately the same technique with the division of the B-3к + B-2п reservoir into four interlayers of uniform thickness was used to determine in these interlayers and their combinations the distributions of acoustic impedances (AI) in each of them. One of the results obtained is shown in Fig. 5.31.

It should be noted scanning a productive section – a reservoir in terms of the shape and AI values has a certain meaning in relation to the concept that explains the influence of geological parameters of a real environment such as lithology, porosity and oil and gas saturation on the distribution of AI values. To a sufficient extent, such features are considered in [13, 14]. Here we note only one of them, namely, that the distribution of the acoustic properties of a specifically studied section against the background of their certain fixed distribution (mostly related to the distribution of the lithological composition) is superimposed by a quite noticeable effect of porosity and the nature of fluid saturation. Under conditions of practically undistorted dynamics of the initial wave field (preserving it during pre-processing), such a concept of the influence of porosity and the nature of fluid saturation works quite well. On this basis, it can be said with high probability under conditions of a

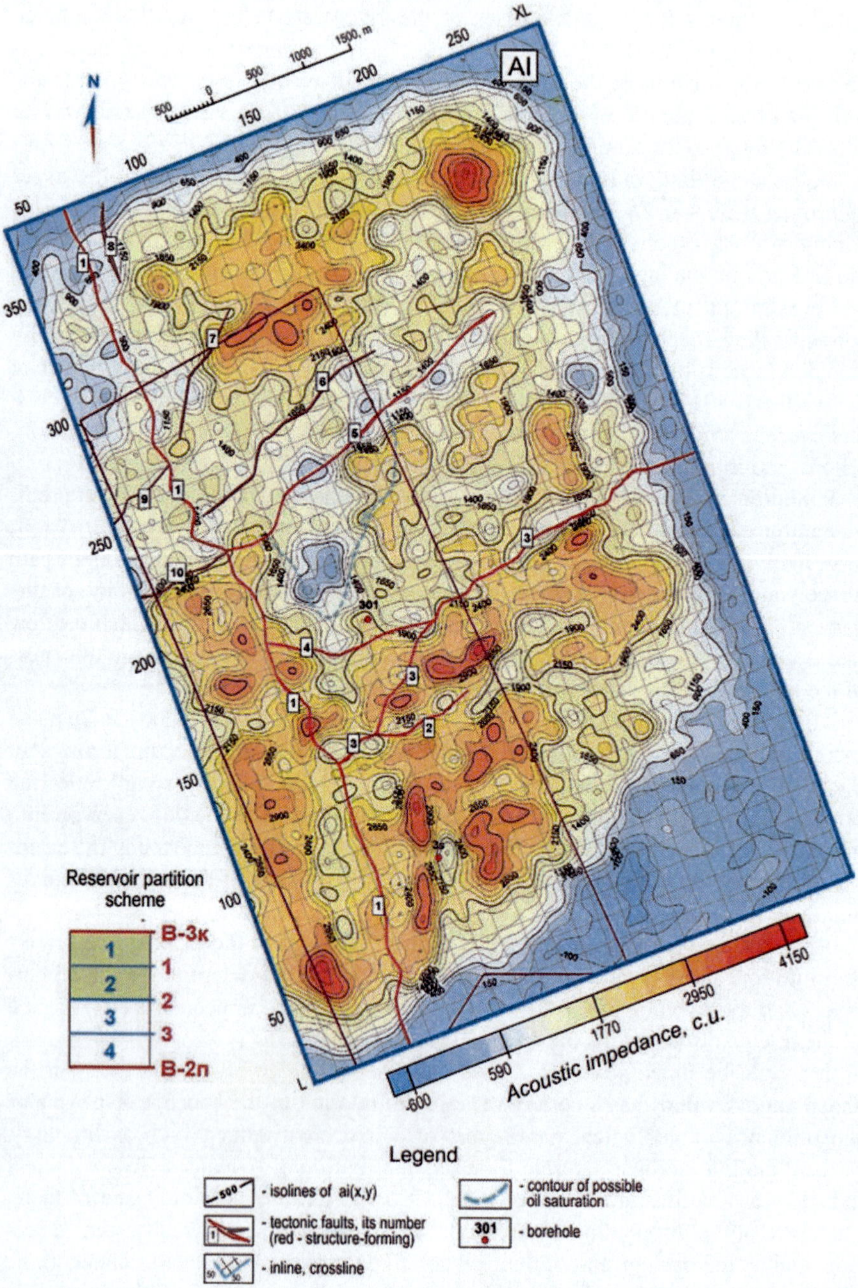


**Fig. 5.31** Acoustic impedance distribution map of the upper half of the reservoir deposits B-3к + B-2п

more or less uniform distribution of AI values, the factor of a significant decrease in AI turns out to be a practical sign of the effect on such a distribution of AI values, to the porosity and the nature of fluid saturation (of course, in this case, it is also necessary to take into account the “structural factor”).

The results of the AI distribution obtained in this way allow us to note the following.

In the northwest of well 301 Savinobor, in the area where the anticline structure is located, bounded from the north by tectonic fault No. 5, for the entire thickness of the B-3к + B-2п formation (in the contour along the isoline −2480), a noticeable decrease is observed in most of this area AI values (despite some isometric shape of AI isolines). In general, this seems to indicate the presence of a clear sign of the influence of porosity and fluid saturation and possible oil saturation on the AI. Such a feature in the distribution of AI values here in Fig. 5.31 – these seem to be possible oil prospects, given the distribution of AI under consideration refers to the upper half of the B-3к + B-2п formation, i.e. the nature of the distribution of AI here refers directly to the reservoir B-3.

The results of the distribution of porosity and oil saturation, presented on fragments of forecast maps in the area of the location of the anticline structure (limited from the north by tectonic fault No. 5 and located to the northwest of well 301) are shown in Fig. 5.32. Considering the procedure for predicting the composition and properties of oil-promising deposits is important, but at the same time quite complicated in the process of prospecting and exploration of oil and gas deposits, we note here only some of the features of this process. Note this process itself is described in detail by us in [13, 14].

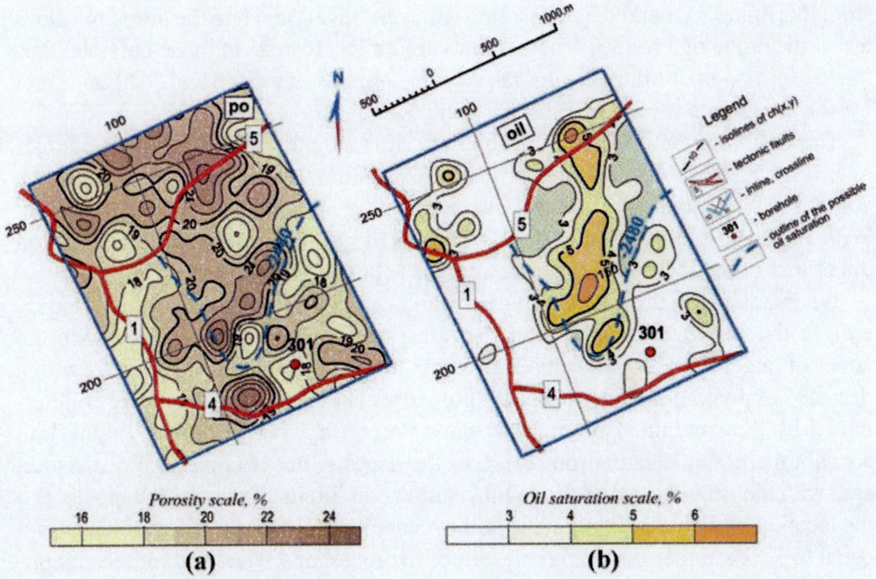


**Fig. 5.32** Predictive maps of the porosity (**a**) and oil (**b**) distribution saturations of the deposits of the upper half of the B-3к + B-2п reservoir in the vicinity of well 301

Seismic data, as is known, are directly related to the distribution of the acoustic properties of the medium, and thus the determination of geological and geophysical parameters is reduced to the search for dependencies between the estimated parameters (lithology, porosity, degree and nature of saturation) and acoustic ones (velocity, density, acoustic impedance, reflections coefficients, etc.). Before direct predicting the desired parameters, the studied section interval, as a rule, is compared with the physical parameters previously obtained for this section interval from well logging data, and then along a given lateral direction in the plane of the seismic section (or 3D seismic cube) for each target interval of the section (formation) the required geological and geophysical indicators are determined.

Determination of the composition and properties of rocks of oil-promising strata in wells is carried out based on the use of well logging data according to the method of functional transformations of geophysical parameters and their integration into information systems in accordance with the methodology and software, as we have shown above in Chap. 4.

In a more specific form, to predict the lithological composition and fluid components from well logging curves, the corresponding dependences of velocity, density, clay content and porosity on acoustic impedance are preliminarily determined [13, 14]. On this basis, using a special system of petrophysical equations, the AI values are recalculated into the desired parameters of porosity and fluid saturation. With this in mind, the relationship that characterizes the increase (decrease) in porosity values with decreased (increased) values of effective acoustic impedance for a particular study interval of the geological section "works".

The distribution of porosity obtained in this way in the upper half of the B-3к + B-2п reservoir shows a slight increase in its values directly in the contour of the anticlinal structure (Fig. 5.32a). The contour of this structure, as can be seen from this figure, is marked by an isoline with a.m. -2480 m. Here the porosity values are in the range of 17–23%. These values are quite adequate to those obtained during the formation of lithological-stratigraphic columns of wells 3 and 301 Savinobor for the considered reservoir (Figs. 5.15 and 5.16).

The results of the oil saturation forecast, shown on a fragment of the prognosis map (Fig. 5.32b), show the most probable distribution of this parameter within the contour of the anticlinal structure, also marked by an isoline with a.m. −2480 m. With regard to the established levels of oil and gas saturation in the considered upper half of the B-3к + B-2п formation, the following can be noted.

When analyzing the desired predictive indicators (reservoir properties), in general, in the considered exploration site, areas were identified associated with the zones of target horizons deposits distribution, that are characterized as areas with a decrease in porosity and permeability properties (FCP) (usually this level is determined by the contour of oil and gas saturation with a range 0–3%). Taking into account the possible saturation nature of the deposits under consideration, associated with the peculiarities of the sedimentation conditions of the target deposits and the location of their distribution, the authors conditionally set the oil saturation level equal to 3% or higher for the mobile part of hydrocarbons. The level for the stationary portion of hydrocarbons, respectively, is also conditionally accepted, equal to

3% and below. On this basis, the contours of signs of oil shows were identified (this approach was used by the authors to determine the contours of oil saturation in the deposits of productive formations). It is also worth noting oil shows in the studied interval of the section are also due to the manifestation of an increased content of silty-sandy rocks (i.e., the presence of a certain lithological factor). This feature of the structure of the section is also well reflected on the constructed lithological-stratigraphic columns and the corresponding initial well logging diagrams (Fig. 5.15 and 5.16).

The distribution of oil saturation shown in Fig. 5.32b, is characterized by a rather complex arrangement of contour shapes, but its maximum values are widely confined to the arched part of the tectonically shielded anticline structure (to its periclinal part), which to a certain extent increases its oil prospects. It also seems zones of deconsolidation of rocks can be associated with the arch of the structure.

### 5.2.7 *Geological Structure Features of the Starooskol Suprahorizon ($D_2st$)*

The regularities of structural forms and tectonic structure in the form of a structural map h(x,y) for the top ($D_2st(k)$) of the Starooskol suprahorizon of the Givetian Stage ($D_2zv$), shown in Fig. 5.33, in general, remain similar (with some structural features) to those for the above horizons. Separate distinctive features of the structure are observed only in the southwestern part of the survey area. The surface morphology of the IIIst(k) ($D_2st(k)$) horizons retains, on the whole, the inherited character of development with the underlying Middle Devonian horizons. The deposits of the Starooskol suprahorizon $D_2st$ are composed of a sandy-aleurite sequence, which is indexed as the "main sequence" and is characterized by negative SP anomalies and minimal GR readings. The sandstones are gray, light gray, inequigranular, well sorted. Quartz siltstones and chlorite-quartz, gray and gray-green, with frequent siderite nodules.

Features and laws of the deposits tectonic structure of the IIIst(k) ($D_2st(k)$) horizons under consideration are largely similar to those for the above-considered deposits – the B-3к + B-2п formation of the horizon ($D_3dzr$ + jr). At the same time, the ranges of changes in the values of isolines h(x, y) ≈ −2430÷ − 2890 m (Fig. 5.33).

As for the overlying formation B-3к + B-2п, a large anticlinal structure can be traced starting from the central part of the prospecting area and increasing in size and hypsometry towards its northern part. Here it turns out to be fragmented by a whole series of tectonic faults of submeridional and sublatitudinal directions. At the same time, tectonic faults of sublatitudinal strike are primarily characterized as low-amplitude.

The brachyanticlinal structure located in the southern half of the study area in the area of the drilled well 3 Savinobor also turns out to be complicated by a number of

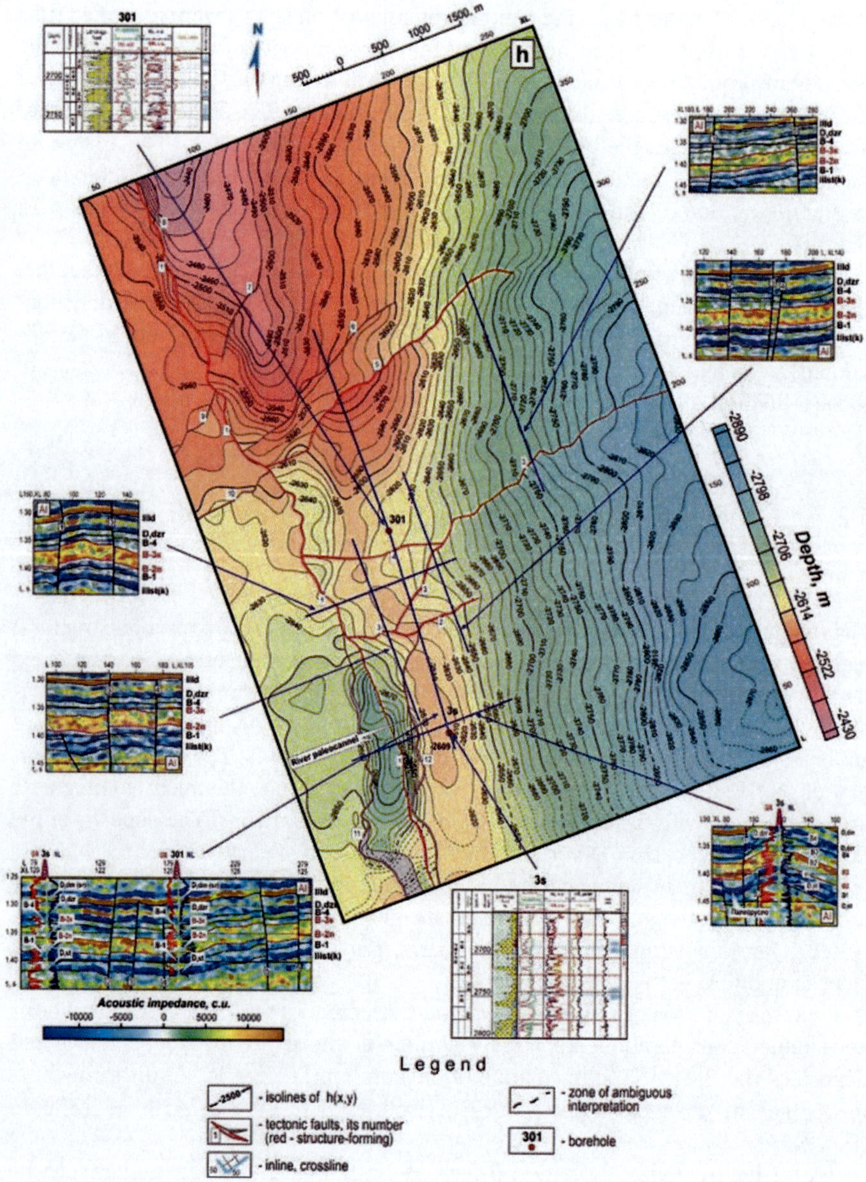


**Fig. 5.33** Structural map along the top of the deposits of the Starooskol suprahorizon $D_2st$ (IIIst(к))

tectonic faults of submeridional and sublatitudinal strike. To the west of well 3, along the contour of submeridional disturbances, the contour of the most probable (in terms of structural features and the nature of the distribution of AI values here) paleochannel of a hypothetical paleoriver identifies itself, which manifested itself at the boundary of the Givetian stage ($D_2$zv) and the Frasnian stage ($D_3$f) of the Middle ($D_2$) and Upper ($D_3$) Devonian (Fig. 5.33). This feature of the geological structure highlights quite confidently in the sections AI and RC presented above.

In the central part of the study area, in the area of well 301 Savinobor, a tectonically screened anticlinal structure is confidently manifested. The wellhead 301 is located relative to the periclinal part of this structure in a southerly direction at a distance of about 900 m (from the axis of the fold of the structure). The sealed structure is limited from the north by tectonic fault No. 5. In the south direction, the anticline structure strongly flattens out, being limited here by low-amplitude tectonic fault No. 4.

For the uppermost part of the deposits of the Starooskol horizon ($D_2$st + 15 ms), seismic facies were classified using the neural network technology (NNT) clustering (implemented in the Stratimagic software package) according to the shape of the curves and their values for the cube of acoustic impedance (AI), wave field (WF) (Fig. 5.34a, b) and reflection coefficients (RC) (Fig. 5.34c). On the classification results obtained for all three forms of seismic data (AI, WF and RC), the zoning of the section heterogeneity is confidently correlated in the shape of the envelope reference curves and their values. To the greatest extent, it finds itself on the map of the distribution of clusters in the cube of acoustic impedance (Fig. 5.34a). Such a heterogeneity of the considered part of the section, i.e., features of the distribution of seismic facies, is due to a significant differentiation in the values of acoustic

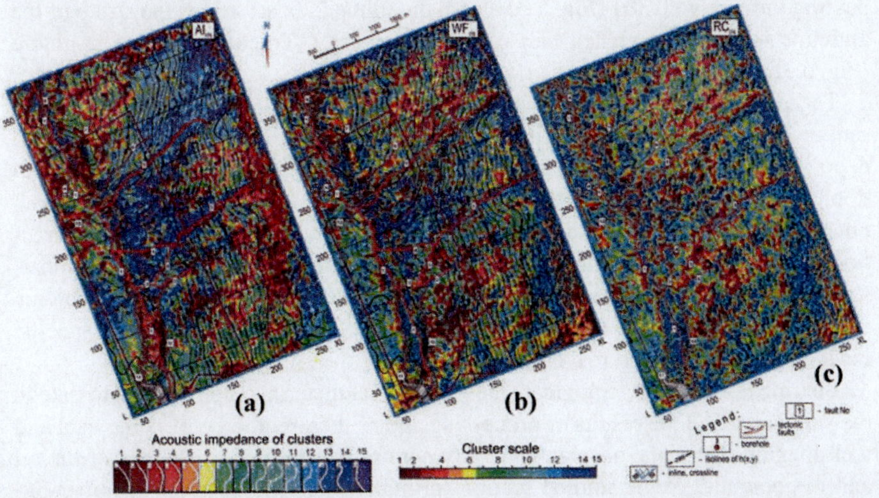


**Fig. 5.34** Distribution map of cube clusters: (**a**) acoustic impedance; (**b**) wave field; (**c**) reflection coefficient for the upper part of the Starooskol suprahorizon ($D_2$st) deposits – relative to the reflecting horizon $D_2$st(к)

impedance (AI) and the corresponding changing shapes of the AI curves (displayed on the scale of reference clusters in AI). Decreased AI values (and changing forms of reference clusters) are most expressed in the form of a certain zone with a violet-blue palette in the central part of the search area. This zone of location of well 301 is enclosed between tectonic faults No. 5 (in the north) and No. 3 (in the south) of the work site. The results of clustering according to the RC data (Fig. 5.34c) to a certain extent detail the results obtained according to the AI data (Fig. 5.34a).

The seismic facies classification results according to the data of the seismic wave field (Fig. 5.34b), despite the interference nature of the record obtained in the conditions of a purely thin-layered section, do not generally contradict the results obtained from the AI and RC data, they have approximately the same character of the manifestation of seismic facies on the studied prospecting site.

### 5.2.8 *Examples of Seismic-Geological Sections for the Frasnian Subdomanik Complex*

When determining the location of the identified structures in the vertical plane of the section along the most informative directions – seismic traverse profiles (formed from 3D cubes of seismic data AI and RC), seismogeological sections were constructed, in the vertical sections of which the desired oil-promising objects are located at various depths (Fig. 5.35a–c).

Such seismogeological sections were built in the interval of tracing the Frasnian subdomanik complex in several directions-profiles. In particular, along: (1) the line of a composite profile passing through wells 3s and 301 (Fig. 5.35a); (2) inline 185 passing through well 301 (Fig. 5.35b); (3) the inline 220, set across the strike of the anticline uplift – the arched part of the tectonically screened anticline structure (Fig. 5.35c).

The presented sections clearly show how the structural features of the studied section are traced. At the same time, the ratios of the structural forms of the studied deposits, the amplitudes of local structures in the productive part of the studied section, the thickness in each of the considered horizons, the identified oil-promising object (in the form of an assumed paleobar (Fig. 5.27a, b)) change in lateral directions. In addition, on the models under consideration (Fig. 5.35) there is an inherited, interconnected character with the distributions of the structural and tectonic elements of the structure listed above, which at the same time characterize specific local areas of the studied real environment.

Bearing in mind the geological and geophysical materials accumulated to date in the study area and the results of processing and reinterpretation of seismic data and well logging materials, one can generally note relatively high prospects for the oil and gas potential of the studied section (primarily deposits of the reservoir B-3к-B-2п Dzhyer-Yaran horizon ($D_3$dzr + jr), as well as a number of other oil-promising objects).

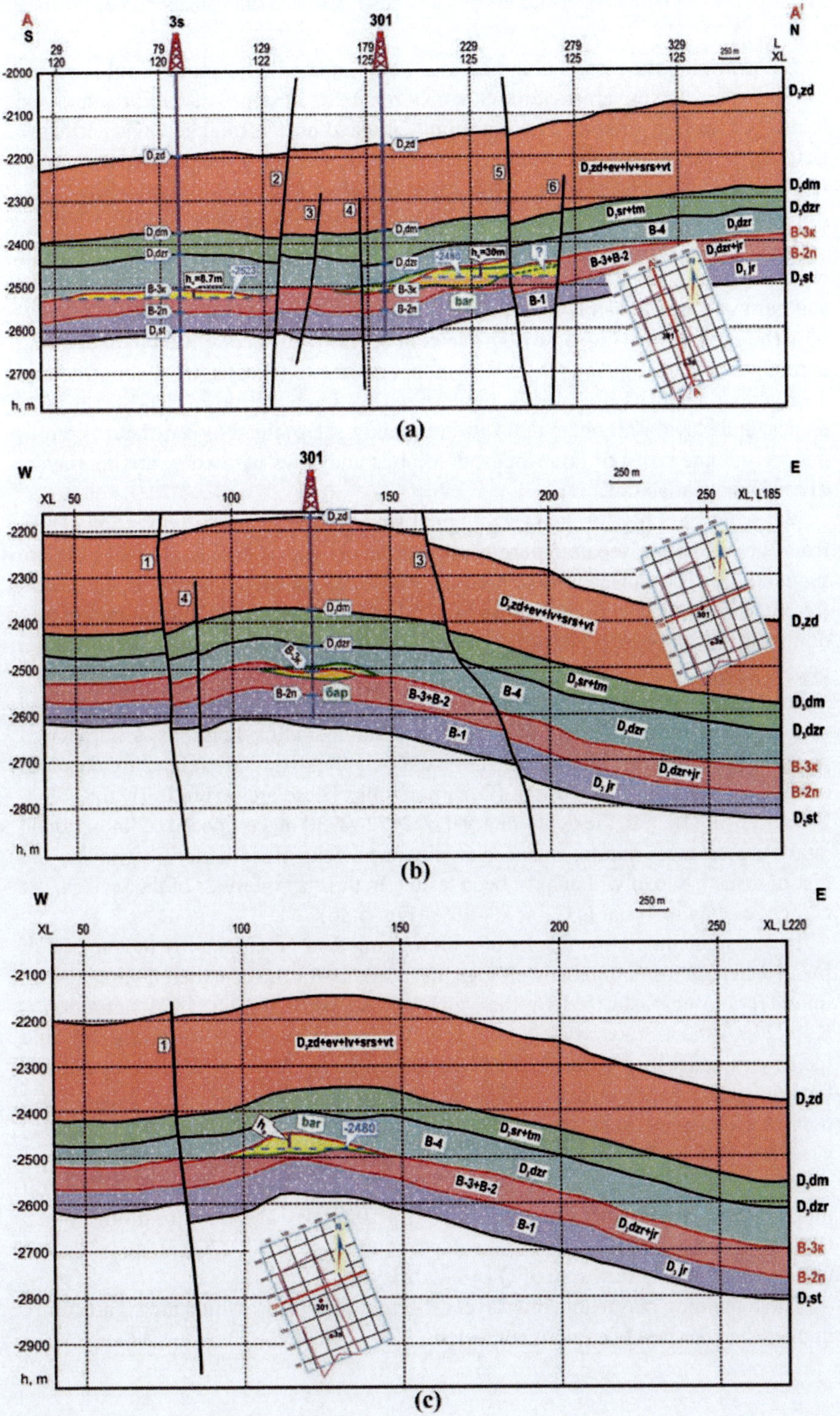

**Fig. 5.35** Seismic and geological sections in the tracing interval of the Frasnian subdomanik complex passing along: **(a)** a composite profile through wells 3s and 301; **(b)** the inline 185 through well 301; **(c)** the inline 220

Summarizing the results obtained as a whole, it should be emphasized in all the studied oil-promising areas of the Savinoborsk field, all stages of interpretation and reinterpretation of seismic data were implemented not on time sections and wave field cubes, but on detailed sections and cubes of geological environment parameters such as sections and cubes of effective reflection coefficients and acoustic impedances.

It should be noted the oil prospects of the exploration area under study can be considered quite high, since in the west this area is located in close proximity to the southern part of the V-Savinobor field, where one of the wells is located – well 9 with the same name (Fig. 5.36). The section of this well is composed of alternating carbonate and terrigenous facies, which are heterogeneous in their structure. Terrigenous deposits are mostly represented by intercalation of sandstones, siltstones, and clays. Carbonate deposits are mainly represented by a different combination (volume ratio) of limestones, dolomites and clays with very rare interlayers of terrigenous material.

We will note only the main geological structure features of the section. Thus, from top to bottom, separate permeable interlayers with a thickness of 4–13 m are sequentially distinguished in the carbonate section – in the sediments of the Zadon ($D_3$zd), Sirachoy ($D_3$srč) and Domanik ($D_3$dm) horizons. These interlayers are predominantly water-saturated with insignificant signs of oil saturation, and having a porosity of $K_p \approx 8–15\%$.

In terrigenous formations (argillites, siltstones, sands, sandstones), represented in the considered interval of the section by subdomanitik horizons, a number of permeable oil-saturated interlayers are fixed, which are directly related to the packs of the Dzhyer ($D_3$dzr) and Yaran ($D_3$jr) horizons. These are layers Iб ($D_3$dzr), B-4, B-3 and B-2 ($D_3$jr). In the depth interval of 2672–2730 m, composed of fine-grained quartz sandstones, quartz, micaceous siltstones, four oil-saturated interlayers are distinguished, two of which have been tested. In the clay intervals of the section, the content of clay material is $C_{cl.} \approx 80–95\%$ (Fig. 5.36).

In accordance with the results of well logging data processing by the method of functional transformations of well logs, the porosity for the considered oil-saturated interlayers is characterized by the values of $K_p \approx 12.1–20.6\%$, oil saturation – $K_o \approx 23.3–80.2\%$, clay content – $C_{cl.} \approx 6.7–37.6\%$. For the same set of permeable interlayers under investigation, the reservoir properties are characterized by the following values: absolute permeability (graph $k_a$) $k_a \approx 0.06–23.28$ mD, relative water permeability (graph $k_w$) $k_w \approx 0.01–1.09$ mD and relative oil permeability (graph $k_o$) $k_o \approx 0.01–5.6$ mD.

When testing the depth interval of 2669–2673 m of formation Iб ($D_3$dzr), an oil inflow was obtained with a flow rate of $Q_{oil} = 10.6$ t/day at the 3 mm fitting. When testing the reservoir B-4 ($D_3$jr) in the depth interval of 2692–2700 m, an oil inflow was obtained with a flow rate of $Q_{oil} = 4.2$ t/day.

The remaining permeable interlayers B-3 and B-2 ($D_3$jr), which are significantly thinner than the first two, were not tested.

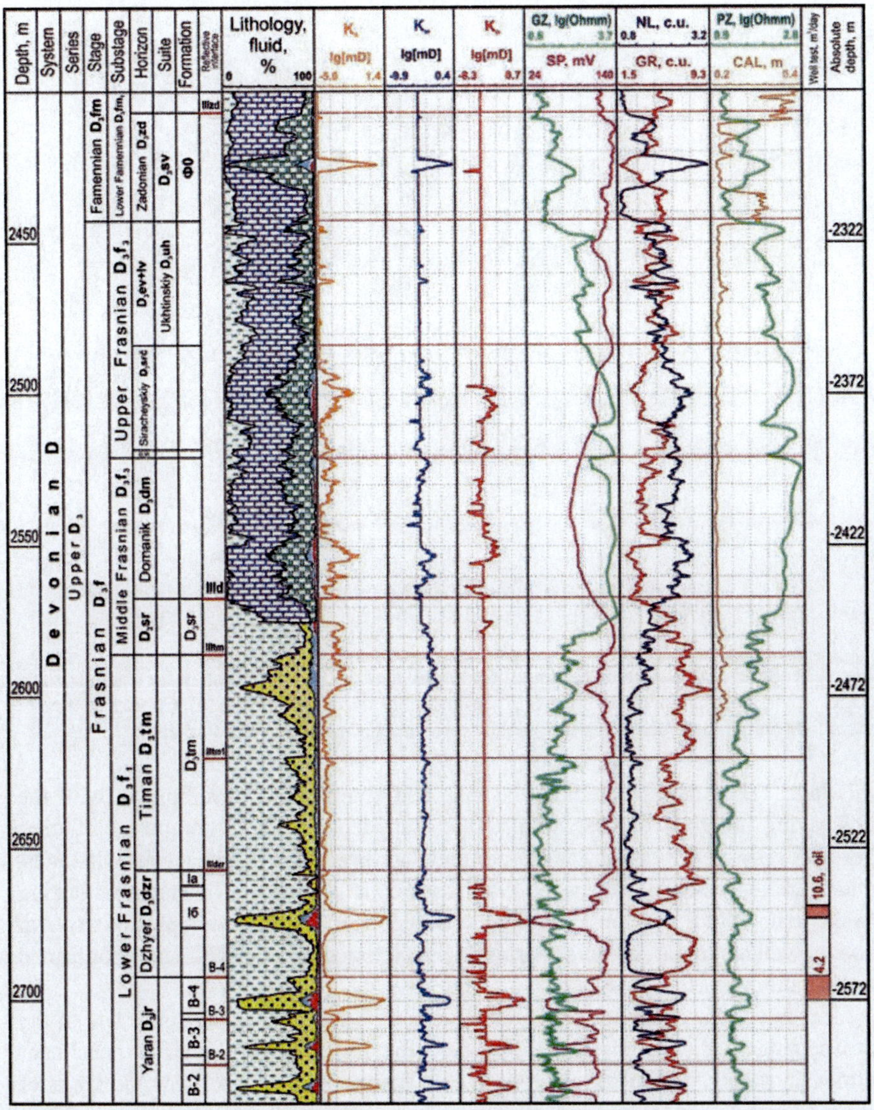


**Fig. 5.36** Lithologic-stratigraphic column with the distribution of the nature of reservoir rock saturation in the well 9 along the interval of the Lower Famennian and Frasnian deposits

## 5.2.9 *Main Geological and Geophysical Parameters of Productive Deposits in the Prospecting Area 3D(S)*

Features and regularities of the Yaran horizon ($D_3jr$) deposits tectonic structure – unit B-3 within the 3D(S) of the southern cube of the Savinobor area are largely similar to the above-considered horizons (Fig. 5.37a). The surface morphology of

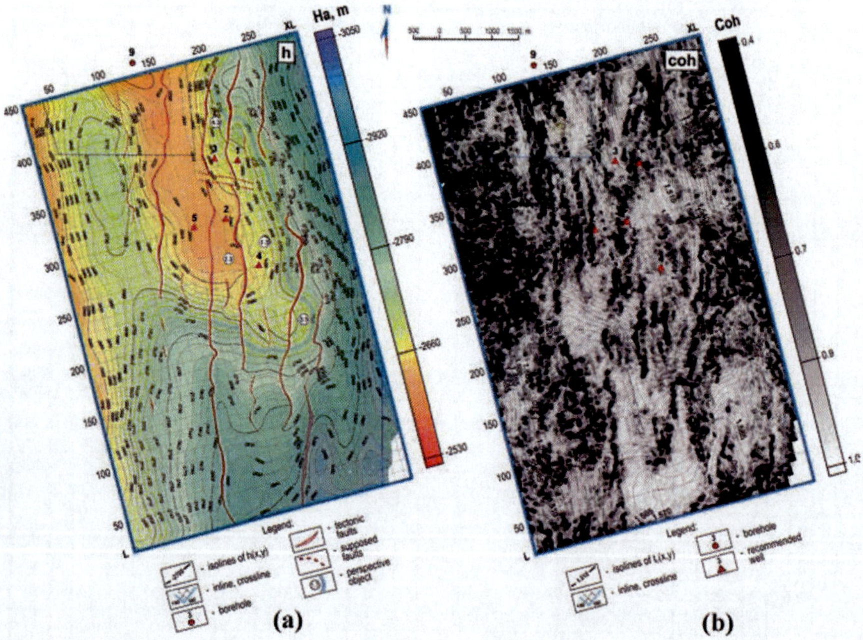

**Fig. 5.37** Structural map along the top of the deposits of the B-3 formation of the Yaran horizon ($D_3$jr) – the reflecting interface of the B-3 within the 3D seismic data cube (**a**). Coherence map of wave field amplitudes cube along the paleoisochronous surface of the B-3 formation (**b**)

this layer (B-3) deposits retains the inherited character of development with the underlying horizons of the Middle Devonian deposits. It is composed of fine-grained, gray-brown quartz sandstones and dark gray, quartz, micaceous siltstones, which are separated by clayey and dense rocks. The deposits of this member belong to the main Middle Devonian-Franian terrigenous oil and gas complex. As shown above, well testing at the neighboring V-Savinobor field of the B-3 unit confirmed its oil saturation (Fig. 5.36).

A relatively large anticlinal structure, traceable throughout the entire site of the prospecting area and complicated by local structures relatively small in their linear dimensions and amplitude, turns out to be a fragmented series of tectonic faults of submeridional and sublatitudinal directions (submeridional faults are generally traced within the 3D(S) seismic data cube). On the structural map along the top of the deposits of the B-3 layer of the Yaran horizon ($D_3$jr) within the 3D(S) cube (Fig. 5.37a), a series of submeridional tectonic faults is confidently traced, separating local tectonically screened structures (designated by us for definiteness by numbers 1.3–4.3) from the pericline of the anticlinal uplift associated with the local structure of the V-Savinobor field in the area of the well 9 (Fig. 5.36). The location of tectonic faults of submeridional strike here is confirmed on the coherence map of the cube of wave field amplitudes along the paleoisochronous surface of the B-3 formation of the Yaran horizon ($D_3$jr) (Fig. 5.37b). There is an irregularity of the seismic record in the form of a background darkening of increased intensity and brightening of the dynamics of the seismic record. Extended lineaments are also observed here in the 3D(S) section of the seismic data cube (the latter one way or another coincide with the identified tectonic faults within the 3D(S) cube of effective acoustic impedance) (Fig. 5.37a, b).

Based on drilling and well logging data for the V-Savinobor field and hypsometry of local tectonically sealed structures, designated for definiteness by numbers 1.3–4.3, we assumed the conditional level of water-oil contact for deposits of the B-3 formation of the Yaran horizon ($D_3jr$) at the absolute mark (a.m.) −2720 m. Four perspective objects (1.3–4.3) (Fig. 5.37) were identified on the structural map along the B-3 formation of the target horizon ($D_3jr$) at this absolute elevation. In this case, tectonic screens are faults of submeridional strike, shown on the target maps by thick red lines. In these local tectonically shielded traps, it seems the main volumes of hydrocarbons are formed.

Based on the use of a structural framework of 2D sections along seismic profiles and 3D cubes of the migrated wave field (primarily cubes of effective acoustic impedance and reflection coefficient associated with 3D(N) and 3D(S) cubes of seismic data of the Savinobor area), drilling results and automated processing of well logging data for the wells of the V-Savinobor field, the results of building detailed 3D models of various geological and geophysical parameters for the main productive horizons were obtained. The elements of such 3D models are industrially productive packs of the Dzhyer ($D_3dzr$) and Yaran ($D_3jr$) sediment complexes. In this case, the formation under study is either a screen-cover for the underlying reservoir, or it is the reservoir itself. In accordance with this, a list of maps has been determined within the search area for each of the considered thin-layer reservoirs (horizons), on the basis of which it is possible to make a certain assessment of the quality of each of the predictive 3D model elements (components) of a thin-layer real environment (including in the 3D(S) area of seismic data cube).

Specifically, a number of predictive models of geological indicators of a thin-layer real environment were built. For horizons characterized as potential screens (i.e. for deposits of the Sargay and upper Timan ($D_3sr + D_3tm$) horizons) within the exploration area of the Savinobor area, a set of only two predictive geological indicators was built: effective acoustic impedance and clay content.

For the productive reservoir of the Yaran ($D_3jr$) horizon, i.e. for the B-3 unit (within the 3D(S) seismic data cube), the following predicted geological indicators are presented in the form of maps below: effective acoustic impedance, porosity, effective thicknesses, clay content, oil-saturated thicknesses and predicted hydrocarbon distribution densities.

The results of conversion of high-resolution seismic data into geological indicators for the main target objects in the exploration area of the Savinobor area (within the 3D(S) seismic data cube) were obtained in a relatively complete volume only for the most promising B-3 unit of the Yaran ($D_3jr$) horizon.

On the distribution maps of the average values of effective **acoustic impedance** (AI) in the sediments of the B-3 pack of the Yaran ($D_3jr$) horizon, a significant heterogeneity of the structure is manifested (Fig. 5.38a). In these deposits, the areas of increased AI values are distributed as follows. The most intense anomalies with elevated AI are confined to the eastern and partly to the western parts of the study area. Approximately in the central part of the area under consideration, a rather extended zone of relatively low AI values is observed. Such a feature in the distribution of AI anomalies is organically "superimposed" by a system of tectonic faults of submeridional strike, which is mainly deep foundation.

The parts of the area with the distribution of low AI values, as is known, are confined to the most clayey intervals of the studied section. They also "fit" well into the indicated system of tectonic faults.

In general, the regularity in the area distribution of low and high values of effective acoustic impedance is quite consistent with the existing ideas about the presence of geological (including acoustic) inhomogeneities in this part of the section, which we previously traced on the sections of acoustic impedance. Here, as noted above (Fig. 5.6), the complex geometry of the Lower Frasnian natural reservoirs of the "filling nested valleys" type and coastal-marine genesis determined, in general, the irregularity of the stratigraphic volumes of the reservoirs in the study area, the seaward nature of the reservoirs of the proposed traps, and the coarsening of the granulometric composition of sandstone layers [12].

In the distribution over the area of porosity studies in the deposits of the horizon (formation) B-3 ($D_3$jr), zoning is highlighted (Fig. 5.38b), which has approximately the same regularity in the distribution over the area as the effective acoustic impedance. At the same time, it can be argued with a high probability the zones of increased porosity (especially those have relatively extended areas in their shape on the area) correspond to relatively increased values of effective acoustic impedance. As well as in the case of the distribution of clay material over the area, here the same

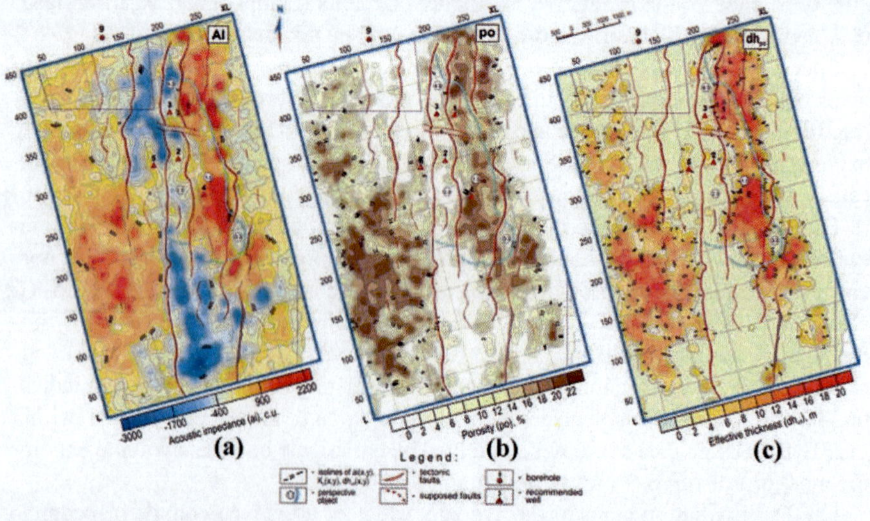


**Fig. 5.38** Maps of acoustic characteristics and predictive parameters in the sediments of the B-3 formation of the Yaran horizon ($D_3$jr): (**a**) acoustic impedance; (**b**) porosity; (**c**) effective thickness

mechanism of formation of sandy sedimentary material took place, with which, in one way or another, the increased reservoir properties of the studied intervals of the section of the real geological environment are associated. For reservoirs of the B-3 unit of the Yaran horizon ($D_3$jr), relatively large zones with increased porosity are located in the western and eastern (largely in its southern half) parts of the study area.

It should be observed the reservoirs with average porosity values varying in the range of 4–20% are most often associated with oil saturation.

In addition, a certain trend is marked, showing the relationship between the location of areas of increased porosity values and increased of effective acoustic impedances.

The map of predicted **effective thicknesses** for reservoir deposits of the B-3 unit of the Yaran horizon ($D_3$jr) is shown in fig. 5.38c. The ranges of variation of this parameter in zones of increased porosity and possible contours of oil and gas saturation for deposits of the B-3 unit are estimated at ≈ 2–18 m.

The general law in the distribution of **clay content** in the deposits of the B-3 unit of the Yaran horizon ($D_3$jr), presented on the corresponding map (Fig. 5.39a), is in good agreement with the distribution of AI for the deposits under consideration. Here, in the deposits of this unit B-3 ($D_3$jr), for areas with increased AI, there is an

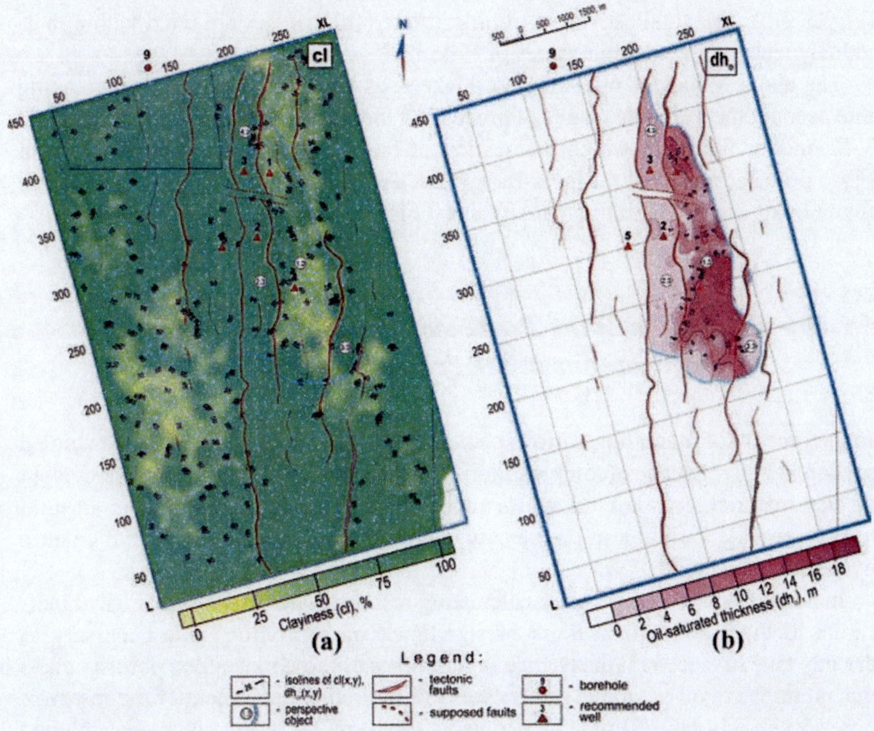


Fig. 5.39 Maps of predictive parameters in the deposits of the B-3 formation of the Yaran horizon (D3jr): (a) clay content; (b) oil-saturated thickness

almost unambiguous correspondence between lower values of clay content. At the same time, the observed zoning in the distribution of clay content values over the area (which is adequate to the distribution of AI values) is obviously due to certain conditions for the formation (sedimentation) of deposits in the considered horizons, as well as the system of location of tectonic faults in the area (which here could play the role of natural barriers on the way spreading paleostreams carrying sedimentary material).

A map of predicted average values of **oil and gas saturated thicknesses** of deposits of the B-3 unit of the Yaran ($D_3$jr) horizon with superimposed contours of oil saturation (at the level of the absolute mark (a.m.) −2720 m) is shown in Fig. 5.39b). (On this map, the minimum oil-saturated thickness is assumed to be ≈ 1 m. This thickness turns out to be the closest in value to the thickness of the so-called elementary layer, which can be determined using seismic data reconstructed applying the HRS-Geo technology software. In particular, the thickness of an elementary layer, equal to 1.5 m, is realized by converting seismic records into acoustic impedance curves based on solving the inverse dynamic seismic problem with a sampling step of seismic records in time $\Delta t = 0.001$ s and an elastic wave propagation velocity in sandy-clayey formations equal to 3000 m/s). The highest values of the parameter under consideration are most likely observed in zones with increased oil saturation, reaching values of ≈ 4–16 m for the deposits of the B-3 unit (Fig. 5.39b). The smallest values of oil-saturated thicknesses are fixed within individual local zones, reaching values of 1–4 m.

The above values of oil-saturated thicknesses were also obtained by us taking into account the available results of processing logging materials for the wells of the V-Savinobor field (for which the section of one of the wells is shown above in Fig. 5.36). The results of the layer-by-layer determination of the desired values were found by us according to the porosity threshold exceeding Kp ≥12%.

### 5.2.10 *Estimation of the Predicted Distribution Density of Hydrocarbons in the Productive Reservoir B-3*

In the process of predicting various geological indicators, it is assumed the studied section is heterogeneous (complex morphology of the pore space, variable content of clay material, complex mineral composition of the matrix, etc.). This, in turn, leads to a wide variation in the density distribution of oil and gas and volumetric hydrocarbon content.

In accordance with this, when calculating resources and various geological indicators in the identified contours of significant oil saturation, it is necessary to directly take into account the change in the values of these indicators: porosity coefficient, effective oil-saturated thicknesses, oil saturation coefficient of the reservoir pore space. This can be implemented using the formulas of the volumetric method in each vertical section of the section using each of the predictive seismic traces.

The heterogeneous distribution of geological indicators, as is known, is due to the complex heterogeneous structure of the target objects of study.

If one of the above-mentioned variables is taken as a basis, namely, the oil-saturated thickness of the prospective horizon, then it should be noted within the volumetric space characterized only by this oil-saturated thickness, in addition to the hydrocarbons themselves (gas saturation), there are also a number of other geological and geophysical indicators. In particular, these are lithological components (the content of sandstones, clays in the corresponding parts of this space), as well as the pore space filled with either hydrocarbons together with formation water, or only hydrocarbons.

To find the share of "pure" oil (or oil only) – hydrocarbons (without fractional content of lithology and water saturation) in linear units of measurement, it is necessary to normalize all constituent components – geological indicators, including the linear interval occupied by oil saturation, in these units of measurement, for example, on a depth scale (i.e. in the same units as the the oil-saturated thickness – in meters). Then, from the total volume of the constituent geological components, everything is not directly related to hydrocarbons should be subtracted. The resulting residual will ultimately characterize the portion of the volume of "pure" oil (*in units of the depth scale at a given point*).

The specified procedure for normalization and recalculation of the volumetric content of various geological components on a depth scale was implemented in the form of an algorithm and a program. Based on their use, it is possible to obtain maps of the predicted distribution density of hydrocarbons ("pure" oil) for reservoirs of target deposits. As one of the examples of such constructions, the authors selected promising objects 1.3–4.3 for deposits of the B-3 pack of the Yaran ($D_3$jr) horizon (Fig. 5.40).

The obtained results of the predicted distribution density of hydrocarbons – "pure" oil for productive formations within the contours of significant oil saturation, in this case, represent a set of cylinders with curvilinear (over the area of work) contours of their vertical walls at the point under consideration. Such hypothetical cylinders of a real geological environment, according to the conditions of the performed calculations, are filled only with hydrocarbons – "pure" oil (as already noted above, without a porous rock matrix of a certain lithology and water saturation). At the same time, the height of the walls of a single curvilinear cylinder is different for each of the objects of study (in the experimental plan, the authors below considered only deposits of the B-3 unit). This height of the walls of the cylinders most often varies in the range of 0.1–0.2 m.

In each of the contours of significant oil saturation, the spatial arrangement of such cylinders implies the presence of sequential stratification: in its lower part there is a cylinder with the largest distribution area, and in the upper part – with the smallest one. At the same time, for each of the objects (oil-promising horizon) within each of the contours of significant oil saturation, its own unique (inherent only to this object) form of the predicted hydrocarbon distribution density can be obtained.

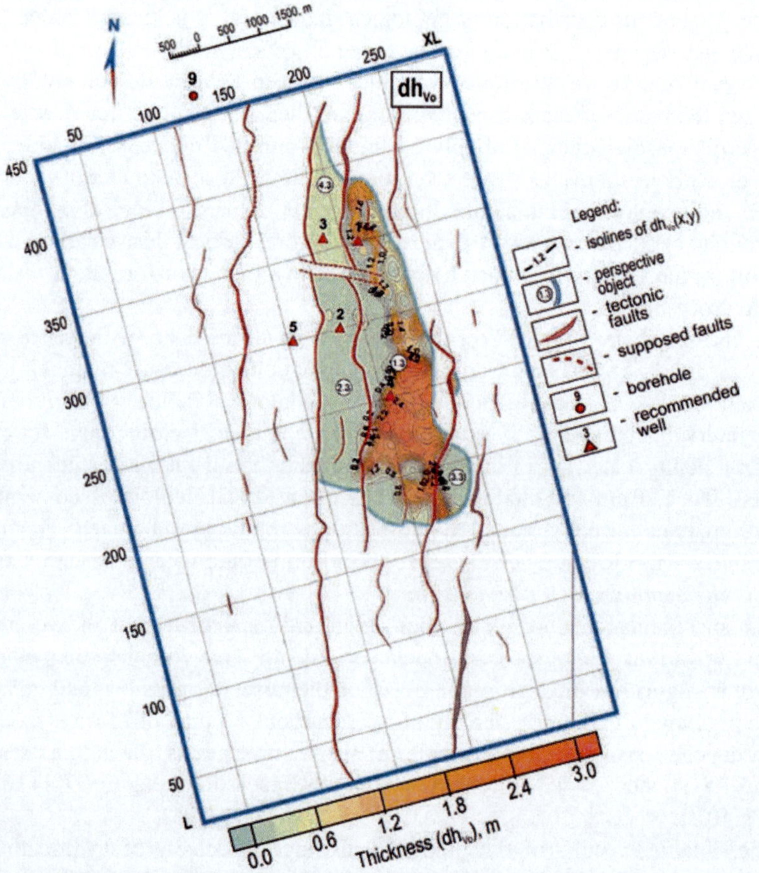


Fig. 5.40 Predicted distribution density map hydrocarbons in the deposits of the B-3 formation of the Yaran horizon ($D_3jr$)

It should also be noted the results in the form of maps of the predicted distribution density of hydrocarbons are of practical interest. It consists in the fact, along with other predictive geological and geophysical indicators (which were discussed above in the previous subsection for the deposits of the B-3 unit of the Yaran ($D_3jr$) horizon), these maps are directly used to determine the most representative oil and gas bearing areas and promising points for laying exploration wells. Specifically, for the considered promising objects of study of the B-3 ($D_3jr$) unit, the following can be noted.

On the map of the predicted density of distribution of hydrocarbons in the productive pack B-3 ($D_3jr$) (Fig. 5.40), in accordance with the scale of the thickness of the "clean" oil reservoir, the range of its change is within 0–2.8 m (with the height of the walls of a single curvilinear cylinder – the reservoir equal to 0.2 m). From the presented results, it can be seen the contours (anomalies relative to the increased value of AI, porosity, effective and oil-saturated thicknesses (Fig. 5.38 and 5.39)) located in the areas enclosed between the lines (Fig. 5.40) are characterized by the highest predicted distribution density of hydrocarbon raw materials:

1. inline: 170 and 300 and crossline: 170 and 240;
2. inline: 310 and 390 and crossline: 200 and 230.

The sizes, configuration and intensity of anomalous zones within these areas are within wide limits. Obviously, this circumstance must be taken into account in combination with those (similar anomalous zones) and for other horizons of the Upper Devonian subdomanik complex of deposits.

The anomalies of the predicted distribution density of hydrocarbons obtained in this way are taken into account directly in combination with other geological indicators, first of all, when determining promising points and designing the location of wells during further opening up of predictive oil-saturated objects in the 3D(S) section of the Savinobor area seismic data cube.

## 5.3 Estimation of Geological Indicators in Oil-Promising Carbonate Deposits of the Middle and Lower Devonian and Upper Silurian Strata (*Kolvin Megaswell and the Northwestern Part of the Khoreyver Depression*)

This subsection presents the results of a comprehensive interpretation of borehole and surface seismic data for one of the areas of the Timan-Pechora oil and gas province. Here, CDP 2D prospecting seismic surveys were carried out at the Kamin area (Fig. 5.41), located in the junction zone of the northeastern part of the Kolvin

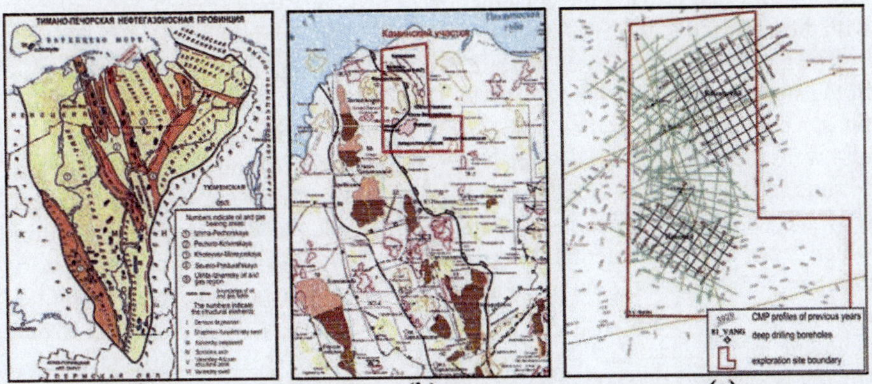


(a) (b) (c)

**Fig. 5.41** Location scheme of the work site: (**a**) scheme of tectonic zoning of the Timan-Pechora province (TPP) (according to S.P. Maksimov); (**b**) copy from the map of the tectonic and oil and gas geological zoning of the TPP (according to the data of JSC "Kolvaneft"); (**c**) scheme of geological and geophysical exploration by 2D seismic survey and drilling of the Kamin area (according to JSC "Naryan-Marseismorazvedka", 2011) megaswell and the northwestern part of the Khoreyver depression of the Timan-Pechora province. The overall goal of the research was to study the geological structure and prepare the work area for prospecting and exploratory drilling, primarily in deposits from the Artinskian stage ($P_1$ar horizon) of the Lower Permian to the Ludlovian stage ($S_2$ld horizon) of the Upper Silurian.

An analysis of the oil content of the territories adjacent to the work area, structural exploration and resource base shows the main prospects in the Kamin area are associated with carbonate Upper Silurian (Gerdyu) deposits, terrigenous Eifelian deposits of the Middle Devonian, Syrachoy (reef-genic) deposits of the Upper Devonian and carbonate Upper Artinskian deposits of the Lower Permian.

The continuation of exploration work, including seismic studies, at the Kamin area is aimed at detailing the geological structure of already known structures and identifying new oil-promising targets for prospecting and exploratory drilling. The main reserves for the growth of hydrocarbon resourses are most likely associated with the possibility of refining the geological model of the structure of hydrocarbon traps, including on the basis of refining the boundaries of deposits along the outer contours of oil potential. First of all, this is due to the Upper Devonian deposits – organogenic structures of Middle Famennian (packs $D_3fm_2$(F6) and $D_3fm_2$(F5)) and Upper Frasnian age (Evlanov-Liven $D_3f_3$($D_3$ev-lv), Syrachoy carbonate $D_3f_3$(III$D_3$srč') and Domanikoid $D_3f_3$($D_3$domanik) horizons), with subdomanik Middle and Lower Devonian (III$D_2$ and III$D_1$) deposits (Kolvin ($D_2$kl) horizon of the Eifelian ($D_2$ef) and Koiven ($D_1$kv) horizon of the Emsian ($D_1$em) stages), as well as with Upper Silurian deposits (the lower part of the Greben horizon ($S_2$gr) of the Przhidolian (IV$S_2$p) and the upper part of the Gerdju horizon ($S_2$gj) of the Ludlovian (IV$S_2$ld) stages). Oil prospects here are associated with the identification of structural-lithological and combined oil traps in marine and coastal-marine, shallow-marine, deltaic, lagoonal sedimentation conditions. Predicting such deposits based on seismic data is a rather difficult task can be solved through the use of modern thin technologies that provide the construction of real environment geological models with the highest possible resolution.

The authors have processed seismic data to obtain high resolution time sections in the studied interval of the wave field. Using a system of functional transformations, a complex interpretation of logging data in the terrigenous-carbonate section of 4 wells of the Kamin exploration area was performed. Using the high-resolution seismic method (HRS-Geo technology), sections of effective acoustic impedance (AI) and reflection coefficients (RC) were obtained – models of a real medium, which depict a detailed internal structure of the geological structures under study. In the course of research, the authors constructed numerous isochrone maps, structural maps, isopach-thickness maps for lithostratigraphic units of the sedimentary cover, including for horizons: $P_2$, $P_1$k, $P_1$ar, $P_1$a + s, $C_{2+3}$, $C_1$s, $C_1$v, $D_3fm_2$(F6), $D_3fm_2$(F5), $D_3f_3$(ev-lv), $D_3$src, $D_3$src', $D_3$vt, $D_3$domanik, $D_3$dm(k), $D_3$dm(p), $D_3$tm, $D_3$dzr, $D_2$ef, $D_1$e, $S_2$p (D-S), $S_2$ld, $S_1$v, $S_1$l, V-O and VI characterizing the structure of the

Paleozoic oil-perspective complexes. On the sections AI and RC, the location of 38 oil-promising objects was determined, which in the study area are characterized by certain, individual zones – the contours of the expected deposits. A detailed assessment was made of the most important geological indicators of the real environment for prospecting and industrial exploration, primarily the material composition and reservoir properties in the deposits of the Artinskian stage ($P_1$ar), the carbonate strata of the Sirachoy ($D_3$srč'), domanikoid ($D_3$domanik), Dzhyer ($D_3$dzr) horizons, the Middle ($D_2$) and Lower ($D_1$) Devonian and Ludlovian ($S_2$ld) Upper Silurian. Oil-promising objects of a structural type were identified in the reservoirs of the horizons $P_1$ar, $D_3$srč', $D_3$domanik, $D_3$dzr, $D_2$, $D_1$, $S_2$ld. In the sediments of the Salyukin horizon ($O_3$sl) of the Upper Ordovician, a paleo-river bed (paleoincision) with contours of coastlines varying in plan was revealed.

As in the analysis of materials for the previous (Savinobor) area (presented in Sect. 5.2), the main processing of seismic data here was carried out using the Western Geco OMEGA 2 processing system. The complex interpretation of seismic data was performed using software tools – the HRS-Geo technology developed by the authors and the interpretation system GeoGraphix R2004.1 v.8.5 of the Landmark company [13, 14]. Seismic data previously obtained on the work area in different field periods in seven sections of the study area were used directly in processing and interpretation. At each of these sites in different field seasons – time periods, various methods of field observations, devices, equipment and layouts of CDP-2D seismic profiles were used. As a result of earlier studies of CDPs at the Kamin and Iev work areas, a set of tasks was outlined, which included an analysis of the state of exploration of the study area, an assessment of the prospects for oil and gas potential, and the preparation of exploration programs. By the time the authors were conducting research, 3 exploratory wells had been drilled within the work site (71 and 81 Vangurei, 1 Yu-Vangurei). In addition to these wells, seismic surveys used materials from well 10 Nyamyurkhit, which is located in the eastern part of the area. Below are just some examples with the results of studies that were used in the joint integrated interpretation of HRS and well logging data.

Regular seismic surveys in the north of the province have been started since 1965, including on the Kolvin megaswell (Fig. 5.41). By the beginning of the 1970s, the Kolvin megaswell had been studied by seismic practically throughout its entire length – 300 km along the strike up to the coast of the Pechora Sea of the Arctic Ocean. The mass introduction of CDP seismic surveys has significantly increased the geological efficiency of seismic explorations, expanded seismic capabilities in detailing the geological section and structural forms of the sedimentary cover. Systematic studies of CDPs in the central part of the Kolvin megaswell have been carried out since 1975.

### 5.3.1 *Processing and Automated Interpretation of Well Log Data*

Processing and automated interpretation of well logging data was carried out to form a detailed geoacoustic model of the medium containing information on the distribution of acoustic velocities, density, lithology, reservoir properties and oil and gas saturation of reservoir rocks in the studied section interval [13, 14].

For wells opened up sediments from Lower Silurian-Lower Devonian to Upper Permian-Triassic, the available well logging data (*recorded in the appropriate LAS-formats*) are represented by a set of standard logging methods. First of all, these are data from radioactive logging methods (GR, NL), electrometry (RL, SP), caliper logging (CAL) and acoustic (SN ($\Delta$T)) logging. For the productive intervals of the well section, test results were also available. It should be noted the neutron logging (NL) data were available only in one of the four wells.

The section along ***well 1 Yu-Vangurei*** (Fig. 5.42a), located in the central part of the Kaminsk area, in the target interval of the section, deposits of the Lower Triassic ($T_1$), Permian (P), Carboniferous (C), Upper ($D_3$), Middle ($D_2$) and partially Lower Devonian $D_1$) are represented.

The upper part of the section in the depth interval of 1950–2420 m (represented by deposits of the Lower Triassic ($T_1$), Upper Permian ($P_2$) and partially by deposits of the upper part of the Kungurian Stage ($P_1$k) of the Lower Permian ($P_1$)) is made up mainly of terrigenous rocks. The formation of these deposits took place in coastal-marine (open shallow-water basin) and continental settings, in particular, mudstone sequences with interlayers of siltstones and sandstones. In this part of the section, permeable interlayers of sandstones and siltstones are fixed in the sediments of the undivided Kazanian + Tartarian ($P_2$kz + t) and Ufimian ($P_2$u) stages. In the first of the stages, three water-saturated interlayers are clearly established (in the depth intervals of 2065–2076, 2108–2119 and 2153–2162 m), separated by clay intervals of the section. In the second of these stages (Ufimian ($P_2$u)) a whole set of water-saturated interlayers is recorded (in the depth intervals of 2222–2266 and 2290–2324 m), separated by both relatively thin and rather thick clayey intervals. These permeable interlayers are characterized by porosity $K_p \approx 10$–23%.

In the depth interval of 2420–4415 m, carbonate deposits of various ages occur predominantly (with rare exceptions) – from the lower part of the Kungurian stage ($P_1$k) of the Lower Permian to the Sirachoi ($D_3$srč + $D_3$srč') and Domanik ($D_3$dm) deposits of the Middle ($D_3f_2$) and Upper ($D_3f_3$) Frasnian. As noted above (in Sect. 5.1), carbonate deposits are formed both in open shallow water shelf conditions (hard, non-stratified, massive, unevenly crystalline limestones) and in coastal-marine and moderate deep-water these (from barrier reef and domanik formations to bituminous clay-siliceous-carbonate formations). Despite the significant interval of the studied section (its length is slightly less than 2000 m), there are not so many areas in the form of reservoirs. A relatively long interval of the carbonate section, which is characterized by the presence of a reservoir, is located in the deposits of the Artinskian stage ($P_1$ar), in the depth interval of 2516–2580 m, where water-saturated carbonates have porosity values $K_p \approx 4$–8%.

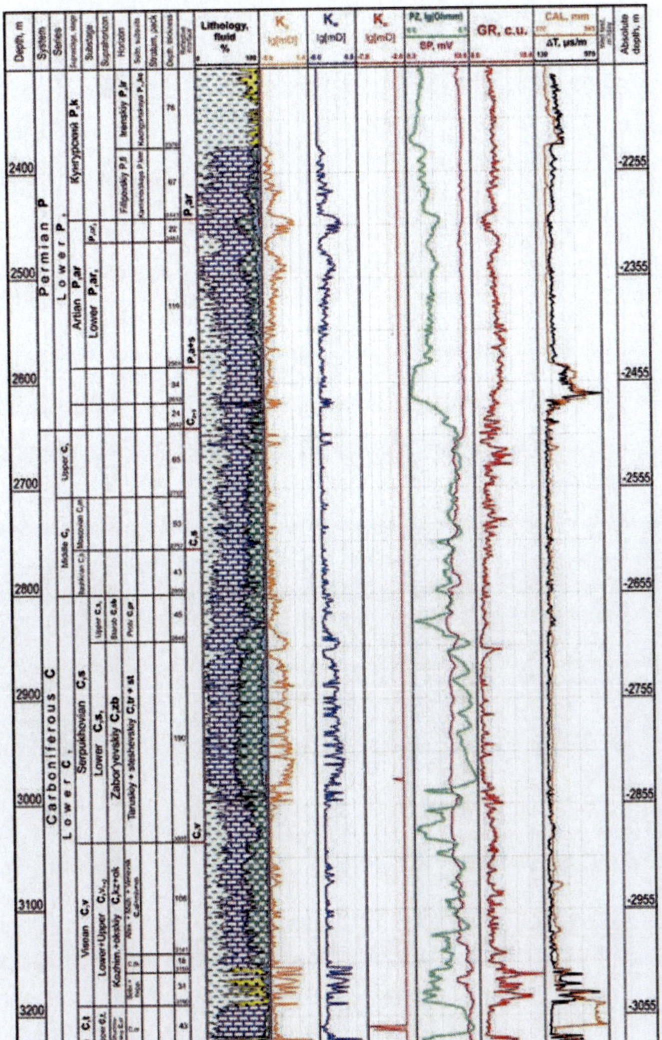


(a)

**Fig. 5.42** Lithological-stratigraphic column with the distribution of the saturation nature of reservoir rocks in well 1 (**a**); sediment correlation scheme for wells 71, 81, 1, and 10 (**b**)

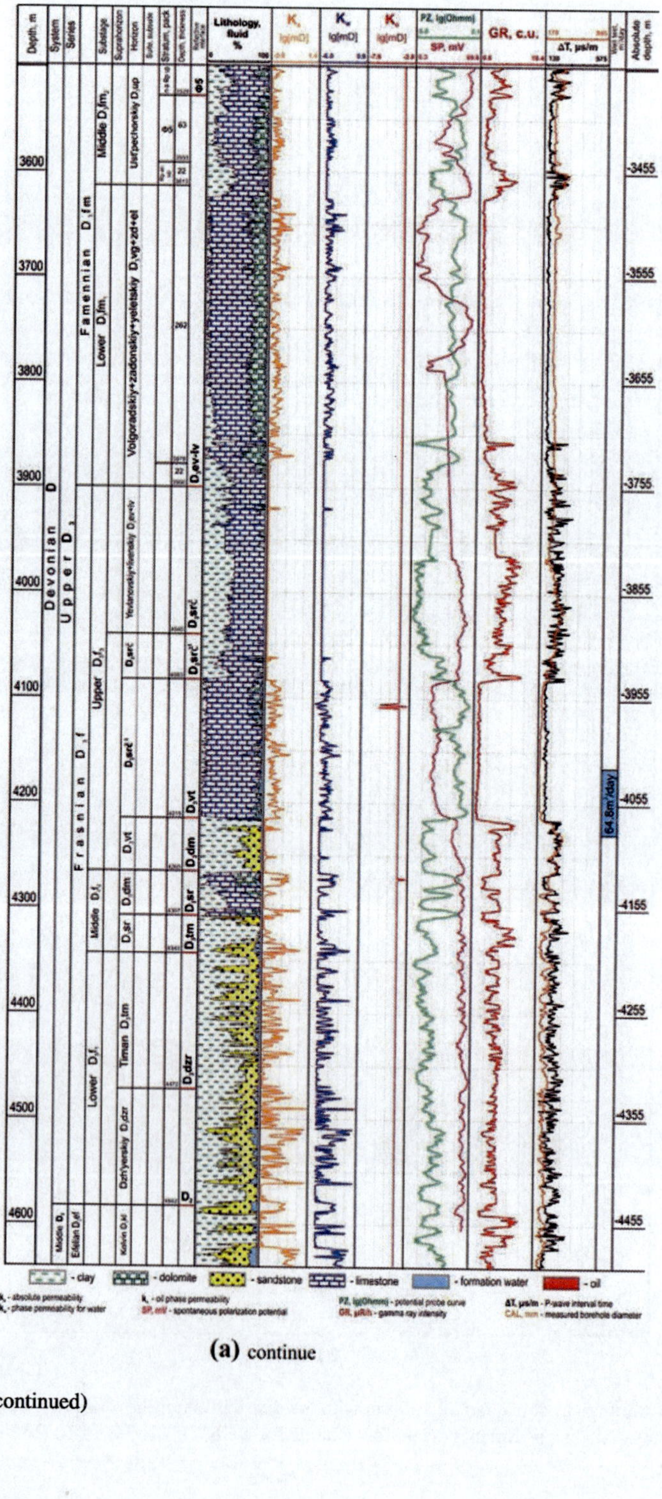


**(a)** continue

**Fig. 5.42** (continued)

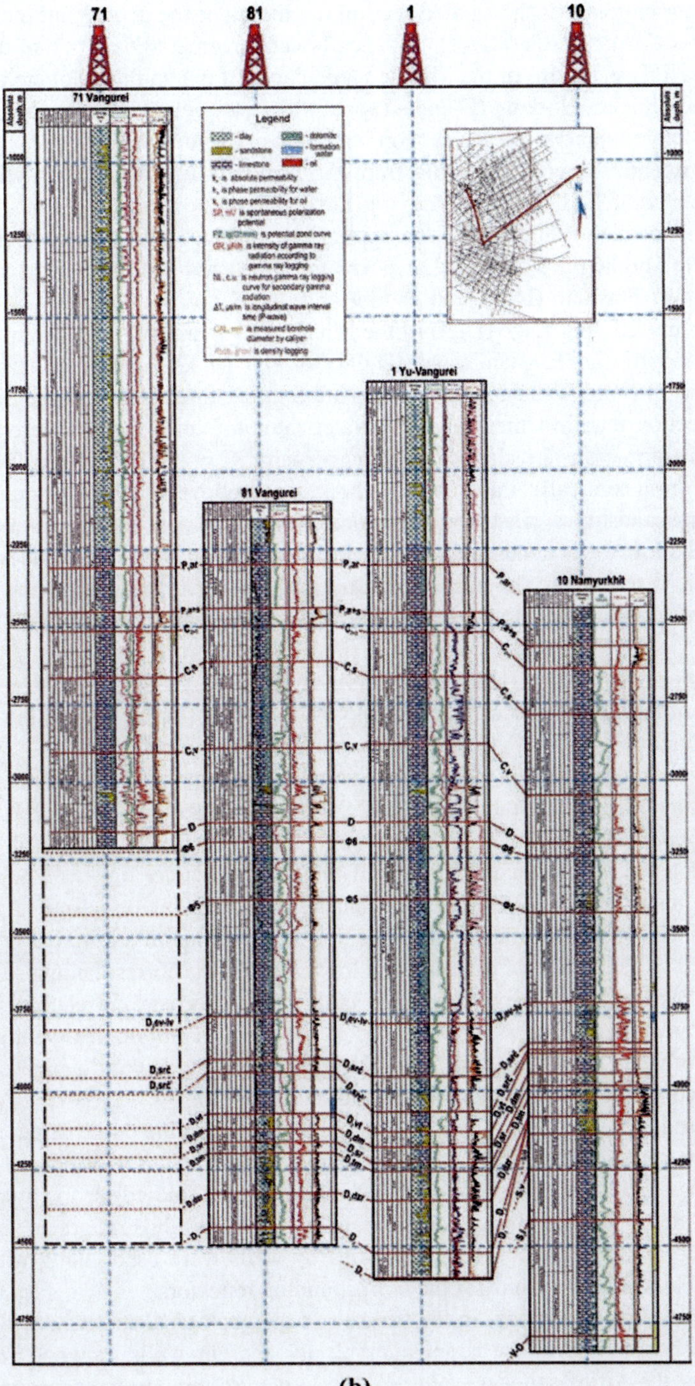


(b)

Fig. 5.42 (continued)

In other intervals of the studied section (including in the deposits of the Middle and Upper Carboniferous ($C_{2+3}$), in the carbonate members F5 and F6 of the Ust'-Pechora Horizon ($D_3$up) of the Middle Famenian ($D_3fm_2$), in the dolomite interlayers of the Sirachoi Horizon ($D_3$srč + $D_3$srč')) there are relatively rare, thin, mostly water-saturated intervals of the section with porosity values $K_p \approx 5$–11%.

In terrigenous deposits, starting from the Bobrikov horizon (C1bb + tlt in the depth interval of 3210–3228 m), from the lower part of the Sirachoy ($D_3$srč + $D_3$srč') and Vetlasyan ($D_3$vt) horizons of the upper Frasnian ($D_3f_3$), the Sargay ($D_3$sr) horizon of the Middle the Frasnian ($D_3f_2$), Timan ($D_3$tm) and Dzhyer ($D_3$dzr) horizons of the lower Frasnian ($D_3f_1$), and ending with the deposits of the Kolvin horizon ($D_2$kl) of the Eifelian stage ($D_2$ef) of the Middle Devonian and the Koivenian horizon ($D_1$kv) of the Emsian stage ($D_1$em) of the Middle Devonian, numerous sandstone-siltstone permeable interlayers are fixed. Almost all permeable intervals of the section manifest themselves as water-saturated interlayers. Heterogeneous sandstone formations are characterized by porosity $K_p \approx 10$–18%. These layers are often hydrodynamically interconnected. As noted above, they are composed of alternating sandstones, siltstones, mudstones (formed in various settings, including coastal shallow water conditions) with a predominance of sandy differences.

The most significant (in terms of thickness and porosity) reservoirs are fixed at depths of: 4186–4209 and 4298–4313 m (in the sediments of the Sirachoy ($D_3$srč + $D_3$srč') horizon); 4435–4465 m (in the lower part of the Sargay ($D_3$sr) and upper part of the Timan ($D_3$tm) horizons); 4535–4540 and 4571–4606 m (in the deposits of the Dzhyer ($D_3$dzr) horizon); 4641–4723 m (in deposits of the Kolvin ($D_2$kl) horizon).

In general, the selected reservoirs according to logging data are estimated as water-saturated. The substantiations made at the same time (*Kuranova T.I., Surina A.I., Voronina L.V.* et al., *"TP NITs Neftegaz", 2011*) practically apply to all oil and gas bearing complexes opened up by 1 Yu-Vangurei borehole.

When analyzing the available well material, linking it with seismic data, one often has to resort to correlation schemes of lithostratigraphic units ("packs", horizons) built either on sets of original well logs or using the corresponding results of their processing and interpretation. We built such a horizon correlation scheme using the results of processing well logs in the form of lithological-stratigraphic columns of wells 71, 81 Vangurei, 1 Yu-Vangurei and 10 Nyamyurkhit together with well log curves of standard logging methods (Fig. 5.42b).

The used lithological-stratigraphic columns of wells in the constructed correlation scheme are set according to their hypsometric levels, that is, in accordance with their inherent absolute marks of geological disjunction of studied section lithostratigraphic units. When identifying the interface between the lithostratigraphic units of the studied section, geological subdivisions by wells were used, along with data from structural constructions for the corresponding reflectors.

In general, lateral changes in the features of the geological structure of the section can be observed in the target intervals for all four wells, ranging from the deposits of the Artinskian stage ($P_1$ar) of the Lower Permian to the deposits of the Upper Devonian – Zelenetsk + Nyumylgsk horizon ($D_3$zl + nm) of the Upper Famennian ($D_3fm_3$). Further, lateral changes in the features of the geological structure of the section are traced only in three wells 81 Vangurei, 1 Yu-Vangurei and 10 Nyamyurkhit, starting from the deposits of the Ust-Pechora horizon ($D_3$up) of the Middle Famenian ($D_3fm_2$) to the bottom of the Dzhyer horizon ($D_3$dzr) of the Lower Frasnian ($D_3f_1$). Only in two wells 81 Vangurei, 1 Yu-Vangurei there is a correlation between the features of the geological structure of the section according to the deposits of the Middle Devonian ($D_2$). Moreover, in the first of the wells (81 Vangurei) only the upper half of the Middle Devonian deposits is opened up. The deposits of the Lower Devonian (its upper part) were discovered only by well 1 Yu-Vangurei.

At the same time, in well 10 Nyamyurkhit, deposits of the Middle ($D_2$), Lower ($D_1$) Devonian and deposits of the Przhidolian stage ($S_2$p) of the Upper series Silurian ($S_2$) are almost completely absent.

Of the most important regularities about the features of sedimentogenesis, the sizes of intervals of promising interlayers and the ratio of their thicknesses, one should first of all note the features of the change in the sediments of the Sirachoi horizon and its carbonate stratum ($D_3$srč + $D_3$srč') between wells 81 Vangurei, 1 Yu-Vangurei and 10 Nyamyurkhit. As can be seen from Fig. 5.42b, the Sirachoi carbonate stratum ($D_3$srč') is the thickest in well 81 Vangurei, significantly decreasing in wells 1 Yu-Vangurei and 10 Nyamyurkhit. In the last of these wells (10 Nyamyurkhit), there is a significant reduction in the Syrachoy horizon and its carbonate stratum ($D_3$srč + $D_3$srč'). In general, in addition to changes in the thicknesses of the Sirachoi horizon, there is a change in its lithological features of the structure. The fully carbonate section in well 81 Vangurei (in the depth interval 4040–4215 m) is partially replaced by terrigenous sediments and clays in wells 1 Yu-Vangurei and 10 Nyamyurkhit. These features of the structure, obviously, can be explained by the strongly changing conditions of sedimentation in the considered part of the section.

The alternation of clayey and sandy sections of the profile from well to well (boreholes 81-1-10) is quite confidently traced in the deposits of the Sargay ($D_3$sr), Timan ($D_3$tm) and Dzhyer ($D_3$dzr) horizons.

Thus, in the above correlation scheme of horizons opened up by boreholes, a characteristic of the geological structure of the studied deposits is given (Fig. 5.42b).

The lithological-stratigraphic columns considered above and the results of well logging data processing in general for existing wells were also used in the analysis of the features of changes in the values of effective acoustic impedance along the profiles of the study area in the context of promising deposits.

### 5.3.2 *Geological Structure of the Section Along the Reference Profiles*

According to the scheme of tectonic and oil and gas geological zoning of the TPP, the main area of the site is confined to the Chernorechen step – a structure of the II order, which is part of the Khoreyver depression and, to a lesser extent, to the eastern side of the Yareyu swell, also a structure of the II order, which makes up the northern part of the Kolvin megaswell (Fig. 5.43).

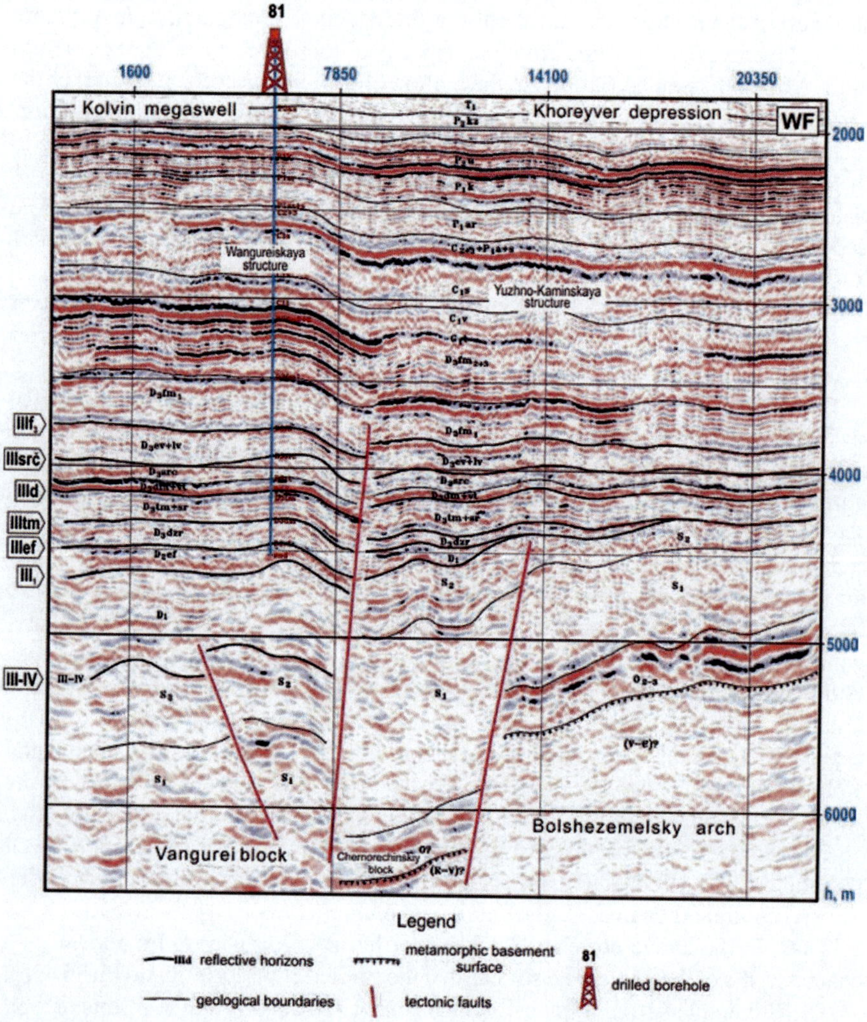


**Fig. 5.43** Seismic-geological section along profile 3512 at the location of well 81 Vangurei (according to JSC "Naryan-Marseismorazvedka", 2002)

**The Khoreyver depression** borders on the Kolvin megaswell in the southwest and the Sorokin swell in the east. The boundary runs along the regional systems of deep faults: V-Kolvin and Varandey. In the north, the depression extends into the Pechora Sea. The dimensions of its continental part are 200 × 120 km. The depression is a large superimposed negative structure with a northwestern orientation and is identified along the Upper Devonian and overlying horizons of the sedimentary cover. According to the basement and Ordovician-Silurian deposits, the southern and central parts of the Khoreyver depression correspond to the Bolshezemel paleo-arch. The arch of the large Bolshezemel swell-like basement ledge is broken by multidirectional tectonic faults, which, in turn, formed the position of local structures within the arch from the Riphean to the early Permian. The top of the arch is located in the area of Sandivey area. Its most submerged sections are confined to the junction zones with the Sorokin and the Yareyu swells.

The Khoreyver depression is characterized by a structural discrepancy between the Ordovician-Lower Devonian deposits, which form the vast buried Bolshezemel dome, and the Upper Devonian-Triassic unconformably overlying them. Above the surface of the pre-Early Frasnian erosion, the dome loses its structural expressiveness, giving way to groups of local uplifts and structural steps.

The Khoreyver depression is asymmetric. Its most submerged part, with the maximum thickness of the sedimentary cover, extends along the junction zone with the Kolvin megaswell. To the east, along the supra-domanik part of the section, there is a rise of horizons.

***The Chernorechen depression*** with dimensions of $170 \times 6 \div 38$ km occupies the most submerged part of the basin in the zone of its junction with the Kolvin megaswell and is determined along the entire section of the sedimentary cover (Fig. 5.43). Its western side is steeper and narrower than the eastern one, and is complicated by tectonic disturbance at the latitude of the Kharyagin-Khanchargin areas. The depression consists of several synclines. The surface of carbonate deposits of the Lower Permian is at the bottom of the depression and lies at depths of $-2425\div-2900$ m. Its northern part is the most submerged. Its amplitude is 100–300 m and increases in the northern direction. Up the section, the structure flattens out.

***The Kolvin megaswell***, located in the eastern part of the Pechoro-Kolvin aulacogen, is one of the most contrasting structures of the TPP. On the surface of the basement, it is a graben with depths from 6.0 km to 8.0 km. The megaswell is bounded by regional faults of northwestern strike, separating it from the west from the Denisov trough – the West Kolvin fault (amplitude 100–250 m in the north of the megaswell along the Upper Paleozoic deposits) and from the east from the Khoreyver depression – the V-Kolvin fault (amplitude up to 100 m) (Fig. 5.43). A series of tectonic faults is also traced, subparallel to regional discontinuous dislocations and within the megaswell itself. First of all, this is the Central-Kolvin deep fault, which divided the territory under consideration along the Lower-Middle Paleozoic deposits into a series of blocks of the same strike. Along these ancient discontinuous dislocations, a graben-like trough was formed, within which, in turn, systems of depressions and protrusions were formed. In the modern structural plan, the Kolvin megaswell, which is one of the most contrasting structures of the first order of the TPP, stretches in a northwestern direction for more than 300 km with a width of 30–40 km. The northern part of the megaswell is established in the waters of the Pechora Sea, where it changes its direction to sublatitudinal.

**The Kamin site** is located in the transitional structural-tectonic zone from the Kolvin megaswell to the Khoreyver depression, which corresponds to the eastern side of the Kolvin megaswell and adjacent blocks of the Bolshezemel arch (Fig. 5.41 and 5.43).

According to seismic data, reef-type anomalies were identified near the licensed area, which, possibly, are single organogenic structures formed on consedimentary

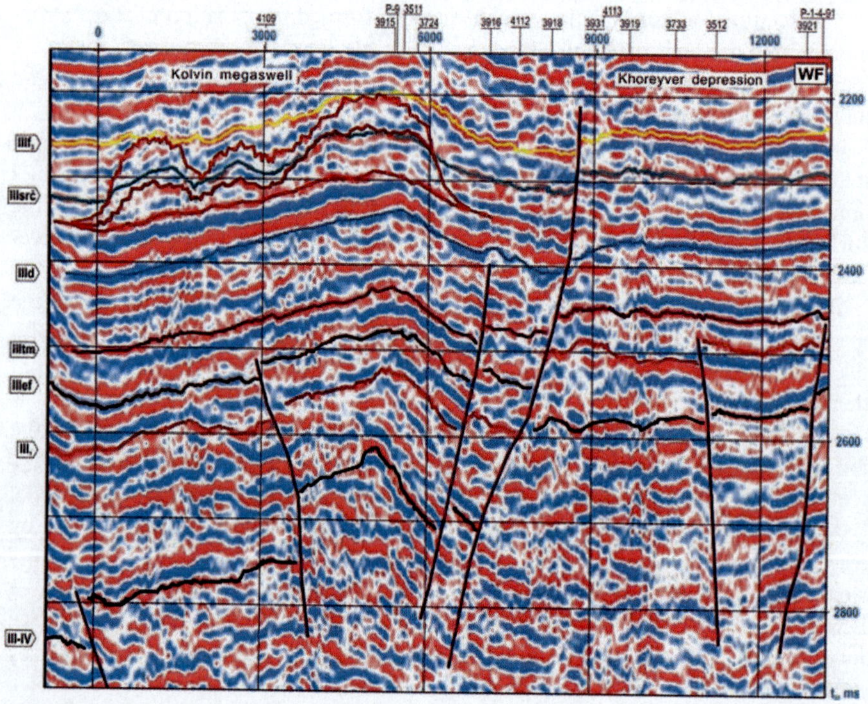


**Fig. 5.44** The nature of the wave field in the junction zone of the Kolvin megaswell and the Khoreyver depression. Reef-type anomaly in the section of the Vangurei structure. Profile 3704 (according to JSC "Naryan-Marseismorazvedka", 2002)

paleo-elevations within the Domanik uncompensated basin (Fig. 5.44). In the previously studied single organogenic structures of the Domanik-Late Frasnian age (Olenya, Seinorogoyakhin, V-Yanemdei, and others), the Domanik bioherm formations are the base for the growth of later (Late Frasnian and Famennian) organogenic structures. The thickness of the Domanik part in the most studied Domanik-Late Frasnian organogenic structures is 40–110 m.

Along the horizons of the lower (III1, III-IV, V reflectors) and intermediate (III2) structural stages, submeridional blocks are distinguished in the area, rising eastward towards the Bolshezemel dome. The boundaries between the blocks are tectonic distortions – faults-throws of submeridional orientation, which are part of the system of the V-Kolvin faults. The greatest amplitude of faults in the area (from 1300 to 50 m) is noted along the fault that limits the Vangurei block from the east.

The band of transverse faults can be traced at the latitude of the Ievskie structures and is the southern limit of the Vangurei block (step) identified as part of the Kolvin trough (megaswell). According to the horizons of the upper and middle structural stages, the fault corresponds to a high-amplitude marginal flexure of the eastern slope of the Kolvin megaswell. Thus, the boundary of the Kolvin megaswell (Kolvin trough) and the Khoreyver depression (Bolshezemel arch) runs along this fault.

Faults of the northwestern orientation prevail in the area, which determines the northwestern orientation of the reflective horizon (RH) isolines and the lines for stopping tracking of the RH. In the south of the site, faults of the meridional and northeastern directions play a decisive role. The southern part of the site is distinguished by a more monoclinal occurrence of horizons within the blocks of the Bolshezemel dome, a small amplitude of faults, and a significantly smaller gradient in the distribution of thicknesses of subdomanik deposits compared to the northern part. This makes it possible to more unambiguously localize the zones of wedging out of subdomanik deposits in the south of the area. It is assumed between the northern and southern parts of the site there is a system of sublatitudinal faults, the movements along which had a shear component, which led to the knee-shaped boundary of the Kolvin megaswell and the Khoreyver depression (Fig. 5.41).

### *5.3.3 Some Characteristic Areas of the Geological Section Structure According to the HRS Data*

The geological structure of the Kamin area is illustrated on separate profiles and their fragments (with elements of interpretation): lines of AI and RC, time section (Figs. 5.45, 5.46, 5.47, 5.48, 5.49, 5.50 and 5.51).

So, according to one of the CDP profiles 3511 of the sublatitudinal direction, the features of the distributions of the effective acoustic impedance (Fig. 5.45a), reflection coefficients (Fig. 5.45b), the wave picture on the migrated time section (Fig. 5.45c) are shown. On these sections, in the central part of the profile (in the area of the vertical section of the well 71 Vangureisk), an anticlinal uplift is manifested, complicated by the zone of the V-Kolvin fault, almost along all correlated reflecting boundaries.

The most complex geological structure is distinguished by deposits that are somehow complicated by tectonic disturbances and regional faults and with which productive and potentially productive horizons are associated. These are, first of all, deposits of the Sirachoi ($D_3$src') and Dzhyer ($D_3$dzr) horizons of the Upper Devonian, the Middle ($D_2$ef) and Lower ($D_1$e) Devonian, and the Gerdju horizon ($S_2$gj) of the Ludlovian stage ($S_2$ld) of the Upper Silurian. Such complexity of the structure of the section is primarily reflected in the complex distribution of the values of effective acoustic impedance (AI) (Fig. 5.45a) and reflection coefficients (RC) (Fig. 5.45b) in the productive deposits under consideration. In some cases, it is not always possible to unambiguously correlate target reflectors and tectonic faults in a thin-layer section.

As a result of the performed correlation of reflecting boundaries and structural constructions, nevertheless, it is possible to identify the main structural elements of the section structure. Here, as noted above, the nature of tracing the reflected waves during the ascent to the Bolshezemel paleoarch is complicated by the influence of a significant number of tectonic faults. The thickness of the Middle and Lower

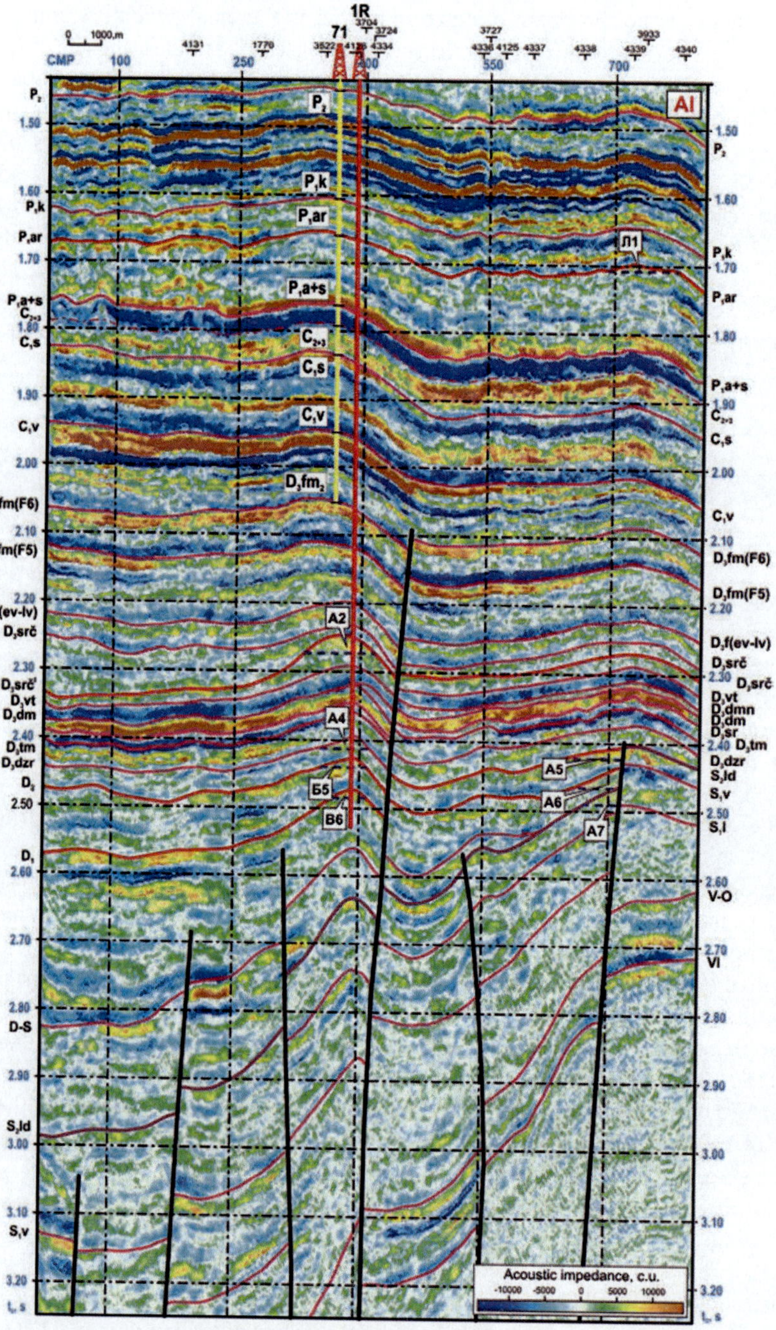

(a)

**Fig. 5.45** Sections of effective acoustic impedance (**a**), reflection coefficient (**b**) and wave field (**c**) along the line 3511 in the tracing interval of target deposits

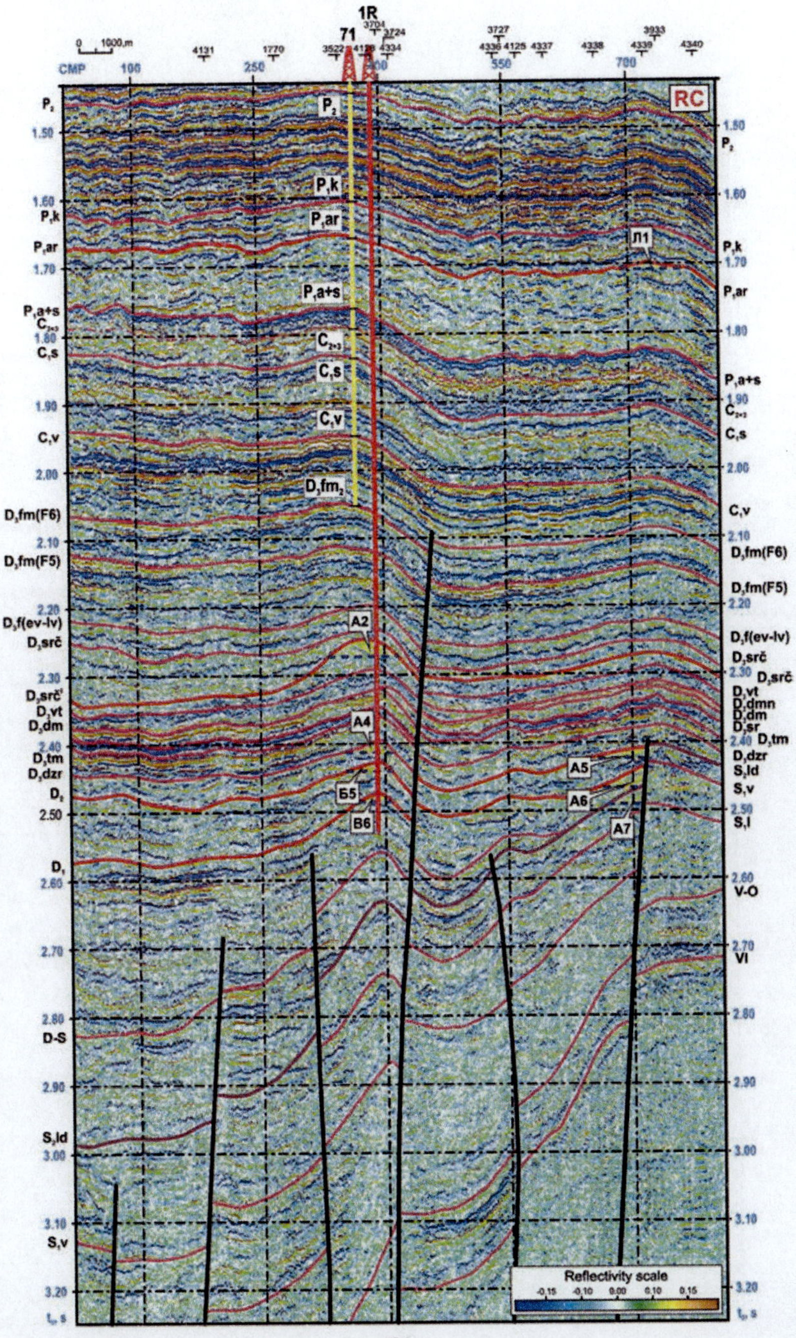

(b)

**Fig. 5.45** (continued)

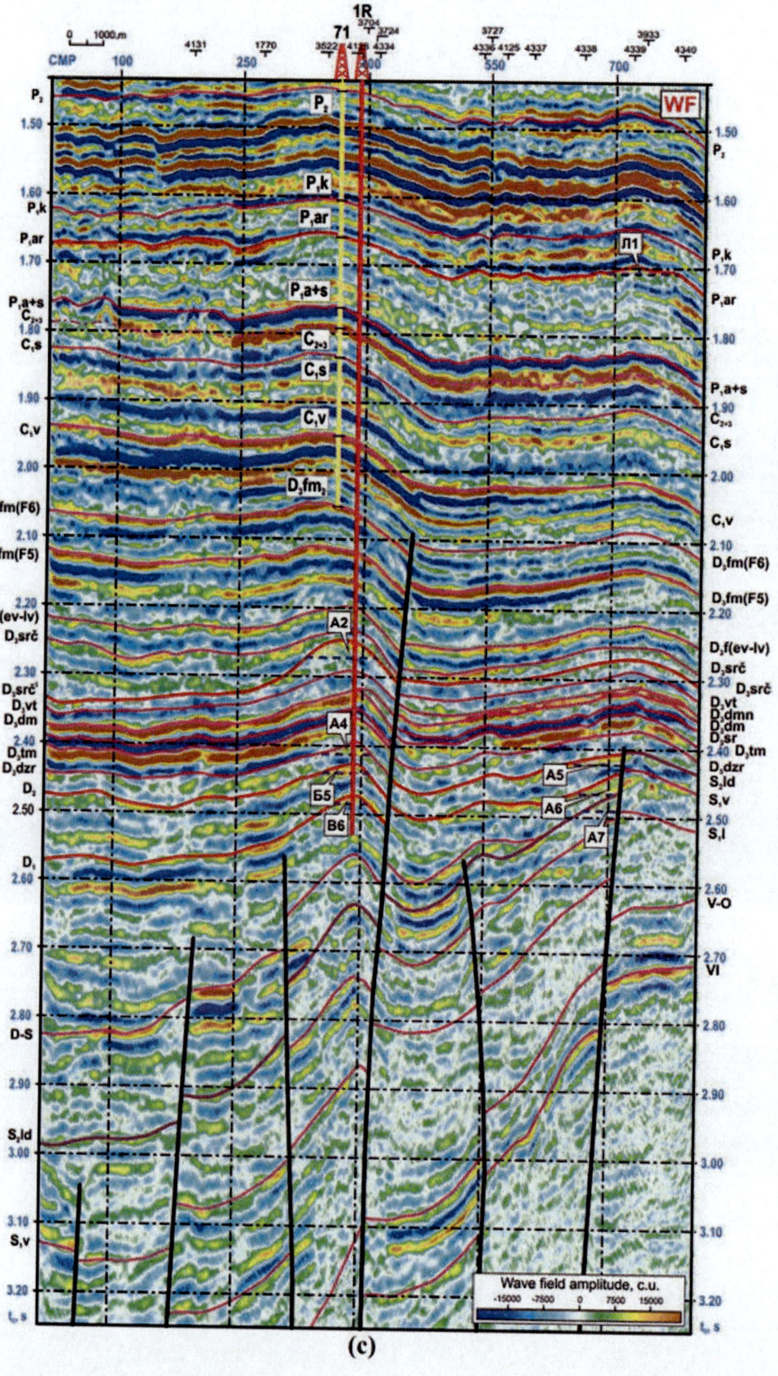


(c)

**Fig. 5.45** (continued)

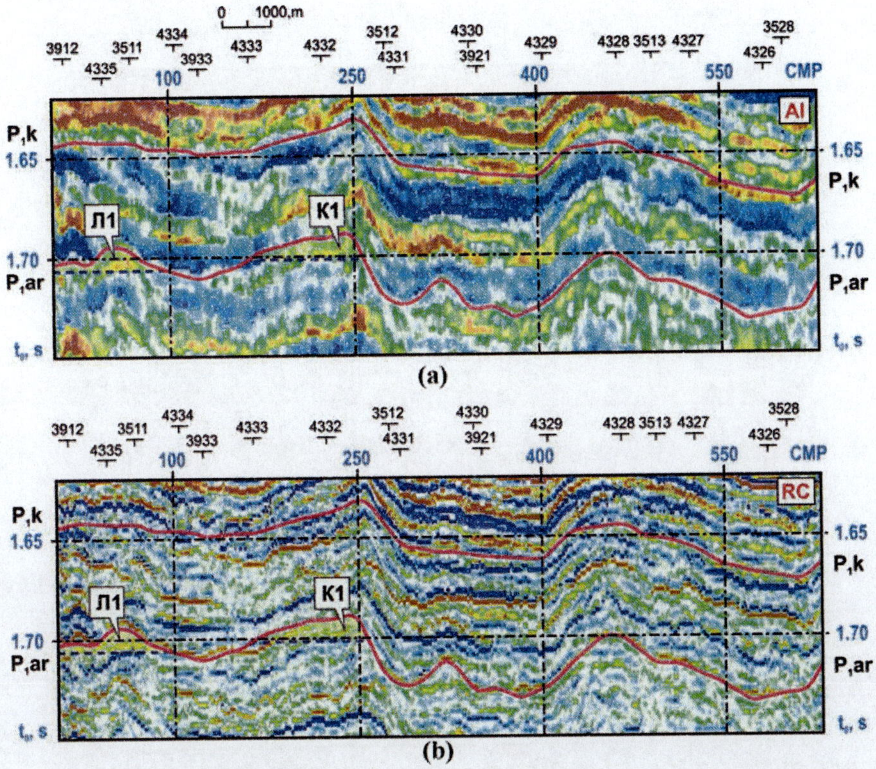


**Fig. 5.46** The manifestation nature of the perspective object JI1, K1 on the Kamin structure (northern and central domes) along the deposits of the Artinskian stage ($P_1ar$) in the sections of line 4339: **(a)** AI; **(b)** RC

Devonian and Silurian deposits is sharply reduced in the east and northeast directions. To the east of the last one indicated in Fig. 5.45 faults, traceability of the Middle and Lower Devonian deposits, as well as the upper part of the Silurian (the deposits of the Pridolian ($S_2p$) stage) stops due to "erosive shearing". At the same time, tectonic faults, in turn, formed the position of local structures at the boundary of the Bolshezemel paleoarch and the Kolvin megaswell from the Riphean to the Early Permian.

The reflectors correlated above on the sections AI and RC are placed on the time section of the CDP along profile 3511 (Fig. 5.45c). These boundaries, which are directly confined to the deposits under consideration, also characterize the possibilities in tracing reflectors using only elements of the wave field structure. At the same time, it is clear these possibilities are significantly limited, since detailed identification and tracing of target reflective boundaries formed on the sections AI (Fig. 5.45a) and RC (Fig. 5.45b), on the time section (Fig. 5.45c) is not always feasible. With the use of elements of the wave field structure, identification of individual reflectors is possible, and it often turns out to be inaccurate. By this circumstance, we emphasize

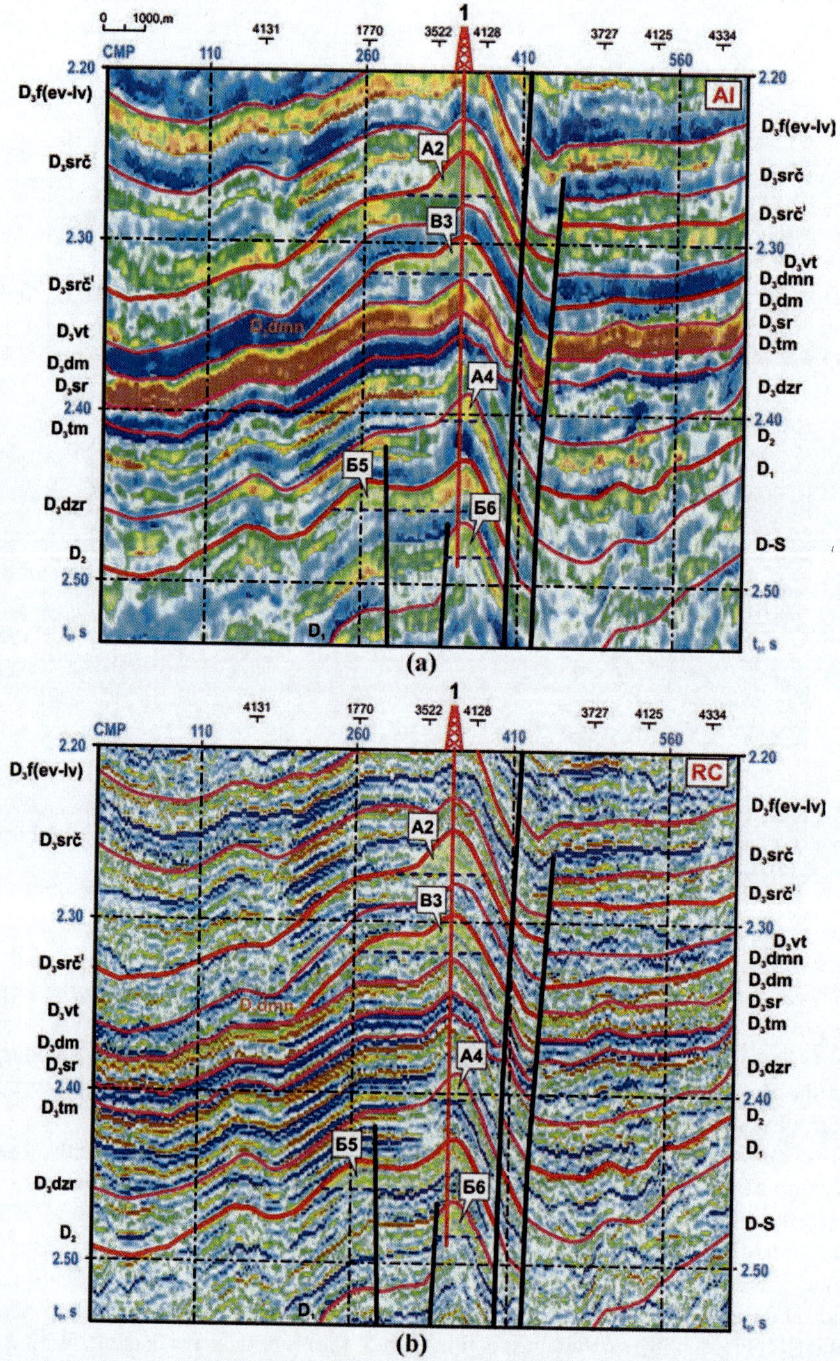


**Fig. 5.47** Prospective object A2 (together with objects B3, Г5 and Б6) on the Vangurei structure along the deposits of the Sirachoy horizon IIID$_3$srč' (object A2) in the sections of profile 4111: **(a)** AI; **(b)** RC

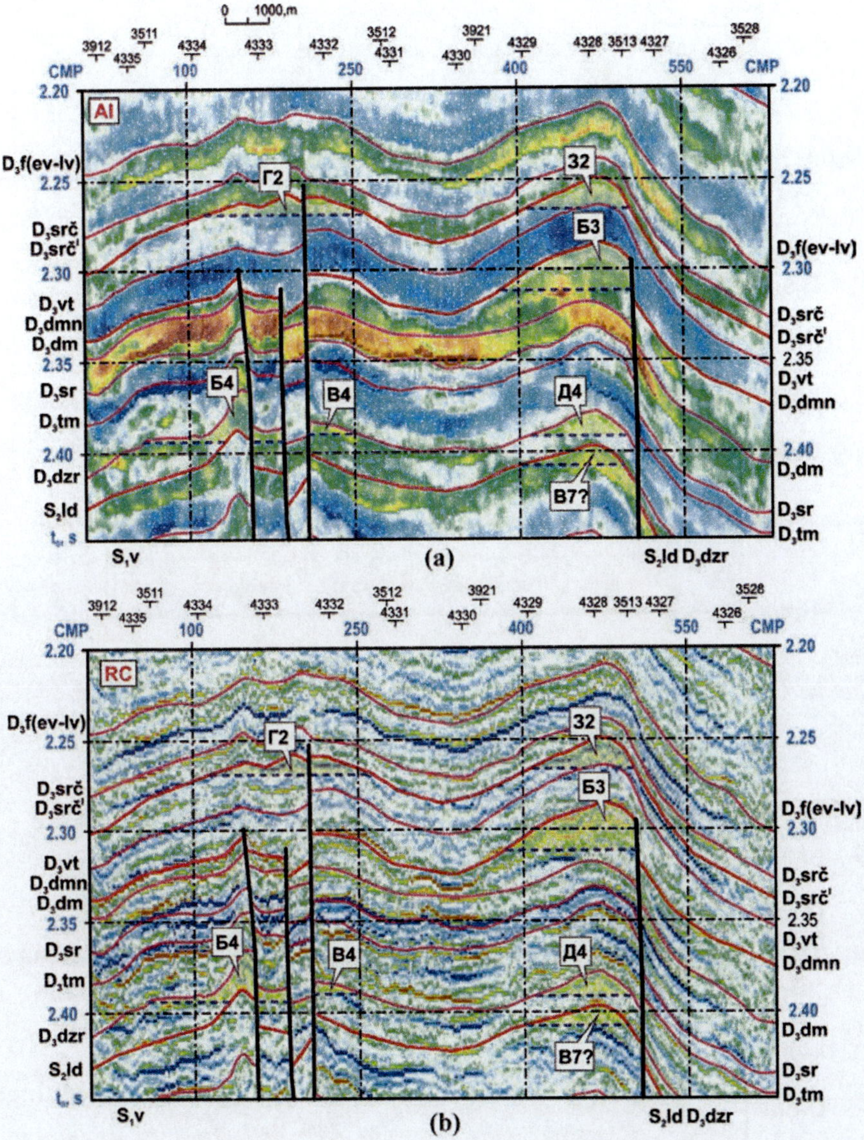


**Fig. 5.48** Prospective objects Г2 and З2 (together with objects Б3, Б4, В4, Д4 and В7) on the Kamin (central dome) and Yu-Kamin structures on the deposits of the Sirachoy horizon IIID3srč' (objects Г2 and З2) in the sections of line 4340: (**a**) AI; (**b**) RC

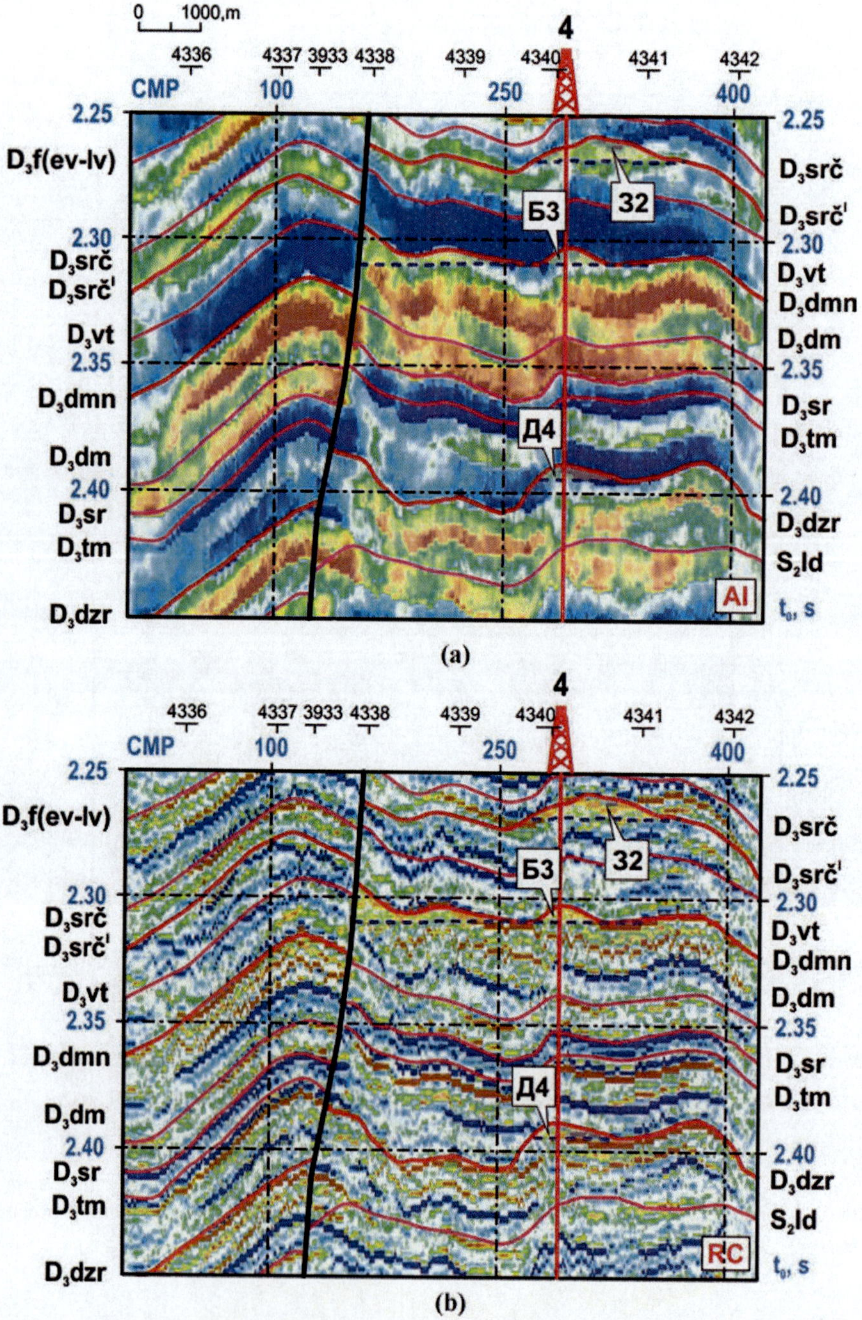


**Fig. 5.49** Prospective object З2 (together with objects Б3 and Д4) on the Yu-Kamin structure along the deposits of the Sirachoy horizon $IIID_3sr\check{c}$' (object З2) on the sections of profile 4328: (**a**) AI; (**b**) RC

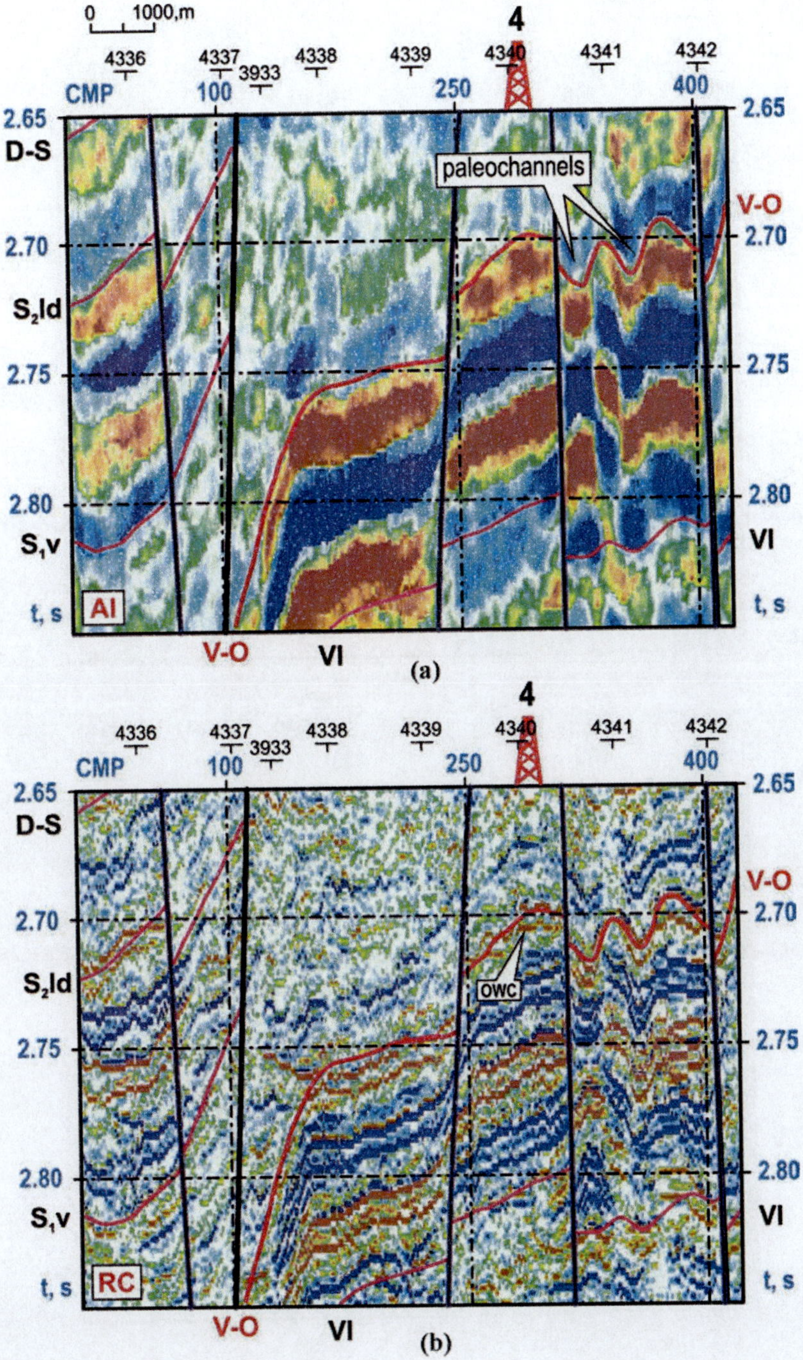


**Fig. 5.50** Fragments of the sections AI (**a**) and RC (**b**) showing the proposed paleochannel incisions of the river system along line 4328

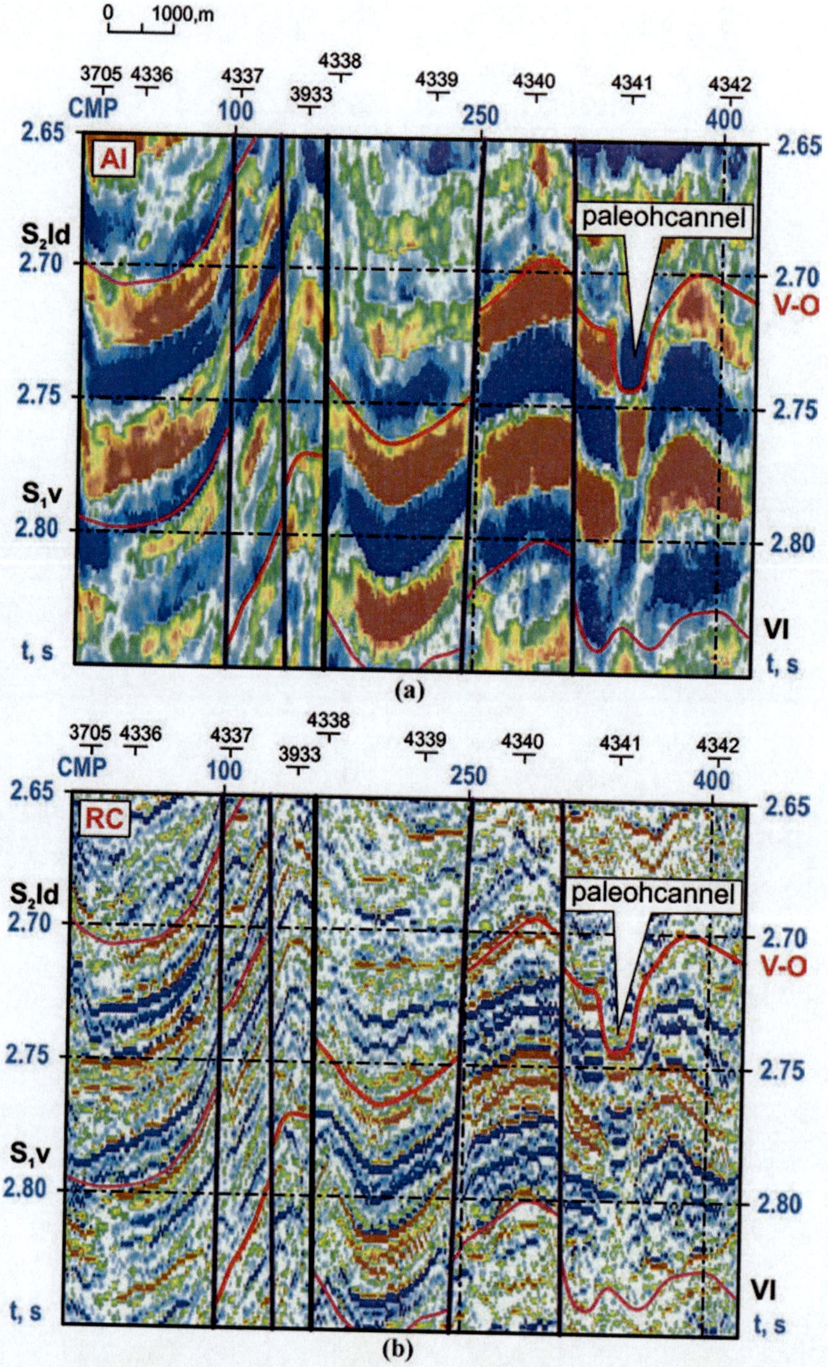


Fig. 5.51 Fragments of sections AI (a) and RC (b) showing the proposed paleochannel incisions of the river system along line 4329

the significant advantage of using high-resolution AI and RC sections over traditional methods for identifying and tracking target reflected waves.

In general, the results obtained for the seismic line (on the example of various types of profile section 3511), we can note the following:

(a) on seismic sections of high resolution, according to the effective acoustic impedance (AI) and reflection coefficients (RC) in the section plane, the heterogeneity of the structure of the studied deposits is most reliably traced, including the features of the manifestation of natural hydrocarbon reservoirs in these heterogeneities;
(b) Based on the distributions of increased and decreased AI values, it is possible to trace zones of facies replacements occurring in isolation, and areas of formation of possible tectonically and lithologically shielded hydrocarbon traps.

The results of the profile interpretation of seismic data for the entire volume of processed sections of CDP-2D seismic data, of course, are the input data for the next stage of processing and automated interpretation – the presentation of the results of structural-tectonic mapping and various geological indicators prediction on an areal basis.

In addition, the obtained results of logging data processing, vertical seismic profiling of deep wells and data on the identification of reflecting horizons in the AI and RC sections make it possible to determine the capability of using the available geological and geophysical materials, perform structural mapping along the target reflectors, identify the most promising areas (zones) of oil and gas accumulation for a number of deposits of the studied section in the Kamin area.

Along the main directions (of the studied sections), fragments of the effective acoustic impedances (AI) and reflection coefficients (RC) are further considered. Based on their use, the locations of promising objects are substantiated, each of which is characterized by its distribution boundaries – the contours of the expected deposits in the sediments of the indicated horizons ($P_1$ar, $D_3$src', $D_3$domanik, $D_3$dzr, $D_2$ef, $D_1$e and $S_2$ld).

Specifically, for each of the horizons of the Paleozoic sediment complex, the following can be marked.

Rationale the presence in the studied section of the distinguished promising objects of a structural type (in particular, objects A1-M1) and to evaluate their dynamic characteristics (features of the object internal structure) in the given vertical sections of the studied medium, fragments of the AI and RC sections were formed for each of the promising objects (Figs. 5.45, 5.46, 5.47, 5.48, 5.49, 5.50 and 5.51). On these sections (specified in certain directions, most often passing objects A1-M1 through their arches), characteristic kinematic (positive structural forms) and dynamic (laterally heterogeneous values of AI and RC) structural features of the proposed prospective objects are confidently distinguished. Thus, in the deposits of the Artinskian stage ($P_1$ar), representative (in terms of the amplitudes of the arch of anticlinal structures) are images of objects Л1 and K1 (Fig. 5.46). Objects K1-Л1 are directly confined to the Kamin local structures, tectonically being low-order structures of the Bolshezemel arch of the Khoreyver depression.

Accounting such a feature of the structure of promising objects in the deposits of the Artinskian stage ($P_1ar$), namely, the zoning of their location in the study area, one can judge with a certain degree of probability about the oil prospects of the identified objects. The screen-seal for these deposits here are dark gray mudstones, silty, unevenly calcareous, dense, in some places platy. These deposits belong to the Lower Karmanov Formation ($P_1krm$) of the Kungurian Stage ($P_1k$) of the Lower Permian. They overlap the target horizon – deposits of the Upper substage ($P_1ar_2$) of the Artinskian stage ($P_1ar$). The latter, in well 7 Khylchuyu, located to the west of the contour of the Kamin prospecting area, turn out to be gas-saturated (as a result of a cased hole tests in the depth interval of 1960–1979 m, a gas influx Qg = 10–15 thousand $m^3$/day was obtained).

To substantiate the presence in the studied section of distinguished promising objects of a structural type (objects A2-Ɜ2 in the deposits of the horizon $D_3$srč') and to evaluate their dynamic characteristics (features of the internal structure of the object) in the given vertical profiles of the studied section, fragments of AI and RC sections for each of the perspective objects were formed (Figs. 5.47, 5.48 and 5.49). It should be noted in the presented fragments of the sections AI and RC, objects A2-Ɜ2 are shown together with a number of promising objects and for other underlying horizons. We also note the joint consideration of promising objects at once for several potentially productive horizons made it possible in the end to determine the choice of recommended wells in the most representative areas of the study area in terms of oil and gas potential. On these sections, objects A2-Ɜ2, overlapping the dome parts of the structures, are confidently distinguished by characteristic kinematic and dynamic (laterally heterogeneous values of AI and RC). Here, in the sediments of the Syrachoy horizon ($D_3$srč), they are the most representative.

Expressed zoning in the location of the contours of objects A2-Ɜ2 in the study area is manifested in the fact they are mostly located in the northern half of the Kamin area. We can also talk about the oil prospects of these objects due to the fact the screen-seal for the deposits under consideration is the terrigenous-carbonate strata, strongly clayey limestones, weakly dolomitized of coastal-marine genesis. These deposits belong to the uppermost part of the Syrachoy horizon ($D_3$srč). They overlap the target horizon, deposits of the reef strata of the Syrachoy horizon ($D_3$srč'). The latter are known to be one of the main productive horizons in the Kamin area. Oil deposits in carbonate structures of this age were found at Tabrovoyakhin, Passedsk, Trebs deposits. Oil shows in the core are noted in the well 51 Yu-Sadayagin.

Something similar in substantiating the presence in the studied section of other promising objects of the structural type (objects A3-B3 in the deposits of the $D_3$domanik horizon) can be seen in the other sections AI and RC, including Figs. 5.47, 5.48 and 5.49. It should be noted in the presented fragments of the AI and RC sections, objects A3-B3 are shown together with a number of promising objects and for other upper and lower horizons. On the presented fragments of sections AI and RC, objects A4-Д4 are also shown together with promising objects. On these sections (along the given control directions crossing the objects A4-Д4 through their arches), the characteristic kinematic and dynamic (laterally heterogeneous values of AI and RC) features of the structure (in conjunction with the geological environment containing them) of each of the promising objects are confidently highlighted.

For prospective objects A4-Д4, something similar can be observed in the deposits of the Dzhyer horizon ($D_3$dzr). Here, these deposits are covered by the terrigenous sequence of the Timan ($D_3$tm) and Sargay ($D_3$sr) horizons. These horizons are known to be represented by mudstones with interbeds of siltstones. Moreover, mudstones are red-brown, chocolate-brown, with gray interlayers with a brownish tint, sometimes crumpled. The siltstones are greenish-gray, irregularly clayey, with thin wavy-lens-shaped layering, the formation of which took place in marine shallow-shelf conditions. In the area, the thickness of these deposits varies from 184 m to 45 m, significantly decreasing in the east direction. Due to the clay composition of the Timan-Sargay deposits and their ubiquitous distribution, they are considered as a regional seal for the underlying oil and gas complexes.

### 5.3.4 *Structural Features of Paleoincisions Based on High-Resolution Seismic Data*

In addition to the above-mentioned promising objects (which are shown in the presented Figs. 5.46, 5.47, 5.48 and 5.49), the authors identified a whole set of objects in the investigated prospecting area (such as A5, B5, Б5, A7, Г7), which are confined to reservoir arched or tectonically screened traps. They are also characterized by their kinematic and dynamic features of the structure (in relation to the surrounding geological environment), zoning and location specifics, formation conditions, possible oil prospects and the structure of the overlying screens-seals. They are not considered here.

Let us dwell only on one more feature that stands out on the interpreted sections of effective acoustic impedance (AI) and reflection coefficient (RC). We are talking about the alleged complexes of buried river systems, which are displayed on the AI and RC sections in the form of a certain set of diagnostic features, according to which, in one way or another, it is possible to identify the desired target objects. The well-known diagnostic features for identifying complexes of buried river systems formed in a sedimentation environment by a paleo-river bed are the following. These are: subparallel and subperpendicular orientation of the accumulative body relative to the strike of the paleoslope of the continental step relief (which determines the energy of the river flow); zonal, relatively tortuous nature of distribution in plan and wedging out of accumulative bodies near the banks of paleo-rivers. At the same time, a single cycle of development of buried river systems (BRS) is characterized by a relative constancy (insignificant displacement) of the coastline, i.e. stable sea level and tectonic position of the region of the river sources. Using these features of the relatively small structure in their linear dimensions accumulative bodies of the river system, a number of possible oil-promising objects associated with such systems were identified in the sections AI and RC.

The supposed paleochannel incisions were identified by the authors directly on the AI and RC sections of a number of seismic profiles. As an example, in Figs. 5.50 and 5.51 show fragments of sections AI and RC along the lines 4328, 4329, which are located almost parallel to each other in the territory of the Bolshezemel arch of the Khoreyver depression. In accordance with the peculiarities of the distribution of AI and RC values on these fragments of the sections, it can be seen the directions of the considered seismic lines are located approximately orthogonal to the formed contour of the river system – the coastlines of the paleo-river channels.

Running a little ahead, we note the above features in the form of paleoincisions are well identified in a fragment of the structural map for these deposits (in Fig. 5.55c). As already pointed out, the buried river system under consideration is confined to the uppermost part of the Ordovician deposits.

Analyzing in general the presented materials on the study of the structural forms of buried river systems (paleoincisions) in the Upper Ordovician sediments, some features of their geological structure should be emphasized.

One of these features is the buried river system under consideration (which has some local distribution in the study area) is directly confined to a number of tectonic faults that have approximately the same area strike as the direction of the axial lines of the paleoincisions channels (Fig. 5.55c). This may indicate a direct relation between the early and late Caledonian stages of tectogenesis (intensive formation of disjunctive dislocations) and the stage of formation of complexly constructed channels of the buried river system at approximately the same period of geological time.

Another feature of the geological structure of the system of paleo-incisions under consideration is the possible oil-perspective objects can be associated both with the channel of the paleo-incision itself and with coastal ledges adjacent to the boundaries (contours) of the paleo-incision. This feature (display on the considered sections of possible oil-promising objects at the level of coastal ledges) is quite confidently displayed on one of the fragments of the sections AI and RC along line 4328 (Fig. 5.50). This figure clearly shows in the upper part of the uplifted block, within the coastal ledge, directly adjacent to the contour of one of the channels of the paleoincision, a possible level of water-oil contact (OWC) is manifested. This is very clearly highlighted in the indicated elevated block on the detailed segment of the values of the reflection coefficients (in the form of their horizontal position) of the RC section (Fig. 5.50b). Almost the same location of the OWC level can be seen on a fragment of the section of effective acoustic impedance (Fig. 5.50a).

The object identified in this way in the form of incisions of river paleovalleys in the conditions of the basin of the Kamin work area, of course, may be of certain oil prospecting interest. Of practical interest, first of all, is the thickness of the channel filling paleoincisions (which are filled, as a rule, with heterogeneous sand and gravel sediments). In addition, in this respect, coastal ledges – blocks (with which high-capacity fractured-cavernous-porous reservoirs eroded by water paleoflows are connected) may be of interest.

Summarizing the results obtained above in such complex seismogeological conditions, which characterize the study area under consideration, one important circumstance should be emphasized: in order to obtain reliable results in the sediments of buried river systems, it is necessary to have original seismic data with good dynamic expression. The technique for identifying the above objects using the HRS-Geo technology allows solving this problem at a fairly high level, as shown in the results obtained above. In order to delineate buried river systems in sediments, it is also necessary to have a denser network of seismic profiles.

### 5.3.5 Prediction of the Main Geological and Geophysical Parameters for Productive and Promising Deposits

Under the conditions of the studied geological section of the northeastern slope of the Kolvin megaswell and the Chernorechen depression in the central part of the Khoreyver depression of the Timan-Pechora province and the Kamin prospecting area, in particular, it makes sense to predict the geological and geophysical parameters of productive and potentially productive deposits based on seismic data mainly for thin-layer real environment models. In accordance with this predicting procedure, seismic data were subjected to characterizing the intervals of the section of those horizons, the deposits on which were considered as oil-promising and top-priority for oil prospecting studies. Such deposits in the study area are deposits from a number of horizons such as, for example, $P_1$ar, $D_3$src', $D_3$domanik, $D_3$dzr, $D_2$, $D_1$ and $S_2$ld. First of all, reservoirs are associated with these deposits, with which, one way or another, oil and gas saturated traps can be associated.

Here we will focus on the characteristics of only some of the marked horizons ($D_3$src', $D_3$dzr, $D_1$) and on only some geological and geophysical parameters.

The sediments of the $D_3$srč' horizon (the carbonate sequence of the Sirachoi horizon III$D_3$srč'), formed chiefly by shallow-water shelf facies, are composed of carbonate (reef-genic) rocks almost free of clay admixture, massive, gray and light gray limestones, inequigranular, unevenly dolomitized and porous-cavernous. The fluid seal for the sediments under consideration, as noted above, is a terrigenous-carbonate stratum – strongly argillaceous limestones, weakly dolomitized, of coastal-marine genesis. These deposits, as a kind of screen-seal, belong to the upper shallow shelf part of the Syrachoy horizon $D_3$srč.

The deposits of the $D_3$srč' horizon are represented by two geological and geophysical parameters – distributions of: effective acoustic impedance (Fig. 5.52a) and porosity (Fig. 5.52b). On the distribution maps of relatively low and high values of effective acoustic impedance (AI) and porosity values in the sediments of the $D_3$srč' horizon, a certain zoning is observed, consisting in the presence of significant heterogeneity in the distributions of AI and reservoir properties – porosity ($K_p$) over the study area. This zoning is obviously associated with certain conditions of sedimentation. At the same time, zones of increased and decreased AI values (and adequate porosity values corresponding to them) in the deposits of the $D_3$srč' horizon are distributed over different parts of the area. Anomalies with the highest AI values are confidently correlated in the central and western parts of the Kamin area.

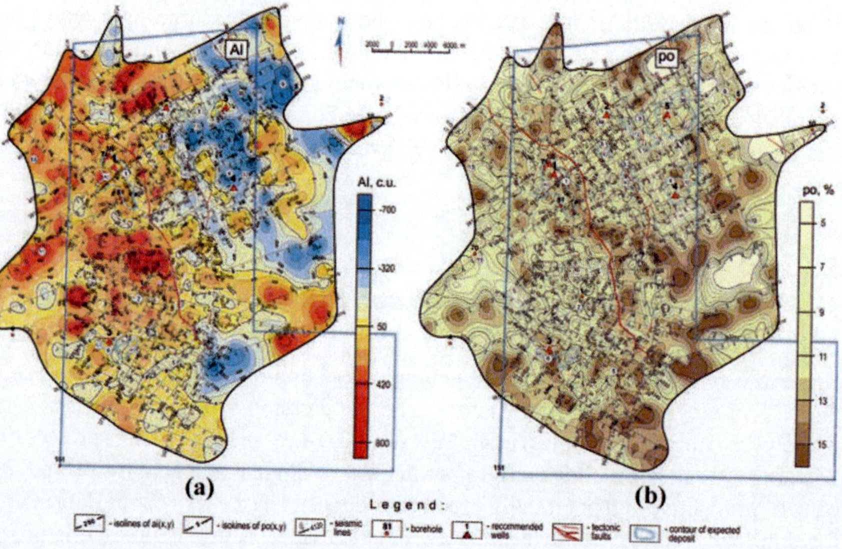


**Fig. 5.52** Distribution maps of acoustic impedance (**a**) and predicted average porosity (**b**) in the deposits of the carbonate strata of the Sirachoy horizon $IIID_3$srč

Anomalies with low AI values are recorded in the eastern and partially in the south-eastern parts of the explored area.

On the materials characterizing the distribution of AI values (Fig. 5.52a) and porosity (Fig. 5.52b) over the deposits of the $D_3$srč' horizon, approximately the following pattern is observed on the area: "relatively lower acoustic impedances (AI) are somehow characterize the increased porosity and vice versa". On the basis of this regularity, it follows from the prediction results obtained the deposits with the highest porosities have the lowest AI values. Judging by the distribution of average porosity values over the area in the sediments of the considered horizon $D_3$srč', they reach values of ≈ 8–15% (Fig. 5.52b).

Analyzing in general the distribution of anomalous porosity (both increased and decreased) over the area, together with the contours of persoective objects, it should be noted when identifying such objects, the structural factor was taken into account first of all. It is clearly seen within the contours of the identified promising objects in the deposits of the $D_3$srč' horizon, the porosity is generally characterized by their average values $K_p \approx 7$–12%.

The deposits of the Dzhyer horizon $D_3$dzr ($IIID_3$dzr) are represented by a uniform alternation of sandstones, siltstones, and mudstones formed in coastal shallow water conditions. The sandstones are light gray, gray, quartz, fine-grained, unevenly calcareous and clayey, with represented thin horizontal and oblique bedding due to the presence of carbonaceous-clayey admixtures. In the middle part of the horizon, there are areas of conglomerate-breccias, consisting of flattened weakly rounded clay fragments cemented by sandy-clay material. Siltstones are gray, brownish, locally calcareous, with a layered texture, sideritized. Mudstones are chocolate-brown, greenish-gray, with gliding surfaces, with remains of esterian shells. At the base of the horizon, thin lenticular interlayers of carbonate rocks are noted, represented by dark gray, silty, clayey limestones.

The screen-cover for the deposits under consideration is the deposits of the terrigenous sequence of the Timan ($D_3tm$) and Sargay ($D_3sr$) horizons. These deposits are represented by mudstones with interlayers of siltstones. Moreover, mudstones are red-brown, chocolate-brown, with gray interlayers with a brownish tint, sometimes crumpled. The siltstones are greenish-gray, irregularly clayey, with fine wavy-lenticular lamination. Due to the clay composition of the Timan-Sargay deposits and their ubiquitous distribution, they are considered as a regional seal for the underlying oil and gas complexes.

Deposits on the considered horizon $D_3dzr$ are characterized by such geological and geophysical parameters in the form of distributions of: acoustic impedance (Fig. 5.53a), porosity (Fig. 5.53b) and clay content (Fig. 5.53c).

On the maps of the distribution of the average effective acoustic impedances (AI) in the deposits of the horizon $D_3dzr$, a significant heterogeneity of the structure is observed (Fig. 5.53a). The area with the highest values of acoustic impedance (AI) is primarily related to the northeastern part of the area. This site of the study area chiefly refers to the Kamin local structures. The largest of all anomalies with elevated AI values is located in the region of the Yu-Kamin local structure.

Areas with relatively low AI values are confined primarily to the central, southern and partially to the northwestern parts of the area (it seems the fields of the area with a distribution of low AI turn out to be confined to the most clay intervals of the studied section, as shown in Fig. 5.53c).

In general, the regularity in the distribution of low and high values of effective acoustic impedance (AI) over the area is consistent with the existing ideas about the presence of geological (including acoustic) inhomogeneities in this part of the section. Here, on the whole, there is a complex geometry of the Lower Frasnian natural

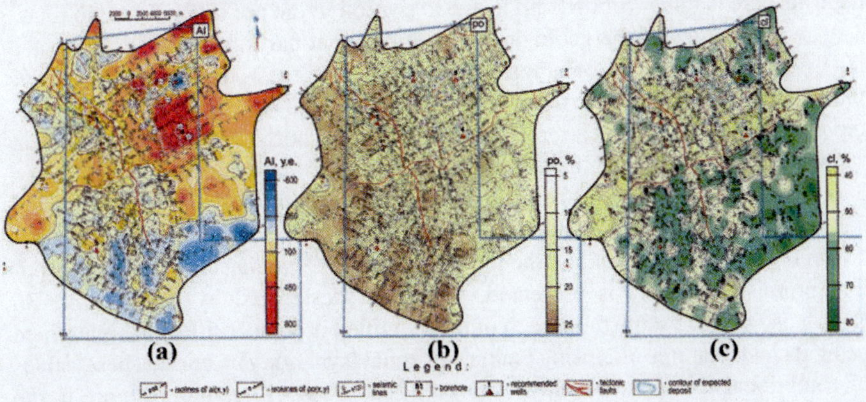


**Fig. 5.53** Distribution maps of acoustic impedance (**a**), predicted average porosity (**b**) and clay content (**c**) in the sediments of the Dzhyer horizon $IIID_3dzr$

reservoirs, represented by a uniform rhythmic interbedding of fine-grained sandstones, siltstones and mudstones with relatively equal vertical ratios and with a thickness of sandstone layers of several meters (Dzhyer deposits in this domain of the section play the role of an intermediate "scattering" stratum).

For the deposits of the $D_3$dzr horizon, as in the previous case (for the sediments of the $D_3$srč' horizon), zoning is observed in the distribution of increased and decreased **porosity** values (Fig. 5.53b), which is largely interrelated with the features of the distribution of AI values over the area. In the contours of possible oil saturation A4-Д4, there is some average distribution of predicted porosity, characterized by values varying in the range of 14–22%. A certain trend is also observed, showing the relationship between the areas of location of promising objects (with average predicted porosities) and zones confined to sites of the area with increased effective acoustic impedances (AI).

The general pattern in the distribution of **clay content** in the deposits of the $D_3$dzr horizon, shown on the map (Fig. 5.53c), mostly agrees well with the distribution of AI for the sediments under consideration. Here, in the deposits of the horizon under review, for areas with increased AI values, there is an almost unambiguous correspondence between lower values of clayiness. At the same time, the observed zoning in the distribution of clay content values over the area (which is adequate to the distribution of AI values) is obviously due to certain conditions for the formation (sedimentation) of deposits of the considered horizon, as well as a certain system of location of tectonic faults in the area (which here could play the role of natural barriers on paths of spreading paleostreams carrying sedimentary material).

For deposits of the Lower series ($D_1$) of the Devonian – the Koiven horizon ($D_1$kv) of the Emsian stage ($D_1$em), the behavior of its top on the Kamin work area in the depth image is characterized by a structural map (Fig. 5.54a). For the Lower Devonian sequence ($D_1$), an isopach map was also constructed – thicknesses between the tops of the Przhidolian stage ($S_2$p) deposits of the Upper ($S_2$) Silurian and Lower ($D_1$) Devonian (Fig. 5.54b). Lower Devonian deposits are known to overlie Upper Silurian rocks with stratigraphic unconformity. The distribution of $D_1$ deposits here is rather limited and is due to erosion in the pre-Early Frasnian time.

The regularities of the geological structure in that part of the Lower Devonian ($D_1$) deposits, which are common in the territory of the Kamin work area, have much in common with those for the overlying deposits of the $D_2$ horizon. A characteristic feature of the structure of deposits along the horizon ($D_1$) is a fairly clear step-like manifestation of the Chernorechen block in the eastern rather extended area of the studied deposits, which is bounded from the west by the V-Kolvin fault, and from the east by the Chernorechen fault.

On the structural map along the top of the Lower Devonian ($D_1$), a set of identified promising objects is presented, which are designated by numbers A6-B6 (Fig. 5.54a). At the same time, each of the identified objects (A6-B6) is highlighted using its own features of closing isolines of equal level h(x,y) = const. These values of conditional contours for objects A6-B6 of the Lower Devonian ($D_1$) are in the range – 4400÷ – 4530 m.

To determine the deposit thickness of the horizon $D_1$, we used the previously constructed isochron map and the structural map along the top of the deposits of the Przhidolian stage ($S_2p$). As a result, an isopach map was built – the thicknesses between the reflecting boundaries $S_2p$ (on the operating sections, this boundary is indicated by the index: III-IV (S-D)) and $D_1$, which is shown in Fig. 5.54b.

A characteristic feature of the deposit structure along the $D_1$ horizon is the most elevated parts of the study area are confined to the eastern boundary where the traceability terminated for the $D_1$ horizon sediments (within the distribution of the Chernorechen block, which gradually sinks in a southeasterly direction).

To the west of the V-Kolvin fault (from the zone of distribution of the Vangurei structure), the deposits of the $D_1$ horizon also tend to subside. The most submerged part of the research area is its southern part (the area of distribution of the Yu-Vangurei, Iev and S-Iev structures).

According to the constructions presented on the isopach map – the thicknesses between the reflecting boundaries $S_2p$ and $D_1$ (Fig. 5.54b), one can make sure in the area of the Chernorechen block, the thickness values $\Delta h_{(S2p\text{–}D1)}$ are significantly lower (changing in the range ≈ 40–240 m) than those for the western half of the reseach area (in the distribution area of the Vangurei, Yu-Vangurei, Iev and S-Iev structures), where the values of $\Delta h_{(S2p\text{–}D1)}$ vary in the range of ≈ 120–500 m and more.

On the formed sections of acoustic impedance (AI) and reflection coefficients (RC) in the deposits of horizon $D_2$, the most representative (with significant amplitudes of the anticlinal structures arch) are images of objects Б6 (Fig. 5.47 on the Vangurei structure), A6 (along profiles 3511, 4335 and 4338 on the northern dome of the Kamin structure) and B6 (along profiles 4327 and 3933 on the Yu-Kamin structure). The proposed deposit in object Б6 is confined to a reservoir dome trap, and in objects A6 and B6 – to tectonically screened traps.

Zoning in the location of the contours of the objects under consideration (A6-B6) in the study area is manifested in the fact the promising object A6 is confined to the northern section of the V-Kolvin fault, objects B6 and B6 to the northern and southern sections of the Chernorechen fault (Fig. 5.54a, b).

Taking into account the features of the geological structure of promising objects A6-B6 in the deposits of the Lower Devonian – horizon $D_1$, the specifics of their distribution in the study area and in the section (dynamic features of the distribution of AI and RC values), one can judge the relatively high oil prospects of the identified objects. The fluid seal (screen-seal) for the sediments under consideration can be layers of mudstones and siltstones with a trough-like layering and clay content increasing up the section, occurring in the lower part of the Middle Devonian ($D_2$) deposits. These deposits were formed in coastal-marine extremely shallow, intertidal conditions (with their possible erosion in the pre-Kolvin time – during the Early Hercynian stage of tectogenesis). At the same time, the Middle Devonian rocks, as is known, unconformably overlie the Lower Devonian and pre-Devonian formations.

The deposits of the Koiven horizon $D_1kv$ are known to be predominantly represented by clastic rocks: sandstones, silty sandstones, and siltstones. In the upper and lower parts of the section, there is a thin uneven interbedding of mixed

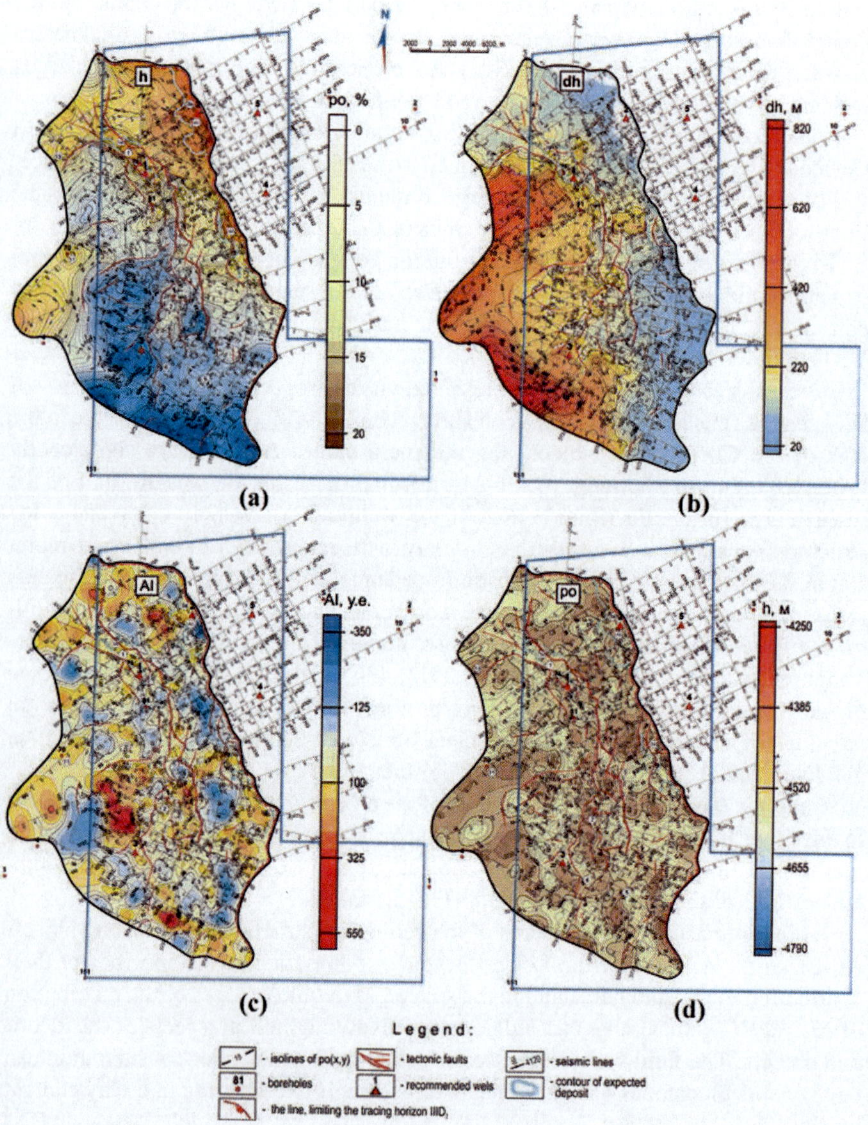


**Fig. 5.54** Structural map along the top of Lower Devonian deposits ($IIID_1$) (**a**), thickness map (isopach) between reflectors ($IVS_2p$-$IIID_1$) (**b**), distribution maps of acoustic impedance (**c**) and predicted average porosity (**d**) in Lower Devonian deposits $IIID_1$

clay-carbonate, clay-carbonate-terrigenous, clay-terrigenous rocks. In addition, the lower part of the section is characterized by the presence of layers and interlayers of red-colored rocks and an increase in the number of fractures of various directions and caverns. Siltstones and silty sandstones are dark gray clayey, unevenly clayey, dense, strong. Sandstones are light gray fine-grained dense, strong. The upper part of the section is represented by light gray quartz sandstones, fine-medium-grained, dense, strong, with microlayers and areas of clay material. The wedging out of the Lower Devonian deposits, as can be seen from the above results of structural constructions, has a structural-tectonic belonging.

In general, the Lower Devonian reservoir stratum is characterized by the predominant development of fractured-porous and fractured reservoirs.

In the neighboring Olenya area, oil inflows were obtained from the Lower Devonian deposits from a number of wells with an oil flow rate of 16–790 $m^3$/day. The reservoirs there are cavernous and fractured dolomites, as well as organogenic dolomitized limestones. In another adjacent Varknavtskaya area, oil inflows from the Lower Devonian deposits (flow rate from 30 to 419 $m^3$/day) were also obtained during testing of a number of wells.

For deposits of horizon $D_1$, the distributions of effective acoustic impedances are shown in Fig. 5.54c, and porosity – in Fig. 5.54d.

On the maps of the distribution of relatively low and high values of effective acoustic impedance (AI), as well as porosity, also for deposits of overlying horizons, there is zoning in the distribution of AI anomalies, consisting in the distribution over the study area of significant heterogeneity in the values of AI and reservoir properties – open porosity ($K_p$). This zoning is most likely due to certain sedimentation conditions. The zones of increased and decreased AI values (and the corresponding adequate porosity values) in that part of the area where deposits of the $D_1$ horizon are traced are distributed over its various parts. In this case, one can observe a certain submeridional direction in the location of anomalies in the AI and porosity. Anomalies with such an arrangement of AI and porosity over the area, of course, may be of some interest in oil exploration. However, for the considered deposits of the $D_1$ horizon, the top-priority (as noted above and for the deposits of the overlying horizons) when prospecting are promising objects identified taking into account the structural factor.

On the presented materials, according to the distributions of AI values (Fig. 5.54c) and porosity (Fig. 5.54d) in the study area, one can note a pattern that is already noted in the deposits of overlying horizons: "relatively lower values of effective acoustic impedance (AI) are characterized by higher porosity values for $D_1$ horizon deposits and vice versa". Following this regularity, it can be observed from the presented forecasting results the deposits, which are basically characterized by increased porosity values, have the lowest AI values. At the same time, the average porosity in general in the deposits of the $D_1$ horizon reach ≈10–18%, and within the identified contours of promising objects A6-B6 of the structural type – ≈13–16% (Fig. 5.54d).

### 5.3.6 *Perspective Objects Contours of the Studied Deposits in the Prospecting Area*

The above results of high-resolution ground seismic data and detailed processing and interpretation of borehole data (well log) make it possible to identify promising objects of a structural type, with which oil and gas deposits can be directly related. Such results are correlated target reflectors on sections of effective acoustic impedance (AI), seismogeological sections built on this basis (examples of such sections are given below), structural maps for target horizons and a number of predictive geological indicators important for prospecting and industrial exploration of oil and gas deposits.

Throughout the list of local structures, within which promising objects were identified in one way or another, i.e. for target horizons ($P_1$ar, $D_3$src', $D_3$dmn, $D_3$dzr, $D_2$kl, $D_1$kv, $S_2$gj and V-O), preliminary special processing and interpretation of HRS and well logging data was performed. According to this list of structures, zones of possible deposits were determined (with indices of supposed traps), the contours of identified promising objects and the areas occupied by these objects were estimated. This information made it possible to identify 38 objects of a structural type, and further analysis based on the use of such information for the entire volume of identified structures to establish the most significant of them. Based on the materials of almost all identified objects, the composition and properties of the deposits of the target section were predicted and, on this basis, the preparation of the most significant objects for exploratory drilling.

When assessing the distribution density over the area of contours of identified promising objects of a structural type, it is convenient to use the results of their imposition in the form of separate sets for successively (top to bottom) occurring horizons (Fig. 5.55a, b). In these figures, on a planned basis, in particular, the alignment of contours along three and four horizons is given. In addition, they clearly reveal the confinement of each of the considered structural forms to a specific part of the study area, as well as their mutual arrangement relative to each other. On Fig. 5.55, and, in particular, data are given for three contours of expected deposits in sediments: the Artinskian stage ($P_1$ar), the reef strata of the Sirachoi horizon ($D_3$src') and the domanikoid horizon ($D_3$domanik). On Fig. 5.55b, data are given for four contours of expected deposits in sediments: the Dzhyer ($D_3$dzr), Kolvin ($D_2$kl), Koiven ($D_1$kv) and Gerdyu ($S_2$gj) horizons.

In general, the information obtained (in Fig. 5.55a, b) makes it easy to navigate when choosing each of the identified objects (structures) both individually for each of the objects, and when analyzing their selected local population.

As can be seen from the above graph, the identified promising objects are distributed over the area mostly within four sites. The first of them is located in the area of wells 71, 81 Vangurei and recommended wells 3R and 1. The second site is located in the northern part of the area in the site of the location of the Kamin local structures, directly at the vicinity of the wellheads of recommended wells 2 and 5. The third site is located in the center of the eastern parts of the work area in the site of

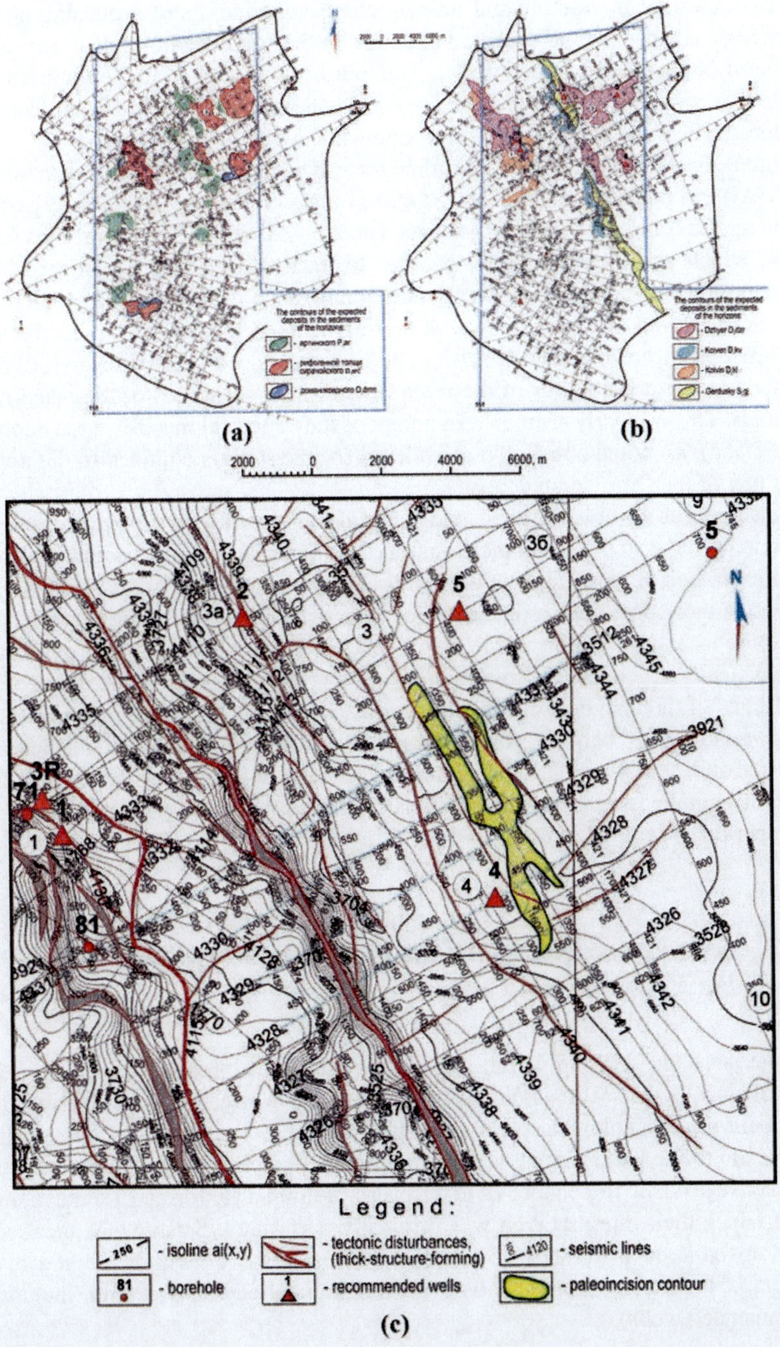


**Fig. 5.55** Combination of the contour location of prospective objects in the deposits of the Artinskian stage ($P_1$ar), the carbonate sequence of the Sirachoy (IIID$_3$srč), domanikoid (IIID$_3$domanik) horizons (**a**) and in the deposits of the Dzher horizon (IIID$_3$dzr), Middle (IIID$_2$), Lower (IIID$_1$) Devonian and Ludlovian Stage (IIIS$_2$ld) (**b**); distribution of the supposed incisions of the river system paleochannels in the sediments of the V-O horizon (**c**)

the Yu-Kamin local structure and linearly elongated contours of promising objects (which are adjacent to the specified structure in the southwest) – in the area of the wellhead of the recommended well 4. The fourth site is located in the southern part of the area, which is directly confined to the site location of the S-Iev local structure, i.e. located near the wellhead of the recommended well 3.

Inferred paleoincisions were identified directly in the sections of acoustic impedance (AI) and reflection coefficient (RC) of a number of seismic profiles. In particular, in fig. 5.50 and 5.51 show fragments of sections AI and RC along profiles 4328, 4329, which are located almost parallel to each other in the territory of the Bolshezemel arch of the Khoreyver depression. In accordance with the peculiarities of the distribution of AI and RC on these fragments of the sections, it can be seen the directions of the considered seismic profiles are placed approximately orthogonally to the formed contour of the river system – the coastlines of the paleo-river channels. This is clearly seen on a fragment of the structural map for these deposits (Fig. 5.55c). As noted above, the considered river system is confined to the uppermost part of the Ordovician deposits.

Analyzing as a whole the presented materials on the study of the structural forms of buried river systems (paleoincisions) in the Upper Ordovician sediments, some features of their geological structure should be emphasized.

One of these peculiarities is the buried river system under consideration (which has some local distribution in the study area) is directly confined to a number of tectonic faults which have approximately the same area strike as the direction of the axial lines of the channels of the paleoincisions (Fig. 5.55c). This may indicate a direct relationship between the intense formation of disjunctive dislocations occurred at the early and late Caledonian stages of tectogenesis, with the formation stage of complexly constructed channels of the buried river system at about the same period of geological time.

### 5.3.7 *Seismogeological Sections in the Areas Where the Recommended Points Are Located*

To determine the location of the identified structures – promising objects in the vertical plane of the section along the most informative directions – seismic profiles were built seismogeological sections, in the vertical sections of which the desired objects are located at different depths. In the same directions, the "initial" sections of effective acoustic impedance (AI) were also preliminarily formed, on the materials of which their interpretation was originally performed. Seismogeological sections are built on a number of seismic profiles, some of which are shown in Fig. 5.56a, b, passing through wells 1 Yu-Vangurei, 81 and 71 Vangurei (including recommended wells).

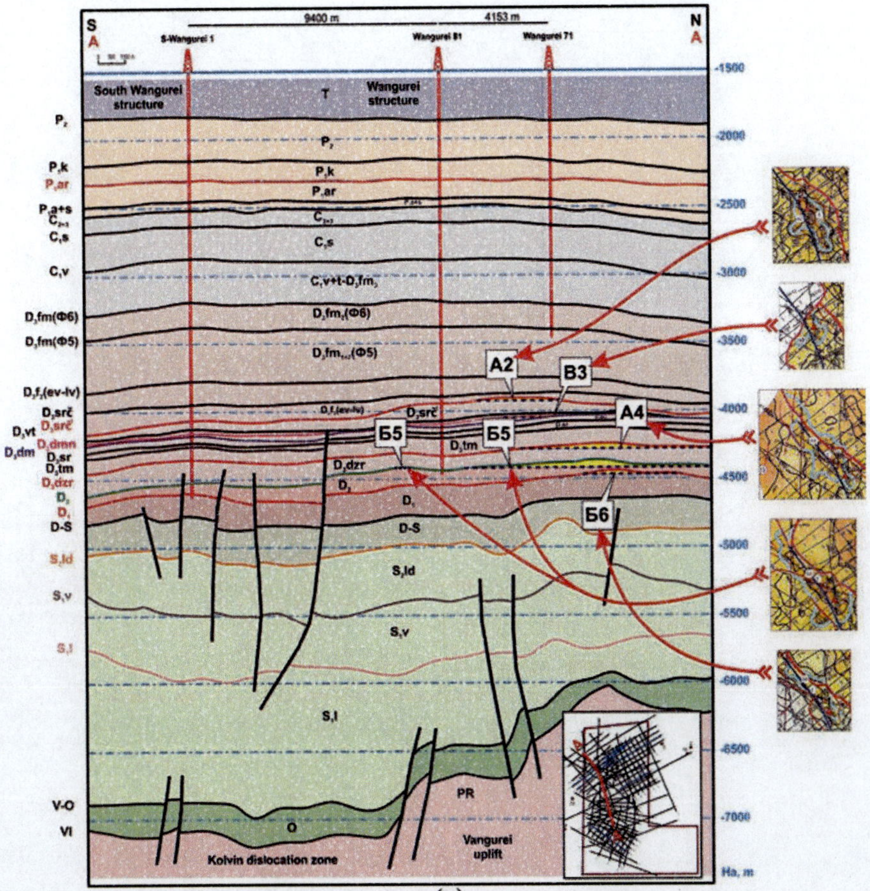


(a)

Fig. 5.56 Seismic-geological sections along: (a) the A-A line through wells 1, 81, and 71; (b) the line 3511 through project well 3R

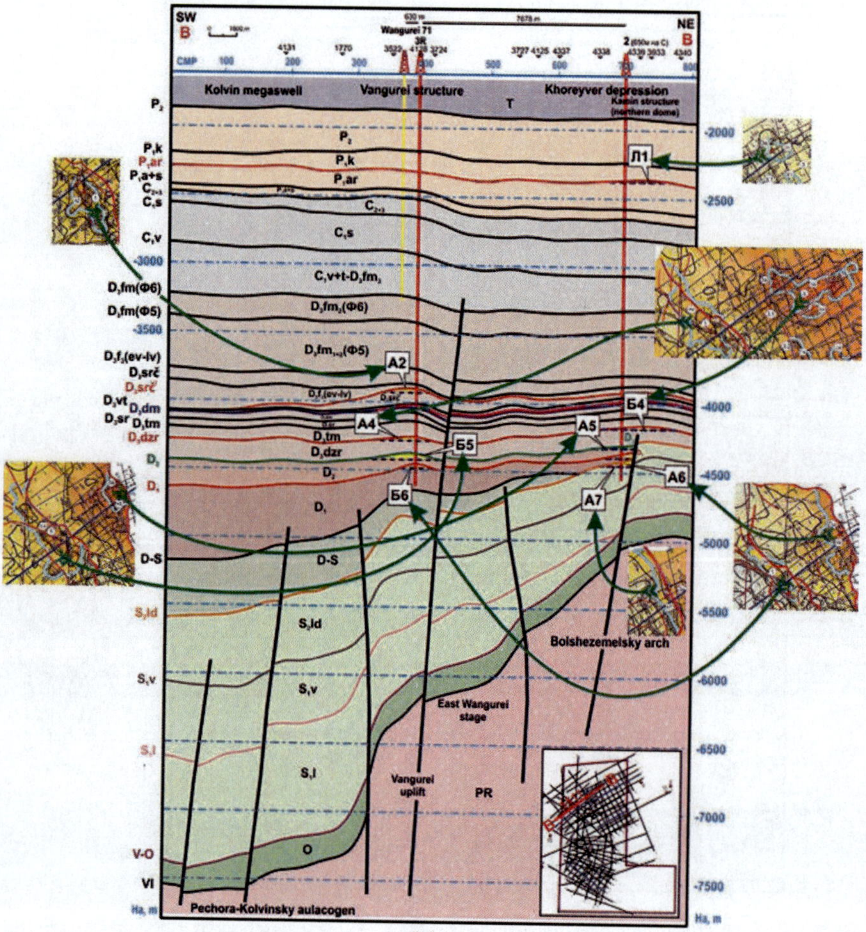


(b)

**Fig. 5.56** (continued)

On the presented sections, purely individual features of the structure of the section are clearly visible (Fig. 5.56). At the same time, the ratios of the structural forms of the studied deposits, the amplitudes of local structures, the location of faults, especially in the lower structural stage, and the thicknesses in each of the considered strata, one way or another, differ from section to section. In addition, the inherited, interconnected nature is clearly visible in the distribution of all the structural and tectonic elements of the structure of the real environment listed above, which at the same time characterize its specific local areas.

The peculiarity of the structure on the studied seismogeological models of the section is also that, according to the main target horizons ($P_1$ar, $D_3$src', $D_3$domanik, $D_3$dzr, $D_2$, $D_1$ and $S_2$ld), almost all encountered promising objects (traps) are presented in the given directions (according to the used 2D seismic profiles) either in the form of anticlinal structures (arched deposits) or in the form of tectonic structural forms (tectonically shielded deposits). At the same time, a number of perspective objects (traps) located in deposits of various horizons are not "covered" by the vertical sections of the recommended wells. In particular, such objects are found on seismogeological models along the A-A line (through sections of wells 1 Yu-Vangurei, 81 and 71 Vangurei) (Fig. 5.56b), as well as along profile 3511 (along sections of well 71 Vangurei and recommended wells 3R and 2R) (Fig. 5.56a).

In that way, the geological and geophysical materials accumulated in the Kamin work area (as a result of processing and interpretation of seismic and field geophysical data), as well as the results of the assessment of oil content in the adjacent areas, show the following. The main prospects in the area under consideration are associated with the Upper Silurian carbonate (Gerdju) deposits, terrigenous Emsian (Koiven) deposits of the Lower and Eifelian (Kolvin) deposits of the Middle Devonian, terrigenous Dzhyer deposits, carbonate domanikoid and Syrachoy (reefgenic) deposits of the Upper Devonian and carbonate Upper Artinskian Lower Permian.

The most obvious hydrocarbon traps within the study area are anticlinal (arched), as well as tectonically and lithologically (wedging out) screened structures accompanying the formation of underlying step-like blocks that formed in the lower structural-tectonic stage during the Early Paleozoic (Caledonian) and Early Hercynian stages of tectogenesis.

The development of numerous disturbances (faults) in various structural zones of the study area, apparently, contributed to the functioning as some controlling factor in the migration and accumulation of hydrocarbons (despite the fact that a significant number of tectonic faults and displacements can significantly complicate the migration of hydrocarbon fluids to traps).

Summarizing the results obtained as a whole, it should be emphasized in the areas of the identified oil saturation contours (*complex shape*), shown in Fig. 5.56a, b, the selected promising points are presented as priorities. In addition to these perspective points, a number of those can also be identified in the Kamin operating area, which may be of practical interest for subsequent prospecting.

## 5.4 The Study of Complex Productive Horizons Using High-Resolution Seismic Data (*Vozei Swell and the Western Part of the Khoreyver Depression*)

This subsection presents a small amount of the results of reinterpretation of CDP 3D seismic data in order to build a detailed tectonic model of the structure of the sedimentary cover within the Yu-Osh and Osokin fields. Using the materials of field surveys obtained here, the authors performed special processing of 3D seismic data in the volume of 174.2 km$^2$ to: (a) obtain 3D cubes and 2D sections of effective acoustic impedance (AI) and reflection coefficients (RC); (b) building a one-dimensional seismic model for a number of wells in the Yu-Osh field; (c) stratigraphic tying of reference boundaries of target horizons (including productive strata of sediments – Zadon $D_3$zd, Starooskol $D_2$st, Kolvin $D_2$kl and Kedrov-Omrin $D_2$kd + om horizons); (d) studying the nature of the layering of deposits of productive strata based on the use of a wave field cube of 3D seismic records (WF), 3D RC and AI cubes; (e) construction of isochrone maps with the application of refined tectonics based on the calculation of interval velocities, plotting dependencies according to well logging and high-resolution seismic data and building on this basis structural maps for target horizons; (f) seismic facies zoning of the study area using 3D cubes of AI and RC by clustering them according to the layers of the main productive and promising deposits.

It should be noted a fairly large amount of geophysical research has been carried out in the study area: gravity, electrical, aeromagnetic and seismic surveys (MRW, CMRW and CDP). Regional geophysical research began in the mid-1950s. Information about all these studies is not provided by the authors, since they do not directly affect the topic considered in this subsection and the results obtained within this theme. We only mark in the most recent period of work (2009–2020), reseaches were carried out in the study area by many geological and geophysical organizations. We are talking about companies with which the authors had operating contacts: OJSC "Severgeofizika" (2006–09), CJSC "NK Nobel Oil", LLC "Nobel Oil (KO)" (2005–20), ZAO "Kolvaneft" (2015), OOO "NAC VNIGRI-Neftegaz" (2015–20) and OAO "CGE" (2013). It should be noted the information on the entire volume of seismic materials and GBS data of these companies is well known. On the issues of preliminary processing and interpretation of these data, the authors of the monograph, in addition, sometimes had to cooperate with the leading specialists of these companies.

The materials of high-resolution seismic (HRS) presented below were obtained exclusively by the authors in the stage of processing seismic data by a special graph and as a result of their geological interpretation (with the exception of the introductory 4 first figures). In addition to the geological significance of the results below for the formation of geological and geophysical models of the studied geological environment (3D), they are also presented by the authors to demonstrate the effectiveness and benefits of using the HRS-Geo technology.

### *5.4.1 Brief Data on the Geological and Geophysical Knowledge of the Study Area*

Brief graphical information on the location of the Yu-Osh field, an overview of the location of hydrocarbon deposits with elements of the tectonic structure, the geological and geophysical knowledge of the study area and the generalized lithological and stratigraphic characteristics of the section are presented in Figs. 5.57 and 5.58.

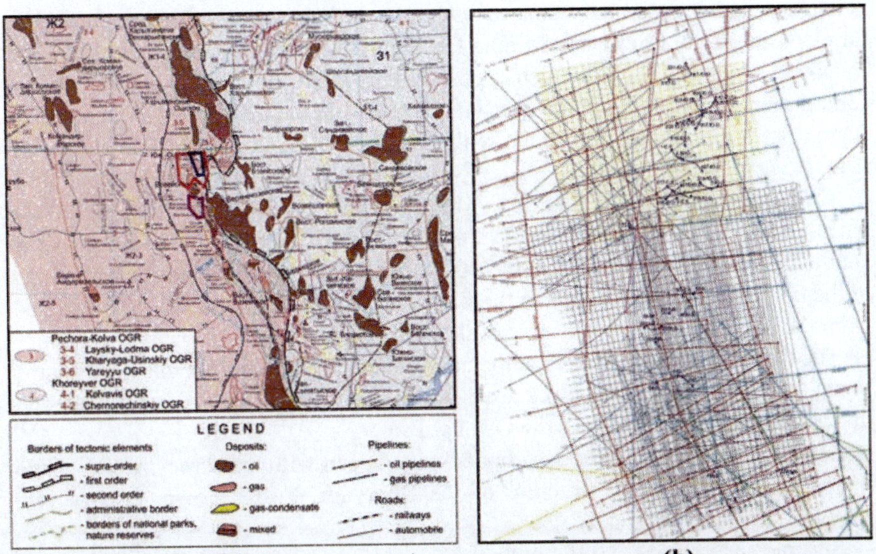


(a) (b)

**Fig. 5.57** Overview placement of hydrocarbon deposits with elements of the tectonic structure of the work area (**a**); Examination of the research area by geophysical methods and drilling (according to JSC "Severgeofizika", 2009) (**b**)

1. The research object in the study area, as noted above, are the Yu-Osh and Osokin oil fields. To the north and northeast, in the immediate vicinity of these fields, are the Osh and Kharyaga fields. In the central part of the area between the Osokin and Yu-Osh-S-Kostyuk blocks, the northern Middle Devonian deposit of the Vozey field is located (Fig. 5.57a). To the southeast are the Kostyuk and Verkhnevozey oil fields.

Tectonically, the study area is located within the Kolvin megaswell of the Pechor-Kolvin aulacogen. The Yu-Osh territory is located in the junction zone of the Vozeya and Kharyaga swells, and the Osokin territory is located at the northern end of the Vozeya swell. From the east, the Kolvin megaswell connects with the large Khoreyver depression through a system of echelon-shaped structures, from the west – with the Denisov depression.

The Vozey swell is a complex uplift and consists of two merged local, echelon-shaped conjugated swells – the Yu-Vozey swell and, attached to it from the north-west, the S-Vosey (Kostyuk) swell. The S-Vozei local swell is a relatively narrow linear anticlinal zone extending from the southeast to the northwest.

According to seismic data, the Yu-Osh area is a complex block located in the zone of an echelon joint: in the north – with the southern periclinal of the Osh swell-like uplift, and in the south – with the northern periclinal of the Kostyuk uplift. From the northeast, the Osh uplift is bounded by a fault zone separating it from the Central Kharyaga uplift, and to the south, by a fault from the Khoreyver depression. On the southern periclinal part of the Osh uplift, the local Yu-Osh structure is isolated, which is separated from the Osh structure by a sublatitudinal trough along the deposits of the Lower and Middle Devonian. As a result of the increase in the thickness of the Middle Devonian deposits to the north, this trough is filled, therefore, the structural plans for the overlying horizons in the Upper Devonian deposits flatten out and a single Osh uplift with local domes (Osh and Yu-Osh) is traced (Fig. 5.57a).

At the initial stage of exploration (1997), before detailed seismic surveys and with limited data on exploratory drilling, the Yu-Osh field was considered as the southern section of the Osh field. In 2000, after refinement of the geological model of the field, the southern section of the field, including well 1 Yu-Osh, was separated from the Osh field along the border of the sublatitudinal fault.

The Osokinsky site is located at the northern end of the Kostyuk uplift within the Vozey swell; S-Osh – Yu-Kostyuk area – at the northern end of the Vozey swell and in the zone of its junction with the Osh uplift.

Seismic exploration in the area under study began with the survey, which made it possible to clarify the structure of the Kharyaga, Yu-Kharyaga, Osh, Vozey, Kostyuk structures (Fig. 5.57b).

Seismic surveys of CDP-2D within the study area began in 1976. Based on the results of these studies, the structure of the Yu-Osh, S-Kostyuk-II, Komavis, S-Komavis, S-Kostyuk and Kostyuk structures was studied and refined.

CDP-3D seismic surveys executed in 2005–2007 at the S-Kostyuk area and the Yu-Osh field were continued in 2007–2008 at the Osokin area (JSC "Severgeofizika"). To build a unified seismic model of the Yu-Osh field, within the framework of these works, a single cube of seismic data was formed for adjacent areas.

Deep drilling within the Yu-Osh field began in 1997. In the near-arch part of the Yu-Osh structure, well 1-Yu.O. was drilled, which opened up carbonate-clay deposits of the Lower Devonian at the bottom (3936 m).

Geological field information on the drilled fund of wells (prospecting, exploration, production) is presented for more than 50 wells. On this basis, an information database has been formed. Materials on geological breakdowns of productive horizons for all available (64) wells are prepared. These geological subdivisions are consistent with seismic data, namely with reflectors – with the corresponding tops, bottoms of the target horizons ($P_1$ar, $P_1$a-s(t), $P_1$a-s(b), $D_3$dm(b), $D_3$tm, $D_3$dzr,) and productive ($D_3$zd, $D_2$st, $D_2$kl and $D_2$kd + om) formations on the sections (inline(s) and crossline(s)) 3D cubes of reflectivities (RC) and acoustic impedances (AI) (the latter – 3D cubes of RC and AI were formed directly by the authors of this work).

2. The lithological and stratigraphic characteristics of the section on the territory under consideration, of course, are based on data from drilling deep wells (Fig. 5.58). In the structure of the section at the study area, deposits of the Cenozoic, Mesozoic and Paleozoic ages, occurring on the Precambrian basement, take part. The deepest well within the study area is the 100-Vozei well, the opened up thickness of the sedimentary cover in which was 4559 m (at the bottom of the Lower Devonian deposits). The characteristics of the underlying sediments are largely presented according to drilling data within the Vozei swell

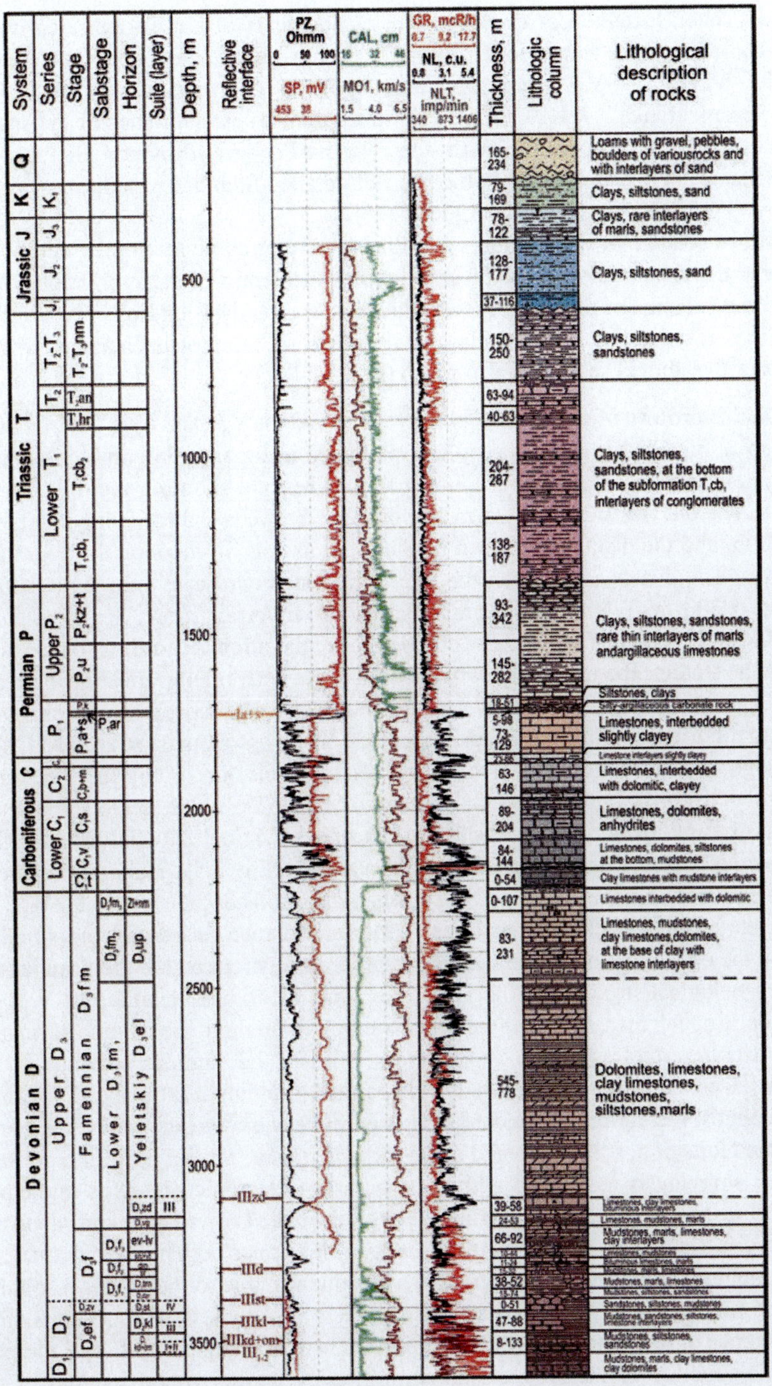

**Fig. 5.58** Summary lithological and stratigraphic section of wells of the Yuzhno-Osh and Osokin oil fields, North-Kostyuk, Vozey and North-Vozey areas (according to JSC "Severgeofizika", 2009)

(51-Vozei, located less than a kilometer southeast of the Osokin area, which exposed deposits from Quaternary to the basement, the actual depth is 4516.4 m, the horizon – V-Є (Vendian-Cambrian)) and seismic survey data (JSC "Severgeofizika", JSC "CGE"). The stratigraphic characteristics of the section are given in accordance with the "Timan-Pechora Province Stratigraphic Scheme" (TPP), compiled on the basis of detailed lithofacies studies carried out by TPO VNIGRI – TP SRC (1996, 1998).

On the scheme of the summary lithological-stratigraphic section, in addition to the traditional vertical sections on stratigraphy, the depths of the stratigraphic main units and subunits of the section, their thicknesses, standard logging curves and the lithological column, a brief lithological description of the rocks characterizing the wells of the studied section is also given (Fig. 5.58).

3. On the territory of the S-Kostyukskaya area and the Yu-Oshskoye field, as noted above, 3D CDP seismic survey was practiced using an orthogonal observation system with overlapping reception points. The total shooting multiplicity was 48. The bin size was 25 × 25 m, the distance between the receiving points was 50 m, and the distance between the shot points was 50 m along the coordinate axes. The distances between the lines along the coordinate axes $\Delta y = 300$ m, $\Delta x = 300$ m. Telemetric seismic stations "Aram Aries" and "Sersel" SN 388 (*JSC "Severgeofizika", 2007*) were used for registration. According to the results of the studies, the necessary constructions were performed (reflectors were correlated, tectonic disturbances were traced, isochron maps, isohypse maps for target horizons, structural maps, sandstone thickness maps $D_2$st and $D_2$kl, seismogeological sections, maps of initial seismic facies according to reflectors in productive intervals in deposits Upper and Middle Devonian, scale 1:25000).

An example of digital processing results of the original 3D CDP seismic data obtained in the area under consideration is shown in Fig. 5.59. Here, on fragments of time sections along several profiles – inline directions of the wave field (WF) 3D cube, the most characteristic features of the interference seismic record structure (along the entire section of the wave pattern) are shown, in which the peculiarities of the geological structure of the Yu-Osh tectonic block were manifested. Without considering such features of the tectonic structure throughout the volume of seismic data, it is nevertheless possible to get some idea of the complex structure of the studied deposits from the presented wave pattern. Running a little ahead, it can be noted almost the entire area under study is complicated by a significant number of tectonic faults of northwestern and submeridional strike. The origin of these disturbances is obviously associated with the processes of tectonic activation in the pre-Lower Frasnian period. As a rule, these faults fade in the Lower Famennian deposits.

Despite the fact the 3D survey makes it possible to trace both high-amplitude and low-amplitude tectonic faults with greater confidence, due to the higher resolution and informative density of the network of profiles, nevertheless, the phenomenon of interference of the seismic record is manifested here to a large extent. The complex tectonic structure of the area is also confirmed by the horizontal sections of the coherence cube near the target reflectors. On such cubes, as a rule, a system of faults

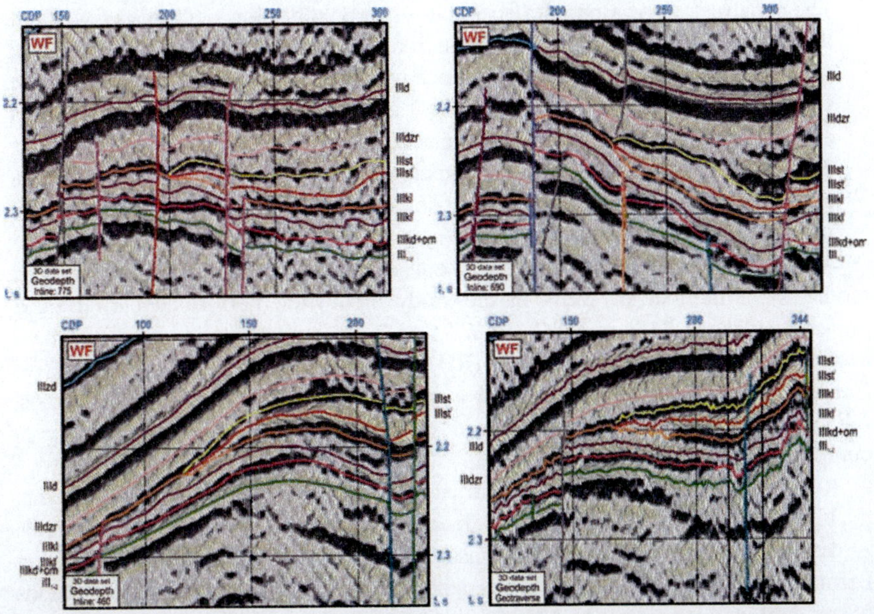


**Fig. 5.59** Stratigraphic wedging of the Starooskol deposits (according to the data of JSC "Severgeofizika", 2009)

distinguished by time sections is traced and it is clearly seen what a complex tectonic zone interpreters have to deal with.

On the CDP time sections themselves, such areas are usually identified by a number of features: a visible shift of the in correlation phase of regular reflected waves; breaks in correlation, complicated in some cases by diffraction phenomena; loss of wave correlation; anomalous attenuation of amplitudes; difficulties associated with the identification of reflected waves on both fault sides (Fig. 5.59). Tectonic faults with an amplitude of 30–50 m are most confidently highlighted (with a break in the axes of reference reflections correlation). Faults with insignificant amplitudes are distinguished by a local change in the dynamic characteristics of reflected waves, such as flexure-like complications of a reflection pack, loss of wave correlation.

It should be noted the reprocessing of original seismic data using the most contemporary software packages, software complexes and interpretation systems such as OMEGA 2.7 (Western Geco), GeoDepth (Power 2D) (Paradigm Geophysical), 3D PSDM, "SeismoScan", DV- 1 DISCOVERY (JSC "CGE", 2013) improves the situation in many respects with tracing the correlation axes of reference reflections and with the selection of tectonic faults of different amplitudes. However, all this powerful arsenal of software does not fundamentally solve one of the important problems of wave seismic exploration – the elimination of wave interference in a seismic record (the solution of which, as is known, is associated with the problem of correctly solving the inverse dynamic seismic problem).

4. The construction of a one-dimensional seismic model for a number of wells of the Yu-Osh field was carried out on the basis of joint visualization of seismic and well logging data. To bind the reference reflectors within the framework of the seismic wave field (WF) structure only, it is usually carried out on the basis of a comparison of the traces of time sections with the ones of synthetic seismograms obtained for a given lithological-acoustic model of the studied section. In this case, special programs for one-dimensional seismic modeling of the wave field are used.

A significant amount of results on the construction of a one-dimensional seismic model for a number of wells in the study area were performed earlier (*JSC "Severgeofizika", 2009 and JSC "CGE", 2013*). Such results are known to be necessary for linking reflective horizons (RH), in this case associated with productive formations $D_3$zd, $D_2$st, $D_2$kl and $D_2$kd + om and reference horizons. On this basis, reflecting boundaries are correctly identified by the selected seismic facies complexes.

An example of the results of such modeling is shown in Fig. 5.60. In this case, the binding of reflecting horizons is carried out along the entire section under study. A detailed division of the section was carried out in the interval of the Lower Permian and Upper-Middle Devonian deposits. The combination of three methods of standard well logging (SN, NLT (NL), GR methods) made it possible to identify sandstone units and subdivide the section according to lithology. Using the possibilities of joint visualization of seismic and well logging data, the logging benchmarks are compared with the corresponding seismic reflectors, such as the top of the Lower Permian carbonates, the top of the Artinskian deposits, the top of pure carbonates of the Lower Permian (the top of the Asselian-Sakmarian deposits), the base of the Domanik deposits (as one of the main reference, well-traced reflective boundaries), the bottom of the Yelets deposits of the Upper Devonian. Further, as a result of a detailed analysis of the wave field, all other reflections were tied, which made it possible to detail the structure of the promising Middle Devonian deposits. When analyzing the wave field, in addition, attention is paid to the dynamics of each positive and negative phase of the oil-promising interval in the Middle Devonian section (extremums of the seismic record closest to the top and bottom of one or another productive formation were selected) (*JSC "Severgeofizika", 2009 and JSC "CGE", 2013*).

The general trends in the distribution of velocity characteristics with depth show a rather monotonous increase in velocity with depth is observed up to the top of the Permian deposits of the Artinskian. For all analyzed wells, a sharp increase in reservoir velocity is observed in the top of the Artinskian stage; in time sections, this boundary corresponds to a dynamically expressed and laterally stable reflection Ia. In this case, the interval of the section from the top of the Artinskian stage to the end of the Permian system is characterized by the highest velocity. The lower intervals of the well section, covering the Carboniferous and Devonian sediments, are generally characterized by a monotonous increase in velocity vs depth. However, in the interval of the section from the Zadon to Starooskol horizons, there is a slight decrease in reservoir velocity.

Stratigraphic binding of the reference boundaries of the reservoirs of the Zadon $D_3$zd, Starooskol $D_2$st, Kolvin $D_2$kl and Kedrov-Omrin $D_2$kd + om productive formations was performed on the basis of the above results of one-dimensional seismic (synthetic) modeling (point **3°**) and a detailed interpretation of the seismic wave field (WF) along various profiles-traverses – CDP time sections formed according to the corresponding inline(s) and crossline(s), which pass directly near the projections of the boreholes of the drilled wells.

A certain advantage in performing the procedure of stratigraphic linking of reflecting boundaries (RB) is obtained when using sections of effective acoustic impedance (AI) and reflection coefficients (RC), since these sections do not have a complex interference pattern of the wave field (WF), which is present on the seismic traces of any of the used time sections (formed from the traces of the cube of seismic records) (Fig. 5.60). It is also important the absence of wave interference makes it possible to obtain a quite adequate model of the acoustic heterogeneity in the sections AI and RC (with a sampling step of the seismic record over time) and, on this basis, apply a more reliable binding of the reflectors to the geological (lithological-stratigraphic) horizons. In turn, the acoustic inhomogeneity of the section is, first of all, the distribution of the product of density and velocity ($\rho \cdot V$) in the medium under study. The distribution ($\rho \cdot V$), one way or another, is due to the presence in the studied reservoirs of such important characteristics as lithological composition, reservoir properties and oil and gas saturation of the rocks of the studied deposits.

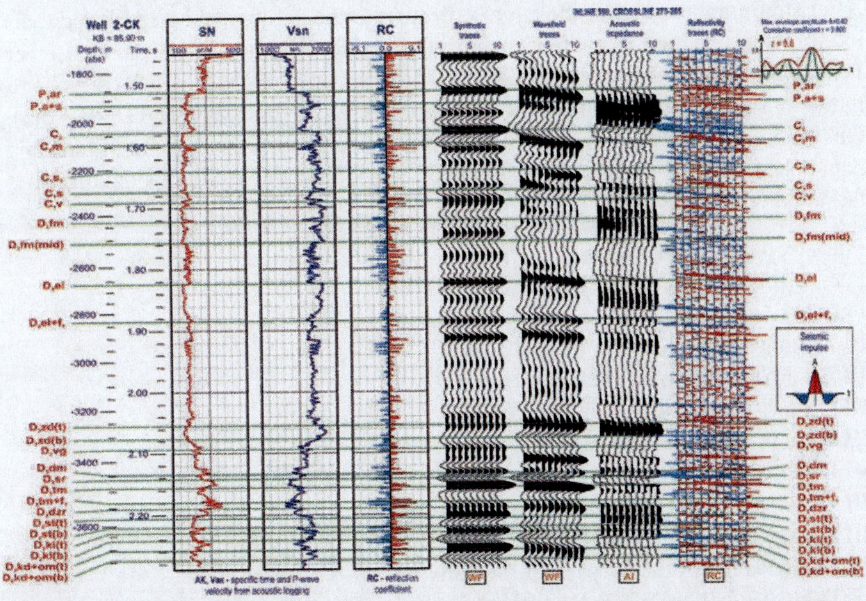

**Fig. 5.60** Linking the geological horizons of thin-layer medium along the well 2-CK to the reflectors of the wave field (WF), acoustic impedance (AI) and reflection coefficients (RC) in the interval of the deposits P1ar-D2kd + om

Using the example of stratigraphic tying of reflectors to geological boundaries, it is shown using the AI and RC sections for each of the target reflectors, the values of $\Delta t$ or $t_0$ are preliminarily calculated from the values of reservoir $V_{fm}$ or average $V_{av}$ velocities and the known depth of the geological boundary in the well. On this basis, it is possible to increase the efficiency of linking stratigraphic and lithological boundaries compared to that to time sections using extremely ambiguous modeling of the seismic wave field. In the latter case, it is necessary to know (or determine) the shape, spectral composition of the seismic wave impulse, as well as the law of their change as this impulse propagates in a real geological environment.

In the case of stratigraphic referencing, in addition, on sections of acoustic impedace, the dynamic properties of both the traces of the AI curve (its increased and decreased values obtained by restoring them from a seismic record with a discretization interval along time) and one of the available curves are taken into account standard logging. At the same time, the latter characterizes either reservoir (SP, GR), or velocity (SN), or density (NL) properties of the studied section (depending on the use of one or another standard well logging method in the vertical section of the model) (Fig. 5.60).

### 5.4.2 *Examples of the Section Geological Structure According to Acoustic Impedance and Reflection Coefficient*

After stratigraphic tying of the target reflectors to the geological horizons and their interpretation on high-resolution sections (AI and RC), 14 horizons from the Lower Permian to the Kedrov-Omrin sediments were correlated, eight of which correspond to the tops and bottoms of productive formations ($D_3zd$, $D_2st$, $D_2kl$ and $D_2kd + om$). On this basis, the correlation of reflectors (RB) over the entire volume of 3D seismic data cubes (AI, RC and WF cubes) was performed for the sediments of following target horizons:

$P_1ar$ is top of the Artinskian,
$P_1a + s(t)$ is top of the Asselian-Sakmarian,
$P_1a + s(b)$ is bottom of the Asselian-Sakmarian,
III $D_3zd(t)$ is top of the Zadon,
III $D_3zd(b)$ is bottom of the Zadon,
III $D_3dm(b)$ is bottom of the Domanik,
III $D_3tm$ is top of the Timan,
III $D_3dzr$ is top of the Dzhyer,
III $D_2st(t)$ is top of the IV pack of the Starooskol,
III $D_2st(b)$ is bottom of the IV pack of the Starooskol,
III $D_2kl(t)$ is top of the IV pack of the Kolvin,
III $D_2kl(b)$ is bottom of the IV pack of the Kolvin,
III $D_2kd + om(t)$ is top of productive I + II packs of Kedrov-Omrin,
III $D_2kd + om(b)$ is bottom of productive I + II packs of Kedrov-Omrin.

At the same time, the study of the sediment layering nature of the formations was carried out on the basis of the use of a time cube of 3D seismic records, a cube of reflectivities (RC) and acoustic impedance (AI).

The main feature of laminated stratigraphic structures, as is known, is the presence of layering. *Layering* is the primary heterogeneity of the sediment, which is expressed in the form of alternation of rocks of different composition. The main element of bedding is a layer – a slab-like body composed of rock of a certain composition, structure, color and limited by two stratifying surfaces separating it from the overlying and underlying layers. The upper surface of the layering (bedding) is the *top*, and the lower one is the *bottom*. The interfaces between the layers, as well as the layering itself, are formed as a result of a change in the sedimentation conditions. If the conditions change quickly, then the boundaries between the layers are clear and sharp, and if the conditions change slowly, then the boundaries will be fuzzy. Its *thickness* is measured by the shortest distance between the top and the bottom. The distance thus measured characterizes the *true* thickness of the layer. Without going further into the features of the structure of various types of bedding (simple flat bedding, discontinuous, ribbon, banded, wave-like, cross bedding, eolian cross bedding, river cross bedding, temporary flow cross layring, coastal-marine, etc.), we note only one important circumstance. It is caused with the possibilities of the applied method of recovering the values of reflection coefficient (RC) and acoustic impedance (AI) from seismic records in relation to determining on their basis the thickness of the studied layers (presence of layering), which are directly formed on the sections and cubes of AI and RC. (i.e., eventually with the possibility of realizing high resolution seismic data). The type and nature of layering are determined most adequately and with less error from high-resolution data, which show thin-layered structures. AI and RC sections, differing from those of the wave field in detail, are most suitable for an unambiguous description of the sediment layering. On all, without exception, sections of AI and RC, characterizing the structure of the geological space within the considered part of the study area (Fig. 5.61, 5.62, 5.63 and 5.64), this property (resolution) is quite well manifested. This pecularity of the structure on the AI and RC ultimately makes it possible to identify the layering detail on the studied tectonic model of the real environment over almost the entire study area.

As mentioned above, when using the HRS method, from the wave field (from seismic traces WF), RC and AI sections are formed, in which the step of reading the RC and AI along the vertical is equal to the sampling step of the seismic record over time. In this case, the practical vertical resolution of studying terrigenous sections is 3–4 m, and for carbonate sections it is 5.5–6.5 m (with a seismic record discretization interval $\Delta t = 0.002$ s and the propagation velocity of elastic oscillations in the terrigenous section $V = 3000–4000$ m/s and in the carbonate section $V = 5500–6500$ m/s, respectively). In the case of applying the quantization step $\Delta t = 0.001$ s, the vertical resolution doubles (i.e., it will be 1.5–2 m and 2.7–3.7 m for the same values of elastic wave propagation velocities). An attempt to determine the thicknesses of a thin-layer model of a real medium using only the elements of the wave field (WF) structure turns out to be quite problematic. Such a

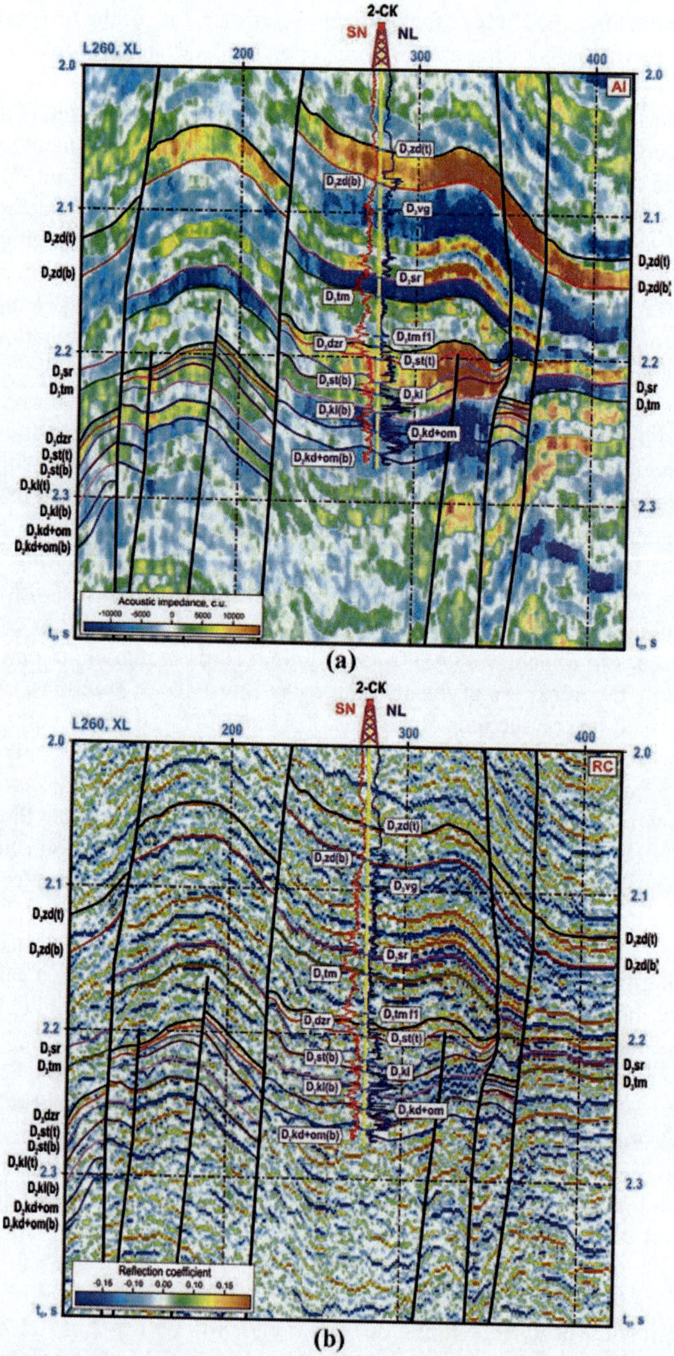


**Fig. 5.61** Detailed binding of geological boundaries along the 2-CK well to the reflectors of the effective acoustic impedances (AI) (**a**) and reflection coefficients (RC) (**b**) along the inline 560 of 3D-cube

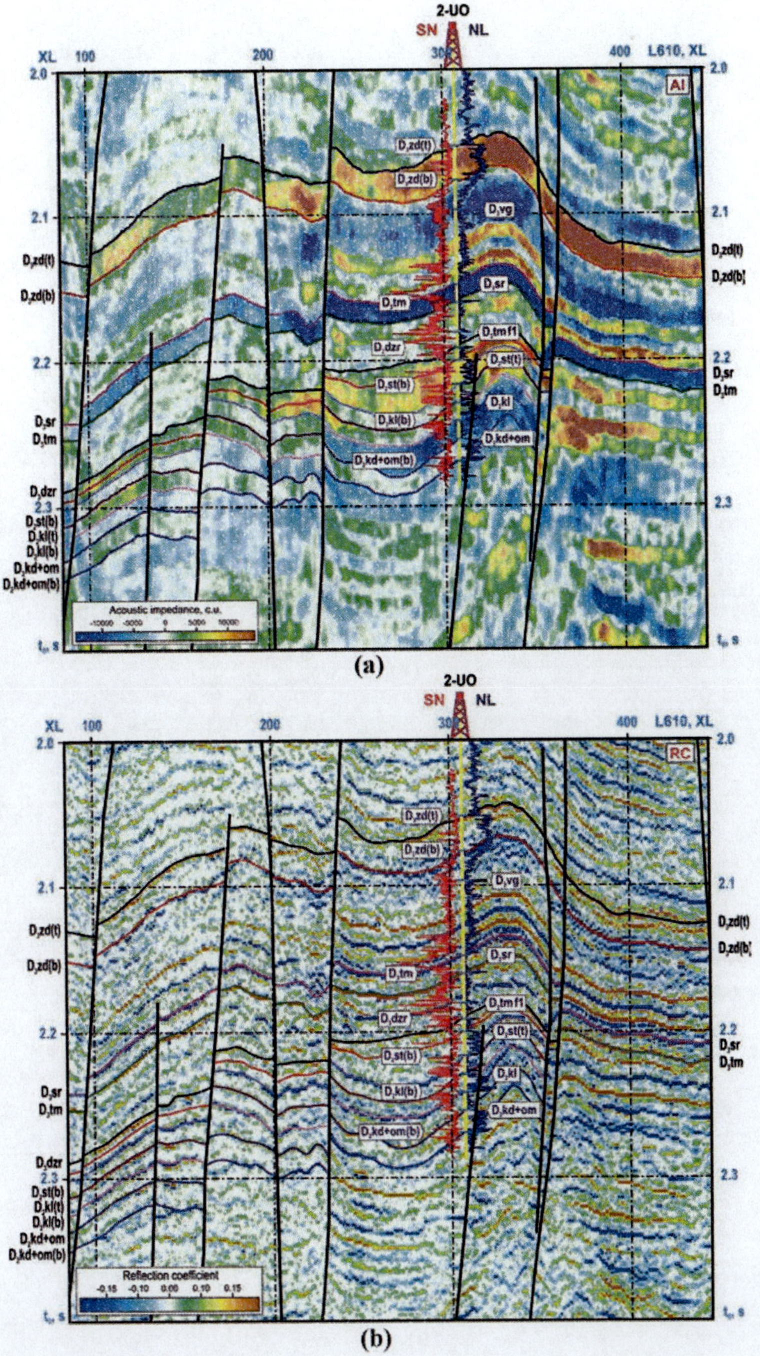


**Fig. 5.62** Detailed tying of geological horizons along the 2-UO well to the reflecting boundaries of the section of effective acoustic impedance (AI) (**a**) and reflection coefficients (RC) (**b**) along the inline 610 of 3D-cube

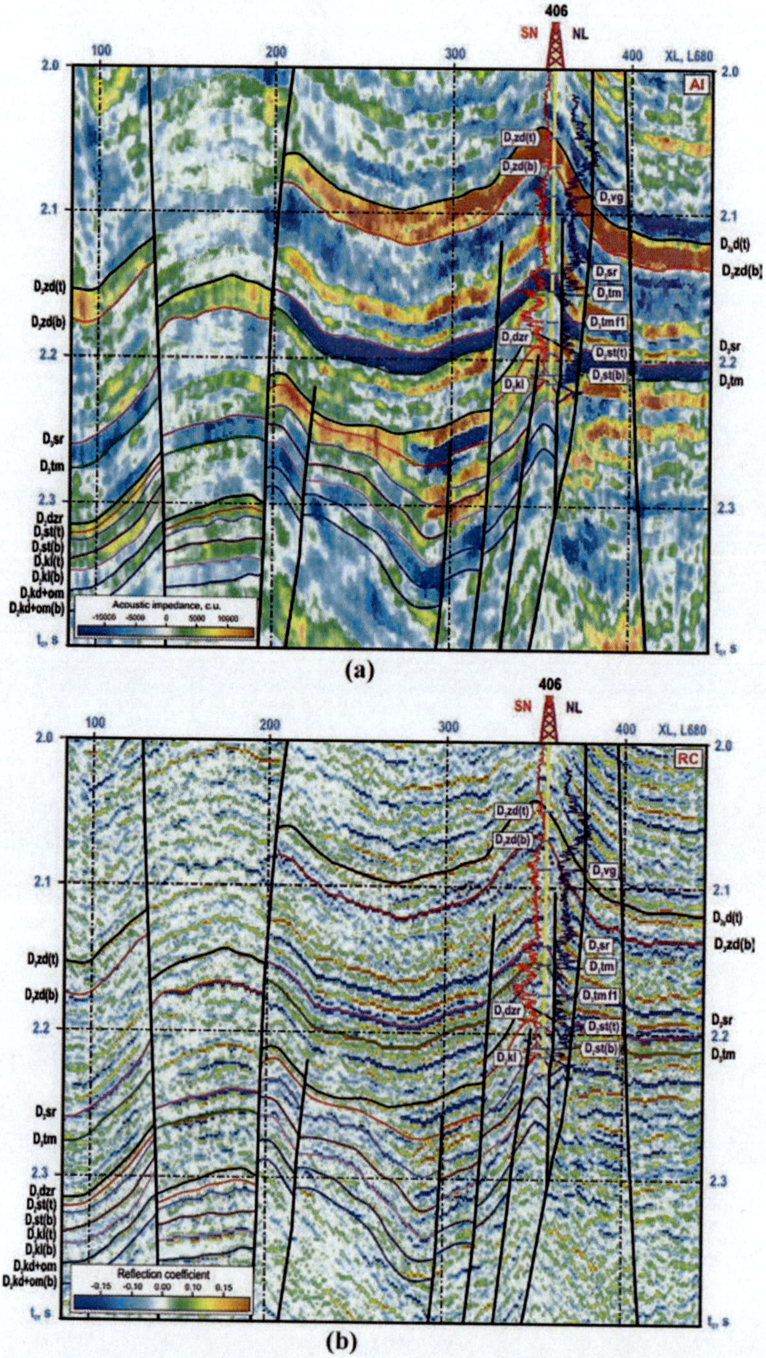


**Fig. 5.63** Geological horizons tying along the 406 well to the reflectors of the section of effective acoustic impedance (AI) (**a**) and reflection coefficients (RC) (**b**) along the inline 680 of 3D-cube

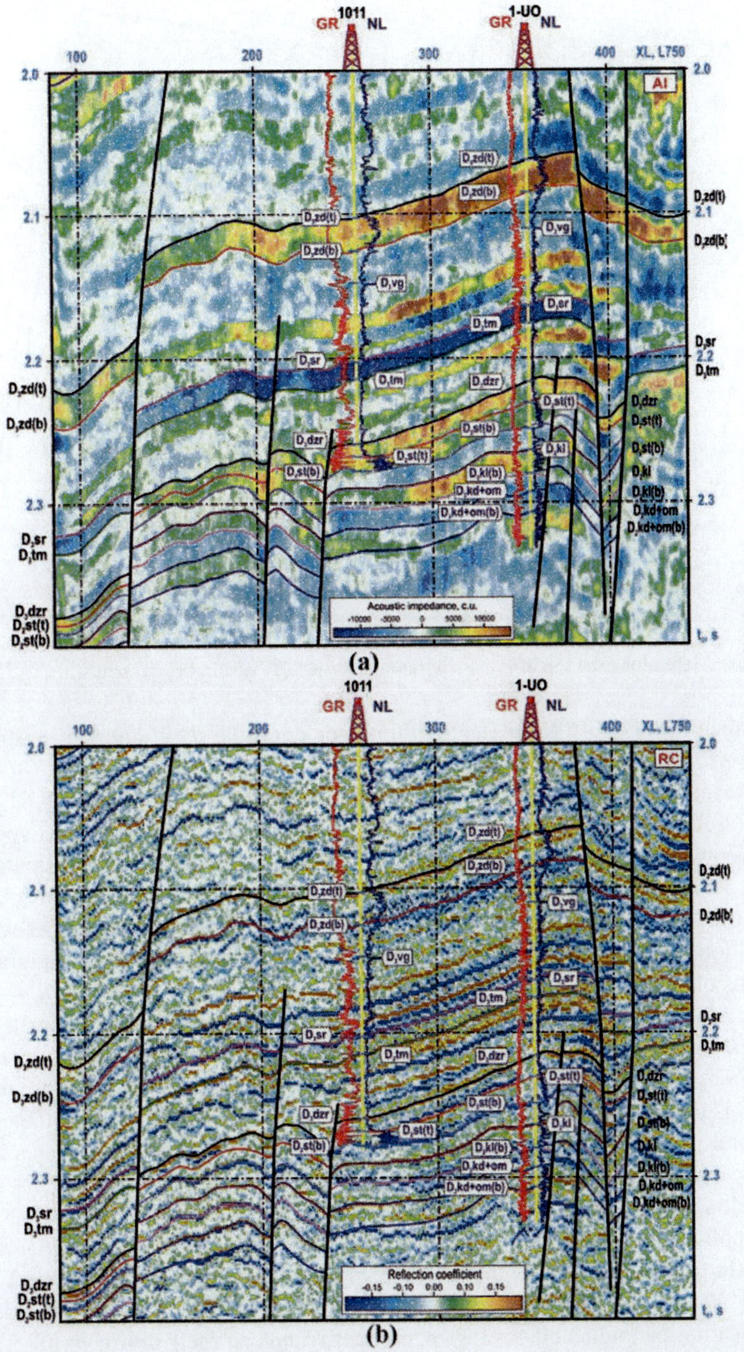


**Fig. 5.64** Geological horizons tying along wells 1-UO and 1011 to the reflectors of the acoustic impedance (AI) (**a**) and reflection coefficients (RC) (**b**) along the inline 750 of 3D-cube

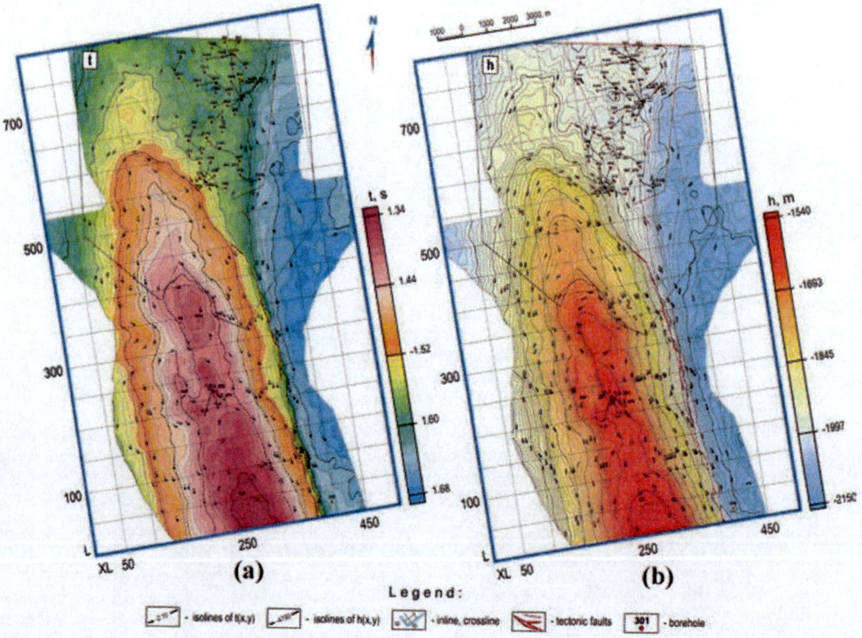


**Fig. 5.65** Maps along the top of $P_1a + s$ horizon deposits: (**a**) isochrone; (**b**) structure

determination of the thicknesses of thin-layer deposits, as is known, introduces strong errors into the results.

As an example of obtaining high-resolution seismic data in Fig. 5.61, 5.62, 5.63 and 5.64 for a number of inline(s) (560, 610, 680 and 750) sections of AI and RC are given with a detailed reference of geological horizons to reflecting boundaries. Along the lines of composite profiles-traverses passing through wells 2 CK, 2 UO, 406, 1011 and 1 UO, fragments of AI and RC sections are given together with a correlation scheme of target reflectors, made according to lithological-stratigraphic columns of the studied wells.

The AI and RC sections clearly show the junction of the Kolvin megaswell with the western edge of the Khoreyver depression through a system of large (high-amplitude) faults and feathering tectonic faults. This feature of the structure of the section is confidently traced from profile to profile and, in general, over the entire study area. As will be shown below (on structural maps), in plan basis, such a complex tectonic conjunction of large structures of the first order occurs along a system of echelon-shaped local structures.

Complications of a significant number of tectonic disturbances, pinching out of layers (in the absence of the phenomenon of interference of seismic records) here on the AI and RC sections are confidently manifested with high resolution. Identification of faults and tracing of productive layers, their thin layering in the lateral direction were largely detailed. The ability to study the internal geological structure of correlating formations (including reservoirs and other prospects) due to the vertical and horizontal resolution of the data has improved in many ways compared to this kind of opportunity based on the use of CDP time sections.

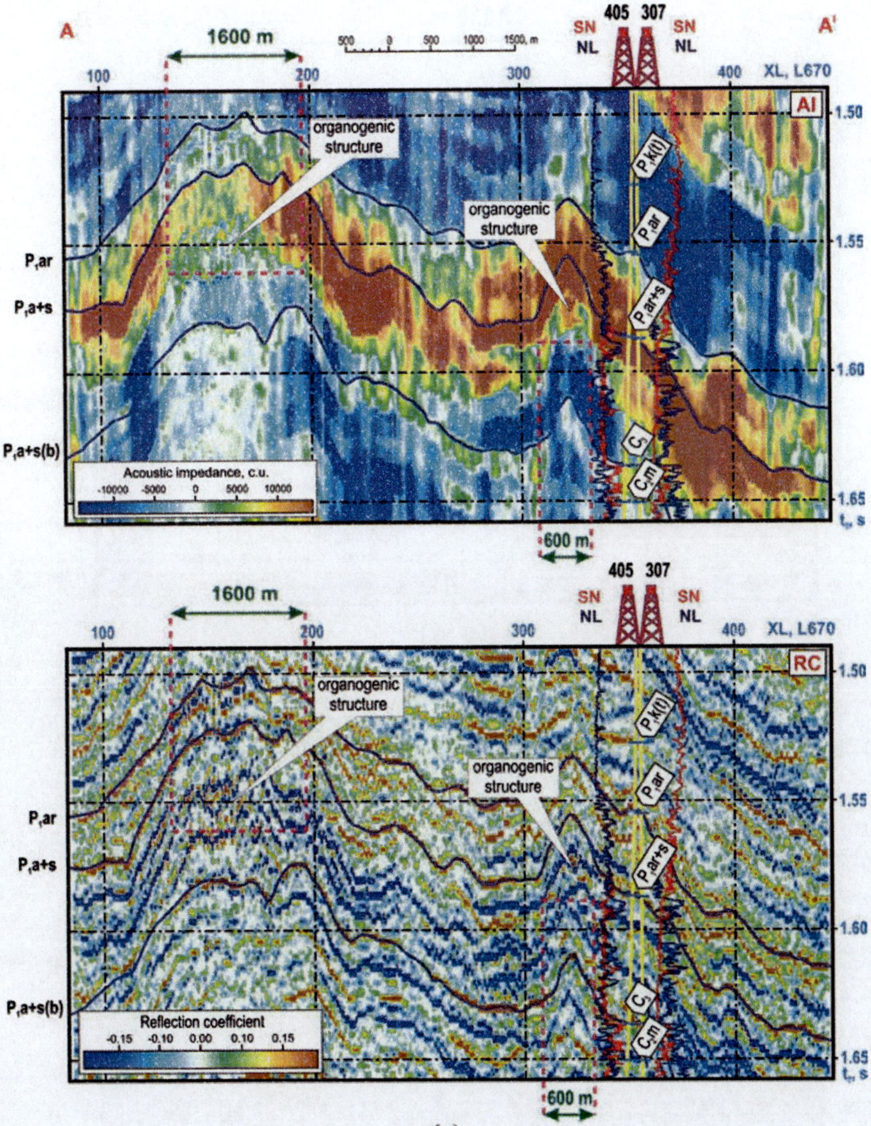


(a)

**Fig. 5.66** Display of the Asselian-Sakmarian organogenic structures on the sections of acoustic impedance (AI) and reflection coefficient (RC) along: (**a**) the inline L670; (**b**) the crossline XL170

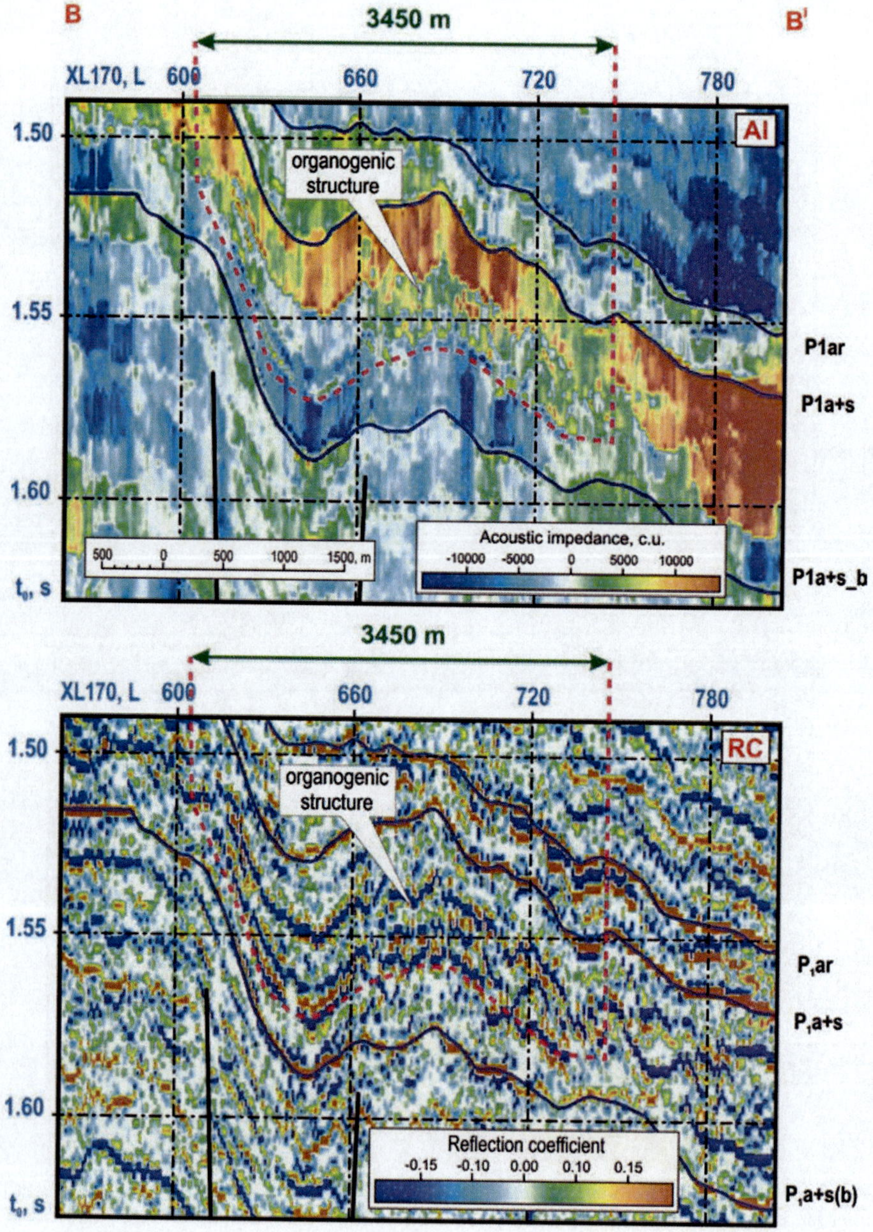


(b)

**Fig. 5.66** (continued)

### 5.4.3 *Possibilities of Studying Perspective Organogenic Structures in the Asselian-Sakmarian Stage ($P_1a$+s) Sediments*

As outlined above, the 3D CDP seismic data using a special graph in the stage of preliminary processing of the original seismic data allowed the authors to form 3D seismic data cubes – 3D WF, AI, and RC 3D cubes and correlate over these data (directly by inline(s) and crossline(s) AI and RC) reflecting boundaries of 14 target horizons (listed above). These stratigraphically identified reflectors (picked up horizons, faults, layers, contacts, surfaces, objects, etc.) along the tops and bottoms of the corresponding productive and prospective formations were the initial data for the construction of the corresponding structural maps and their derivatives (maps of isochrones, isopachs, thicknesses, structural maps and boundary surfaces for the formations with in-situ information content of the predictive assessment of the main geological indicators: lithology, porosity, oil saturation, etc.). The input information for the construction of isochrone maps was the observed times ($t_0$) of the reflectors stratified with the boundaries of the geological section, as well as the polygons of tectonic faults (for deposits of those horizons where these faults are established).

This subsection provides an example of the use of HRS data for a detailed study of the organogenic structure in the deposits of the Asselian-Sakmarian stage ($P_1a + s$) of the Lower series ($P_1$) of the Permian, namely, the upper part of these deposits.

The Lower series of $P_1$ is known to be composed of limestones and dolomites of the undivided Asselian-Sakmarian stage and upper-lying clay-carbonate rocks of the Artinskian stage, mudstones, siltstones with thin interlayers of limestones and marls of the Kungurian that. The reflecting horizon $I_{1\text{-}2}$ ($P_{1\text{-}2}$) is confined to the base of the Upper Permian and the top of the Lower Permian deposits. The top of the pure carbonates of the Lower Permian, the top of the Asselian-Sakmarian deposits, is confined to the reflector Ia + s ($P_1a + s$). The Asselian-Sakmarian formations themselves are represented by gray limestones, slightly clayey, biomorpho-detrital, algal interlayers, irregularly recrystallized, not stratified, in some interlayers cavernous and porous. In general, the thickness of the Assel-Sakmarian deposits varies by area from 70 to 130 m.

The use of the HRS-Geo technology software tools and the interpretation system GeoGraphix Discovery™ R2004.1, v.8.5 (Landmark Graphics Corporation) made it possible to perform the necessary structural constructions using sections specified in the inline(s) and crossline(s) directions of 3D cubes AI and RC. The software and technological tools of the GeoGraphix Discovery™ R2004.1, v.8.5 system, as you know, are a full-featured interpretation system for estimating and managing topographic, seismic and well data on the Microsoft Windows® platform. The products of this system are the packages: SeisVision™ (*an integrated system for in-depth seismic 2D/3D interpretation*), IsoMap™ (*an application for interpretive mapping – all kinds of surface modeling, gridding and isolines in the GESXplorer™*

*product suite*) and GeoAtlas™ (*a comprehensive system for graphical editing and visualization of maps and sections*).

Preparation, loading and operation of these system modules ultimately provide:

1. Correlation of target reflectors and assessment of the convergence of time lines $t_0(x)$ in vertical sections of the section at the sections of the intersection of profiles according to inline(s) and crossline(s) of 3D cubes AI and RC.
2. Machine correlation of target reflectors in the range of studied deposits of the sedimentary cover at the stage of simultaneous construction and refinement of an adequate geological model of the real environment.
3. Construction of isochron maps, velocity model of the environment and structural maps along the target reflecting boundaries of productive and prospective sediment complexes using the IsoMap™ and GeoAtlas™ packages of the system.
4. Identification and localization of tectonic faults along the CDP sections for characteristic inline and crossline 3D profiles of the AI and RC cube using the identified inhomogeneous zones on the coherence maps of the wave field amplitude cube along the paleoisochronous surfaces of the target horizons.
5. Mapping the distribution of AI values for target horizons and relatively thin interlayers of productive complexes with elements of lithofacies variability, relative changes and direction of sedimentation processes.

On the maps obtained in this way: the isochrone map (Fig. 5.65a) and the structural map (Fig. 5.65b), over the top of the deposits of the Asselian-Sakmarian stage $P_1a + s$, a brachianticline structure is confidently traced with the highest amplitude in the area of the southern boundary of the 3D cube (on the site of the Kostyuk-Osokin swell). In the northwest of the Yu-Osh prospecting area, the Kostyuk-Osokin swell gradually flattens out, turning into a complexly built pericline of the brachian-anticlinal structure. This complex periclinal is confined to a relatively small zone, which has an approximate limit for inline profiles ≈ 650–690 and for crossline profiles ≈ 140–210 of AI and RC 3D seismic cubes (Fig. 5.65).

The obtained general patterns in displaying the distributions of AI and RC values in the planes of the AI and RC sections over the entire analyzed set of traverse profiles for these inline(s) and crossline(s) allowed the authors to identify and identify on this basis organogenic structures (one of them is relatively large, and the other is rather small in its linear dimensions).

An example of the display of these formations on the sections of AI and RC is shown in Fig. 5.66a, b. These figures, in particular, demonstrate fragments of sections of AI and RC along one of the sublatitudinal profiles with inline 670 (Fig. 5.66a) and along the submeridional profile with crossline 170 (Fig. 5.66b). The internal structure of the identified organogenic structures in the presented sections of the sections manifests itself in the form of a more chaotic recording of the values of AI and RC (this is especially clearly seen in the distribution of the RC values) in relation to the rest of the studied formation. In addition, there is a characteristic increase in the thickness of the Asselian-Sakmarian strata analyzed within each of the structures under consideration. In addition, there is also a slope character of the structure of the section, primarily due to the presence of a complex periclinal of the Kostyuk-Osokin swell.

In order to determine the approximate dimensions of the area occupied by these organogenic structures in the study area, maps of the coherence of the distribution of acoustic impedance values for two time levels were constructed using given horizontal sections (slices): (a) at time $t_0$ = 1560 ms (to determine the approximate contour of a larger organogenic structure) (Fig. 5.67a); (b) at time $t_0$ = 1588 ms (to determine the approximate contour of a relatively small organogenic structure) (Fig. 5.67b). At the same time, the selected approximate contours (boundaries) of the distribution of such structures are characterized by a darker palette in relation to the surrounding areas of the real geological environment.

As noted in one of the papers [6], the interest in the oil and gas content of the deposits forming the slopes of the marginal reef and bank formations is based on the idea of the existence of pre-reef clastic talus aprons. On the whole, studies of sections of slope facies in various regions of the Timan-Pechora province have shown they are composed to varying degrees of argillaceous limestones with rare interlayers of clay-free carbonate rocks and layers enriched in carbonate fragments, sometimes terrigenous material. The rocks are usually dense, usually without pore-type reservoirs. Oil shows and small deposits with off-balance reserves are discovered in carbonate and clay-carbonate fractured low-capacity reservoirs of extremely limited distribution.

For a better study of the formation patterns of organogenic structures, their classifications, lithofacies zonality under various conditions of sediment accumulation in areas adjacent to structures, the characteristics of frame ones consolidating the basal parts of reefs – cementers, etc., one could recommend a large number of publications on reef issues. The authors managed to get acquainted with the main elements of this topic in the works [4, 6, 12, 15].

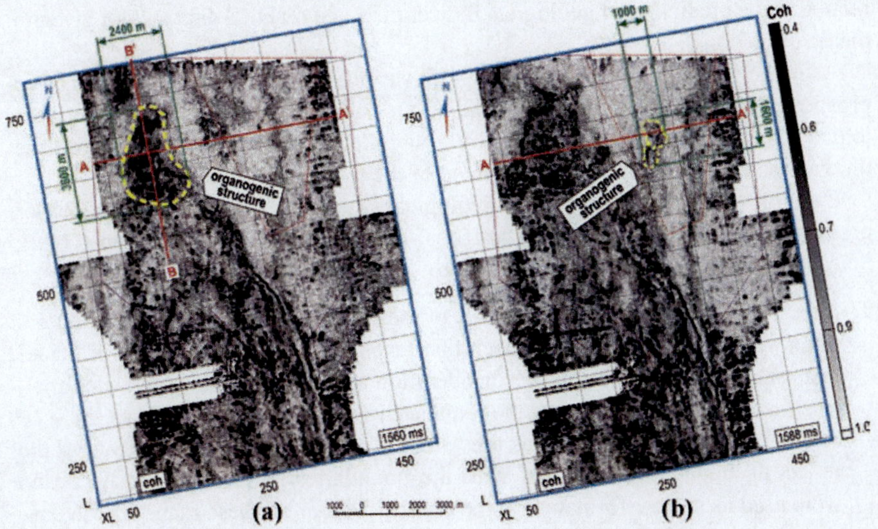


**Fig. 5.67** Display of the Asselian-Sakmarian organogenic structures on the acoustic impedance coherence map at times: (**a**) 1560 ms; (**b**) 1588 ms.

In order to somehow indicate the conditions for the formation of the organogenic structures discussed above as some unique natural objects of the geological past, the authors considered it possible to give an example of reef objects as a comparative material, which is directly related to real geological outcrops (Fig. 5.68). This example, with elements of the detailed internal structure of biohermal objects, but for their different type of formation, was obtained from a completely different geological region. But the most interesting thing that follows from this example is the presented real picture of the internal structure of these objects (Fig. 5.68) can be largely (with a good degree of probability) correlated with the internal structure of the models of organogenic structures in the AI and RC sections (Fig. 5.66). In turn, such a positive correlation of real geological material (representation of structures in the picture of in-situ outcrops of geological sections) and the results of converting seismic data into environmental parameters in the form of sections of the acoustic impedance AI and reflection coefficients RC seems to be very impressive.

At the same time, we would like to note: such a positive result of the comparative material, as shown in Figs. 5.66 and 5.68, only when using the structure of the seismic wave field (WF) in this kind of case it is practically impossible to provide.

### 5.4.4 *Seismic Facies Distribution Features in the Sediments of the Zadon Horizon ($D_3zd$)*

In this subsection, as well as in the remaining final sections 5.4.5, 5.4.6, and 5.4.7, a small volume of materials on pre-processing and partial interpretation of high-resolution seismic (HRS) data is presented in order to demonstrate the effectiveness and benefits of using HRS-Geo technology tools (at the same time, the features related to the prediction of geological indicators based on HRS data are not given in these subsections).

As in the examples discussed above with various well and seismic data for other prospecting areas, here at the site of the Yu-Osh and Osokin fields, the seismogeological conditions are characterized by a significant variety of structure and the uniqueness of the geological development of the territory.

Examples with the features of the geological structure of sediments on models of productive and promising layers (horizons) are considered below: Zadon ($D_3zd$), Starooskol ($D_2st$), Kolvin ($D_2kl$) and Kedrov-Omrin ($D_2kd + om$).

1. First, we note, based on the materials of these deposits, all the necessary operations were performed to build isochrone maps, structural maps, maps of the AI distribution in each layer under consideration and perform a seismic facies analysis (using AI and RC data) in the sequence in which it is shown above (in Sect. 5.4.3). It should be added to this the values of the observed times $t_0(x,y,z)$ of the reflecting boundaries stratified with the boundaries of the geological section were used as input information in the construction of all these maps. These constructions also included polygons of tectonic faults for those horizons where

(a)

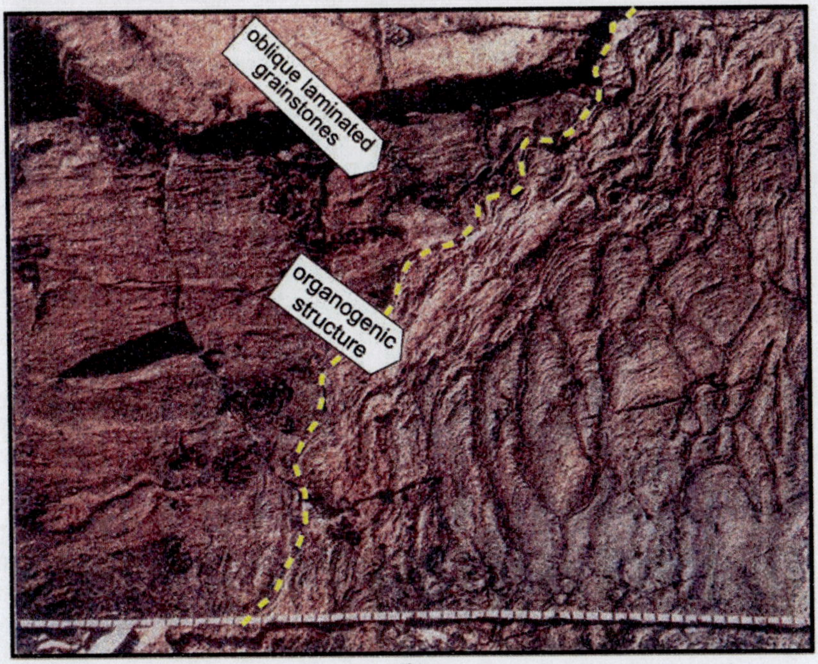


(b)

**Fig. 5.68** Fragments of bioherm structures in geological outcrops: (**a**) horizontal boundary of a large bioherm built by columnar stromatolites; (**b**) flank of a small (3 m) bioherm built by stromatolites. The sediments adjacent to the bioherms are cross-bedded grainstones. Taltelley Formation, Early Proterozoic, left arm of Great Slave Lake, NW Canada (Carbonate Depositional Environments. AAPG Memoir, 33, 1983) these faults are confidently traced. Identifing and tracing of tectonic faults was carried out on the basis of tracing fault surfaces (fault planes) simultaneously with the correlation of target reflectors.

As will be shown below, almost the entire area under study is complicated by a significant number of tectonic faults of northwestern and submeridional strike. The initiation of these faults is directly related to the processes of tectonic activation in the pre-Lower Frasnian geological period. At the same time, most of the faults are attenuated in the lower Famennian deposits. In general, the 3D seismic survey made it possible to trace both high- and low-amplitude tectonic faults with a high degree of confidence. This was obtained due to the high resolution and informative density of the profile network and the use of sections and cubes of AI and RC with a bin size of 25 × 25 m. The complex tectonic structure of the area is recorded both in the area of the drilled cluster wells (to a large extent in the north-eastern and southern parts of the research area) and in areas not covered by drilling (mainly in the western and central parts of the area).

Note also the velocity fields were calculated for each horizon. To construct velocity fields, average velocities are usually calculated from well depths and isochron times. Then the statistical dependences of average velocities on time are determined. To find the dependence $V_{av} = f(t_0)$, we used the depths of several horizons, for which a single law of velocity change can be adopted, i.e. the velocity field in depth was approximated by a piecewise linear dependence. The calculated dependencies $h_{abs} = f(t_0)$ for different coverage intervals of the studied section (primarily for deposits of productive formations $D_3$zd, $D_2$st, $D_2$kl and $D_2$kd + om), together with the values $t_0(x,y)$ of the isochron maps of the target reflectors, were directly involved in the construction of structural maps.

2. We note one more feature that is important to take into account and implement in the detailed geological interpretation of seismic data obtained in complex seismogeological conditions. This implies the presence of a sufficiently strong lithofacies variability in the most significant productive deposits. This feature of the section structure predetermines, as is known, the need to formulate and solve the problem of reconstructing the sedimentation environment on the basis of seismic facies analysis. Well logging data and traditional wave seismic data are usually used as input data in such a problem.

To solve this problem, the authors first used the method of cluster analysis (pattern recognition) to directly determine the zones of facies replacement using high-resolution seismic data and digital models of various geological indicators in the form of corresponding distributions of three-dimensional grids. As a tool for solving such a problem of facies zoning, the authors used the cluster data analysis program implemented in the STATISTICA software package (APP). A distinctive feature of this method is the user specifies the desired number of finite clusters and a parameter that reflects the convergence of the iterative process. In this case, the k-means algorithm forms a set of k clusters located at the greatest possible distance from each other. These metrics are defined both in one-dimensional and in multidimensional space. The most direct way to calculate distances between objects in a multi-dimensional space is to calculate the Euclidean distance, which is the most popular metric in cluster analysis. The need to have a hypothesis about the most probable number of clusters (i.e., the most optimal choice of the number k) is based on the results of previous studies and theoretical considerations. The same problem, in principle, can be solved by conducting a hierarchical cluster analysis with a randomly selected sample of observations.

The use of the cluster analysis method allowed the authors to obtain positive results on the facies zoning of the reseach area (in particular, in the exploration areas of Western Siberia) using the HRS data (AI and RC). The results obtained at the same time, together with the established predictive geological indicators, allowed the authors to outline the optimal places for further prospecting, to outline promising areas for opening up oil-promising objects of a structural-lithological type with deep holes in a complex of studies directly related to the development of oil and gas resources.

3. As a whole, it should be borne in mind that standardless classification allows, on a formal basis, to really reduce the description of a sufficiently large set of objects to a small number of objects groups related to each other by some regular property, for example, by "similarity" to each other, to some one "typical" object or for a given number of classes that are homogeneous in the sense of a given class metric. On this basis of pattern recognition, as is known, it is possible to more effectively jointly interpret a significant number of geological and geophysical parameters by combining them.

In this situation, the authors used one of the most popular in our country and abroad software package Stratimagic by Paradigm (*Malyarova T.N., Ptetsov S.N., Ivanova N.A., 2004–2007*). In this complex, the most effective is the classification of traces according to their shape using three main methods: self-organizing neural network technology (NNT), hybrid (Hybrid) and hierarchical with a randomly selected sample of observations. The obtaning results are a sequence of color coded model facies traces reflecting the heterogeneity of the seismic data. Based on these data, maps of seismic facies are formed.

Those who have tested the specified complex for classifying sets of traces according to their shape using neural network technology (NNT), as the authors of this development note, allows you to determine, analyze and interpret the detailed characteristics of reservoirs are not detected by other methods. First of all, this is the ability of NNT to identify stable dependencies in the conditions of encountered interference, reflect all the subtleties of the seismic wave field and, on this basis, improve the accuracy of the results when evaluating reservoirs and geometrizing their boundaries (*Malyarova T.N., 2007*). At the same time, to control the process of automatic classification and highlight individual zones on the seismic facies map, a correlation curve is used (for areas with different slopes on this curve), as well as a correlation map and an assessment of the diversity of signal forms in a particular zone under study. A significant decrease in the quality of the correlation can serve as a sign of the boundary of different (additional) sedimentation zones. The authors also note after the formation of a seismic facies map, the selected seismic facies classes can be identified with well logs, such as SP, and, taking into account the facies determined by electrometric models (*for example, according to the method of*

*Muromtsev V.S., 1984 or Akhiyarov A. V.* et al., *2009*), draw a conclusion about the conditions of sedimentation under which the productive interval was formed.

4. In order to determine various seismic facies zones and their attributes under conditions of general subsidence of the northeastern margin of the Russian platform and the accumulation of significant masses of marine sediments of different facies, the authors of this work used maps (digital models) of effective acoustic impedance (AI), reflectivities (RC) as input information. For the genetic interpretation of the classification results, the allocation of zones of complex shape, confined to various areas of facies replacement of reservoirs, the authors planned to use directly well logging data – electrometric SP models for the total number of wells drilled in the study area, as well as analogues of these electrometric models – fragments of curves AI, directly confined to the thickness of each of the productive layers. However, this last part of the work (genetic interpretation of the calculated seismic facies clustering results) turned out to be very significant in its scope (due to the presence of a large number of wells) and, moreover, it went beyond the framework of the work program, which the authors did not carry out, adhering to this works of only the initial stage of research
5. Next, we dwell in somewhat more detail on the materials of the Zadon horizon $D_3zd$. The deposits of this horizon include the upper productive carbonate unit, which contains the most thick carbonate layers. At the Yu-Osh field, an oil deposit is associated with it. The thickness of the pack is 34–50 m. The top of the Zadon sediments is confidently identified according to well logging data by a decrease in the level of GR, an increase in NL, and an increase in RL.

The lower part of the section of the Zadon horizon is composed of gray, dark gray to black, bituminous limestones, with interlayers of marls and mudstones (borehole 72 Vozey). Further along the section, the deposits are represented by uneven interbedding of mudstones, marls, and nodular clayey limestones. Limestones are light gray, spherical-clotty, organogenic-detrital, granular, unevenly porous-cavernous or gray, dark gray, slightly bituminous, clayey. Sediments in the upper part of the section are expressed by uneven alternation of limestones, marls and mudstones. Those are overlain by argillaceous limestones of the Yelets horizon, which are a cover for productive deposits of the Zadon horizon.

The study area is located in the east of the Kolvin megaswell, which is separated by a fault in the interval of pre-Upper Devonian deposits and flexure along the overlying strata from the Khoreyver depression located to the east.

The results of structural constructions along the top of the deposits of the Zadon horizon are presented in the form of an isochrone map (Fig. 5.69a) and a structural map (Fig. 5.69b). Figure 5.70a–c shows maps of the area distribution of the average values of acoustic impedance (AI) between the reflecting boundaries $D_3zd(t)$ and $D_3zd(b)$ and the distribution of seismic facies obtained as a result of classifying trace sections according to their shape using neural network technology (NNT) according to the values of AI and RC, respectively. It can be seen from the presented materials both the results of structural constructions (Fig. 5.69), and the acoustic parameters of the environment and seismofacies models (Fig. 5.70) for the deposits of the Zadon horizon are distinguished by high detail and accuracy of constructions.

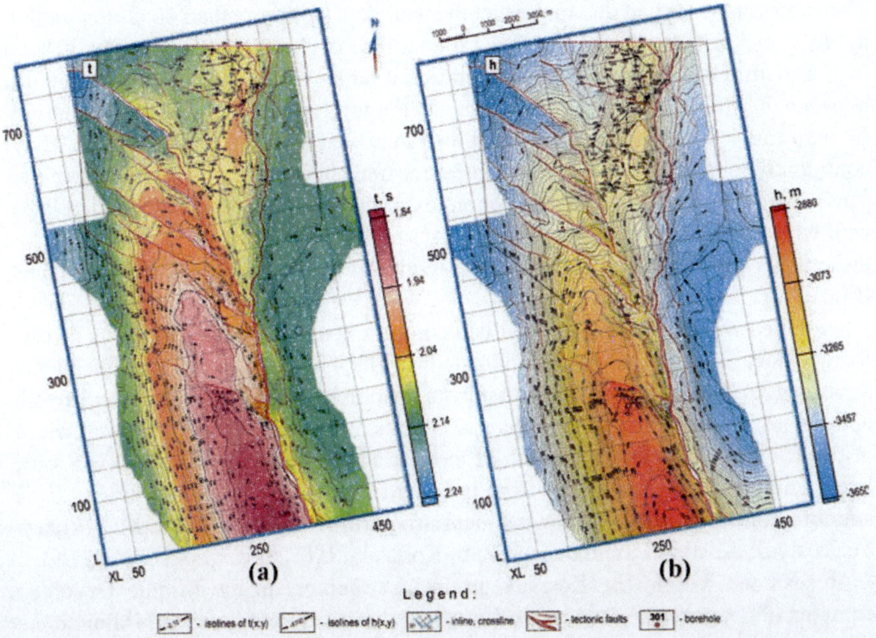


**Fig. 5.69** Maps along the top of the $D_3$zd horizon deposits: (**a**) isochrone; (**b**) structure

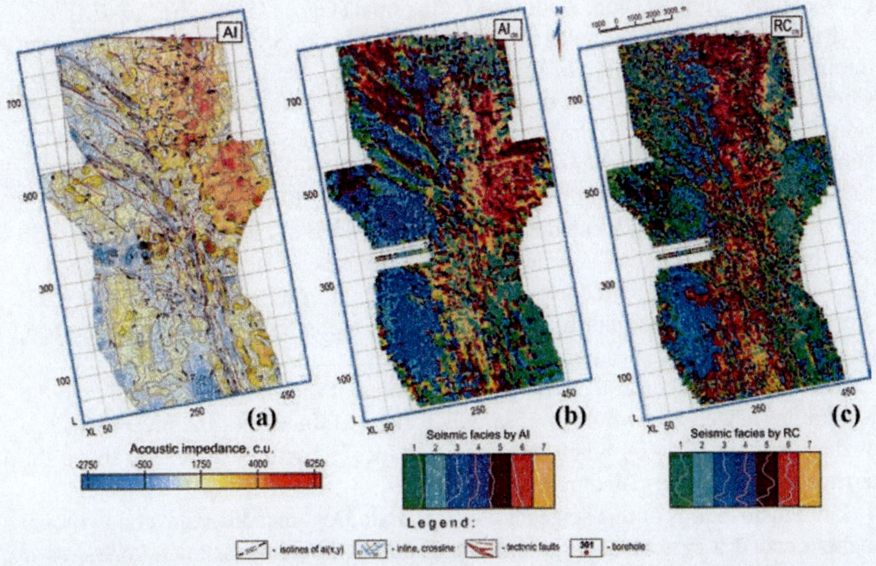


**Fig. 5.70** Map of mean AI between reflectors D3zd(t) and $D_3$zd(b) (**a**). Maps of the distribution of seismic facies according to the clustering of the AI (**b**) and RC (**c**) fields in the deposits of the $D_3$zd horizon

The latter can be argued due to the fact the well data (for more than 50 claster wells) are fully correlated with the identified top surface of the target horizon $D_3$zd(t), as well as with related data: an almost meridional fault (V-Kolvin pre-Frasnian a fault between the structures of the first order) and the framework of numerous feathering (in plan and in the spatial orientation of the fault fissure planes) tectonic faults. They confidently separate the echelon structures located chiefly in the northern half of the Yu-Osh area (Fig. 5.69). In turn, the complex framework of tectonic faults correlates well with local anomalies in the distributions of AI and seismic facies (SF) values – these faults are most often traced in all extended zones of AI, SF and RC anomalies (Fig. 5.70).

As noted above, the Vozey swell is a complex one and actually consists of two merged local, echelon-shaped conjugated swells – the Yu-Vozey swell and the S-Vozey (Kostyuk) swell attached to it from the northwest. The S-Vozey local swell, better known as the Kostyuk group of structures, is a relatively narrow (total width of the swell is 12–14 km; the width of local structures does not exceed 2–2.5 km), linear anticlinal zone extending from the south-southeast to north-northwest. This anticline zone includes the following local structures (from south to north): Z-Vozey, Yu-Kostyuk, Kostyuk, Srednekostyuk, S- Kostyuk (*JSC "Severgeofizika", 2009*).

Researchers within the Kostyuk group of structures in the Middle Devonian complex of deposits divide tectonic faults into three different types according to the time of their occurrence and the nature of faults: (1) Middle Devonian normal faults associated with the formation of grabens under extensional conditions, (2) Late Devonian faults associated with microinversion movements of short-term compression episodes, (3) reverse faults associated with the inversion uplift of the study area ($C_1$v-T) under compression conditions (collisional stage) (*JSC "CGE", 2013*).

The Middle Devonian faults are faults oriented from SSE to the NWW and cut through the entire thickness of the Middle Devonian deposits (except for its top), where faults are interrupted at the boundary with the surface of the Predjerian erosion. Faults formed within the S-Vozeisky paleograben in the area under consideration, five local paleograbens, an echelon-like shifted relative to each other in latitude. From south to north, respectively: first, the Kostyuk local paleograben (boreholes 71 B., 50 K., located near the southern edge of the contour of the search area, but beyond it); then – a series of three paleograbens successively displaced to the west – Yu-Osokin, Srednekostyuk (boreholes 1, 2 Osokin) and S-Kostyuk-I (borehole 68 B.). To the north of the latter, there is the S-Kostyuk-II paleograben, far displaced to the east (borehole 55 B) (Fig. 5.69).

The modern structural-tectonic plan of the area of the Osokin structure (and the Kolvin megaswell as a whole) was largely formed at the end of the inversion stage of development ($C_1$v-T), when the swell and local structures appeared above the territory of the former paleograben.

Fault dislocations of this stage and, in part, Late Devonian microinversion movements created a system of large and small horsts, half horsts, microhorsts, as if "squeezed out" from paleograbens, throughout the Kolvin megaswell. As a result, the Middle Devonian deposits of the Kostyuk group of structures were also broken through by reverse fault dislocations – in terms of a new "network" of reverse faults, the old "network" of fault was superimposed.

As a result, the Yu-Osh and Osokin structures were formed in the area under consideration, complicated by smaller steps and microblocks. Fault faults are predominantly oriented submeridionally (from SSE to NNW); there are sublatitudinal dislocations in areas of the echelon-shaped junction of paleograbens. All horsts in plan are well correlated with local paleograbens. The Osokin structure has the smallest dimensions in terms of area.

Accounting the results of drilling wells on the Osokin structure and data on the complex structure of the Middle Devonian deposits obtained from seismic data and expert evaluation of the structural-tectonic model for the top of the $D_3$zd, $D_2$st, $D_2$kl formations, the geological structure in the area of this structure was refined (*JSC "Severgeofizika", 2009, JSC "CGE", 2013*). On this basis, the correlation of the main reflectors and the model of tectonic faults in the area of the Osokin field were refined.

With regard to the distributions of AI values and seismic facies (SF) (Fig. 5.70), one can notice these results, despite the lack of their genetic interpretation, show a high level of agreement with the features that were noted above in connection with many refinements of the structural-tectonic construction and historical development of sediments of the Zadon ($D_3$zd) horizon. Almost all the detailed features of the tectonic faults structure in terms of the time of their occurrence, traceability in a real environment and various lengths find their manifestation in the deposits under consideration. Such pecularities of the structure on the presented seismic facies are confidently differentiated here according to the shapes of the reference curves AI and RC (Fig. 5.70b, c). These maps show the influence of tectonic processes, various formation and occurrence conditions, structural features of the formation structure, as well as expressed zoning in the distribution of high and low values (AI and RC), reflecting changes in the composition and properties of the studied formation.

The formation deposits, as noted above, are dense limestones, often bituminous, consistent in thickness and area. Layer $D_3$zd has been opended up by all wells drilled in the field. It is represented by deposits of a carbonate-argillaceous strata of the fill, which characterize the transition from the conditions of a deep-water basin to a shallow-water shelf. The $D_3$zd deposit is known to be confined to a fractured zone formed as a result of the action of tensile forces in the area of large tectonic faults: from the east, a meridional tectonic fault with an amplitude of 20 to 80–100 m and a group of parallel tectonic faults of northwestern strike with an amplitude of 30–40 m, from the south – limiting the Komavis structure. All of the above indicates a limited deposit type, which is associated with fracture zones within a carbonate formation with a complex reservoir type containing syngenetic bitumen. It is known one oil deposit has been identified in the reservoir, its oil content has been established based on the results of testing wells, for which anhydrous oil flows have been obtained.

### 5.4.5 Seismic Facies Distribution Features in the Sediments of the Starooskol Suprahorizon ($D_2st$)

The Starooskol suprahorizon ($D_2st$) of the Givetian Stage ($D_2zv$) is composed of sandstones interbedded with siltstones and mudstones. Sandstones are brownish-gray, oil-saturated, fine-grained, quartz (wells 1 Osokin, 68 Vozey, 2 Yu-Osh). The siltstones are light gray with a greenish tint, quartz, thinly bedded, dense, with thin interlayers of clayey-carbonaceous material. Mudstones are gray, irregularly clayey, platy, with coalified plant remains (borehole 68 Vozey). The deposits of the Starooskol suprahorizon (IV pack) are associated with a commercial oil deposit. The thickness of these deposits is on average about 40 m. It should be noted the top of the Starooskol suprahorizon is well identified according to well logging data using a set of methods: increased readings of NL (NLT), low readings of GR, increased SN, and increased resistivity. At the top of the reservoir, there is erosion typical of weathering crust deposits, which stratigraphically corresponds to the regional pre-Frasnian erosion.

From geological sources (JSC "Severgeofizika", 2009) it is known in the western direction (towards well 64 Vozey) the thicknesses of the Middle Devonian deposits decrease due to the distance from the axial part of the semigraben, a similar decrease in thicknesses, but only due to erosion in the Starooskol part of the section also occurs to the east, towards the Bolshezemel paleoarch. Reduced thicknesses of the Starooskol sediments are observed to the south (Kostyuk paleograben), where there was a local pre-Dzhyer paleouplift, which in turn was the zone of maximum manifestation of the pre-Dzhyer erosion. The Starooskol deposits are completely wedged out to the south and east, in close proximity to the Khoreyver depression. Located in the arch of the well 1 Osokinsk, opened up the top of the Starooskol sediments at an absolute depth of −3267 m. The local structure within the horst is outlined by the isohypse −3280 m, the amplitude of the structure is 25 m, the dimensions are 1.2 × 1.0 km. The magnitude of displacement along reverse faults at the boundaries of the Srednekostyuk block is: in the north – 30 m, in the south – up to 15 m, and in the west – 30 m.

The results of structural constructions along the top of the deposits of the Starooskol suprahorizon $D_2st$ are presented in the form of an isochrone map (Fig. 5.71a) and a structural map (Fig. 5.71b). Distribution maps of average values of acoustic impedance (AI) between reflectors $D_2st(t)$ and $D_2st(b)$ and distribution of seismic facies obtained as a result of trace classifying according to their shape using neural network technology (NNT) according to AI and RC values are given in Fig. 5.72a–c, respectively.

As for the deposits of the Zadon horizon, for the deposits of the Starooskol suprahorizon, in the features of structural constructions (Fig. 5.71), a largely inherited nature of their manifestations is observed. These features, along with the acoustic parameters of the environment and models of seismofacies (Fig. 5.72) for the

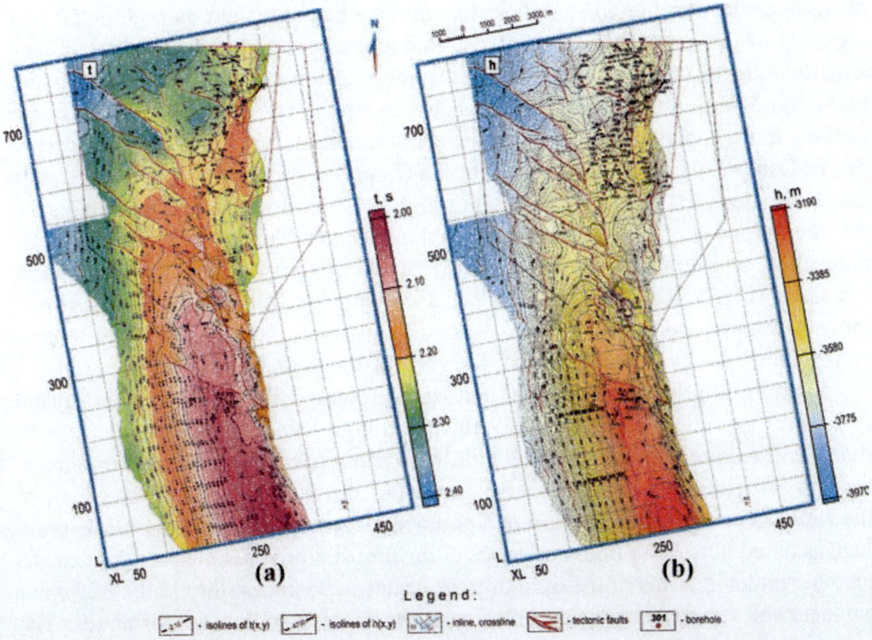


**Fig. 5.71** Maps along the top of the $D_2st$ horizon deposits: (**a**) isochrone; (**b**) structure

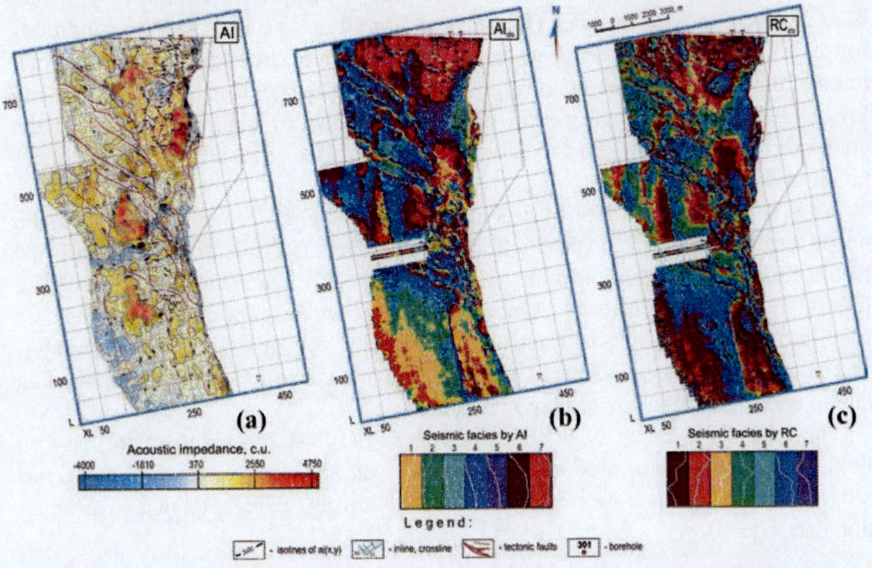


**Fig. 5.72** Map of mean AI between reflectors $D_2st(t)$ and $D_2st(b)$ (**a**). Maps of the distribution of seismic facies according to the clustering of the AI (**b**) and RC (**c**) fields in the deposits of the $D_2st$ horizon

deposits of the considered suprahorizon, are also distinguished by high detail and accuracy of determination. A confident analogy with the previous structural constructions can be traced in almost the entire set of tectonic faults and the meridional break (the V-Kolvin pre-Frasnian fault between structures of the first order). These tectonic faults separate the echelon structures located mainly in the northern part of the Yu-Osh prospecting area (Fig. 5.71). The complex framework of tectonic faults also correlates well with local anomalies in the distributions of AI values and seismic facies (SF) (Fig. 5.72) – they are traced in all extended zones of AI, SF and RC anomalies, while highlighting the main features of the structural forms of local structures (Fig. 5.71). All reverse fault dislocations, Late Devonian microinversion movements, systems of large and small horsts, half horsts, microhorsts and paleograbens in the deposits of the Starookol reservoir also manifest themselves.

In terms of the distributions of AI and seismic facies (SF) values using the AI and RC values as initial data (Fig. 5.72), there is a high level of correspondence with those features are manifested in relation with many peculiarities of the structural-tectonic structure and historical development of the Starooskol deposits ($D_2$st) of the suprahorizon, despite the absence of a genetic interpretation of the SF anomalies, as noted above. As in the previous case, almost all the detailed features of the tectonic faults in terms of the time of their occurrence, traceability in the real environment and various lengths are reflected in the deposits of the $D_2$st formation. The classification of seismic facies according to the shape of the reference models AI and RC (Fig. 5.72b, c) shows these results generally complement each other and, in addition, they are well aligned with the features of complex structures (Fig. 5.71a, b) and identified AI anomalies (Fig. 5.72a). On all these maps, various conditions for the formation of target objects and their occurrence, structural and tectonic features of the Starooskol reservoir, highlighted zoning in the distribution of high and low values (AI and RC) are confidently traced, which, in addition to changing the composition and properties of the studied reservoir, also reflect the nature sedimentation of deposits.

As is known, one oil deposit was identified in the $D_2$st formation. It is the main one in terms of reserves and production at the field and is associated with an uplifted tectonic block. The reservoir rocks, as noted above, are represented by sandstones, less often by silty sandstones. According to the core descriptions, sandstones are gray, brownish-gray, fine-medium-grained, quartz and quartz-micaceous, clayey and calcareous to varying degrees, well sorted, with horizontal and inclined bedding due to uneven distribution of clay material.

The oil-bearing capacity of the reservoir in the $D_2$st formation of the Yu-Osh field, which is the main production oilfield object, has been established by almost all wells drilled in it based on the results of testing, development and production data.

### 5.4.6 Seismic Facies Distribution Features in the Sediments of the Kolvin Horizon ($D_2kl$)

The Kolvin horizon ($D_2kl$) is represented by the eponymous formation, the Kolvin suite of the Eifelian Stage ($D_2ef$) of the Middle Devonian. The upper part of the suite is composed of siltstones, mudstones with limestone interbeds. This part of the Kolvin horizon was called the clay-carbonate pack: in addition to mudstones, it contains many limestones that form layers up to several meters thick and contain remains of marine fauna. The thickness of the unit is up to 15–30 m. The base of the suite is composed of interbedded quartz sandstones, siltstones and mudstones (1 Osokin, 64 Vozey). Mudstones are dark gray, unevenly silty, platy, thinly bedded. Limestones are dark gray, finely crystalline, clayey. Sandstones are light gray, fine-grained, quartz, dense, brownish-gray in interbeds, unevenly (porous layers up to 1.3 m thick are found) saturated with oil (well 68 Vozey). The deposits of the Kolvin horizon (according to the local nomenclature, it stands out as a productive unit III) is associated with a commercial oil deposit. The thickness of the Kolvin horizon is 52–69 m.

Deposits, represented by interbedding of quartz sandstones and siltstones, are identified according to well logging according to increased NL readings, increased resistivities, and lower GR readings. At the bottom of the productive horizon, there is a clayey-siltstone pack of the Omrin horizon, which is also well identified by well logging.

The results of structural constructions along the top of the deposits of the Kolvin horizon ($D_2kl$) are presented in the form of an isochrone map (Fig. 5.73a) and a structural map (Fig. 5.73b). Figure 5.74a–c shows maps of the area distribution of the average values of acoustic impedance (AI) between the reflecting boundaries $D_2kl(t)$ and $D_2kl(b)$ and the distribution of seismic facies, obtained as a result of trace classifying according to their shape using neural network technology (NNT) according to the values of AI and RC, respectively. From these materials, in particular, it can be seen the results of structural constructions (Fig. 5.73), acoustic parameters of the environment and seimofacies models (Fig. 5.74) for the deposits of the Kolvin horizon are distinguished by high detail and accuracy of constructions. Well data (for more than 50 cluster wells) are linked to the top of the target horizon $D_2kl(t)$, and all faults and numerous feathering tectonic faults organically fit into the structural and tectonic features of the this horizon top structure and, in particular, the presented area distributions acoustic parameters (AI) and seismic facies (SF). Tectonic faults confidently separate the echelon-shaped structures located primarily in the northern half of the Yu-Osh prospecting area. Almost all Late Devonian microinversion movements, which formed a system of large and small horsts, half horsts, and microhorsts throughout the Kolvin megaswell, form a modern structural-tectonic plan in the prospecting area, in which the main place is given to the Yu-Osh and Osokin structures. As noted above, faults are predominantly oriented submeridionally (from SSE to NNW), and sublatitudinal dislocations are also present in areas of the echelon-shaped junction of paleograbens. All local paleograbenhorsts are

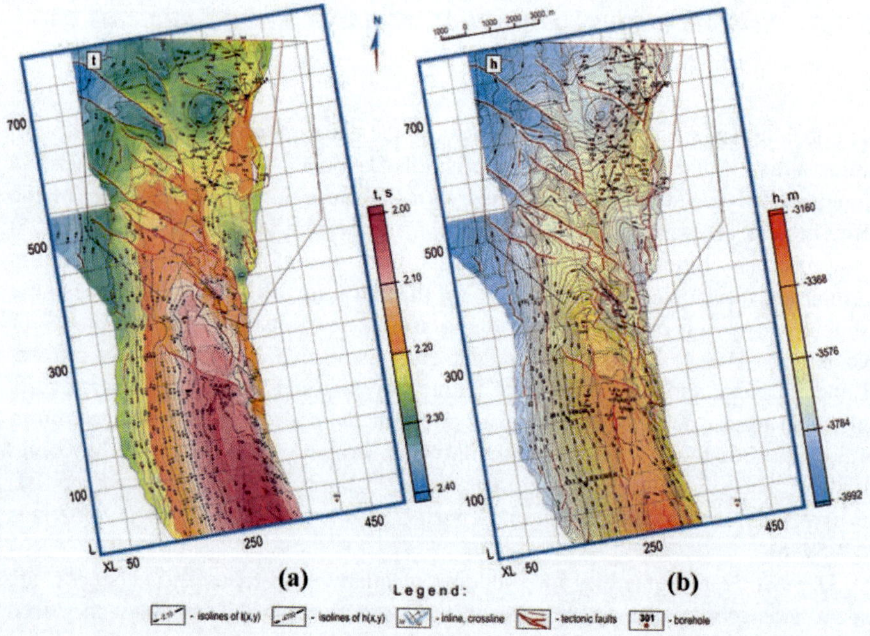


**Fig. 5.73** Maps on the top of the deposits of the $D_2kl$ horizon: (**a**) isochrone; (**b**) structure

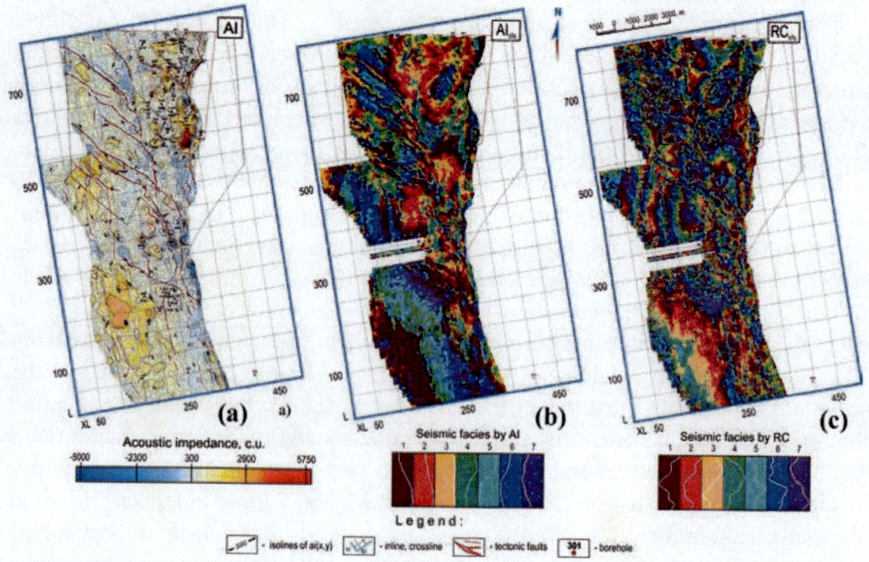


**Fig. 5.74** Map of mean AI between reflectors $D_2kl(t)$ and $D_2kl(b)$ (**a**). Maps of the distribution of seismic facies according to the clustering of the AI (**b**) and RC (**c**) fields in the deposits of the $D_2kl$ horizon

well correlated with pageohorsts in plan view. The Osokin structure is also characterized by smaller sizes in area (Fig. 5.73).

Once again, it should be noted the northern end of the Vozey swell, complicated by a system of tectonic faults such as faults, forms unevenly high blocks along the Devonian deposits. The faults do not connect with each other and represent four echelon-shaped branches of the northwest strike. The northernmost fault in this system is the most extended and high-amplitude (amplitude reaches 160–180 m) in all deposits in the section of the Middle-Upper Devonian. The greatest amplitude of disturbances is observed in the central part of each branch (towards the end and beginning of the branches, the amplitude decays). In each block, within the northern end of the Vozey swell, additional disturbances of various lengths and amplitudes are traced. Some of them are feathering in relation to major faults. In the southern part of the blocks formed by these faults, the S-Kostyuk-II structure was identified. To the north, the Komavis and S-Komavis near-fault structures were traced. They are identified by all traced reflecting horizons in the deposits of the Upper-Middle Devonian.

In general, the Yu-Osh, Komavis, S-Komavis, S-Kostyuk-II structures form a complex Osh-Komavis zone, which has the shape of a wedge directed to the south and limited by tectonic faults from the east and a fall of the structure to the west. The separation of the two internal structural zones of this wedge, eastern and western, runs along a trough-shaped trough in the central and northern parts of the area and a low-amplitude inflection in the southern part of the area.

The dimensions of the Komavis structure are 3.7 × 0.8 km, the amplitude is 40 m. The dimensions of the S-Komavis structure are 2.2 × 0.5 km, the amplitude is 20 m. The dimensions of the Yu-Osh structure along reflector $III_{1\text{-}2}$ ($D_{1\text{-}2}$) are 4.6 × 1.2 km with an amplitude of up to 120 m. The dimensions of the entire Osh-Komavis block are estimated as follows: according to reflector $III_{2\text{-}3}$ ($D_{2\text{-}3}$) within closed isohypse −3620 m is 8.5 × 3.5 km with an amplitude of about 120 m; along the reflecting horizon III kl ($D_2$kl) within the closed isohypse −3720 m are 10.5 × 3.5 km with an amplitude of about 160 m.

In the distribution of acoustic impedance (AI) values, as well as in the distributions of seismic facies (SF) in the prospecting area (Fig. 5.74), in general, a relatively high level of compliance with the features characterized the structural and tectonic structure and the development of Kolvin deposits ($D_2$kl) horizon. These features of the structure on the presented seismic facies obtained using the shapes of the reference curves of AI and RC are clearly manifested here. The maps clearly show the influence of tectonic processes, various conditions of formation and occurrence, structural features of the structure of the target reservoir, as well as expressed zoning in the distribution of high and low values (AI and RC), reflecting changes in the lithological composition and, obviously, reservoir properties of the deposits of the Kolvin horizon. At the same time, it should be noted the distributions of AI, RC and SF mutually complement each other and detail the internal structure of the deposits under consideration, despite the lack of a detailed genetic interpretation of the manifested local anomalous zones (AI, RC and SF) (Fig. 5.74).

Two oil reservoirs are identified in the reservoir, separated from each other by a tectonic fault, identified according to the reinterpretation of 3D seismic. The reservoirs are represented by sandstones and siltstones. According to the core descriptions, sandstones are gray, brownish-gray, fine-medium-grained, quartz, strong, medium-platy, sometimes there are clay fragments or gravel-sized quartz grains oriented along the bedding. Reservoir opended up by 11 wells.

At the base of the productive Kolvin horizon, there is a clayey-siltstone pack of the Omrin horizon, which is well identified according to well logging data.

### 5.4.7 *Seismic Facies Distribution Features in the Sediments of the Kedrov-Omrin Horizon ($D_2kd+om$)*

The Kedrov and Omrin horizons ($D_2$kd + om) are represented in the volume of the Vozeyshor formation. In the top of the Vozeishor formation, deposits are represented by interlayer of mudstones, unevenly calcareous and silty, with limestones and siltstones (borehole 64 Vozey). In the middle part, the deposits are composed of light gray quartz sandstones, silty and argillaceous interbeds, with siltstone interbeds (hole 68 Vozey). In the well 1 Osokin horizon sediments are dominated by mudstones with rare sandstone interbeds. Mudstones with interlayers of siltstones are noted at the base of the deposits.

The thickness of the deposits varies from 26 m (407 ST2) to 61 m (well 100 Vozey).

The Omrin horizon $D_2$om conformably overlies the Kedrov horizon. In the local nomenclature, it was called the clay-aleurite pack. The lower part of the horizon is composed of alternating gray microgranular thin limestones with dark gray mudstones. There are remains of marine fauna. The upper part of the horizon is represented by a thin interbedding of mudstones, siltstones, and fine-grained sandstones. The total thickness of the horizon is 18–26 m.

The Kedrov horizon $D_2$kd unconformably overlies the Lower Devonian deposits. Its section from below begins with a thin (3–10 m) clayey-siltstone pack. The stratum lies above, represented by rhythmic interbedding of sandstones, siltstones and mudstones, which, according to the local stratigraphic scale, is called "pack I + II". The rocks are gray, light gray and brownish gray. Sandstones of coarse, medium and fine grained quartz with clayey cement, form layers up to 6 m thick, containing porous varieties. Quartz siltstones compose layers up to 1.2 m thick. Mudstones are thin-layer dark gray with plant detritus, forming layers and interlayers up to 0.6 m thick. The thickness of the interval I + II is 40–55 m. The total thickness of the Kedrov horizon is 43–65 m.

The structural plans of the Lower Paleozoic and Lower-Middle Devonian deposits coincide. With minor changes, which are in particular related to the amplitude of the folds, they persist up to the top of the Devonian deposits. Only the depression to the west of the Yu-Osh structure along the Lower Devonian and Kedrov-Omrin sediments ($D_2kd$ + om) has an amplitude of more than 80 m, with its reduction along the top of the Starooskol horizon to 15 m, and by the Timan time it is completely compensated. In the modern structure of the Permian-Carboniferous deposits, the Yu-Osh structure disappears (Fig. 5.65) and graben-like depressions are divided into Devonian strata by the S-Kostyuk-Komavis and Kostyuk-Osokin swells. Slightly north of the latitude of the northern end of the Osokin structure, the flexure bifurcates, located on the eastern slope of the Kostyuk-Osokin swell. One branch of the flexure remains tied to the eastern slope of the swell, while the other continues with a deviation to the east, the border of the Kolvin swell with the Khoreyver depression.

The results of structural constructions along the top of the deposits of the Kedrov-Omrin horizon ($D_2kd$ + om) are presented in the form of an isochrone map (Fig. 5.75a) and a structural map (Fig. 5.75b). Distribution maps of average values of acoustic impedance (AI) between reflectors $D_2kd$ + om(t) and $D_2kd$ + om(b) and distribution of seismic facies obtained as a result of classifying trace sections by their shape using neural network technology (NNT) by AI and RC values are shown in Fig. 5.76a–c, respectively.

As for sediments of overlying horizons, for deposits of the Kedrov-Omrin horizon, in the features of structural constructions (Fig. 5.75), a largely inherited nature of their manifestations is observed. These features, along with the acoustic parameters of the environment and seismofacies models (Fig. 5.76), for the deposits of the formation under consideration are distinguished by high detail and accuracy of

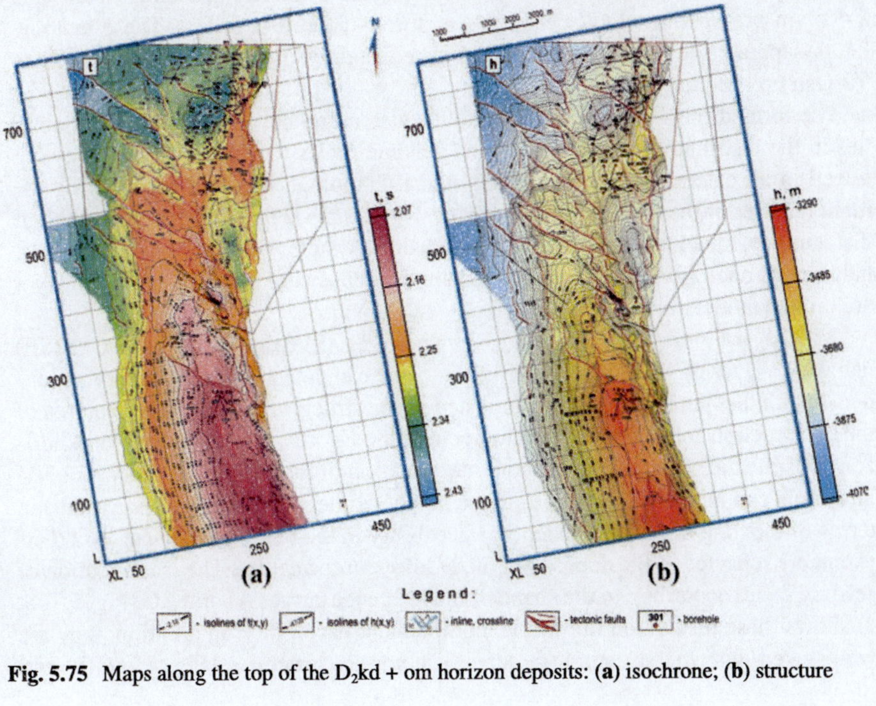


**Fig. 5.75** Maps along the top of the $D_2kd$ + om horizon deposits: (**a**) isochrone; (**b**) structure

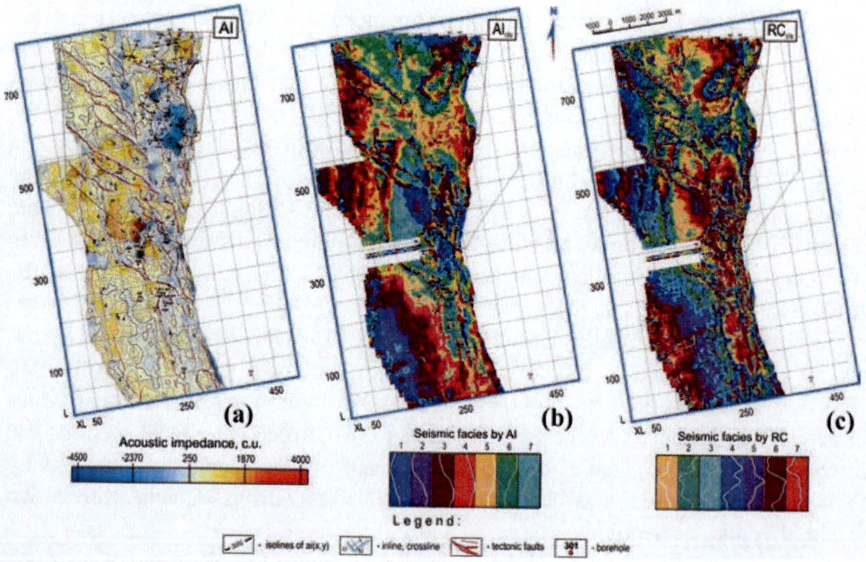


**Fig. 5.76** Map of mean AI between reflectors $D_2$kd + om(t) and $D_2$kd + om(b) (**a**). Maps of the distribution of seismic facies according to the clustering of the AI (**b**) and RC (**c**) fields in the deposits of the $D_2$kd + om horizon

determination. There is also a confident analogy with the previous structural constructions for almost the entire set of tectonic faults and the meridional fault (the V-Kolvin pre-Frasnian fault between structures of the first order). These tectonic faults separate the echelon structures located mainly in the northern part of the Yu-Osh prospecting area (Fig. 5.75).

The formed framework of tectonic faults also correlates well with local anomalies in the distributions of AI values and seismic facies (SF) (Fig. 5.76) – they are traced in all extended zones of AI, SF and RC anomalies, while highlighting the main features of the structural forms of local structures (Fig. 5.75). All reverse fault dislocations, Late Devonian microinversion slips, systems of large and small horsts, half horsts, microhorsts and paleograbens in the deposits of the Kedrov-Omrin layer are quite confident here.

With regard to the distribution of AI values (Fig. 5.76a) and seismic facies (SF) using the values of AI and RC as initial data, one can note a high level of correspondence with the features that are manifested in the structural and tectonic features of the construction and historical development of the Kedrov-Omrin ($D_2$kd + om) horizon. Despite the absence of a genetic interpretation of the SF anomalies (which has already been noted above), almost all the details of the structure of tectonic faults in terms of the time of their occurrence, traceability in the real environment, and their extent are reflected in the deposits of the $D_2$kd + om formation. The classification of seismic facies according to the shape of the reference curves AI and RC (Fig. 5.76b, c) shows these results, on the whole, complement each other. In addition, they are well linked with the structural features of complex structures (Fig. 5.75a, b) and formed AI anomalies (Fig. 5.76a). On all these maps, various target objects, the conditions of their occurrence, and specific structural and tectonic features of the structure of the Kedrov-Omrin layer are confidently traced. They show zoning in the distribution of high and low values of AI and RC, as well as information on changes in the composition and properties of the studied reservoir and on the conditions and nature of sedimentation of the deposits of the Kedrov-Omrin horizon.

With regard to the oil content of the target deposits, we note once again two deposits of oil were found in the Yu-Osh field in the Middle Devonian deposits: in the sandstones of the Starooskol suprahorizon (IV pack) and sandstones of the Kolvin horizon (III pack) of the Middle Devonian. Type of deposits – flat sheet, arched, tectonically and lithologically screened.

The Eifelian Stage ($D_2ef$) includes the Kedrov-Omrin and Kolvin horizons. The Kedrov-Omrin deposits are presented from bottom to top: a clayey pack 2–10 m thick, a sandy pack I-II 26–49 m thick, and a clayey-silty pack separating it from the Kolvin deposits. In the Kedrov-Omrin deposits I + II, a sandstone unit was opened up in wells: 1, 2, 407 Yu.O.; 2 S.K.; and 72 Vozey. This pack was tested only in wells 1 Yu.O. and 2 S.K. The fluid inflow was: 0.075 $m^3$ of drilling mud and 0.01 $m^3$ of oil per hour (well 1 Yu.O) and 0.11 $m^3$ of slightly gassed mud per 1 h (well 2 C.K). In wells 2, 407 Yu.O. Kedrov-Omrin sandstones were tested together with Kolvin depositions.

On the eroded surface of the Lower Devonian sediments, the Middle Devonian deposits ($D_2$) occur transgressively, which are the main oilfield object in the explore area and are represented in the Givetian and Eifelian stages.

Thus, the commercial oil and gas potential at the Yu-Osh and Osokin fields was revealed within the Middle Devonian-Lower Frasnian oil and gas complex in the Starooskol suprahorizon $D_2st$ (Givetian stage) and the Kolvin suite $D_2kl$ (Eifelian stage) of the Middle Devonian, as well as within the Middle Frasnian-Tournaisian oil and gas complex – in the deposits of the Zadon horizon $D_3zd$ (Famenian stage of the Upper Devonian). Middle Devonian oil deposits, as is well known, have also been discovered in the Usinsk, Kharyagin, and Vozey fields of the Kharyaga-Usinsk oilfield. The oil and gas potential of the Upper Devonian carbonates of various genesis has also been established at the Kharyagin and Vozey fields.

## 5.5 Summary

1. Using the software tools of the HRS-Geo technology, studies were carried out on three prospecting sites (areas) of the Timan-Pechora Province (TPP), located in the northeastern outskirts of the Russian platform. The seismic materials obtained here are characterized by a number of significant features in which the influence of valuable tectonic activity is manifested, which, in turn, determines the specifics of the geological structure of the study area, the formation of hydrocarbon traps and oil and gas potential. Based on the results of the studies per-

formed, the geological structure of local uplifts was refined for all productive and promising objects of study (horizons) of the Paleozoic. The problem of predicting the composition and properties of target deposits using HRS data is partially solved.

2. At the first of these sites – Savinobor, located in the southern part of the Mechayu-Pashnin swell and the Tebuk step of the Izhma-Pechora saddle of the TPP, according to the results of the formed 3D cubes of effective acoustic impedance (AI) and reflection coefficients (RC) – models of the real environment, the geological structure of local uplifts has been refined. The results of reinterpretation of structural constructions are analyzed taking into account the drilling data of two wells. Deposits of productive and promising horizons were studied based on the classification of seismic facies using the shapes and values of reference curves – clusters of 3D AI and RC cubes in the deposits of horizons – II-III (C-D), $D_3$zd, B-3 k + B-2p and $D_2$st(k). Oil-perspective objects of structural, tectonically and lithologically screened types were identified in the deposits of the upper part of the Famennian horizon (II-III (C-D)), the upper part of the Zadon horizon ($D_3$zd) and in the reservoir beds of the B-3к + B-2п units of the Dzhyer-Yaran horizon deposits ($D_3$dzr + jr).
3. In the junction zone of the northeastern part of the Kolvin megaswell and the northwestern part of the Khoreyver depression, on the territory of the Kamin prospecting area, a detailed assessment is given of the most important geological indicators of the real environment, primarily the material composition and reservoir properties in the deposits of the Artinskian stage ($P_1$ar), the carbonate strata of the Sirachoi ($D_3$srč'), Domanikoid ($D_3$domanik), Dzhyer ($D_3$dzr) horizons, Middle ($D_2$) and Lower ($D_1$) Devonian, and Ludlovian ($S_2$ld) Upper Silurian. Oil-promising objects of a structural type were identified in the reservoirs of these horizons. Oil prospects here are associated with the identification of structural-lithological and combined oil traps formed in marine and coastal-marine, shallow-marine, deltaic and lagoonal sedimentation conditions.
4. In the central part of the Kolvin megaswell of the Pechora-Kolvin aulacogene and in the junction zone of the large Vozeya and Kharyagin swells within the Yu-Osh and Osokin fields, using 3D cubes and 2D sections of the AI and RC, the bedding character of the deposits along the tops and bottoms of the target Artinskian ($P_1$ar), Asselian-Sakmarian ($P_1$as), Dzhyer ($D_3$dzr), Zadon ($D_3$zd) horizons, Starooskol ($D_2$st) suprahorizon, Kolvin ($D_2$kl) and Kedrov-Omrin ($D_2$kd + om) horizons is studied. Based on these horizons, using the AI and RC data, detailed structural constructions were made and the conditions for the formation of sediments were determined by classifying and predicting various groups of facies in order to identify zones of facies replacement, zones of heterogeneities, which are one of the main directions for studying the detailed internal structure and lithofacies variability (zonality) of these productive and prospective deposits.

## References


1. Botneva, T. A., & Shulova, N. S. (1985). Geneticheskiye osobennosti neftey Timano-Pechorskoy provintsii (Genetic features of the oils of the Timan-Pechora province). *Geologiya nefti i gaza (Geology of Oil and Gas), 10*, 45–50.
2. Viskunova, K. G. (1997). *Seysmostratigrafiya i perspektivy neftegazonosnosti ordoviksko-nizhnedevonskogo karbonatnogo kompleksa Timano-Pechorskoy provintsii* (Seismic stratigraphy and oil and gas potential of the Ordovician-Lower Devonian carbonate complex of the Timan-Pechora province). Avtoref. disser. na soisk. uch. stepeni kand. geol.-miner. nauk. RAN, VNIGRI. Sankt-Peterburg.
3. Yermakova, Shch. L., & Lanskikh, N. A. (2006). Kharakternyye cherty stroyeniya i perspektivy neftegazonosnosti nizhnepermskikh karbonatnykh otlozheniy yuzhnoy chasti Denisovskoy vpadiny i Kolvinskogo megavala (Characteristic features of the structure and prospects of oil and gas potential of the Lower Permian carbonate deposits of the southern part of the Denisov depression and the Kolvin megaswell). Geologiya i neftegazonosnost' Timano-Pechorskogo basseyna. Sb-k nauch.trudov – Kirov: KOGUP "Kirovskaya oblastnaya tipografiya" (Collection: Kirov: KOGUP "Kirov Regional Printing House") (pp. 115–118).
4. Zhemchugova, V.A. (2014). *Rezervuarnaya sedimentologiya karbonatnykh otlozheniy* (Reservoir sedimentology of carbonate deposits). M., OOO "EAGE Geomodel" (LLC "EAGE Geomodel") (p. 232).
5. Leshchenko, V. Y. (1973). Rol' gidrodinamicheskikh faktorov pri formirovanii i razrushenii zalezhey nefti i gaza v Timano-Pechorskoy provintsii (The role of hydrodynamic factors in the formation and destruction of oil and gas deposits in the Timan-Pechora province). *Geologiya nefti i gaza (Geology of Oil and Gas), 12*, 31–35.
6. Menner, V. V. (1989). *Litologicheskiye kriterii neftegazonosnosti paleozoyskikh tolshch severo-vostoka Russkoy platform* (Lithological criteria for oil and gas potential of the Paleozoic strata of the northeast of the Russian Platform) (p. 133). M., Nauka.
7. Novikov, A. A., Kapustin, I. N., & Kiryukhin, L. G. (1985). Evolyutsiya struktur vostochnykh rayonov Vostochno-Yevropeyskoy platformy i ikh neftegazonosnost' (Evolution of the structures of the eastern regions of the East European platform and their oil and gas potential). M., VNIIOENG. Obzornaya informatsiya. Seriya: Neftegazovaya geologiya i geofizika (Overview information. Series: Oil and gas geology and geophysics) (p. 70).
8. Peremyshlennikova, N. V., Petrenko, Y. L., & Surina, A. I. (2006). Nekotoryye zakonomernosti razmeshcheniya kollektorov v sredneordoviksko-nizhnedevonskom i domanikovo-turneyskom karbonatnykh rezervuarakh na severe Khoreyverskoy vpadiny i svyazannyye s nimi perspektivy neftenosnosti (Some patterns of reservoir location in the Middle Ordovician-Lower Devonian and Domanik-Tournaisian carbonate reservoirs in the north of the Khoreyver depression and the oil potential associated with them). Geologiya i neftegazonosnost' Timano-Pechorskogo basseyna. Sb-k nauch.trudov – Kirov: KOGUP "Kirovskaya oblastnaya tipografiya" (Collection: Kirov: KOGUP "Kirov Regional Printing House") (pp. 36–44).
9. Petrenko, Ye. L. (2006). Perspektivy neftegazonosnosti rifosoderzhashchikh otlozheniy domanikovo-turneyskogo kompleksa yuzhnoy chasti Denisovskoy vpadiny (Prospects for the oil and gas potential of reef-bearing deposits of the Domanik-Tournaisian complex in the southern part of the Denisov depression). Geologiya i neftegazonosnost' Timano-Pechorskogo basseyna. Sb-k nauch.trudov – Kirov: KOGUP "Kirovskaya oblastnaya tipografiya" (Collection: Kirov: KOGUP "Kirov Regional Printing House") (pp. 96–103).
10. Romanov, I. Ye., & Teplov, Ye. L. (1988). Rol' regional'nykh razlomov v formirovanii zalezhey UV v Timano-Pechorskoy provintsii (The role of regional faults in the formation of hydrocarbon deposits in the Timan-Pechora province). Geologiya nefti i gaza, № 12 (Geology of Oil and Gas, No. 12) (pp. 30–33).
11. Safonov, A. S., Kondrat'yeva, O. O., & Fedotova O. V. (2011). Poisk neantiklinal'nykh lovushek uglevodorodov metodami seysmorazvedki (Search for non-anticlinal hydrocarbon traps using seismic survey methods). M., Nauchnyy mir (Scientific world) (p. 512).

12. Teplov, Ye. L., Kostygova, P. K., & Larionova, Z. V. (2011). *Prirodnyye rezervuary neftegazonosnykh kompleksov Timano-Pechorskoy provintsii* (Natural reservoirs of oil and gas complexes of the Timan-Pechora province). GUP RK TP NITS, SPb: OOO "Renome" (GUP RK TP NIC, St. Petersburg: Renome LLC) (p. 286).
13. Trofimov, V. L., Khaziyev, F. F., & Trofimova, A. V. (2019). *Tekhnologiya vysokorazreshayushchey seysmiki pri poiskakh i razvedke neftegazovykh zalezhey* (High-resolution seismic technology in prospecting and exploration of oil and gas deposits). M., EAGE Geomodel LLC (p. 343).
14. Trofimov, V. L., Khaziev, F. F., & Trofimova, A. V. (2022). *Oil and gas reservoir prospecting and exploration. High-resolution seismic (HRS) techniques and technology*. Springer. https://doi.org/10.1007/978-3-030-84389-2
15. Fortunatova, N. K., Shvets-Teneta-Guriy, A. G., & Gumarov, R. K.. (2000). *Sedimentologicheskoye modelirovaniye karbonatnykh osadochnykh kompleksov* (Sedimentological modeling of carbonate sedimentary complexes) (p. 249). NIA-Priroda, M.
16. Khain, V. Ye. (2001). *Tektonika kontinentov i okeanov* (Tectonics of continents and oceans). (p. 606). M., Nauchnyy mir (Scientific world).

# Conclusion

1. Further improvement of the numerical algorithm for solving the inverse dynamic seismic problem (IDSP) predetermined the consideration of the searching process for the acoustic impedance (AI) and reflection coefficient (RC) models using objective functions (vector of objective functions), in which various types of residuals between real and model data are iteratively computed. Moreover, such discrepancies are estimated from the energy (amplitudes) and the shape of the seismic signal. It is for this kind of residuals that in Chap. 1 we analyzed the information content and sensitivity of the amplitude (energy) and phase spectra of the seismic record. Such an analysis is performed in the frequency domain in which the contribution of the medium model parameters perturbations (velocity of elastic waves, density of the medium, and wave attenuation) affects the result of seismic inversion, which is proportional to the frequency of the waves. Previously, the authors did not consider these issues in deep. They are studied here (in Chap. 1) for the first time. The results of modeling and numerical experiments have shown the main contribution of the medium model parameters modifications (wave propagation velocity, medium density and wave absorption coefficient) is reflected in the low-frequency band of the amplitude spectrum. In this case, the phase spectrum responds to such perturbations to the greatest extent, in its high-frequency band.
2. Successful extraction of geo-environment parameters from HRS data requires adequate preparation of field observation materials, taking into account the features and objectives of solving the problem. It includes: preliminary special processing of seismic survey data; assessment of the quality and quantity of information in the original seismic data; optimal parametrization of the environment model; parametrization of the main characteristics of the procedure for solving the inverse dynamic seismic problem (IDSP) by a numerical method. The parametrization of the medium model based on the use of the amplitude and phase spectra characteristics of the original digital data makes it possible to preliminary determine the initial and final details of the seismic data inver-



V. L. Trofimov et al., *Theory and Practice of High Resolution Seismic HRS-Geo Technology Application*, https://doi.org/10.1007/978-3-031-41590-6

sion results, i.e., in general, the results of the inversion of the IDSP. It is also advisable to preliminarily study the features of residual noise on a seismic record (after performing data pre-processing), determine the structure of this noise, and build regularization functions on this basis.

3. Based on the use of well-known theoretical positions for predicting the oil and gas potential of the subsurface, presented as a system of geostructural, lithological, stratigraphic principles and criteria for the formation and regularities of oil and gas accumulations placement, i.e. principles of system analysis, substantiates the geological basis and specifics of the complex geological interpretation of high-resolution acoustic characteristics (HRS) together with the geological indicators of the real medium extracted from these characteristics. On specific examples, it is shown the selection of weak seismic signals in the original wave field against the background of a high level of noise and reflected waves from strong interfaces is one of the main tasks of the seismic data complex interpretation. Features and examples of complex geological interpretation of data based on the use of HRS-Geo technology, production geophysics and drilling data are demonstrated. The features of processing and interpretation of materials using data from HRS-Geo and well logging technology are analyzed sequentially by developing and applying a special (optimal) processing graph and a set of seismic record regularization procedures in the process of prospecting and exploration of oil and gas deposits.
4. For the joint use of high-resolution seismic (HRS) and well logging data, an interpretive system was created as part of the technology, which is based on the principles of continuous functional data transformations for studying well profiles, in which the physical parameters of the section, extracted from well logging data, are optimally matched with corresponding high-resolution seismic parameters obtained using HRS-Geo technology. The main features of the development and improvement of the methodology for processing and interpreting data from standard well logging methods according to the technique of integrating geophysical data by their functional transformations into information systems are considered. These features of log data transformation are aimed at quantitative and qualitative study of well sections with complex lithological composition. Positive results of well logging data interpretation were obtained due to the wide application of the log normalization method and their functional transformations, elimination of distortions in the original log curves, data integration when identifying complex reservoirs, assessing the porosity and density of various lithology rocks, determining factors that affect geophysical parameters, influences bound water on resistivity when estimating saturation of shale reservoirs, accounting for porosity when interpreting electrical resistivity, and the effect of fluid composition on readings of normalized porosity and elastic wave velocity curves.
5. The effectiveness of the developed HRS-Geo technology was studied by the authors on the materials obtained in rather difficult seismo-geological conditions, at one of the polygons located in the northeastern margin of the Russian platform, on the territory of the Timan-Pechora Province (TPP). It is character-

ized by a number of significant features, primarily due to its high tectonic activity, which determines the specifics of its geological structure and oil and gas potential.

The researches were carried out on the basis of seismic and well logging data for three prospecting areas.

At the first of these sites, Savinobr, located in the southern part of the Mechayu-Pashnin swell and the Tebuk step of the Izhma-Pechora saddle of the TPP. Here, according to the results of the generated 3D cubes of effective acoustic impedance (AI) and reflectivities (RC) – models of the real environment with a detailed internal structure of the studied geological section, the geological structure of local uplifts was refined. The results of structural constructions reinterpretation are analyzed accounting the drilling data of two wells. Deposits of productive and promising formations were studied based on the classification of seismic facies using the shapes and values of reference curves – clusters of 3D cubes AI and RC. Oil-perspective objects of structural, tectonically and lithologically shielded types have been identified in the deposits of the upper part of the Famennian, the upper part of the Zadon and in the reservoirs of the Dzhyer-Yaran horizons.

6. In the junction zone of the northeastern part of the Kolvin megaswell and the northwestern part of the Khoreyver depression, on the territory of the Kamin prospecting area, a detailed assessment of the most important geological indicators of the real environment is given, primarily the material composition and reservoir properties in the sediments of the Devonian and Upper Silurian strata. Oil-perspective objects of a structural type were identified in the reservoirs of the target formations. Oil prospects in this area are associated with the identification of structural-lithological and combined oil traps in marine and coastal-marine, shallow-marine, deltaic and lagoon sedimentation conditions.
7. In the central part of the Kolvin megaswell of the Pechera-Kolvin aulacogen and in the junction zone of the large Vozeya and Kharyagin swells within the Yuo-Osh and Osokin fields, using 3D cubes and 2D sections of the AI and RC, the layering nature of deposits in productive and promising formations was studied. For these horizons, detailed structural constructions were made and various sedimentation environments were considered by seismic facies classification in order to identify facies replacement zones, heterogeneities zones of the geological section in the process of studying the detailed internal structure and lithofacies variability of target deposits.

# Appendix: To the Relevance and Novelty of the Author's Development

In this appendix, the authors limited themselves to mentioning only the main achievements, the distinguishing feature of which is novelty and relevance. Such achievements seem to be very important for the formation of adequate, thin-layer geological models of the real environment of high quality in the processing and complex interpretation of seismic data. It should be noted these features of the developments made by the authors were the material for writing original works, including two books (the first of them is Book 1 [1, 2]), the second is this monograph – Book 2. These achievements are aimed at obtaining great advantages over the traditional, currently available, methods of constructing seismogeological models of the real environment, in which only the principles of "wave" seismic exploration are usually used.

Let us briefly dwell on the distinctive features of these achievements in solving specific seismogeological problems and evaluating the results obtained. In particular, we are talking about the following distinctive features and methods for solving problems:

1. The numerical method for solving the inverse dynamic seismic problem (IDSP), which implements the separation of seismic information associated with the source of elastic oscillations and the environment of the studied geological section, is aimed at studying real media, is focused on studying the dynamic parameters of a seismic record, and realizing the maximum possible resolution of these parameters. On this basis, the construction of 2D sections and 3D cubes of effective acoustic impedance (AI) and reflection coefficients (RC) is implemented, having a vertical resolution equal to or close to the seismic record sampling step along time using appropriate observing system and digital data processing.
2. To search for a model of the environment by means of multicriteria optimization, a method of dynamic regularization in the process of solving the IDSP has been developed. Unlike traditional approaches to regularization of the inverse problem, where the regularization parameter is usually estimated and set (as a


V. L. Trofimov et al., *Theory and Practice of High Resolution Seismic HRS-Geo Technology Application*, https://doi.org/10.1007/978-3-031-41590-6

rule, this is one number for the entire algorithm), in the HRS-Geo technology, according to the obtained estimates of the signal and noise characteristics, sets of regularization functions are formed (Fig. 2.18), which are dynamically applied in the course of solving the inverse problem, ensuring the stability of the solution search process and the detail of the desired result.

The searching process for a model of AI and RC in the algorithm for solving the IDSP is studied in detail using objective functions (a vector of objective functions), in which various types of residuals between real and model data are iteratively calculated. Such residuals are estimated from the energy (amplitudes) and the shape of the seismic signal. For these discrepancies, the information content and sensitivity of the amplitude (energy) and phase spectra of the seismic record are analyzed. The analysis is performed in the frequency ranges, in which the contribution of the medium model parameters perturbations (velocity of elastic waves, density of the medium, and wave attenuation) affects the result of seismic record inversion, which is proportional to the frequency of the waves. Numerical experiments have shown the contribution of the medium model parameters modifications is displayed in the low-frequency band of the amplitude spectrum. The phase spectrum responds to such modifications to the greatest extent mainly in its high-frequency region.

3. To obtain a stable (correct) solution to the inverse problem, various methods of regularization are used. However, in order to estimate the regularization parameter in traditional algorithms or in the construction of regularization functions (in HRS-Geo technology), it is necessary to determine the type, evaluate the level and statistical characteristics of the noise contained in the seismic record. To describe the wave process, a more detailed convolutional seismic trace model (2.2) was developed and presented. In accordance with this concept of the seismic trace model, noise also participates in the process of wave propagation in space and along time (i.e., they are of a wave nature) and are also subject to the same action of an elementary pulse sounding the real environment. This effect consists in the fact a seismic impulse affects the primary noise that has arisen, which transforms it through filtering: the original noise changes its shape and frequency composition, and participates in the general interference process, complicating the overall wave pattern. In this case, the interferences turn out to be constituent elements of a more complex interference phenomenon described by convolution. The representation of this process in the form of a mathematical model (2.2) makes it possible to estimate the contribution of random primary noise to interference noise (changed by an impulse during wave propagation in space), which, in turn, is included as an element of a seismic record in the general convolutional model, where the components are already the coefficients reflections and a medium-probing pulse.
4. The structure, type, level and statistical characteristics of noise are important categories in technologies for solving inverse problems of geophysics. How accurate knowledge about them determines the result of solving the IDSP. The presented monographs define the roles of noise and useful signals at different

stages of seismic data processing. One and the same signal can be a useful signal at one stage of work and a noise at another [1, 2].

A new type of noise has been introduced into the noise classification: interference noise for seismic waves in a thin layer section [1, 2]. The source of this noise is a seismic impulse probing the real medium. At the stage of preprocessing of the wave field (for example, when filtering the wave field from noise), a seismic pulse that forms a primary wave field is a useful signal and a carrier of information about the real medium ("bearer"). When solving the IDSP, the seismic impulse turns into a strong noise, which limits the detail of the solution and which must be removed (significantly attenuated) from the processed record.

5. The concepts of an elementary layer and an elementary boundary are introduced for the unambiguous definition of a model of a thin-layer acoustic medium in the development and use of the technology for obtaining seismic data with high spatial resolution.

   A criterion has been developed for evaluating the contribution of various components of the geo model, signals and noise of the seismic record to the wave field and its various transformations, represented by the "interference matrix of contributions" (ICM). The ICM is a two-dimensional table where the elements represent a contribution quantification of the considered factor to the seismic record. A joint analysis of the information content of the thin-layer section elements into an interference wave field also makes it possible to study the contributions of local responses from lithology, porosity, and water and gas saturation to the seismic wave field and to the results of its various transformations.
6. It is shown the solution of the "***detection of weak signals***" is directly related to the problem of "***solving the inverse dynamic seismic problem (IDSP)***". On this basis, the nature and mechanism of the formation of a seismic wave field is explained and an adequate solution is found in the problem of separating heterogeneous information contained in reflected waves. Such an approach to the analysis of a seismic wave field structure makes it possible to understand the nature of reflected waves that are weak in intensity, which, depending on the values of the reflection coefficients that change along the section, "respond" appropriately to low-contrast seismoacoustic inhomogeneities.
7. A special system of petrophysical equations has been developed to determine the lithological composition and the nature of fluid saturation of reservoir rocks. With an integrated approach to the analysis of the results obtained, the extraction of the maximum possible amount of reliable geological and geophysical information from seismic data is achieved for the purpose of searching the oil-perspective objects, assessing their characteristics for calculating hydrocarbon reserves and resources, optimal placement of exploration and production wells and, in general, to substantiate the strategy for the development of oil and gas deposits. Before predicting the desired parameters, the studied section interval is compared with the physical parameters previously obtained for this section

interval from well logging data, and then the desired geological and geophysical indicators are determined for each point of the target interval (formation) according to the specified lateral direction in the plane of the seismic section (or 3D seismic cube).

8. Adequate preparation of seismic data includes: preliminary data processing using HRS-Geo and well logging technology based on the development and application of a special (optimal) processing graph and a set of seismic record regularization procedures; assessment of the quality and quantity of information in the original seismic data; optimal parametrization of the environment model; parametrization of the main characteristics of the procedure for solving the inverse dynamic seismic problem by a numerical method.
9. To calculate the signal to noise ratio of a reflectivity thin-layer section representing graphs of discontinuous functions, the authors developed and applied a method for estimating the signal/noise ratio based on the change in signal energy in the power spectrum and the value of the regularization parameters when solving the IDSP with respect to the S/N value of the initial field. The signal-to-noise ratio is estimated to section the reflectivity distribution of the tested seismic data processing versions using the formula (7) presented in [1] (or formula (4.7) in [2]).
10. The existing ambiguity in determining the parameters of the medium (lithology, porosity and oil saturation) in the context of each of the deep wells using hole measurement methods is presented as a special "geological interference" of different characteristics – geological indicators of the real environment. Based on the set of used parameters located in a certain system, the material composition, porosity, content of bound water and useful capacity are determined, the oil and gas saturation of reservoirs is estimated, and a number of very important characteristics of the geological section are calculated. On the one hand, on this basis, a thin-layer geological model is formed in the vertical section of the well, lithological-stratigraphic columns and detailed geoacoustic well models are built. And on the other hand, on the same basis, the authors, according to the seismic data of the HRS, build predictive lithological-stratigraphic columns in the inter-well space with a discretization step of seismic data, using exclusively HRS data with a relatively high probability of forecasting.
11. In accordance with the methodology for predicting geological indicators extracted from high-resolution seismic data (HRS) for those intervals are considered as oil-perspective and priority for oil exploration, the main set of indicators is determined, such as oil and gas saturation, porosity, effective thicknesses, clay content, carbonate content, oil-saturated thicknesses, oil saturation coefficient and a number of others. The practice of generating and applying numerous results of complex geological interpretation of HRS data allowed the authors to choose the most optimal value (level) of a very important parameter that is necessary to determine the contour of the mobile part of hydrocarbons. This level was chosen accounting the obtained results of well logging data processing and the available well test results. In the process of numerous experience and analysis of the predicted indicator (oil and gas saturation), along with the

possible nature of saturation of the target deposits, associated with the peculiarities of the sedimentation conditions and the location of their distribution, for the mobile part of hydrocarbons, the authors conditionally set the level of oil saturation equal to 3% and higher. The level for the stationary part of hydrocarbons, respectively, is also conditionally accepted, equal to 3% and below. On this basis, the contours of signs of oil shows were identified, and later we used the same approach to determine the contours of oil saturation in the sediments of most productive horizons. It also takes into consideration the presence in the section of the corresponding lithological composition and reservoir properties of the studied deposits, which are confirmed in the corresponding intervals of the section of the lithological-stratigraphic columns of existing wells. The identified contours of the mobile part of hydrocarbons are largely taken into account when predicting all other formed predictive geological indicators.

12. Interpretation of reflecting boundaries and a system of faults when forming of a tectonic and formation model of sedimentation is based on the sections and cubes of acoustic impedance (AI) and reflection coefficient (RC) obtained as a result of solving the inverse dynamic seismic problem. That is, the sections themselves AI and RC are obtained regardless of the presence of the original boundaries and faults. While in traditional methods for calculating sections and cubes of AI and RC, it is necessary to have these data (boundaries and faults) – i.e. first one has to interpret the wave field and then apply these results (boundaries and faults) as input to the wave field inversion. In addition, in traditional technologies, inversions are required as input data, acoustic parameters obtained from the well logging. In the HRS-Geo technology, since AI and RC are calculated regardless of the available boundaries, faults and data on the well, then the model of tectonic faults and horizons of thin-layer layers is determined from AI and RC, and not vice versa. This model turns out to be much more accurate and detailed, since thin-layer sections of RC and AI are interpreted, rather than a smoothed wave field. Thus, AI and RC turn out to be independent characteristics during interpretation (do not depend on the interpretation of boundaries, horizons, layers, and faults). This gives high importance to the results obtained using the HRS-Geo technology and shows their advantage over traditional technologies.
13. In the complex interpretation of HRS data, the parameter that directly characterizes the regularity of the distribution density of oil and gas and volumetric hydrocarbon content distribution in the study area is of great importance. The need for such a parameter is explained by the fact in the process of predicting various geological indicators, it is assumed the studied geological section is heterogeneous (complex morphology of the pore space, variable content of clay material, complex mineral composition of the rock matrix, etc.). This leads to a wide variation in the distribution density of oil and gas and volume content of hydrocarbons.

It is also known when calculating resources and various geological indicators in the identified contours of significant oil saturation, the change in the values of these indicators is directly taken into account, namely, the porosity coefficient, effective oil-saturated thicknesses, and the oil saturation coefficient of the pore space of the reservoir. According to the formulas of the volumetric method, this is realized in each vertical profile of the section along each of the predictive seismic traces. The nonuniform distribution of geological indicators is due to the complex heterogeneous structure of the target objects of study.

If one of the above-mentioned variables is taken as a basis, namely, the oil-saturated thickness of the prospective horizon, then it should be noted within the volumetric space characterized only by this oil-saturated thickness, in addition to the hydrocarbons themselves (oil and gas saturation), there are also a number of other geological and geophysical indicators. In particular, these are lithological components (for a terrigenous section, this is the content of sandstones and clays in the corresponding parts of this space), as well as the pore space filled with either hydrocarbons together with formation water, or only hydrocarbons.

To find the share of "pure" oil – hydrocarbons (without fractional content of lithology and water saturation) in linear units of measurement, it is necessary to normalize all constituent components – geological indicators, including the linear interval occupied by oil saturation, in these units of measurement, for example, on a depth scale (i.e. in the same units as the oil saturated thickness – in meters). Then, from the total volume of the constituent geological components, everything is not associated with hydrocarbons should be subtracted. The resulting residual will ultimately characterize the fraction of the volume of "clean" oil (*in units of the depth scale*).

The use of the described approach (hydrocarbons distribution density maps over the area) to the calculation of hydrocarbon resources and reserves gives more reliable and accurate results. The specified procedure for normalization and recalculation of the volumetric content of various geological components on a depth scale was implemented in the form of an algorithm and a program. Based on their use, maps of the predicted distribution density of hydrocarbons ("pure" oil) are obtained for the studied deposits (reservoirs, horizons, etc.). In general, The obtained results of the predicted distribution density of hydrocarbons – "pure" oil for the considered elements of the section within the contours of significant oil saturation are a set of cylinders with curvilinear (in the area of work) contours of their vertical walls. Such hypothetical cylinders of a real geological environment, according to the conditions of the performed calculations, are filled only with hydrocarbons – "pure" oil (without a porous rock matrix of a certain lithology and water saturation). The height of the walls of a single curvilinear cylinder is different for each of the research objects. This height of the cylinder walls is usually taken in the range of 0.1–0.2 m. In each of the contours of significant oil saturation, the spatial arrangement of such cylinders has a sequential layering: in its lower part there is a cylinder with the largest distribution area, and in the upper part – with the smallest one. At the same time, for each of the objects (oil-promising horizon) within each of the contours of significant oil saturation, its own unique (inherent only to this object) form of the predicted hydrocarbon distribution density is obtained.

The results obtained in the form of the predicted hydrocarbons distribution density maps are directly used for:

(a) a more accurate assessment of reserves and resources in a volumetric way at each point in the area;
(b) determining not only the integral characteristics of hydrocarbon resources and reserves in the form of their total quantity by volume and mass, but differential ones – the distribution of oil and gas over the area;
(c) selection of the optimal point for opening up a deposit when searching for hydrocarbons and operating a well.

14. When solving linear inverse problems of geophysics by the classical method of least squares, the quality of the solution is usually estimated by its covariance matrix, the diagonal elements of which represent the parameter variances [3]. The variance of the parameters depends on the variance of the observational errors and the diagonal elements of the $\mathbf{G}^T\mathbf{G}$ matrix. The solution obtained by the Baykus-Hilbert method is interpreted as smoothed, and the degree of deviation of the smoothing kernel from the delta function determines the resolution of the original data. In this case, the *quality* of the solution is characterized by the resolution, which is determined by one or another functional from the smoothing kernel. If the number of observations $n$ is greater than the number of required parameters $k$, then the $\mathbf{G}^T\mathbf{G}$ matrix is nonsingular, i.e. exists an inverse matrix $(\mathbf{G}^T\mathbf{G})^{-1}$. Conversely, if the number of parameters is greater than the number of observations, then exists a matrix $(\mathbf{G}\mathbf{G}^T)^{-1}$. Thus, there exists a solution for the limiting case $\alpha = 0$. It is possible both matrices $\mathbf{G}^T\mathbf{G}$ and $\mathbf{G}\mathbf{G}^T$ turn out to be degenerate (singular).

Since the matrices $(\mathbf{G}\mathbf{G}^T)^{-1}$ and $(\mathbf{G}\mathbf{G}^T + \alpha\mathbf{I})^{-1}$ (the addition of a small positive term to the diagonal (positive) elements of the matrix is used to obtain a stable solution of ill-conditioned linear systems) are positive-definite, the difference $s_i(\alpha) - s_i(0)$ is always positive. The minimum value of $s_i$ corresponds to $\alpha = 0$, $s_i(0) = 1 - \mathbf{g}_i^T(GG^T)^{-1}\mathbf{g}_i$, and the maximum value corresponding to $\alpha \to \infty$ is equal to 1. The variance of the $i$-th parameter is $\rho_i^2 = \sigma_y^2\mathbf{h}_i^T\mathbf{h}_i = \sigma_y^2\mathbf{g}_i^T(\mathbf{G}\mathbf{G}^T + \alpha\mathbf{I})^{-2}\mathbf{g}_i$, which shows the variance decreases with increasing $\alpha$.

For $\alpha = 0$, the variance is equal to $\sigma_y^2\mathbf{g}_i^T(\mathbf{G}\mathbf{G}^T)^{-2}\mathbf{g}_i$, and for $\alpha \to \infty$, $\rho_i^2 \to 0$.

The quality of the solution in this case will be the better, the smaller $\rho_i^2$ and $s_i$. But it is impossible to decrease $\rho_i^2$ and $s_i$ simultaneously: a decrease in one value leads to an increase in the other. Therefore, when constructing a solution (i.e., when choosing the value of $\alpha$), one has to look for a compromise between variance and resolution. The curves of the solution variance dependence on the resolution characteristic (as shown in Fig. 2.29 in Chap. 2) are called *trade-off curves*. In practice, they will have a similar shape for any other choice of the resolution characteristic$s_i$.

According to the data [3], there is no formalized way to achieve such a goal, i.e., the choice of the value $\alpha$, although, in principle, they offer such an approach. Since it is usually required the variance of the parameters does not exceed a certain set value $\delta$, a point is found on the trade-off curve (and the value $\alpha_\delta$ corresponding to this point) in which $\rho^2 = \delta$ (Fig. 2.29). The value $\alpha \geq \alpha_\delta$ is taken as $\alpha$.

In the HRS-Geo technology of high-resolution seismic, due to the use of multi-objective optimization and successive regularization, a more detailed study of the noise structure and special data processing, the "compromise" curve has a different shape and a different pattern (Fig. 2.30). This dependence monotonically increases with the deterioration of resolution, and then the variance stabilizes (Fig. 2.30). At the same time, within certain limits, a decrease in the solution dispersion occurs not due to a deterioration in the detail of the resulting model (Fig. 2.30, Table 2.3) (within the limits of applicability of the described inversion method).

15. In the process of further numerical algorithm improvement for solving the inverse dynamic seismic problem (IDSP), rationale and searching for a medium model in the form of a AI and RC distribution, the authors solved a number of local tasks, without which it is practically impossible to solve the main problem – converting a seismic record into acoustic parameters of a real geological section. In particular, local tasks were solved, which are distinguished by their relevance and novelty:

    15.1 Criteria for assessing the information content of the amplitude and phase spectra of seismic signals have been developed. The sensitivity of the spectra determines the potential detail of seismic data, i.e. their information content contained in the original seismic data. Figure 1.1 and formulas (1.13)–(1.21) theoretically substantiate the algorithm for assessing the information content in the spectral components of a seismic record. Sections 1.3, 1.4, 1.5, 1.6, 1.7, and 1.8 prove the assumptions made about the possibility of informativeness estimates on model examples for thin-layer media, in Sects. 2.1, 2.2 and 2.3 their applicability to real experiments on measuring the characteristics of the wave field is shown.

    15.2 Changes in the parameters of models and noise are formed, the influence on the spectra the wave velocity, medium density and wave absorption and their combinations is considered and evaluated (Tables 1.1, 1.2, 1.3 and 1.4, Figs. 1.4, 1.5, 1.6, 1.7, 1.8, 1.9, 1.10, 1.11, 1.12, 1.13, 1.14, 1.15, 1.16, 1.17, 1.18, 1.19, 1.20, 1.21, 1.22, 1.23, 1.24, 1.25, 1.26, 1.27 and 1.28). Frequency intervals are identified and the distribution of the contributions of the medium parameters and noise over the frequency ranges of the spectra is estimated.

    15.3 The features and mechanism of the absorbing layer thickness influence on the structure of the seismic trace, the spectra of the wave field were studied (Tables 1.5, 1.6 and 1.7, Figs. 1.30, 1.31, 1.32, 1.33, 1.34, 1.35 and 1.36). The influence of the absorbing layer is shown: in particular, it does not stop below the location level of the object under study, which absorbs seismic waves, complicating the wave picture. The dependence of the absorption effect on the ratio between the thickness of the absorbing layer and the wavelength is studied and presented. The presented dependence of the absorption effect on the ratio of the thickness of the absorbing layer and the wavelength shows at thicknesses much smaller than the seismic wavelength, the absorption contribution to the seismic trace is smaller and with a decrease in the formation thickness, the absorption effect decreases. At thicknesses greater than the wavelength, the absorbing effect increases and the formation absorbs more energy from the information carrier. When the thickness of the actively absorbing layer changed from 13.4 to 261.8 m, the change in the amplitudes (energy) of the traces ranged from 6% to 53%. The absorption effect leads to a narrowing of the frequency band, especially from the high frequencies, and this is well presented in the graphs of the amplitude spectrum. In the phase spectrum, the absorption effect appears in the form of expanded horizontal intervals.

15.4 The manifestation in the amplitude-phase spectra of the seismic trace resolution property (Fig. 1.43) reveals the disappearance of the spectrum components with a significant increase in the absorption coefficient. In the phase spectrum, you can find areas where there is no set of thin-layer beds of a certain thickness. In general, spectrum analysis makes it possible to find operating frequency bands that are generally different for the amplitude and phase spectra.

15.5 The influence of the medium model discretization step on the characteristics of the seismic trace and its display on the spectral components was studied (Figs. 1.46 and 1.47). An increase in the sampling interval of the model leads to a narrowing of the operating frequency band. This phenomenon manifests itself most clearly in the phase spectrum.

15.6 The influence of the random noise level on the characteristics of seismic traces and spectra is considered (Figs. 1.49 and 1.50). An increase in the signal-to-noise ratio is manifested in the amplitude spectrum by an increased intensity in the high-frequency band. The unwrapped phase spectrum responds to this in a more regular way: the phase spectrum becomes more monotonically flat with increasing frequency. This creates the prerequisites for filtering noise in the frequency domain without losing the components of the useful signal.

15.7 The representations of the main classical components of the IDSP solution in the HRS-Geo technology are considered: (a) parametrization of the medium model (Sect. 2.1), (b) parameterization of the seismic data inversion process (Sect. 2.2) and (c) an example of solving the IDSP (Sect. 2.3).

Thus, the obtained results of solving inverse dynamic seismic problems by a numerical method using multicriteria optimization, regularization of seismic data, and estimation of dependencies that improve the resolution of reconstructed acoustic data are of great (fundamental) importance for solving various problems of geophysics in general.

## References


1. Trofimov, V. L., Khaziyev, F. F., & Trofimova, A. V. (2019). *Tekhnologiya vysokorazreshayushchey seysmiki pri poiskakh i razvedke neftegazovykh zalezhey* (High-resolution seismic technology in prospecting and exploration of oil and gas deposits) (p. 343). M., EAGE Geomodel LLC.
2. Trofimov, V. L., Khaziev, F. F., & Trofimova, A. V. (2022). *Oil and gas reservoir prospecting and exploration. High-Resolution Seismic (HRS) techniques and technology*. Springer (Hardcover Book, eBook). https://doi.org/10.1007/978-3-030-84389-2
3. Yanovskaya, T. B., & Porokhova, L. N. (2004). *Obratnyye zadachi geofiziki* (Inverse problems of geophysics) (p. 213). Izdatel'stvo SPbGU (St. Petersburg State University Publishing House).

# Index




V. L. Trofimov et al., *Theory and Practice of High Resolution Seismic HRS-Geo Technology Application*, https://doi.org/10.1007/978-3-031-41590-6